POWER GEOMETRY IN ALGEBRAIC AND DIFFERENTIAL EQUATIONS

North-Holland Mathematical Library

VOLUME 57

ELSEVIER
Amsterdam – Lausanne – New York – Oxford – Shannon – Singapore – Tokyo

Power Geometry in Algebraic and Differential Equations

Alexander D. Bruno
Keldysh Institute of Applied Mathematics
Moscow

2000
ELSEVIER
Amsterdam – Lausanne – New York – Oxford – Shannon – Singapore – Tokyo

ELSEVIER SCIENCE B.V.
Sara Burgerhartstraat 25
P.O. Box 211, 1000 AE Amsterdam, The Netherlands

First edition 2000

Library of Congress Cataloging in Publication Data
A catalog record from the Library of Congress has been applied for

ISBN: 0 444 50297 1

Transferred to digital printing 2005

My uncle... used to wonder what became of all the unsuccessful tinkers, and gunsmiths, and shoemakers, and blacksmiths, and engineers...

Mark Twain. My watch

Preface

Power Geometry is a new calculus developing the differential calculus and aimed at the nonlinear problems. Its main concept consists in the study of nonlinear problems not in the original coordinates, but in the logarithms of these coordinates. Then to many properties and relations, which are nonlinear in the original coordinates, some linear relations can be put in correspondence. The algorithms of Power Geometry are based on these linear relations. They allow to simplify equations, to resolve their singularities, to isolate their first approximations, and to find either their solutions or the asymptotics of the solutions. After the first step of such simplifying transformations, the power solutions or the power asymptotics of solutions are easily found. After the multiple application of these algorithms, the solutions or the asymptotics of solutions containing the multiple logarithms and exponents may be easily obtained. This approach allows to compute also the asymptotic and the local expansions of solutions. Algorithms of Power Geometry are applicable to equations of various types: algebraic, ordinary differential and partial differential, and also to systems of such equations. These algorithms include the simplifying algorithms of both types cited in [Bruno 1998a]: the transformations of coordinates and the transformations of equations. Power Geometry is an alternative to Algebraic Geometry, Differential Algebra, Group Analysis, Nonstandard Analysis, and other disciplines. The first study on Power Geometry was the memoir by Newton [1711]. An elementary introduction to Power Geometry for the algebraic and ordinary differential equations is expounded in Chapters I and II respectively of the book [Bruno 1979a].

This book contains the more advanced presentation for all types of equations. The effectiveness of the algorithms is demonstrated on some complicated problems from various fields of science (Robotics, Celestial Mechanics, Hydrodynamics, Thermodynamics). At present, there are many nonlinear problems which may be solved by these algorithms (and by them only). It is demonstrated that these algorithms give also the classical results, which were obtained earlier intuitively. The expounding of material is detailed, and it is explained by great number of examples and figures.

The difference between the present English edition of the book and the original Russian one ("Power Geometry in algebraic and differential equations". Fizmatlit, Moscow, 1998) is in the following. Section 22 in Chapter 5 and two chapters are added, namely, Chapters 7 and 8; the list of the literature is considerably increased; in some places, the expounding of the material in Chapters 1–6 was changed; a considerable number of misprints and inaccuracies found in the Russian edition were corrected. I express my gratitude to Lev M. Berkovich, Michel Hénon, Vladimir Yu. Petrovich, Victor P. Varin, Mikhail M. Vasiliev, who had pointed out some of these defects of the Russian edition. Special thanks to Prof. M. Hénon, who had maid a draft translation of Chapters 1 and 2 by his own initiative, and had discovered many misprints there (but not all of them). Special thanks also to Dr. V.P. Varin both for his competent translation and for the multitude of remarks he had made to the Russian text. I carefully examined the translation and have made some changes in the final version. I hope that in spite of all our efforts, the English text still contains a sufficient number of misprints and inaccuracies, the disclosure of which might serve as a test of comprehension of the expounded material. I would be grateful for any comments on this English edition, which I ask to direct to the following e-mail:

bruno@spp.keldysh.ru

I acknowledge good cooperation with Elsevier, especially with Drs. Arjen Sevenster, Ms. Claudette van Daalen, and Ms. Titia Kraaij.

Moscow, January 2000 Alexander D. Bruno

Contents

Chapter 0

Introduction

1. Concepts of Power Geometry

Many problems in mechanics, physics, biology, economics and other sciences are reduced to nonlinear equations or to systems of such equations. The equations may be algebraic, ordinary differential or partial differential; and systems may comprise the equations of one type, but may include equations of different types. The *solutions* to these equations and systems subdivide into *regular* and *singular* ones. Near a regular solution the implicit function theorem or its analogs are applicable, which gives a description of all neighboring solutions. Near a singular solution the implicit function theorem is inapplicable, and until recently there had been no general approach to analysis of solutions neighboring the singular one. Although different methods of such analysis were suggested for some special problems.

The purpose of the book is to supply a general purpose set of algorithms for analysis of singularities applicable to all types of equations. At present, the usual way of development of mathematical sciences may be depicted as the following sequence:

$$\begin{aligned}&\text{Concept} \to \text{Theory} \to \text{Algorithm} \to \\ &\qquad\qquad \text{Software} \to \text{Computation} \to \text{Application},\end{aligned} \tag{1.1}$$

where some elements may be in plural and feedbacks play a major part [Bruno 1998a]. This book comprises all elements of the sequence (1.1).

The main concept of Power Geometry is to study the properties of solutions to an equation through the power exponents of its monomials. For instance, to the polynomial

$$f(X) = \sum f_Q X^Q, \quad Q \in \mathbf{S}, \tag{1.2}$$

where $X = (x_1, \ldots, x_n)$, $Q = (q_1, \ldots, q_n)$, $X^Q = x_1^{q_1} \ldots x_n^{q_n}$, there corresponds the set $\mathbf{S} \subset \mathbb{R}^n$ of the vector power exponents Q, for which the coefficients $f_Q \neq 0$. Together with the set $\mathbf{S}$ (the *support* of Polynomial (1.2)), we consider its convex hull $\mathbf{\Gamma}$ (the *Newton polyhedron* of Polynomial (1.2)) and faces $\mathbf{\Gamma}_j^{(d)}$ of the polyhedron $\mathbf{\Gamma}$ in the space $\mathbb{R}^n$, and also their normal cones $\mathbf{U}_j^{(d)}$ in the dual space $\mathbb{R}_*^n$ that is the space of logarithms of coordinates x_i. To each face $\mathbf{\Gamma}_j^{(d)}$ there corresponds the *truncation* of Polynomial (1.2)

$$\hat{f}_j^{(d)}(X) = \sum f_Q X^Q, \quad Q \in \mathbf{S} \cap \mathbf{\Gamma}_j^{(d)}. \tag{1.3}$$

It is the first approximation to the Polynomial (1.2) in the part of X-space that corresponds to the normal cone $\mathbf{U}_j^{(d)}$ in the $(\ln X)$-space $\mathbb{R}_*^n$. In that part of the X-space, the first approximation to a solution to the equation $f(X) = 0$ is the solution to the truncated equation $\hat{f}_j^{(d)}(X) = 0$. The power transformations of coordinates, which are linear transformations of logarithms of coordinates, induce the linear transformations in spaces $\mathbb{R}^n$ and $\mathbb{R}_*^n$, which allows to reduce the number of coordinates in a problem, especially for the truncated equations. The linear transformations in $\mathbb{R}^n$ and $\mathbb{R}_*^n$ make Power Geometry a geometry in the Klein's sense (see [Klein 1872]).

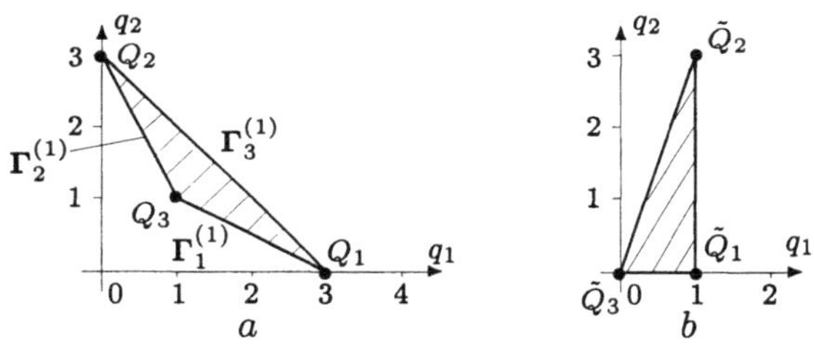

Figure 0.1: Supports and Newton polyhedra for equations (1.4) (a) and (1.8) (b).

Example. Consider a plane (i.e. $n = 2$) algebraic curve, which is called the *folium of Descartes* and is defined by the equation

$$f(X) \stackrel{\text{def}}{=} x_1^3 + x_2^3 - 3x_1x_2 = 0. \tag{1.4}$$

Let us study solutions to this equation near the origin $x_1 = x_2 = 0$ and at infinity, where the implicit function theorem can not be applied. The set $\mathbf{S}$ of power exponents of the equation consists of three points: $Q_1 = (3, 0)$, $Q_2 = (0, 3)$, $Q_3 = (1, 1)$. Their convex hull $\mathbf{\Gamma}$ is the triangle (Fig. 0.1, a). The points Q_1 and Q_3 lie in its edge $\mathbf{\Gamma}_1^{(1)}$. Hence, to it there corresponds the truncated equation

$$\hat{f}_1^{(1)}(X) \stackrel{\text{def}}{=} x_1^3 - 3x_1x_2 = 0. \tag{1.5}$$

Its nontrivial solution $x_2 = x_1^2/3$ is the first approximation to the branch $\mathcal{F}_1$ of solutions to Equation (1.4) (see Fig. 0.2). The branch $\mathcal{F}_1$ passes through the point $x_1 = x_2 = 0$. The points Q_2 and Q_3 lie on the edge $\mathbf{\Gamma}_2^{(1)}$; to it there corresponds the truncated equation $x_2^3 - 3x_1x_2 = 0$. Its nontrivial solution $x_1 = x_2^2/3$ is the first approximation to the second branch $\mathcal{F}_2$ passing through the origin. To obtain subsequent approximations to the branch $\mathcal{F}_1$, let us make the power transformation

$$\begin{cases} y_1 = x_1^2 x_2^{-1}, \\ y_2 = x_1^{-1} x_2, \end{cases} \quad \text{with inverse} \quad \begin{cases} x_1 = y_1 y_2, \\ x_2 = y_1 y_2^2. \end{cases} \tag{1.6}$$

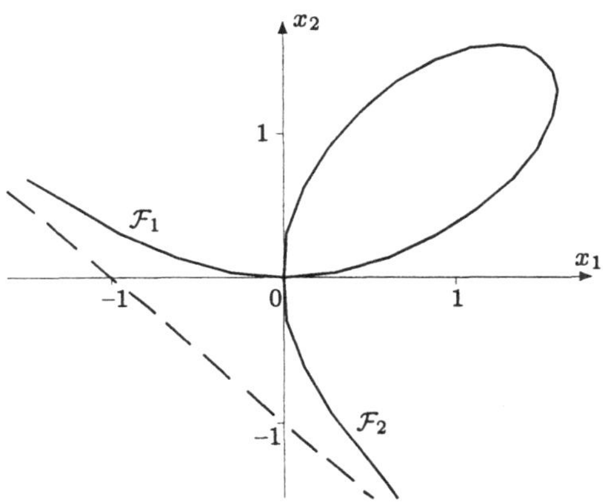

Figure 0.2: The folium of Descartes.

Then Equations (1.4) and (1.5) are transformed into

$$f \stackrel{\text{def}}{=} y_1^2 y_2^3 (y_1 + y_1 y_2^3 - 3) = 0, \quad \hat{f}_1^{(1)} \stackrel{\text{def}}{=} y_1^2 y_2^3 (y_1 - 3) = 0. \tag{1.7}$$

Cancelling $y_1^2 y_2^3$, we obtain the complete equation

$$y_1 + y_1 y_2^3 - 3 = 0 \tag{1.8}$$

and its truncated equation $y_1 - 3 = 0$. For Equation (1.8), the support and polyhedron are shown in Fig. 0.1, *b*. The root $y_1 = 3$ of the truncated equation is simple. Hence in the neighborhood of the point $y_1 = 3$, $y_2 = 0$ the implicit function theorem is applicable to the complete Equation (1.8), which allows to obtain $y_1 - 3$ as a power series of y_2. Here it may be done explicitly: $y_1 = 3/(1 + y_2^3)$. Substituting this expression in right Formulae (6), we obtain the parametric representation for the branch $\mathcal{F}_1$:

$$x_1 = 3y_2/(1 + y_2^3), \quad x_2 = 3y_2^2/(1 + y_2^3).$$

The points Q_1 and Q_2 lie on the edge $\mathbf{\Gamma}_3^{(1)}$, and to it there corresponds the truncated equation $x_1^3 + x_2^3 = 0$. Its only real solution $x_1 + x_2 = 0$ is the first approximation to the asymptote $x_1 + x_2 = -1$ of the folium of Descartes, shown in Fig. 0.2. To obtain the asymptotic expansion for its branches that go into infinity we need to apply the power transformation $z_1 = x_1$, $z_2 = x_1^{-1}x_2$.

Exercise. Plot the support and the Newton polyhedron for the left Equation (1.7).

2. Historical remarks

Power Geometry is based upon the three concepts: the Newton polyhedron, the power transformation and the logarithmic transformation. The crucial points of their development are as follows.

I. The *Newton polyhedron.* For $n = 2$, approximately in 1670 Newton [1711] suggested to use one edge of the "Newton open polygon" [Bruno 1979a] of a polynomial $f(x, y)$ to find the branches of solutions to the equation $f(x, y) = 0$ near the origin $x = y = 0$, where the polynomial f has no constant or linear terms. Puiseux [1850] was already using all the edges of the Newton open polygon and had given a rigorous substantiation to the solution of the problem by this method. Liouville [1833] was using this approach to find the rational solutions $y = y(x)$ to the linear ordinary differential equation

$$a_0(x)d^n y/dx^n + \ldots + a_{n-1}(x)dy/dx + a_n(x) = 0,$$

where $a_i(x)$ are polynomials. Briot and Bouquet [1856] were using an analog to the Newton open polygon to find solutions $y(x)$ to the nonlinear ordinary differential equation $dy/dx = f(x, y)/g(x, y)$ near the point $x = y = 0$, where polynomials f and g vanish. A survey of other applications of the Newton (open) polygon was made by Chebotarev [1943]. The survey should be completed by mentioning the dual open polygons [Bruno 1979a] (see for example the Frommer open polygon [Frommer 1928]) and by attempts to reduce the solution of a system of algebraic equations of three or more coordinates to the plane Newton open polygon (see Section 10 of Chapter 2).

Sintsov [1898] to obtain expansions $y(x)$, $z(x)$ of the branches of solutions to an algebraic system of equations $f_1(x, y, z) = f_2(x, y, z) = 0$ near the origin $x = y = z = 0$, where the polynomials f_i have no constant or linear terms, was considering (in our terminology) intersections of normal cones $\mathbf{U}_{ij}^{(1)}$ of the edges $\mathbf{\Gamma}_{ij}^{(1)}$ of Newton polyhedra $\mathbf{\Gamma}_i = \mathbf{\Gamma}(f_i)$, $i = 1, 2$, with the plane $p_3 = -1$ in dual space $\mathbb{R}_*^3$ (see Sections 3, 9, 10 of Chapter 1 and Section 10 of Chapter 2).

Shestakov [1960, 1961] to find the asymptotics of solutions to the system of ordinary differential equations $dx_i/dt = \varphi_i(X)$, $i = 1, \ldots, n$, was considering the supports of polynomials φ_i in $\mathbb{R}^n$. For such a system, there was suggested in [Bruno 1962, 1965], the writing in the form

$$d \ln x_i/dt = f_i(X), \quad i = 1, \ldots, n, \tag{2.1}$$

its support and Newton polyhedron in $\mathbb{R}^n$, and the normal cones of the faces in $\mathbb{R}^n_*$.

Mikhailov [1963, 1965, 1967a, b] studying the properties of solutions $u(X)$ to the linear partial differential equation $f(D)u = 0$, where $D = (\partial/\partial x_1, \ldots, \partial/\partial x_n)$ and $f(Y)$ is a polynomial in Y, considered the Newton polyhedron $\mathbf{\Gamma}(f)$ of the operator polynomial $f(Y)$. Gindikin [1973] considering an analogous problem introduced the term "Newton polyhedron", which became generally accepted. Formerly these polyhedra were called characteristic. Kushnirenko [1975a, b] was considering the Newton polyhedra for usual polynomials.

The Newton polyhedron was introduced in [Bruno 1994] for arbitrary differential polynomials both with ordinary and partial derivatives.

II. The *power transformation.* It was used by Newton [1711] and all his followers in the simplest form $y = x^\alpha z$. Weierstrass [1902] suggested the sequence of transformations $y = xz$ and $x = zy$ analogous to the σ-process in Algebraic Geometry. Power transformations in general form $\log X = \alpha \log Y$ were suggested in [Bruno 1962, 1965] for systems (2.1), and in [Bruno 1976, 1979a] for systems of algebraic equations. Power transformations of different types of coordinates, i.e. parameters, independent and dependent variables, were proposed in [Bruno 1996b].

III. *Logarithmic transformations* in the context of Power Geometry were introduced in [Bruno 1997a], where the term "Power Geometry" also appeared. Although logarithmic transformations themselves were always used.

Besides the local and asymptotic properties of solutions, the Newton polyhedra allow to study the global properties of solutions, as it was shown by D.N. Bernshtein, A.G. Kushnirenko, A.G. Khovanskii and others.

3. A brief survey of the book

This approach to nonlinear problems in coordinates X lead to linear problems in coordinates $\log X = (\log x_1, \ldots, \log x_n)$. If the dimension of the problem $n \leq 3$, then the corresponding linear problem may be solved graphically. If $n > 3$, then to solve the problem one needs special algorithms and computer programs described in Chapter 1. Generally, the book pays a great attention both to description of algorithms of Power Geometry and to examples of their application. The applications are mainly related to problems in the Theory of Mechanisms, Celestial Mechanics, Hydrodynamics and Thermodynamics. Applications to problems in other sciences are also possible.

The theory and algorithms developed in the book should be considered only as the first steps on the way of using the concepts of Power Geometry. Here there is a great field for activity, although the "linear" mentality in the logarithmic space is still uncommon among the specialists in nonlinear analysis. As to expounding material, the book classifies as something between a textbook and a monograph, because its statements and proofs are substantiated by large number of examples and figures.

Chapter 1 gives some information in the theory of linear inequalities in $\mathbb{R}^n$ along with their geometric interpretation, which is necessary for analysis of nonlinear problems in other chapters. Algorithms of solution of systems of linear inequalities are set forth, as well as their modifications for the purposes of Power Geometry, and the corresponding computer programs are described. The problems with finite or infinite set $\mathbf{S}$ are considered, along with the problems described by a single equation or by a system of equations. The necessary material in the theory of linear transformations is given as well.

Chapter 2 deals with truncations and power transformations of a single algebraic equation (Sections 1–4) and of systems of such equations (Sections 5–8). The methods of successive resolving of a singularity, which allow to find all branches of the algebraic manifold near the singularity and at infinity, are cited. This approach is particularly advanced for the algebraic curves. The problems of Robotics are considered.

Chapter 3 deals with systems of ordinary differential equations (2.1). An algorithm is designated for them, which allows to find asymptotics of solutions tending to a stationary point or to infinity. In addition, along with truncations and power transformations, the logarithmic transformations of coordinates are also introduced, which allow to find asymptotics of solutions with multiple logarithms of coordinates. It is shown, that with the help of generalized power transformations many systems that are common in applications may be simplified.

Chapter 4 deals with a Hamiltonian system of ordinary differential equations with m degrees of freedom near the origin or near infinity. It is supposed that the Hamiltonian function is a polynomial or a Laurent series. The truncated systems of such a Hamiltonian system are studied. It is shown, that not all truncated systems are Hamiltonian ones. An algorithm allowing to find all truncated systems, which are Hamiltonian ones, is suggested. The problems of Celestial Mechanics are considered.

In Chapter 5, the solutions to a reversible four-dimensional system of ordinary differential equations (2.1) with two small parameters are studied near the stationary point. The system appeared in the problem of surface water waves after its reduction on the center manifold. The basic first approximation to the system is singled out using the Newton polyhedron. Then the number of its parameters is reduced to one by a power transformation. The resulting system is studied near the stationary point with the help of the normal form for all values of the parameter. The theory of structure of the normal form is developed for the resonant cases with pure imaginary eigenvalues. By a power transformation of the normal form and by an isolation of the first approximations to the resulting system, the new local families of periodic solutions and of quasi-periodic solutions are found. The results are applied to the initial Hydrodynamical problem.

In Chapter 6, the technique of truncations and power transformations is generalized on arbitrary differential equations (ordinary and partial), and also on systems of arbitrary equations including mixed systems of algebraic and differential equations. It is shown how to find asymptotics of their solutions. In Section 6 with the help of this approach the strict substantiation of the boundary layer theory is given for the problem of the flow around a flat plate. Previously the mechanical and physical heuristics were used for the purpose.

In Chapter 7, a quasi-homogeneous partial differential equation without parameters is considered. It is shown how to find all its quasi-homogeneous (self-similar) solutions by the support of the equation with the help of Linear Algebra computations. The simplifications of such an equation are studied with the help of power and logarithmic transformations. It is shown that these transformations allow to reduce the order of the quasi-homogeneous ordinary differential equation, and that for such an equation the boundary value problems may be simplified. Generalizations of these results for a quasi-homogeneous system of differential equations are formulated. In examples, equations of combustion process without a source and with a source are considered.

In the concluding Chapter 8, the present state of Power Geometry is summed up. A classification of the complexity levels of problems of Power Geometry is proposed. The classification consist of four levels, and it is based on the complexity of the geometric objects corresponding to a problem in the space of power exponents. It is also given a comparative survey of these objects, and the based on them methods of analysis of solutions for systems of algebraic equations, for systems of ordinary differential equations, and for systems of partial differential equations. The publications where the methods of Power Geometry were effectively used are cited.

All Chapters are fairly independent from each other, and one can read them in any order.

In the double indexing of formulae, theorems, lemmas and remarks the first number designates Section (in Chapter), the second one is the internal number inside Section. In the figure and table indexing the first number designates the chapter. For the sake of brevity, vectors are written as matrix-rows; but in matrix operations they are supposed to be matrix-columns. For a matrix α, its transposed matrix is denoted as α^*.

In Bibliography, the Russian publications are indicated as such. In other publications, the language of the title coincides with the language of the publication.

Author thanks his students and colleagues Alexander B. Aranson (figures and tables), Vladimir Yu. Petrovich (typesetting), Victor P. Varin (LaTeX make-up) for preparation of the Russian version of the book. Special gratitude to Professor Ahmadjohn Soleev, who was the initiator of the writing of the book. It was supposed, that he would be a co-author of the book, but he failed to get to Moscow from Samarkand and to participate in writing it. A financial support from the Russian Foundation for Basic Research was helpful in writing the book (Grants 93–01–16045 and 96–01–01411), as well as in its publication (Grant 96–01–14122).

Chapter 1

The linear inequalities

1. Principal definitions and properties

Let us recall the principal definitions, which are common in the theory of linear inequalities, and their geometric interpretations (see [Goldman and Tucker 1956, Tucker 1956, Chernikov 1968, Pshenichnyi 1980, Brøndsted 1983] and examples in Section 1 of Chapter I [Bruno 1979a]).

Let $\mathbb{R}^n$ be the vector space of $Q = (q_1, \ldots, q_n)$, and $\mathbb{R}^n_*$ be the dual vector space of $P = (p_1, \ldots, p_n)$, such that the scalar product

$$\langle P, Q\rangle \stackrel{\text{def}}{=} p_1 q_1 + \ldots + p_n q_n$$

is defined. Let $\mathbf{S}$ be a set in $\mathbb{R}^n$. Consider its *inner hulls*

$$\{Q\colon\ Q = \lambda_1 Q_1 + \ldots + \lambda_s Q_s,\ \ Q_j \in \mathbf{S}\} \tag{1.1}$$

for various coefficient values $\lambda_j \in \mathbb{R}$.

1. λ_j are arbitrary, then (1.1) is the *linear hull* of the set $\mathbf{S}$; it is denoted as $\operatorname{Lin}\mathbf{S}$.

2. $\lambda_1 + \ldots + \lambda_s = 1$, then (1.1) is the *affine hull* of the set $\mathbf{S}$; it is denoted as $\operatorname{Aff}\mathbf{S}$.

3. $\lambda_j \geq 0$, then (1.1) is the *conic hull* of the set $\mathbf{S}$; it is denoted as $\operatorname{Con}\mathbf{S}$.

4. $\lambda_1 + \ldots + \lambda_s = 1$ and $\lambda_j \geq 0$, then (1.1) is the *convex hull* of the set $\mathbf{S}$; it is denoted as $\operatorname{Cnv}\mathbf{S}$.

The following inclusions are obvious

$$\operatorname{Lin}\mathbf{S} \supset \operatorname{Aff}\mathbf{S} \supset \operatorname{Cnv}\mathbf{S}, \quad \operatorname{Lin}\mathbf{S} \supset \operatorname{Con}\mathbf{S} \supset \operatorname{Cnv}\mathbf{S}.$$

If the set $\mathbf{S}$ coincides with its hull, then in cases 1–4 it is respectively:

1) *linear subspace*;
2) *linear manifold*;
3) *convex cone*;
4) *convex set.*

When a vector $P \in \mathbb{R}^n_*$ and a constant c are fixed, the equation

$$\langle P, Q\rangle = c \tag{1.2}$$

defines in $\mathbb{R}^n$ a *hyperplane* $\mathbf{H} = \{Q\colon \langle P, Q\rangle = c\}$ orthogonal to the vector P. It contains the point $Q = 0$ if and only if $c = 0$. In $\mathbb{R}^n$ it isolates the *negative halfspace*

$$\mathbf{H}^- = \{Q\colon \langle P, Q\rangle \le c\}. \tag{1.3}$$

For the set $\mathbf{S} \subset \mathbb{R}^n$ and the fixed vector $P \in \mathbb{R}^n_*$, $P \neq 0$, the hyperplane (1.2) is called *supporting* to $\mathbf{S}$, if $c = \sup\langle P, Q\rangle$ over $Q \in \mathbf{S}$, it is denoted as $\mathbf{H}_P$. The corresponding *halfspace* (1.3) is also called *supporting*, and it is denoted as $\mathbf{H}^-_P$. If $\mathbf{H}_P = \mathbf{H}_{-P}$, then $\mathbf{S} \subset \mathbf{H}_P$.

Consider now the *outer hulls* of the set $\mathbf{S}$.

1. The intersection of all supporting hyperplanes $\mathbf{H}_P = \mathbf{H}_{-P}$, which include the set $\mathbf{S}$ and the origin $Q = 0$. It is called the *linear hull* of the set S, and it is denoted as LIN S.

2. The intersection of all supporting hyperplanes $\mathbf{H}_P = \mathbf{H}_{-P}$, which include the set $\mathbf{S}$. It is called the *affine hull* of the set $\mathbf{S}$, and it is denoted as AFF S.

3. The intersection of all supporting to $\mathbf{S}$ halfspaces $\mathbf{H}^-_P$, which have the constant $c = 0$ in the writing (1.3). It is called the *conic hull* of the set $\mathbf{S}$, and it is denoted as CON S.

4. The intersection of all supporting to $\mathbf{S}$ halfspaces $\mathbf{H}^-_P$ is called the *polyhedron hull* of the set $\mathbf{S}$ and is denoted as CNV S.

The following inclusions are obvious

$$\mathrm{LIN}\,\mathbf{S} \supset \mathrm{AFF}\,\mathbf{S} \supset \mathrm{CNV}\,\mathbf{S}, \quad \mathrm{LIN}\,\mathbf{S} \supset \mathrm{CON}\,\mathbf{S} \supset \mathrm{CNV}\,\mathbf{S}.$$

The topological *closure* of the set $\mathbf{S}$ is denoted as $\overline{\mathbf{S}}$. Let us now establish the relations between the inner and the outer hulls of the set $\mathbf{S}$:

1) $\mathrm{Lin}\,\mathbf{S} = \overline{\mathrm{Lin}}\,\mathbf{S} = \mathrm{LIN}\,\mathbf{S}$;
2) $\mathrm{Aff}\,\mathbf{S} = \overline{\mathrm{Aff}}\,\mathbf{S} = \mathrm{AFF}\,\mathbf{S}$;
3) $\mathrm{Con}\,\mathbf{S} \subset \overline{\mathrm{Con}}\,\mathbf{S} = \mathrm{CON}\,\mathbf{S}$;
4) $\mathrm{Cnv}\,\mathbf{S} \subset \overline{\mathrm{Cnv}}\,\mathbf{S} = \mathrm{CNV}\,\mathbf{S}$.

Here the linear space and the linear manifold are closed sets.

The intersection of the set $\mathbf{S}$ with its supporting hyperplane $\mathbf{H}_P$ is called the *boundary subset* and is denoted as

$$\mathbf{S}_P = \mathbf{S} \cap \mathbf{H}_P. \tag{1.4}$$

The boundary subsets $\mathbf{S}_P$ of the set $\mathbf{S}$ form a structure [Bruno 1962, 1973b], if $\mathbf{S}_P \cap \mathbf{S}_R$ means the set-theoretic intersection, $\mathbf{S}_P \cup \mathbf{S}_R$ means the least boundary subset that includes $\mathbf{S}_P$ and $\mathbf{S}_R$, and the order relationship means inclusion $\mathbf{S}_P \supset \mathbf{S}_R$. The unit in the structure is the set $\mathbf{S}$ itself, and zero is the empty set $\varnothing$. If the set $\mathbf{S}$ is convex, then its boundary subset $\mathbf{S}_P$ is also convex, and it is called the *face*. The operations of taking the inner convex hull and of isolating the boundary subset commute, i.e.

$$\mathrm{Cnv}\,(\mathbf{S}_P) = (\mathrm{Cnv}\,\mathbf{S})_P. \tag{1.5}$$

In general, this property does not hold for the outer convex hull, and there is only the inclusion

$$\mathrm{CNV}\,(\mathbf{S}_P) \subset (\mathrm{CNV}\,\mathbf{S})_P. \tag{1.6}$$

Besides,

$$\mathbf{S}_P = \mathbf{S} \cap (\mathrm{Cnv}\,\mathbf{S})_P = \mathbf{S} \cap (\mathrm{CNV}\,\mathbf{S})_P. \tag{1.7}$$

Example 1.1. Let $n = 2$ and the set $\mathbf{S}$ consist of integer points $Q = (q_1, q_2)$ with $q_1 \geq 0$, $q_2 \geq 1$, and the point $(0,0)$. Then $\mathrm{Cnv}\,\mathbf{S} = \{Q\colon\ q_1 \geq 0,\ q_2 > 0$ and point $Q = 0\}$, but $\mathrm{CNV}\,\mathbf{S} = \{Q\colon\ q_1, q_2 \geq 0\}$. For $P = (0, -1)$ we have $\mathbf{S}_P = \{0\}$, $(\mathrm{Cnv}\,\mathbf{S})_P = \{0\}$, $\mathrm{CNV}\,(\mathbf{S}_P) = \{0\}$, and $(\mathrm{CNV}\,\mathbf{S})_P = \{Q\colon\ q_1 \geq 0,$ $q_2 = 0\}$. $\square$

Below the words *poly-*, *multi-* and *finite-* are used as synonyms. Let us recall, that the *dimension* d of a linear subspace $\mathbf{L}$ is equal to the number of linearly independent over $\mathbb{R}$ vectors lying in $\mathbf{L}$. Consequently, $\mathbf{L}$ is fully determined either by d vectors $Q_1, \ldots, Q_d \in \mathbf{L} \subset \mathbb{R}^n$ as $\mathrm{Lin}\,\{Q_1, \ldots, Q_d\}$, or by $n-d$ vectors $P_1, \ldots, P_{n-d} \in \mathbb{R}^n_*$ as the orthogonal complement to $\mathrm{Lin}\,\{P_1, \ldots, P_{n-d}\}$, i.e. as the set of solutions Q to the system $\langle P_j,\, Q\rangle = 0$, $j = 1, \ldots, n-d$. Here vectors $Q_1, \ldots, Q_d$ form a *basis* of the linear subspace $\mathbf{L}$, and vectors $P_1, \ldots, P_{n-d}$ form a *basis* of its normal subspace. Similarly, a linear manifold $\mathbf{M}$ is determined either by $d+1$ vectors $Q_1, \ldots, Q_{d+1} \in \mathbb{R}^n$ as $\mathrm{Aff}\,\{Q_1, \ldots, Q_{d+1}\}$, or by $n-d$ vectors $P_1, \ldots, P_{n-d} \in \mathbb{R}^n_*$ and constants $_1, \ldots, c_{n-d} \in \mathbb{R}$ as the set of solutions Q to the system of equations

$$\langle P_k,\, Q\rangle = c_k, \quad k = 1, \ldots, n-d.$$

Here $\mathbf{M} = Q_0 + \mathbf{L}$, where Q_0 is an arbitrary vector in $\mathbf{M}$; $\mathbf{L}$ is the linear subspace, and $+$ means the vector summation of sets, i.e. taking the sums of vectors from each addend. Here d is the *dimension* of the manifold $\mathbf{M}$. Generally, the *dimension* of a convex set $\mathbf{S}$ is equal to the dimension of its affine hull $\mathrm{Aff}\,\mathbf{S}$, and it is denoted as $\dim \mathbf{S}$.

There is a one-to-one correspondence between the boundary subsets $\mathbf{S}_P$ and $(\mathrm{Cnv}\,\mathbf{S})_P$ due to the property (1.5). Faces $(\mathrm{Cnv}\,\mathbf{S})_P$ and $(\mathrm{CNV}\,\mathbf{S})_P$ have their own dimensions. Each face $(\mathrm{Cnv}\,\mathbf{S})_P$ lies in a face $(\mathrm{CNV}\,\mathbf{S})_R$ of the same dimension, but in general not every face $(\mathrm{CNV}\,\mathbf{S})_R$ contains a face $(\mathrm{Cnv}\,\mathbf{S})_P$ of the same dimension. Thus, in Example 1.1 for $P = (0, -1)$ we have $\dim(\mathrm{CNV}\,\mathbf{S})_P = 1$; this face contains only zero-dimensional face $(\mathrm{Cnv}\,\mathbf{S})_P = \{0\}$.

If a cone $\mathbf{C}$ is the conic hull of a finite number of vectors $Q_1, \dots, Q_s \in \mathbb{R}^n$, then it may be given as the intersection of a finite number of halfspaces

$$\langle P_k, Q\rangle \le 0, \quad k = 1, \dots, t. \tag{1.8}$$

The reverse is also true. Such a cone is called the *polyhedral* one.

A cone $\mathbf{C}$ is called the *forward* one, if it does not contain a linear subspace (or a linear manifold) of a positive dimension. A polyhedral forward cone may be given as the cone hull of the minimal number of vectors $Q_1, \dots, Q_r$, which form its *skeleton*. None of the vectors of the skeleton is lying in the cone hull of other vectors of the skeleton. An arbitrary polyhedral cone $\mathbf{C}$ may be represented as the sum of the forward cone $\mathbf{N}$ and the linear subspace $\mathbf{L}$:

$$\mathbf{C} = \mathbf{N} + \mathbf{L};$$

here the cone $\mathbf{N}$ and its skeleton are determined up to vectors from $\mathbf{L}$. A *cone* $\mathbf{C}^* \subset \mathbb{R}^n$ is called the *dual* to the cone $\mathbf{C} \subset \mathbb{R}^n$, if $\langle P, Q\rangle \le 0$ for all $P \in \mathbf{C}^*$ and all $Q \in \mathbf{C}$. Note, that $\mathbf{C}^{**} = \overline{\mathbf{C}}$.

The inner convex hull of a finite number of points $Q_1, \dots, Q_s$ is called the *polyhedron* or *polytope*. A polyhedron $\mathbf{\Gamma}$ is also may be given as the intersection of a finite number of halfspaces of the form

$$\langle P_k, Q\rangle \le c_k, \quad k = 1, \dots, t. \tag{1.9}$$

Conversely, the intersection of a finite number of halfspaces of the form (1.9) is called the *polyhedral set*, it may be represented as the sum of a polyhedron and a cone. The boundary subsets of the polyhedral set are its *faces* $\mathbf{\Gamma}_j^{(d)}$, where d is the dimension of the face, and j is its number. Their number is finite. Zero-dimensional faces $\mathbf{\Gamma}_j^{(0)}$ are *vertices*, one-dimensional faces $\mathbf{\Gamma}_j^{(1)}$ are *edges* etc. If the dimension of a polyhedral set equals n, then it may be given as the intersection of the lowest number of halfspaces of the form (1.4), the boundary of each of them passes through its own hyperface $\mathbf{\Gamma}_k^{(n-1)}$.

Faces $\mathbf{\Gamma}_j^{(d)}$ of a polyhedral set $\mathbf{\Gamma}$ form a structure [Bruno 1962, 1973b], if $\mathbf{\Gamma}_j^{(d)} \cap \mathbf{\Gamma}_k^{(e)}$ means the set-theoretic intersection, $\mathbf{\Gamma}_j^{(d)} \cup \mathbf{\Gamma}_k^{(e)}$ means the least face that includes $\mathbf{\Gamma}_j^{(d)}$ and $\mathbf{\Gamma}_k^{(e)}$, and the order relationship means inclusion $\mathbf{\Gamma}_j^{(d)} \supset \mathbf{\Gamma}_k^{(e)}$. The unit in the structure is the polyhedral set $\mathbf{\Gamma}$, and zero is the empty set $\varnothing$. It is possible to draw the structural diagram of the faces. The faces of the same dimension are placed in one line of the diagram, and the face

$\mathbf{\Gamma}_i^{(d)}$ is connected by dash with the face $\mathbf{\Gamma}_j^{(d+1)}$, if $\mathbf{\Gamma}_j^{(d+1)} \supset \mathbf{\Gamma}_i^{(d)}$ (Fig. 1.1). Here each face is the intersection of some hyperfaces $\mathbf{\Gamma}_j^{(l-1)}$, where $l = \dim \mathbf{\Gamma}$.

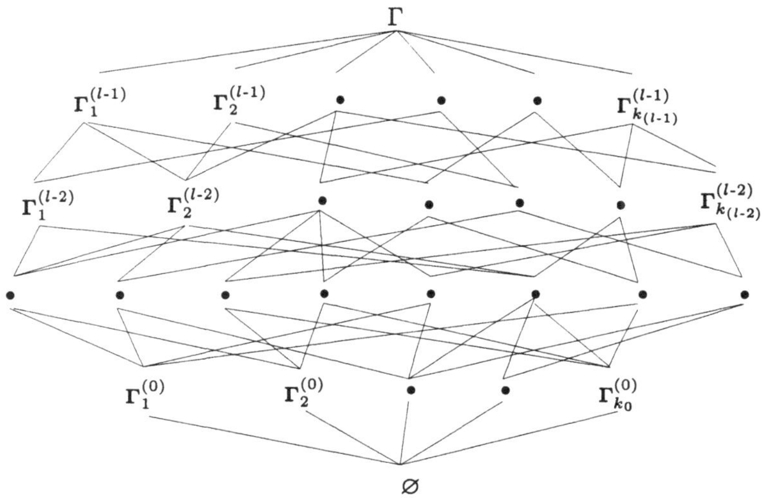

Figure 1.1: The structural diagram of the faces of a polyhedral set.

A polyhedron, i.e. a compact polyhedral set, is the convex hull of its vertices, and each its face is the convex hull of the vertices belonging to that face. Therefore a polyhedron may be defined by its vertices, and each its face may be defined by its own vertices.

Let $\mathbf{\Gamma}$ be a polyhedron of a dimension l and k_d be the number of its faces of dimension d. The *Euler formula* holds

$$\sum_{d=-1}^{l} (-1)^d k_d = 0, \quad k_{-1} = k_l = 1, \quad l \geq 0. \tag{1.10}$$

A face $\mathbf{\Gamma}_j^{(d)}$ that differs from the whole polyhedron $\mathbf{\Gamma}$ and from the empty set $\varnothing$ is called the *proper* one; it coincides with the intersection of all hyperfaces $\mathbf{\Gamma}_k^{(l-1)}$, which include $\mathbf{\Gamma}_j^{(d)}$. For $d \leq l-3$ there are no less then $l-d$ of such hyperfaces, for $d = l-2$ there are exactly two (i.e. $l-d$), and for $d = l-1$ there is exactly one (i.e. $l-d$).

A polyhedron $\mathbf{\Gamma}$ of a dimension l is called the *simple* one, if for $d = 0, 1, \ldots, l-1$ each its face $\mathbf{\Gamma}_j^{(d)}$ belongs exactly to $l-d$ of its hyperfaces $\mathbf{\Gamma}_k^{(l-1)}$. The *Dehn—Sammerville equations* [Brøndsted 1983] for the simple polyhedron are valid

$$\sum_{d=0}^{i}(-1)^d \frac{(l-d)!}{(l-i)!\,(i-d)!}\, k_d = k_i, \quad i = 0, 1, \ldots, l. \tag{1.11}$$

When $i = l$ formula (1.11) coincides with the Euler formula (1.10). Formulae (1.11) are useful for checking the number of faces obtained by means of an algorithm. For example, when $l = 3$ if one of the three numbers k_0, k_1, k_2 is given, then the other two are determined from formulae (1.11). Namely, $k_0 = 2k_2 - 4$, $k_1 = 3k_2 - 6$.

The most simple polyhedron is the *simplex*, the number of vertices k_0 of which is greater by one then its dimension l: $k_0 = l + 1$. For $l = 0, 1, 2, 3$ a simplex is respectively a point, a segment, a triangle, a tetrahedron. Each d-dimensional face of a simplex again is a simplex. The number of d-dimensional faces of a l-dimensional simplex is $k_d = (l+1)!/[(d+1)!\,(l-d)!]$.

A forward cone is called the *simplicial* one, if the convex hull of the vertices of its skeleton vectors and zero point is a simplex. Each of its faces is also a simplicial cone. For a l-dimensional simplicial cone the number of d-dimensional faces equals $l!/[d!\,(l-d)!]$.

2. The normal and tangent cones

In Section 1 there was exposed the theory of closed objects defined by non-strict inequalities for the coefficients λ_j in (1.1), and for the scalar product in (1.3), (1.8) and (1.9) as well. In fact analogous objects are defined in an obvious way with strict inequalities, and also with mixed inequalities. In particular, we will encounter convex cones given in the form (1.1), where some $\lambda_j \geq 0$, and others $\lambda_j > 0$, or in the form (1.8), where some of the inequalities are strict. Such a cone may not be closed and may not include some (or all) vectors of its skeleton.

Let $\mathbf{S}$ be a given set in $\mathbb{R}^n$ and its boundary subset $\mathbf{S}' \stackrel{\text{def}}{=} \mathbf{S}_{P'}$ corresponding to the vector $P' \in \mathbb{R}^n_*$ be chosen (see (1.4)). The set of all vectors $P \in \mathbb{R}^n_*$ for which $\mathbf{S}_P = \mathbf{S}_{P'} \stackrel{\text{def}}{=} \mathbf{S}'$ is called the *normal cone* of the boundary subset $\mathbf{S}'$. It is denoted as $\mathbf{U}(\mathbf{S}; \mathbf{S}')$ and determined by the relations

$$\mathbf{U}(\mathbf{S}; \mathbf{S}') = \{P\colon \langle P, Q' \rangle = \langle P, Q'' \rangle, \ \ Q', Q'' \in \mathbf{S}';$$
$$\langle P, Q' \rangle > \langle P, Q \rangle, \ \ Q \in \mathbf{S} \setminus \mathbf{S}'\}. \tag{2.1}$$

The normal cones corresponding to different boundary subsets of the set $\mathbf{S}$ do not intersect, and the union of all normal cones complemented by the point $P = 0$ forms a convex cone, which is called the *cone of the normals* of the set $\mathbf{S}$ and denoted as $\mathbf{U}[\mathbf{S}]$. The closures of the normal cones $\overline{\mathbf{U}}(\mathbf{S}; \mathbf{S}_P)$ form a structure dual to the structure of the boundary subsets $\mathbf{S}_P$.

Problem 1. For a given set $\mathbf{S} \subset \mathbb{R}^n$ find all boundary subsets $\mathbf{S}_P$ and their normal cones $\mathbf{U}(\mathbf{S}; \mathbf{S}_P)$.

According to the property (1.5)

$$\mathbf{U}(\mathbf{S}; \mathbf{S}_P) = \mathbf{U}(\operatorname{Cnv} \mathbf{S}; (\operatorname{Cnv} \mathbf{S})_P),$$

therefore if we solve Problem 1 for the set $\operatorname{Cnv} \mathbf{S}$ instead of $\mathbf{S}$ and take advantage of the first equality (1.7), then we obtain the solution of Problem 1. With respect to computations, it is simpler to solve Problem 1 for a closed set $\mathbf{S}$. Therefore it is more convenient at first to solve Problem 1 for the set $\operatorname{CNV} \mathbf{S}$, which is the closure of the set $\operatorname{Cnv} \mathbf{S}$. Then from the faces $(\operatorname{CNV} \mathbf{S})_P$ one has to extract those which contain faces $(\operatorname{Cnv} \mathbf{S})_R$ of the same dimension, and for each face $(\operatorname{Cnv} \mathbf{S})_R$ to compose its normal cone as the union of all those normal cones $\mathbf{U}(\operatorname{CNV} \mathbf{S}; (\operatorname{CNV} \mathbf{S})_P)$ that correspond to faces $(\operatorname{CNV} \mathbf{S})_P$ including the face $(\operatorname{Cnv} \mathbf{S})_R$ and not including any larger face $(\operatorname{Cnv} \mathbf{S})_S \supset (\operatorname{Cnv} \mathbf{S})_R$.

Firstly, in Sections 3–5 we shall describe methods of solution of Problem 1 for a set $\mathbf{S}$ consisting of a finite number of points $Q_1, \ldots, Q_s$, i.e.

$$\mathbf{S} = \{Q_1, \ldots, Q_s\}. \tag{2.2}$$

In that case the convex hulls $\operatorname{Cnv} \mathbf{S}$ and $\operatorname{CNV} \mathbf{S}$ coincide. It is a closed polyhedron $\mathbf{\Gamma}$. Its boundary $\partial\mathbf{\Gamma}$ consists of a finite number of faces $\mathbf{\Gamma}_j^{(d)}$. Besides there is a one-to-one correspondence between the boundary subsets $\mathbf{S}_P$ of the set $\mathbf{S}$ and the faces $\mathbf{\Gamma}_P$ of the polyhedron $\mathbf{\Gamma}$. Namely, to each face $\mathbf{\Gamma}_j^{(d)} = \mathbf{H}_P \cap \mathbf{\Gamma}$ there corresponds its boundary subset $\mathbf{S}_j^{(d)} = \mathbf{S} \cap \mathbf{\Gamma}_j^{(d)} = \mathbf{H}_P \cap \mathbf{S}$. Their normal cones coincide: $\mathbf{U}(\mathbf{\Gamma}; \mathbf{\Gamma}_j^{(d)}) = \mathbf{U}(\mathbf{S}; \mathbf{S}_j^{(d)})$; we denote them as $\mathbf{U}_j^{(d)}$. The faces $\mathbf{\Gamma}_j^{(d)}$ (and the boundary subsets $\mathbf{S}_j^{(d)}$) form a structure, and the closures $\overline{\mathbf{U}}_j^{(d)}$ of the normal cones $\mathbf{U}_j^{(d)}$ form the dual structure.

Let $\tilde{Q}$ be an internal point of the face $\mathbf{\Gamma}_j^{(d)}$ of the polyhedral set $\mathbf{\Gamma}$, i.e. $\tilde{Q}$ does not lie in a face of smaller dimension. If $d = 0$, then $\tilde{Q} = \mathbf{\Gamma}_j^{(0)}$. We call the cone $\operatorname{Con}(\mathbf{\Gamma} - \tilde{Q}) \subset \mathbb{R}^n$ the *tangent cone of the face* $\mathbf{\Gamma}_j^{(d)}$ [Bruno 1979a], and we denote it as $\mathbf{T}_j^{(d)}$. It is easy to see that it is the dual one to the normal cone $\mathbf{U}_j^{(d)}$ of the same face. If the cone $\mathbf{U}_j^{(d)}$ is defined by a skeleton $N_1, \ldots, N_l$ and is non-closed, then the tangent cone $\mathbf{T}_j^{(d)}$ is given by the inequalities:

$$\mathbf{T}_j^{(d)} = \{Q : \langle N_i, Q \rangle \leq 0, \ i = 1, \ldots, l\}.$$

Conversely, if a closed cone $\mathbf{T}_j^{(d)}$ is defined by a basis $B_1, \ldots, B_k$ of its linear subspace and the vectors of its skeleton $R_1, \dots, R_l$, then

$$\mathbf{U}_j^{(d)} = \{P\colon \langle P, B_i\rangle = 0, \ \ i = 1, \dots, k; \ \ \langle P, R_m\rangle < 0, \ \ m = 1, \dots, l\}.$$

Let the dimension of the whole polyhedron $\mathbf{\Gamma}$ be equal to n. Then the normal cone $\mathbf{U}_j^{(n-1)}$ of the hyperface $\mathbf{\Gamma}_j^{(n-1)}$ is a ray, orthogonal to the hyperface $\mathbf{\Gamma}_j^{(n-1)}$ and directed outside of the polyhedron $\mathbf{\Gamma}$. The tangent cone $\mathbf{T}_j^{(n-1)}$ is a half-space, which is bounded by the hyperplane passing through the origin $Q = 0$ and parallel to the hyperface $\mathbf{\Gamma}_j^{(n-1)}$. The normal cone $\mathbf{U}_j^{(n-2)}$ is a two-dimensional sector, situated in the plane orthogonal to the face $\mathbf{\Gamma}_j^{(n-2)}$. The tangent cone $\mathbf{T}_j^{(n-2)}$ is the intersection of two half-spaces. The normal cone $\mathbf{U}_j^{(1)}$ of the edge $\mathbf{\Gamma}_j^{(1)}$ is a part of its orthogonal hyperplane; the tangent cone $\mathbf{T}_j^{(1)}$ is n-dimensional and contains only one straight line, passing through the origin $Q = 0$ and parallel to the edge $\mathbf{\Gamma}_j^{(1)}$. The normal cone $\mathbf{U}_j^{(0)}$ of the vertex $\mathbf{\Gamma}_j^{(0)}$ is a n-dimensional polyhedral forward cone. The tangent cone $\mathbf{T}_j^{(0)}$ of the vertex $\mathbf{\Gamma}_j^{(0)}$ is also a n-dimensional polyhedral forward cone. From this geometrical interpretation it follows that since the dimension of the face $\mathbf{\Gamma}_j^{(d)}$ equals d, the dimension of the normal cone $\mathbf{U}_j^{(d)}$ equals $n - d$, and the dimension of the tangent cone $\mathbf{T}_j^{(d)}$ always equals n.

If $\dim \mathbf{\Gamma} = l < n$, then the tangent cones $\mathbf{T}_j^{(d)}$ lie in the linear subspace of dimension l, parallel to $\mathbf{\Gamma}$, and their dimension equals l. Here a tangent cone $\mathbf{T}_j^{(d)}$ can be represented in the form of the sum of a linear subspace $\mathrm{Lin}\,(\mathbf{\Gamma}_j^{(d)} - \tilde{Q})$ and a forward cone of dimension $l - d$.

The faces $\mathbf{\Gamma}_j^{(d)}$, their normal cones $\mathbf{U}_j^{(d)}$ and their tangent cones $\mathbf{T}_j^{(d)}$ can only be given by means of vertices of the polyhedron $\mathbf{\Gamma}$. Moreover, the normal cone $\mathbf{U}_j^{(d)}$ can be given using only edges lying in the face $\mathbf{\Gamma}_j^{(d)}$ and adjacent to it. Let the edges $\mathbf{\Gamma}_1^{(1)}, \dots, \mathbf{\Gamma}_k^{(1)}$ lie in the face $\mathbf{\Gamma}_j^{(d)}$, and the edges $\mathbf{\Gamma}_{k+1}^{(1)}, \dots, \mathbf{\Gamma}_l^{(1)}$ be adjacent to the face $\mathbf{\Gamma}_j^{(d)}$. Let us denote R_i the unit vector of the edge $\mathbf{\Gamma}_i^{(1)}$, with the vectors $R_{k+1}, \dots, R_l$ directed from the face $\mathbf{\Gamma}_j^{(d)}$. Then the normal cone is written as

$$\mathbf{U}_j^{(d)} = \{P\colon \langle P, R_i\rangle = 0, \ \ i = 1, \dots, k;$$

$$\langle P, R_i\rangle < 0, \ \ i = k+1, \dots, l\}, \qquad (2.3)$$

and the tangent cone as

$$\mathbf{T}_j^{(d)} = \{Q\colon Q = \gamma_1 R_1 + \cdots + \gamma_k R_k + \delta_{k+1} R_{k+1} + \cdots + \delta_l R_l, \ \delta_i \geq 0\}. \quad (2.4)$$

3. Graphical solution of Problem 1

If $n = 2$, then we can plot the points $Q_1, \dots, Q_s$ of the set (2.2) in the plane (q_1, q_2) and draw their convex hull $\mathbf{\Gamma}$. Then in the plane (p_1, p_2) we have to draw the normal rays to the edges of the open polygon $\partial\mathbf{\Gamma}$. The decomposition of the open polygon $\partial\mathbf{\Gamma}$ into edges and vertices and the corresponding decomposition of the plane (p_1, p_2) into rays and sectors between them give the solution to Problem 1. This is described in detail in Section 1, Chapter I of [Bruno 1979a].

If $n = 3$, then we can represent the points Q_j of the set (2.2) in one (or two) projection(s) of $\mathbb{R}^3$ on a plane, draw their convex hull, that is a projection of the polyhedron $\mathbf{\Gamma}$, and extract its vertices $\mathbf{\Gamma}_j^{(0)}$, edges $\mathbf{\Gamma}_j^{(1)}$ and faces $\mathbf{\Gamma}_j^{(2)}$. In order to understand the disposition of the normal cones $\mathbf{U}_j^{(d)}$, we must take two planes

$$\mathbf{H}_\pm = \{P\colon \langle P, H \rangle = \pm 1\}$$

and plot on them the intersections with the normal cones. Here we can take as vector H, for instance, $H = (1, 0, 0)$ or $H = (1, 1, 1)$.

Example 3.1. Let $n = 3$, $s = 5$, and the set $\mathbf{S}$ consists of the points

$$\begin{aligned} &Q_1 = (1,1,1), \quad Q_2 = (4,0,0), \quad Q_3 = (0,4,0), \\ &Q_4 = (0,0,4), \quad Q_5 = (2,0,2). \end{aligned} \tag{3.1}$$

The convex hull $\mathbf{\Gamma}$ of the set $\mathbf{S}$ is a tetrahedron, which is shown on Fig. 1.2. It consists of four vertices Q_1–Q_4, six edges and four ordinary faces. Here the point Q_5 lies on the edge which joins the vertices Q_2 and Q_4.

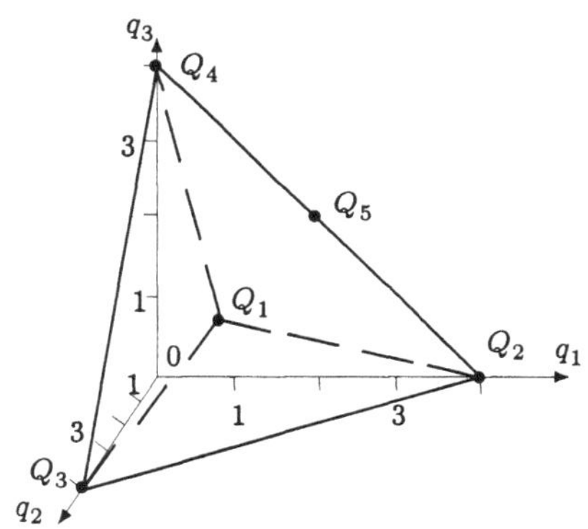

Figure 1.2: The tetrahedron $\mathbf{\Gamma}$ of Example 3.1.

To each element $\mathbf{\Gamma}_j^{(d)}$ there corresponds its normal cone $\mathbf{U}_j^{(d)}$, defined by the inequalities (2.1), where for Q, Q' and Q'' we can take only vertices. Thus, the vertex $\mathbf{\Gamma}_1^{(0)} = \{Q_1\}$ has the normal cone

$$\mathbf{U}_1^{(0)} = \{P\colon \langle P, Q_1\rangle > \langle P, Q_j\rangle,\ j = 2,3,4\} = \{P\colon \langle P, Q_j - Q_1\rangle < 0\} =$$
$$= \{P\colon 3p_1 - p_2 - p_3 < 0,\ -p_1 + 3p_2 - p_3 < 0,\ -p_1 - p_2 + 3p_3 < 0\} =$$
$$= \{P\colon 3p_1 < p_2 + p_3,\ 3p_2 < p_1 + p_3,\ 3p_3 < p_1 + p_2\}.$$

The edge $\mathbf{\Gamma}_1^{(1)} \supset \{Q_1, Q_2\}$ has the normal cone

$$\mathbf{U}_1^{(1)} = \{P\colon \langle P, Q_1\rangle = \langle P, Q_2\rangle,\ \langle P, Q_1\rangle > \langle P, Q_3\rangle,\ \langle P, Q_1\rangle > \langle P, Q_4\rangle\} =$$
$$= \{P\colon 3p_1 = p_2 + p_3,\ 3p_2 < p_1 + p_3,\ 3p_3 < p_1 + p_2\} =$$
$$= \{P\colon 3p_1 = p_2 + p_3,\ 2p_2 < p_3,\ 2p_3 < p_2\}.$$

The face $\mathbf{\Gamma}_1^{(2)} \supset \{Q_1, Q_2, Q_3\}$ has the normal cone

$$\mathbf{U}_1^{(2)} = \{P\colon \langle P, Q_1\rangle = \langle P, Q_2\rangle = \langle P, Q_3\rangle > \langle P, Q_4\rangle\} =$$
$$= \{P\colon 3p_1 = p_2 + p_3,\ 3p_2 = p_1 + p_3,\ 3p_3 < p_1 + p_2\} =$$
$$= \{P\colon 2p_1 = 2p_2 = p_3 < 0\}.$$

For the vector $H = (1,1,1)$ the intersections of the normal cones $\mathbf{U}_j^{(d)}$ with the planes $\mathbf{H}_-$ and $\mathbf{H}_+$ are shown in Fig. 1.3 with the coordinates p_2, p_3. Here the intersections are labeled as the normal cones themselves. The corresponding boundary subsets are

$$\mathbf{S}_j^{(0)} = \{Q_j\},\ j = 1,2,3,4;\ \mathbf{S}_{k-1}^{(1)} = \{Q_1, Q_k\},\ k = 2,3,4,$$

$$\mathbf{S}_4^{(1)} = \{Q_2, Q_3\},\ \mathbf{S}_5^{(1)} = \{Q_2, Q_4, Q_5\},\ \mathbf{S}_6^{(1)} = \{Q_3, Q_4\};$$

$$\mathbf{S}_1^{(2)} = \{Q_1, Q_2, Q_3\},\ \mathbf{S}_2^{(2)} = \{Q_1, Q_2, Q_4, Q_5\},\ \mathbf{S}_3^{(2)} = \{Q_1, Q_3, Q_4\},$$

$$\mathbf{S}_4^{(2)} = \{Q_2, Q_3, Q_4, Q_5\}.\square \tag{3.2}$$

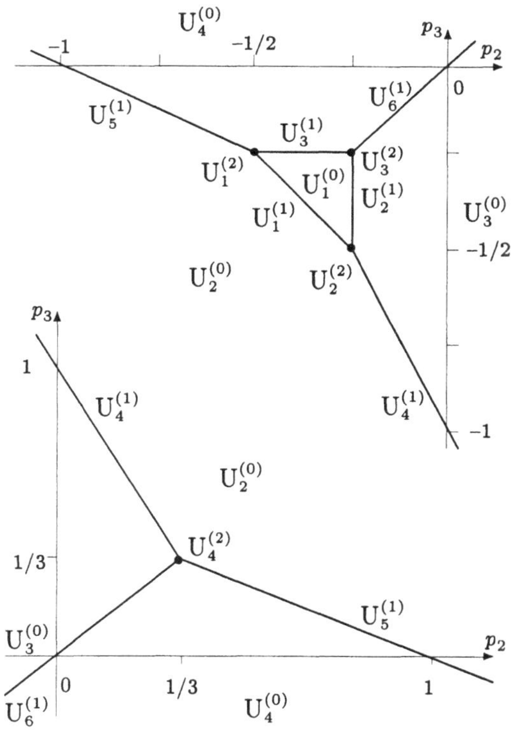

Figure 1.3: Intersections of the normal cones $\mathbf{U}_j^{(d)}$ of Example 3.1 with the planes $p_1 + p_2 + p_3 = -1$ (above) and $p_1 + p_2 + p_3 = 1$ (below) in the coordinates p_2, p_3.

4. The Motzkin–Burger algorithm

In this section we present one of the methods of computation of the skeleton of a cone given by half-spaces, i.e. a method to obtain solutions to a system of linear homogeneous inequalities of the form

$$l_j(P) \stackrel{\text{def}}{=} a_{j1}p_1 + a_{j2}p_2 + \ldots + a_{jn}p_n \leq 0, \quad j = 1, \ldots, m. \tag{4.1}$$

We note that the simplex method is an algebraic one of solving the problems of linear programming. With its help, one can solve the problems relating to finite systems of linear inequalities, and find one or another particular solution. However, the simplex method and the other methods of solving the problems of linear programming can not determine the totality of the solutions (in the form of skeletons) to the finite system of inequalities (4.1).

To solve this problem, Motzkin and Schoenberg [1954] proposed a method of double description, which was not substantiated, but only illustrated descriptively with the help of geometric considerations. A rigorous algebraic presentation and justification of that method was given by Burger [1956]. Below we present briefly the Motzkin–Burger algorithm. For details, see the book [Chernikov 1968].

Let the rank of the matrix (a_{jk}) in the system (4.1) be r, the cone of its solutions be $\mathbf{C}$ and the maximal linear subspace in $\mathbf{C}$ be $\mathbf{L}(\mathbf{C})$. Let a basis for $\mathbf{L}(\mathbf{C})$ be

$$\mathbf{B}(\mathbf{C}) = \{B^1, \ldots, B^s\}, \quad s = n - r$$

(if $r = n$, then $\mathbf{B}(\mathbf{C})$ is the empty set), and a *fundamental system of solutions* to the system (4.1) be

$$\mathbf{N}(\mathbf{C}) = \{N^1, \ldots, N^t\}, \tag{4.2}$$

i.e. every solution P to the system (4.1) is represented in the form

$$P = \sum_{i=1}^{t} \lambda_i N^i + \sum_{j=1}^{s} \mu_j B^j,$$

where $\lambda_i \geq 0$ (if the system (4.1) has no fundamental solution, then $\mathbf{N}(\mathbf{C})$ is the empty set). Supplementing the system (4.1) by the inequality

$$l(P) \stackrel{\text{def}}{=} a_1 p_1 + \ldots + a_n p_n \leq 0, \tag{4.3}$$

we obtain the system

$$l_j(P) \leq 0, \quad j = 1, 2, \ldots, m, \quad l(P) \leq 0, \tag{4.4}$$

whose cone of solutions $\mathbf{C}'$ is a part of the cone of solutions $\mathbf{C}$ to the system (4.1) and, in general, differs from it.

The problem of determining at least one basis $\mathbf{B}(\mathbf{C}')$ of the maximal linear subspace $\mathbf{L}(\mathbf{C}')$ of the cone $\mathbf{C}'$ and at least one fundamental system $\mathbf{N}(\mathbf{C}')$ of solutions to the system (4.4) has a great practical interest. To that problem reduces the finding of a general formula of solutions to the system (4.1). In fact, supplementing the trivial inequality $0p_1 + \ldots + 0p_n \leq 0$ (for which $\mathbf{C} = \mathbb{R}^n_*$, i.e. $\mathbf{B}(\mathbf{C}) = \{E_1, \ldots, E_n\}$ and $\mathbf{N}(\mathbf{C}) = \varnothing$) by other inequalities of the system (4.1) one by one and solving each time the obtained system, we can find the sets $\mathbf{B}(\mathbf{C})$ and $\mathbf{N}(\mathbf{C})$ for the system (4.1).

Theorem 4.1 [Chernikov 1968] *1. If* $s \geq 1$ *and*

$$l(B^i) = 0, \quad i = 1, 2, \ldots, s,$$

then $\mathbf{L}(\mathbf{C}) = \mathbf{L}(\mathbf{C}')$, *and therefore* $\mathbf{B}(\mathbf{C}) = \mathbf{B}(\mathbf{C}')$; *if* $s > 1$ *and, for instance,* $l(B^1) \neq 0$, *then the* $s - 1$ *elements*

$$B^i l(B^1) - B^1 l(B^i), \quad i = 2, \ldots, s,$$

form a basis $\mathbf{B}(\mathbf{C}')$ *of the space* $\mathbf{L}(\mathbf{C}')$. *For* $s = 1$ *and* $l(B^1) \neq 0$, *and also for* $s = 0$, *the set* $\mathbf{B}(\mathbf{C}')$ *is empty.*

2. a) If $s \geq 1$ *and*

$$l(B^i) = 0, \quad i = 1, 2, \ldots, s,$$

and also if $s = 0$, *then the system* $\mathbf{N}(\mathbf{C}')$ *is composed of those vectors* N^i *of the set* (4.2) *for which* $l(N^i) \leq 0$, *and in the case* $t > 1$ *also of those vectors whose indices (but not their order) differ from each other*

$$N(k, k') = N^k |l(N^{k'})| + N^{k'} |l(N^k)| \tag{4.5}$$

with $l(N^k)/l(N^{k'}) < 0$, *for each of which there exist* $r - 2$ *linearly independent inequalities of the system* (4.1) *(i.e. the* $r - 2$ *inequalities* $l_j(N(k, k')) \leq 0$ *with linearly independent forms* $l_j(P)$*), reducing to equalities for* N^k *and for* $N^{k'}$ *as well (for* $r = 2$ *this condition drops off).*

b) If, for instance, $l(B^1) \neq 0$, *then the system* $\mathbf{N}(\mathbf{C}')$ *is composed of the elements*

$$B^0, \; -N^k l(B^0) + B^0 l(N^k), \quad k = 1, 2, \ldots, t,$$

where B^0 *is an arbitrary element of* $\mathbf{L}(\mathbf{C})$ *with* $l(B^0) < 0$.

3. Among the inequalities of the system (4.1), *reducing to equalities by element* N^k *as well as by element* $N^{k'}$ *of the system* $\mathbf{N}(\mathbf{C})$, *there exist* $r - 2$ *linearly independent inequalities if and only if no one of the other elements of the system* $\mathbf{N}(\mathbf{C})$ *reduces all inequalities* (4.1) *to equalities.*

The algorithm described in Theorem 4.1 is considerably shortened if the rank r of the system (4.1) coincides with the number n of its unknowns, because with that condition the system has not nonzero solutions, which would change into equalities all inequalities, and so the space $\mathbf{L}(\mathbf{C})$ is the origin. In that case, Theorem 4.1 is transformed into the following theorem.

Theorem 4.2 [Chernikov 1968] *If* $N^1, \ldots, N^t$ $(t > 1)$ *is a fundamental system* $\mathbf{N}(\mathbf{C})$ *of solutions to the system of linear inequalities* (4.1) *of rank* n *over* $\mathbb{R}^n_*$ *and* (4.3) *is an arbitrary linear inequality over* $\mathbb{R}^n_*$, *then those elements* N^i *for which* $l(N^i) \leq 0$, *and those elements* (4.5) *with* $l(N^k)/l(N^{k'}) < 0$ *(differing from each other at least by one of the indices* k, k'*), for each of which there exist* $n - 2$ *linearly independent inequalities* $l_j(P) \leq 0$, *reducing to equalities for* N^k *and for* $N^{k'}$, *form a fundamental system of solutions to the system of inequalities* (4.4). *If such elements do not exist, then the system* (4.4) *has no fundamental solution.*

Among the inequalities of the system (4.1) *of rank* n, *reducing to equalities by element* N^k *as well as by element* $N^{k'}$ *of the system* $\mathbf{N}(\mathbf{C})$, *there exist* $n-2$ *linearly independent inequalities if and only if no one of the other elements of the system* $\mathbf{N}(\mathbf{C})$ *reduces all inequalities* (4.1) *to equalities.*

5. Algorithmic solution of Problem 1

Let $\mathbf{S}'$ be a subset of the set $\mathbf{S}$ from (2.2), not necessarily a boundary one. We determine the cone

$$\overline{\mathbf{U}}(\mathbf{S}') = \{P\colon \langle P, Q_i\rangle = \langle P, Q_j\rangle,\ \ Q_i, Q_j \in \mathbf{S}';$$

$$\langle P, Q_j\rangle \geq \langle P, Q_k\rangle,\ \ Q_k \in \mathbf{S}\setminus\mathbf{S}'\} \tag{5.1}$$

as the closure of the cone (2.1). In a similar way, we denote as $\overline{\mathbf{U}}_l^{(d)}$ the closure of the normal cone $\mathbf{U}_l^{(d)}$. Obviously,

$$\overline{\mathbf{U}}_l^{(d)} = \overline{\mathbf{U}}(\mathbf{S}_l^{(d)}).$$

Lemma 5.1 *Let $\mathbf{S}'$ and $\mathbf{S}''$ be subsets of the set $\mathbf{S}$. Then*

$$\overline{\mathbf{U}}(\mathbf{S}'\cup\mathbf{S}'') = \overline{\mathbf{U}}(\mathbf{S}')\cap\overline{\mathbf{U}}(\mathbf{S}'').$$

Proof. Let Q_i and Q_j run through $\mathbf{S}'$; Q_k and Q_l run through $\mathbf{S}''$; and Q_m run through $\mathbf{S}\setminus(\mathbf{S}'\cup\mathbf{S}'')$. Then $\overline{\mathbf{U}}(\mathbf{S}')$ is given by the relations

$$\langle P, Q_i\rangle = \langle P, Q_j\rangle,\ \ \langle P, Q_i\rangle \geq \langle P, Q_k\rangle,\ \ \langle P, Q_i\rangle \geq \langle P, Q_m\rangle,$$

and $\overline{\mathbf{U}}(\mathbf{S}'')$ by the relations

$$\langle P, Q_k\rangle = \langle P, Q_l\rangle,\ \ \langle P, Q_k\rangle \geq \langle P, Q_i\rangle,\ \ \langle P, Q_k\rangle \geq \langle P, Q_m\rangle.$$

For the intersection $\overline{\mathbf{U}}(\mathbf{S}')\cap\overline{\mathbf{U}}(\mathbf{S}'')$ these two systems are satisfied simultaneously, i.e. the middle inequalities reduce to equalities. We obtain the relations

$$\langle P, Q_i\rangle = \langle P, Q_j\rangle = \langle P, Q_k\rangle = \langle P, Q_l\rangle,\ \ \langle P, Q_i\rangle \geq \langle P, Q_m\rangle,$$

which determine the cone $\overline{\mathbf{U}}(\mathbf{S}'\cup\mathbf{S}'')$. □

Theorem 5.1 *Let $\mathbf{S}'$ be a subset of the set $\mathbf{S}$. Then:*

$$\overline{\mathbf{U}}(\mathbf{S}') = \overline{\mathbf{U}}_l^{(d)}, \tag{5.2}$$

where $\mathbf{\Gamma}_l^{(d)}$ is the smallest face containing all points of $\mathbf{S}'$; if $\mathbf{S}'$ does not lie entirely in any face, then $\overline{\mathbf{U}}(\mathbf{S}') = \{0\}$.

Proof. Let $\mathbf{\Gamma}_l^{(d)}$ be the smallest face containing all points of $\mathbf{S}'$. Then $\mathbf{S}' \subset \mathbf{S}_l^{(d)}$, therefore

$$\overline{\mathbf{U}}(\mathbf{S}') \supset \overline{\mathbf{U}}(\mathbf{S}_l^{(d)}) = \overline{\mathbf{U}}_l^{(d)}.$$

If $\mathbf{S}' = \mathbf{S}_l^{(d)}$, then the equality (5.2) is satisfied.

We consider the case where $\mathbf{S}' \neq \mathbf{S}_l^{(d)}$ and we assume that in that case the assertion of the theorem is not true, i.e. there exists a vector $P_0 \in \overline{\mathbf{U}}(\mathbf{S}')$ such that $P_0 \notin \overline{\mathbf{U}}_l^{(d)}$. Then the hyperplane $\mathbf{H}_{P_0}$ supporting to $\mathbf{S}$ contains $\mathbf{S}'$ and does not contain at least one point of $\mathbf{S}_l^{(d)}$. But the supporting hyperplanes intersect $\mathbf{\Gamma}$ only along faces, and the mentioned supporting hyperplane $\mathbf{H}_{P_0}$ intersects $\mathbf{\Gamma}$ along a face $\mathbf{\Gamma}_{l_1}^{(d_1)}$ containing $\mathbf{S}'$. Here $\mathbf{\Gamma}_{l_1}^{(d_1)} \supset \mathbf{S}'$ and $\mathbf{\Gamma}_{l_1}^{(d_1)} \not\supset \mathbf{\Gamma}_l^{(d)}$. Therefore the intersection $\mathbf{\Gamma}_{l_1}^{(d_1)} \cap \mathbf{\Gamma}_l^{(d)}$ is a face, which is smaller than $\mathbf{\Gamma}_l^{(d)}$ and contains $\mathbf{S}'$. But this contradicts the hypothesis that the face $\mathbf{\Gamma}_l^{(d)}$ is the smallest containing $\mathbf{S}'$.

Assume that $\mathbf{S}'$ does not lie in any face. We suppose that $\overline{\mathbf{U}}(\mathbf{S}')$ contains a vector $P \neq 0$. Then $\mathbf{S}'$ lies in the supporting hyperplane $\mathbf{H}_P$, corresponding to the vector P. But every supporting hyperplane intersects with $\mathbf{\Gamma}$ along a face. In particular, $\mathbf{S}'$ lies in the face $\mathbf{H}_P \cap \mathbf{\Gamma}$, which contradicts the initial hypothesis. □

Among points of the set $\mathbf{S}$ of the form (2.2), the vertices play a special role: every face $\mathbf{\Gamma}_j^{(d)}$ is the convex hull of its vertices. If

$$\dim \mathbf{\Gamma} = n, \tag{5.3}$$

then in the dual space $\mathbb{R}_*^n$ the normals N_i to the hyperfaces $\mathbf{\Gamma}_i^{(n-1)}$ play an analogous role. Namely, the skeleton of each normal cone consists of normal vectors N_i. Here if $N_{i_1}, \ldots, N_{i_k}$ are the skeleton of a normal cone $\mathbf{U}_j^{(d)}$, then the rank of the matrix

$$T = (N_{i_1} \ldots N_{i_k}), \tag{5.4}$$

where vectors N_i are considered as matrices-columns, equals $n - d$. If $\mathbf{\Gamma}_i^{(d)} \supset \mathbf{\Gamma}_j^{(d')}$, then $\overline{\mathbf{U}}_i^{(d)} \subset \overline{\mathbf{U}}_j^{(d')}$ and the skeleton of the cone $\mathbf{U}_i^{(d)}$ constitutes a part of the skeleton of the cone $\mathbf{U}_j^{(d')}$. Let $r = k_{n-1}$ be the number of hyperfaces $\mathbf{\Gamma}_i^{(n-1)}$. On the basis of Theorem 5.1, one can compose the following table of correspondence between the points $Q_j \in \mathbf{S}$ and the normals N_i to the hyperfaces $\mathbf{\Gamma}_i^{(n-1)}$.

We set $\theta_{ij} = +$ if N_i belongs to the skeleton of the cone $\overline{\mathbf{U}}(Q_j)$, and $\theta_{ij} = -$ in the opposite case, and we write the values θ_{ij} in a table with r lines (corresponding to the N_i) and s columns (corresponding to the Q_j). We call the table of the values θ_{ij} the *table of correspondence* of the set $\mathbf{S}$ and denote it as $M(\mathbf{S})$. We call the totality of the values θ_{ij} for $i = i_1, \ldots, i_k$ and $j = j_1, \ldots, j_l$ a *minor* and we denote it as $M^{i_1,\ldots,i_k}_{j_1,\ldots,j_l}$. We call this minor *positive* if all its elements θ_{ij} are pluses. Finally, we call a positive minor *maximal* if for each fixed $i \neq i_1, \ldots, i_k$ there is a minus among the values $\theta_{ij_1}, \ldots, \theta_{ij_l}$ and for each fixed $j \neq j_1, \ldots, j_l$ there is a minus among the values $\theta_{i_1 j}, \ldots, \theta_{i_k j}$.

Theorem 5.2 *a) Let* $\mathbf{S}^{(d)}_i = \{Q_{j_1}, \ldots, Q_{j_l}\}$ *be a boundary subset of the set (2.2), and the skeleton of its normal cone* $\mathbf{U}^{(d)}_i$ *consists of the vectors* $N_{i_1}, \ldots, N_{i_k}$. *Then the minor* $M^{i_1,\ldots,i_k}_{j_1,\ldots,j_l}$ *of the table* $M(\mathbf{S})$ *is maximal positive.*

b) Conversely, if $M^{i_1,\ldots,i_k}_{j_1,\ldots,j_l}$ *is a maximal positive minor of the table* $M(\mathbf{S})$, *then the subset* $\{Q_{j_1}, \ldots, Q_{j_l}\}$ *of the set (2.2) is a boundary subset, let us say* $\mathbf{S}^{(d)}_i$, *and the vectors* $N_{i_1}, \ldots, N_{i_k}$ *form the skeleton of its normal cone* $\mathbf{U}^{(d)}_i$.

Proof. a) Let $\mathbf{\Gamma}^{(d')}_j$ be the smallest face containing the point Q_{j_1}. Obviously, $\mathbf{\Gamma}^{(d')}_j \subset \mathbf{\Gamma}^{(d)}_i$. According to Theorem 5.1

$$\overline{\mathbf{U}}(Q_{j_1}) = \overline{\mathbf{U}}^{(d')}_j \supset \overline{\mathbf{U}}^{(d)}_i.$$

Therefore all vectors $N_{i_1}, \ldots, N_{i_k}$ belong to the skeleton of the cone $\overline{\mathbf{U}}(Q_{j_1})$, i.e. all $\theta_{i_1 j_1}, \ldots, \theta_{i_k j_1}$ are $+$. In a similar way, it can be shown that $\theta_{i_1 j_2}, \ldots, \theta_{i_k j_2}$ are $+$, and so on. In that way, moving along $j = j_2, \ldots, j_l$, we find that the minor $M^{i_1,\ldots,i_k}_{j_1,\ldots,j_l}$ is positive. We show now that it is maximal. Assume that for some $j' \neq j_1, \ldots, j_l$ all values $\theta_{i_1 j'}, \ldots, \theta_{i_k j'}$ are $+$, i.e. the vectors $N_{i_1}, \ldots, N_{i_k}$ belong to the skeleton of the cone $\overline{\mathbf{U}}(Q_{j'})$. But then $Q_{j'}$ lies in the intersection of the supporting hyperplanes for these vectors; since their intersection with the polyhedron $\mathbf{\Gamma}$ is a face $\mathbf{\Gamma}^{(d)}_i$, then $Q_{j'} \in \mathbf{\Gamma}^{(d)}_i$. But this contradicts the condition that $\mathbf{S}^{(d)}_i = \{Q_{j_1}, \ldots, Q_{j_l}\} \not\ni Q_{j'}$. Therefore, for each $j' \neq j_1, \ldots, j_l$, among the values $\theta_{i_1 j'}, \ldots, \theta_{i_k j'}$ there is a $-$. Assume now that for some $i' \neq i_1, \ldots, i_k$ all values $\theta_{i' j_1}, \ldots, \theta_{i' j_l}$ are $+$. Then the vector $N_{i'}$ belongs to the skeleton of the cone $\mathbf{U}^{(d)}_i$, which contradicts the hypothesis that this skeleton consists of the vectors $N_{i_1}, \ldots, N_{i_k}$. The assertion a) is proved.

b) Since the minor $M^{i_1,\ldots,i_k}_{j_1,\ldots,j_l}$ is positive, the vectors $N_{i_1}, \ldots, N_{i_k}$ belong to the skeletons of each of the cones

$$\overline{\mathbf{U}}(Q_{j_1}), \ldots, \overline{\mathbf{U}}(Q_{j_l}),$$

i.e. to the skeleton of the cone $\overline{\mathbf{U}}(\mathbf{S}')$, where $\mathbf{S}' = \{Q_{j_1}, \ldots, Q_{j_l}\}$. Therefore, $\overline{\mathbf{U}}(\mathbf{S}') \supset \overline{\mathbf{C}}$, where $\overline{\mathbf{C}}$ is the convex closed conical hull of the vectors $N_{i_1}, \ldots, N_{i_k}$. According to Theorem 5.1

$$\overline{\mathbf{U}}(\mathbf{S}') = \overline{\mathbf{U}}(\mathbf{S}^{(d)}_i) = \overline{\mathbf{U}}^{(d)}_i,$$

where $\mathbf{\Gamma}_i^{(d)}$ is the smallest face containing $\mathbf{S}'$. Since the minor $M_{j_1,\dots,j_l}^{i_1,\dots,i_k}$ is the maximal one, then $\mathbf{S}_i^{(d)} = \mathbf{S}'$ and the cone $\overline{\mathbf{C}}$ exhausts the cone $\overline{\mathbf{U}}_i^{(d)}$. □

Remark 5.1. If in the table of correspondence we keep only the columns corresponding to vertices, then Theorem 5.2 is also true for this modified table.

We note the following **properties of the table of correspondence**.

1. The ith line corresponds to the hyperface $\mathbf{\Gamma}_i^{(n-1)}$; the hyperface $\mathbf{\Gamma}_i^{(n-1)}$ contains those and only those points Q_j for which there is a plus in the ith line, i.e. $\theta_{ij} = +$. The vector N_i forms the skeleton of the normal cone $\mathbf{U}_i^{(n-1)}$ of the hyperface $\mathbf{\Gamma}_i^{(n-1)}$.

The proof follows from the properties of the normal cones.

2. The jth column corresponds to the point Q_j; the point lies in those and only those hyperfaces $\mathbf{\Gamma}_i^{(n-1)}$ for which $\theta_{ij} = +$. Therefore it lies in the intersection of these hyperfaces.

The proof follows from the properties of polyhedra.

3. A point Q_j is a vertex if and only if there are no other columns having pluses in all lines for which there is a plus in the jth column.

Proof. Assume that the pluses in the jth column coincide with some pluses of a kth column in the lines with numbers $l_1, \dots, l_m$, and in addition the kth column has pluses in the lines with numbers $l_{m+1}, \dots, l_t$. Since

$$\overline{\mathbf{U}}(Q_k) = \left\{P\colon\ P = \sum_{j=1}^{t} \lambda_i N_{l_i},\ \lambda_i \geq 0,\ i = 1, \dots, t\right\},$$

then, obviously, $\overline{\mathbf{U}}(Q_j) \subset \overline{\mathbf{U}}(Q_k)$, $\overline{\mathbf{U}}(Q_j) \neq \overline{\mathbf{U}}(Q_k)$, and according to the properties of the normal cone the point Q_j is not a vertex.

Assume now that the pluses in the jth column do not coincide with some pluses of any other column. Then $\overline{\mathbf{U}}(Q_j)$ is not a subset of any $\overline{\mathbf{U}}(Q_i)$, $i = 1, \dots, s$, $i \neq j$, and according to the properties of the normal cones the point Q_j is a vertex. □

4. Let $Q_{j_1}, \dots, Q_{j_{d+1}}$ be various vertices; they correspond to the columns with numbers $j_1 \neq \dots \neq j_{d+1}$. If only in k lines with numbers $i_1, \dots, i_k$ (where $k \geq n-d$) there are only pluses in the intersection with the mentioned columns, and the matrix (5.4) formed by the vectors $N_{i_1}, \dots, N_{i_k}$ has a rank $n-d$, then the face $\mathbf{\Gamma}_i^{(d)}$, whose normal cone has the skeleton formed by the vectors $N_{i_1}, \dots, N_{i_k}$, is spanned on the points $Q_{j_1}, \dots, Q_{j_{d+1}}$.

Proof. By the property 1 the convex hull of the set $\{Q_{j_1}, \dots, Q_{j_{d+1}}\}$ coincides with the intersection $\mathbf{\Gamma}_{i_1}^{(n-1)} \cap \dots \cap \mathbf{\Gamma}_{i_k}^{(n-1)}$, where $k \geq n-d$, and it constitutes a face $\mathbf{\Gamma}_i^{(d')}$. Its normal cone has a skeleton $\{N_{i_1}, \dots, N_{i_k}\}$. By the assumption the matrix (5.4) has a rank $n-d$, hence the dimension of that cone is $n-d$. Therefore, $d'=d$. □

5. Let a face $\Gamma_i^{(d)}$ be spanned on the points $Q_{j_1},\dots,Q_{j_{d+1}}$ and the skeleton of its normal cone be formed by the vectors $N_{i_1},\dots,N_{i_k}$. If for some j' all values $\theta_{i_1 j'},\dots,\theta_{i_k j'}$ are $+$, then $Q_{j'}\in\Gamma_i^{(d)}$.

The proof follows from the preceding property.

Let a numeration of points $Q_1,\dots,Q_s$ and normal vectors $N_1,\dots,N_r$ be fixed. Then the boundary subsets $\mathbf{S}_i^{(d)}=\{Q_{j_1},\dots,Q_{j_l}\}$ and the normal cones $\mathbf{U}_i^{(d)}=\{P\colon\ P=\lambda_1 N_{i_1}+\dots+\lambda_k N_{i_k},\ \lambda_i>0\}$ can be given by the sets of indices $[j_1,\dots,j_l]$ and $[i_1,\dots,i_k]$ respectively. The answer to the problem of extracting all boundary subsets and their normal cones is the enumeration of all pairs

$$\mathbf{S}_i^{(d)}=[j_1,\dots,j_l],\quad \mathbf{U}_i^{(d)}=[i_1,\dots,i_k]. \tag{5.5}$$

Now for a finite set (2.2) we describe an algorithm, which gives the solution (5.5) to Problem 1. The algorithm consists of the following steps:

1) for each $j=1,\dots,s$ writing out a system of the linear inequalities which define the cone $\overline{\mathbf{U}}(Q_j)$;

2) solving these systems of inequalities;

3) composition of the table of correspondence $M(\mathbf{S})$;

4) construction of the structural diagram of the faces and the dual structural diagram.

We assume that (5.3) is satisfied. To the system

$$\langle P, Q_k-Q_1\rangle\le 0,\quad k=1,\dots,s, \tag{5.6}$$

which determines the cone $\overline{\mathbf{U}}(Q_1)$, we can obtain the solution in the form of the conical hull of the vectors N_i with the help of the Motzkin–Burger algorithm, i.e. applying Theorem 4.2, we obtain the list of normal vectors $N_1,\dots,N_l$ and we fill the first column of the table of correspondence. Then we find the skeleton of the cone $\overline{\mathbf{U}}(Q_2)$, i.e. we add the normals $N_{l+1},\dots,N_{l+k}$ to the normals $N_1,\dots,N_l$, and we fill the second column of the table of correspondence. Thus, consecutively in j, we find the skeleton of the cone $\overline{\mathbf{U}}(Q_j)$, and we fill the jth column of the table of correspondence. At the sth step we obtain the enumeration of all normals $N_1,\dots,N_r$, and we terminate the composition of the table of correspondence.

Firstly, according to Property 1 of the obtained table of correspondence, we isolate all hyperfaces $\Gamma_i^{(n-1)}$ ($i=1,\dots,k_{n-1}=r$) with the skeletons N_i of the cones $\mathbf{U}_i^{(n-1)}$, and we fill the second line of the structural diagram. The pairwise intersections of the faces $\Gamma_i^{(n-1)}$ ($i=1,\dots,k_{n-1}$) give all faces $\Gamma_i^{(n-2)}$ ($i=1,\dots,k_{n-2}$). For instance, let $\mathbf{S}_i^{(n-1)}=[1,\dots,l,l+1]$ and $\mathbf{S}_j^{(n-1)}=$

$[1, \dots, l, m]$, $m \neq l+1$; then

$$\mathbf{S}_k^{(n-2)} = \mathbf{S}_i^{(n-1)} \cap \mathbf{S}_j^{(n-1)} = [1, \dots, l],$$

and the skeleton of the normal cone $\mathbf{U}_k^{(n-2)}$ will be $[i, j]$. In that way we find all faces $\mathbf{\Gamma}_i^{(n-2)}$ $(i = 1, \dots, k_{n-2})$, and we write them in the third line of the structural diagram. With the help of each intersection of at least three faces $\mathbf{\Gamma}_i^{(n-2)}$ we find the faces $\mathbf{\Gamma}_i^{(n-3)}$ $(i = 1, \dots, k_{n-3})$ with their normal cones $\mathbf{U}_i^{(n-3)}$ $(i = 1, \dots, k_{n-3})$ and we write all $\mathbf{\Gamma}_i^{(n-3)}$ in the fourth line of the structural diagram of the faces. For instance, let

$$\mathbf{S}_1^{(n-2)} = [1, 2, \dots, l], \quad \mathbf{S}_2^{(n-2)} = [2, \dots, l, l+1],$$
$$\mathbf{S}_3^{(n-2)} = [2, \dots, l, m], \quad m \neq l+1,$$
$$\mathbf{U}_1^{(n-2)} = [r, s], \ \mathbf{U}_2^{(n-2)} = [s, t], \ \mathbf{U}_3^{(n-2)} = [r, t];$$

then

$$\mathbf{S}_1^{(n-3)} = [2, \dots, l], \quad \mathbf{U}_1^{(n-3)} = [r, s, t].$$

We note that in that case the rank of the matrix $T = (N_r N_s N_t)$, formed by the normals belonging to the skeleton of the normal cone $\mathbf{U}_1^{(n-3)}$, must be equal to three. Exactly in the same way, we find the faces $\mathbf{\Gamma}_i^{(n-4)}$ $(i = 1, \dots, k_{n-4})$, and so on.

Assume that all faces $\mathbf{\Gamma}_i^{(d+1)}$ with their normal cones $\mathbf{U}_i^{(d+1)}$ are found. We find all faces $\mathbf{\Gamma}_i^{(d)}$ $(i = 1, \dots, k_d)$ as intersections of k $(k \geq n-d)$ faces $\mathbf{\Gamma}_i^{(d+1)}$ $(i = 1, \dots, k_{d+1})$. Here the skeleton of the normal cone $\mathbf{U}_i^{(d)}$ is the union of k $(k \geq n-d)$ skeletons of normal cones $\mathbf{U}_i^{(d+1)}$, and the cone $\mathbf{U}_i^{(d)}$ has a dimension $n-d$, i.e. the rank of the matrix T (see property 4), formed by the normal vectors belonging to the skeleton of the normal cone $\mathbf{U}_i^{(d)}$, equals $n-d$. Thus, we continue this process until we have extracted all vertices $\mathbf{\Gamma}_i^{(0)}$ with the normal cones $\mathbf{U}_i^{(0)}$ $(i = 1, \dots, k_0)$ and we fill in the last but one line of the structural diagram. As it was already mentioned, we obtain the complete structural diagram of the faces and the boundary subsets by writing $\mathbf{\Gamma}$ in the first line and the empty set $\varnothing$ in the last line. To verify the correctness of the extraction of the faces, we apply the Euler formula (1.10). And that is the end of the extraction of all boundary subsets $\mathbf{S}_i^{(d)}$ and skeletons of their normal cones $\mathbf{U}_i^{(d)}$.

We will represent the results of these computations in the form of the *table of the boundary subsets*, organized as follows (see Table 1.2 below). Each line of the table corresponds to one face $\mathbf{\Gamma}_i^{(d)}$, beginning with the hyperfaces $\mathbf{\Gamma}_i^{(n-1)}$ and ending with the vertices $\mathbf{\Gamma}_i^{(0)}$. Here the first column contains the dimensions d of the faces $\mathbf{\Gamma}_i^{(d)}$; the second column contains their numbers i; the third column contains the numbers j of the points $Q_j \subset \mathbf{S}_i^{(d)}$; the fourth column contains the numbers k of the normals N_k, forming the skeleton of the normal cone $\mathbf{U}_i^{(d)}$ (all hyperfaces $\mathbf{\Gamma}_m^{(l-1)}$ including $\mathbf{\Gamma}_i^{(d)}$ can be reconstituted from these normals N_k), the last column contains the numbers m of the faces $\mathbf{\Gamma}_m^{(d+1)}$ of the dimension $d+1$, the intersection of which is the given face $\mathbf{\Gamma}_i^{(d)}$.

If in the resulting structural diagram of the faces, instead of the faces $\mathbf{\Gamma}_i^{(d)}$ we write their normal cones $\mathbf{U}_i^{(d)}$, given by means of skeletons, then as a result we obtain a dual structural diagram (see Figs. 1.4 and 1.5 below).

If (5.3) is not satisfied and $\dim \mathbf{\Gamma} = l < n$, then in $\mathbb{R}_*^n$ there is a linear subspace $\mathbf{L}$, normal to $\mathbf{\Gamma}$, with $\dim \mathbf{L} = n - l$. Let the vectors $B_1, \dots, B_{n-l}$ form a basis in $\mathbf{L}$. Each normal cone $\mathbf{U}_j^{(d)}$ can be given in the form

$$\mathbf{U}_j^{(d)} = \left\{ P : P = \sum_{k=1}^{n-l} \mu_k B_k + \sum_{i=1}^{r} \lambda_i N_{n-l+i}, \quad \lambda_i > 0 \right\},$$

where μ_k are arbitrary real numbers, and N_i are defined up to terms of the form $\sum \mu_k B_k$. To eliminate this ambiguity, we consider in $\mathbb{R}_*^n$ a linear subspace $\mathbf{N}$, transverse to $\mathbf{L}$. For definiteness we can take $\mathbf{N}$ orthogonal to $\mathbf{L}$. In that case $\mathbf{N} = \mathrm{Lin}\,(\mathbf{S} - Q_1)$. If we require the inclusions $N_i \subset \mathbf{N}$, then the normals N_i are uniquely determined. Here the properties of the table of correspondence (between B_k, N_i and Q_j) are preserved, but the hyperface is understood to be a face of dimension $l - 1$.

The algorithm is as follows. We compute all $\overline{\mathbf{U}}(Q_j)$ according to Theorem 4.1 of the Motzkin–Burger algorithm. We obtain a basis $B_1, \dots, B_{n-l}$ in $\mathbf{L}$ and vectors of the skeletons $\widetilde{N}_1, \dots, \widetilde{N}_r$. We will look for the vectors

$$N_j = \widetilde{N}_j + \sum_{k=1}^{n-l} \sigma_{ik} B_k, \quad j = n-l+1, \dots, n-l+r,$$

satisfying the condition $N_j \subset \mathbf{N}$, i.e.

$$0 = \langle B_i, N_j \rangle = \langle B_i, \widetilde{N}_j \rangle + \sum_{k=1}^{n-l} \sigma_{ik} \langle B_i, B_k \rangle,$$

$$i = 1, \dots, n-l; \; j = n-l+1, \dots, n-l+r.$$

For each j we obtain a system of linear equations for $\sigma_{j1}, \dots, \sigma_{j,n-l}$, from which they are uniquely determined. Now we fill the table of correspondence and from it we build the structural diagram. Here the first $n - l$ lines of the table correspond to the vectors of the basis $B_1, \dots, B_{n-l}$. In these lines, all

$\theta_{ij} = +$. All properties of the table of correspondence are preserved. In that case the table of the boundary subsets begins with the hyperfaces $\mathbf{\Gamma}_i^{(l-1)}$, and its fourth column contains the numbers k of the basis vectors B_k and of the normals N_k forming the skeleton of the normal cone $\mathbf{U}_i^{(d)}$.

Example 5.1 (continuation of Example 3.1). We take (3.1) as set $\mathbf{S}$. The convex hull $\mathbf{\Gamma}$ of the set $\mathbf{S}$ is a tetrahedron, represented in Fig. 1.2. We will extract algorithmically all boundary subsets $\mathbf{S}_j^{(d)}$ and normal cones $\mathbf{U}_j^{(d)}$. We list the system of linear inequalities, which define $\overline{\mathbf{U}}(Q_1)$:

$$\begin{aligned} \langle P, Q_2 - Q_1 \rangle &= 3p_1 - p_2 - p_3 \le 0, \\ \langle P, Q_3 - Q_1 \rangle &= -p_1 + 3p_2 - p_3 \le 0, \\ \langle P, Q_4 - Q_1 \rangle &= -p_1 - p_2 + 3p_3 \le 0, \\ \langle P, Q_5 - Q_1 \rangle &= p_1 - p_2 + p_3 \le 0. \end{aligned} \tag{5.7}$$

We will solve this system according to the Motzkin–Burger algorithm exactly, as it was described in Section 4 (see [Chernikov 1968, Bruno and Soleev 1991a]). As a result of the computation we obtain the skeleton $N_1 = (-1,-2,-1)$, $N_2 = (-1,-1,-2)$, $N_3 = (-2,-1,-1)$ of the cone $\overline{\mathbf{U}}(Q_1)$. In the same way we build the systems of linear inequalities of the type (5.7) for the points $Q_2, \dots, Q_5$, and we find for each $\overline{\mathbf{U}}(Q_j)$ its skeleton. The results of these computations are shown in Table 1.1.

Table 1.1: Table of correspondence of the set (3.1).

	Q_1	Q_2	Q_3	Q_4	Q_5
$N_1 = (-1,-2,-1)$	+	+	−	+	+
$N_2 = (-1,-1,-2)$	+	+	+	−	−
$N_3 = (-2,-1,-1)$	+	−	+	+	−
$N_4 = (1,1,1)$	−	+	+	+	+

From Table 1.1 we find the faces $\mathbf{\Gamma}_i^{(d)}$ of the polyhedron $\mathbf{\Gamma}$, their boundary subsets $\mathbf{S}_i^{(d)}$ and their normal cones $\mathbf{U}_i^{(d)}$, shown in Table 1.2.

In Table 1.2, the numbering of the faces $\mathbf{\Gamma}_i^{(d)}$ corresponds to the numbering (3.2) in Example 3.1. The structural diagram of the faces of the polyhedron $\mathbf{\Gamma}$ for the set (3.1) is shown on Fig. 1.4.

Substituting in this diagram, instead of the boundary subsets, the skeletons of their normal cones, we obtain its dual structural diagram of the normal cones (Fig. 1.5). □

Table 1.2: Boundary subsets of the set (3.1).

d	i	j	k	m
2	1	1, 2, 4, 5	1	1
2	2	1, 2, 3	2	1
2	3	1, 3, 4	3	1
2	4	2, 3, 4, 5	4	1
1	1	1, 2	1, 2	1, 2
1	2	1, 4	1, 3	1, 3
1	3	2, 4, 5	1, 4	1, 4
1	4	1, 3	2, 3	2, 3
1	5	2, 3	2, 4	2, 4
1	6	3, 4	3, 4	3, 4
0	1	1	1, 2, 3	1, 2, 4
0	2	2	1, 2, 4	1, 3, 5
0	3	4	1, 3, 4	2, 3, 4
0	4	3	2, 3, 4	4, 5, 6

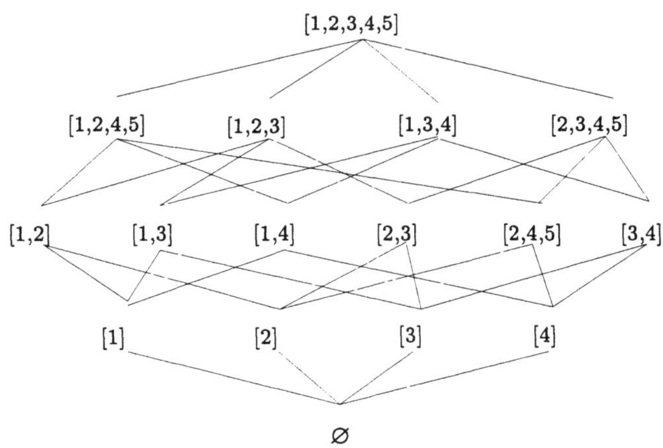

Figure 1.4: The structural diagram of the boundary subsets $\mathbf{S}_j^{(d)}$ of Example 3.1.

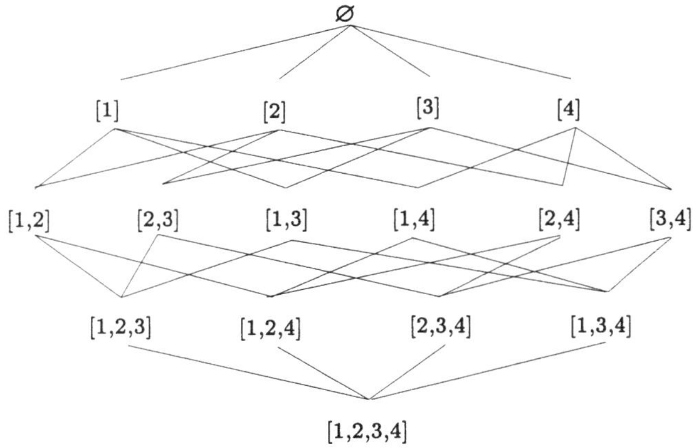

Figure 1.5: The structural diagram of the normal cones $\mathbf{U}_j^{(d)}$ of Example 3.1.

The algorithm proposed in this section for the computation of a polyhedron is based on the computation of all cones $\overline{\mathbf{U}}(Q_j)$. One can organize the computation of the polyhedron in a different way. For instance, if $\dim \mathbf{\Gamma} = n$, then one can compute $\overline{\mathbf{U}}(Q_{j_1}, \ldots, Q_{j_n})$ for all collections $Q_{j_1}, \ldots, Q_{j_n}$ for which the rank of the matrix

$$(Q_{j_2} - Q_{j_1} \ldots Q_{j_n} - Q_{j_1})$$

equals $n - 1$. For such collections the normal vector $\widetilde{N}$ is uniquely determined as the solution to a system of linear equations. There remains only to compute the values $\langle \widetilde{N}, Q_l \rangle$ for all $l \neq j_1, \ldots, j_n$ and to compare them with $\langle \widetilde{N}, Q_{j_1} \rangle$.

If $\langle \widetilde{N}, Q_{j_1} \rangle \geq \langle \widetilde{N}, Q_l \rangle$, then $\widetilde{N}$ is normal to some hyperface $\mathbf{\Gamma}_k^{(n-1)}$. Points $Q_{j_1}, \ldots, Q_{j_n}$ and those of the points Q_l for which $\langle \widetilde{N}, Q_{j_1} \rangle = \langle \widetilde{N}, Q_l \rangle$ are in the hyperface.

If $\langle \widetilde{N}, Q_{j_1} \rangle < \langle \widetilde{N}, Q_l \rangle$ for an index l, then the points $Q_{j_1}, \ldots, Q_{j_n}$ do not lie in one hyperface.

In this way Theorem 5.2 allows to obtain vectors N_k and the corresponding hyperfaces $\mathbf{\Gamma}_k^{(n-1)}$ all at once. Then, moving down the structural diagram, we can determine the remaining $\mathbf{\Gamma}_i^{(d)}$. Here mainly the systems of linear equations are to be solved. But the algorithm based on the computation of $\overline{\mathbf{U}}(Q_j)$ is more standardized (it contains less exceptional cases, requiring a separate program) and requires less item-by-item examination (s points instead of $s! / [\, n! \, (s-n)! \,]$ combinations).

6. Cone of the problem

By the algorithm proposed in the previous section we can also solve for a finite set $\mathbf{S}$ the

Problem 1′. Let in $\mathbb{R}^n_*$ be given an open convex cone $\mathbf{K}$ (*cone of the problem*), and in $\mathbb{R}^n$ be given a set $\mathbf{S}$. From the set $\mathbf{S}$ it is necessary to extract only those boundary subsets $\mathbf{S}_P$ for which the normal cones $\mathbf{U}_P$ intersect the cone of the problem $\mathbf{K}$.

Let the set $\mathbf{S}$ consist of a finite number of points $Q_1, \ldots, Q_s$. Then $\mathbf{S}_P = \mathbf{S}_j^{(d)}$ and $\mathbf{U}_P = \mathbf{U}_j^{(d)}$. Let the cone of the problem $\mathbf{K}$ be given by the skeleton $W_1, \ldots, W_\sigma$ and also be described by the system of inequalities

$$\mathbf{K} = \{P: \langle P, K_k \rangle < 0, \quad k = 1, \ldots, t\}. \tag{6.1}$$

The skeleton of the cone $\mathbf{K}_j^{(d)} \stackrel{\text{def}}{=} \mathbf{U}_j^{(d)} \cap \mathbf{K}$ consists of the normals N_l to the hyperfaces and of the vectors D_m lying in the boundary of the cone $\mathbf{K}$ and not being normals to hyperfaces.

The following algorithm is proposed for the solution of this problem in the case (5.3). Firstly, according to algorithm of Section 3 or Section 5 we find all boundary subsets and their normal cones, including all normal vectors for the set $\mathbf{S}$. Then, using the description (6.1) of the cone of the problem $\mathbf{K}$, we extract those of the normal vectors which lie in the cone $\mathbf{K}$ or in its boundary $\partial\mathbf{K}$. Let these normals be

$$N_1, \ldots, N_r. \tag{6.2}$$

If $\mathbf{K}_j^{(d)} = \mathbf{U}_j^{(d)} \cap \mathbf{K}$ is not empty, then the boundary subset $\mathbf{S}_j^{(d)}$ lies in one of the subsets $\mathbf{S}_k^{(n-1)}$ for $k = 1, \ldots, r$. Therefore the other boundary subsets (not lying in $\mathbf{S}_1^{(n-1)} \cup \ldots \cup \mathbf{S}_r^{(n-1)}$) can further be dropped, and for each of the remaining subsets $\mathbf{S}_j^{(d)}$ we can find the skeleton for $\mathbf{K}_j^{(d)}$, using the Motzkin–Burger algorithm. From these vectors of the skeleton we must extract those that belong to the set (6.2), and the additional $D_{r+1}, \ldots, D_\tau$ that lie in the boundary of the cone $\mathbf{K}$ and are not normals to hyperfaces. For each of the vectors D_k we compute all scalar products $\langle D_k, Q_j \rangle$ for $j = 1, \ldots, s$, and we find their maximum c_k on $Q_j \in \mathbf{S}$. In the kth line of the table of correspondence we put a + in the jth column if $\langle D_k, Q_j \rangle = c_k$, and a − if $\langle D_k, Q_j \rangle < c_k$.

According to the results of these computations, we compose the table of correspondence in the same way as in Section 5. That table has all properties of tables of correspondence, except the first, and the first property holds for the vectors N_j. On the basis of these properties, successively, as in Section 5, we build the structural diagram. Since the structural diagram in that case consists only of those faces of the polyhedron $\mathbf{\Gamma}$ for which the normal cones intersect the cone of the problem $\mathbf{K}$, then the Euler formula (1.10) does not hold for it.

If (5.3) is not satisfied and $\dim \mathbf{\Gamma} = l < n$, then at first the basis $B_1, \ldots, B_{n-l}$ of the linear subspace $\mathbf{L}$ normal to $\mathbf{\Gamma}$, and the normals $N_{n-l+1}, \ldots, N_{n-l+r}$ lying in $\mathrm{Lin}\,(\mathbf{S} - Q_1)$ are computed, as described in Section 5. Let $\mathbf{L}_1$ be the maximal linear subspace in the closure of the cone of the problem $\mathbf{K}$, $\mathbf{L}_2 = \mathbf{L} \cap \mathbf{L}_1$, $\mathbf{N}_2$ be the orthogonal complement to $\mathbf{L}_2$, and $\mathbf{K}_2 = \mathbf{K} \cap \mathbf{N}_2$. The intersection $\mathbf{L} \cap \mathbf{K}_2$ has a skeleton

$$N_0^1, \ldots, N_0^{l_0}; \tag{6.3}$$

each of the intersections $(\mathbf{L} + N_i) \cap \mathbf{K}_2$ has a skeleton consisting of the vectors (6.3) and the vectors

$$N_i^1, \ldots, N_i^{l_i}, \quad i = n - l + 1, \ldots, n - l + r. \tag{6.4}$$

We extract those values of the index i for which the sets (6.4) are not empty. Let them be $i = n - l + 1, \ldots, n - l + \tau$. If the cone $\mathbf{K}_j^{(d)}$ is nonempty, then the boundary subset $\mathbf{S}_j^{(d)}$ lies in one of the subsets $\mathbf{S}_i^{(l-1)}$ for $i - n + l = 1, \ldots, \tau$. Therefore we can drop the other boundary subsets, and for each of the remaining ones find the skeleton for the intersection $\mathbf{K}_j^{(d)} \cap \mathbf{N}_2$, using the Motzkin–Burger algorithm. From these vectors of the skeleton we must extract those that belong to the sets (6.3) and (6.4), and the additional vectors D_k that lie on the boundary of the cone $\mathbf{K}_2$ and are not normals to the hyperfaces $\mathbf{\Gamma}_j^{(l-1)}$. Now each cone $\mathbf{K}_j^{(d)}$ is given by a basis $B_1, \ldots, B_\omega$ of the subspace $\mathbf{L}_2$ and by the skeleton of the cone $\mathbf{K}_j^{(d)} \cap \mathbf{K}_2$, consisting of the vectors (6.3) and the vectors (6.4) for some values i and some vectors D_k. Here the vectors (6.4) with one index i are present or absent all together. The result of these computations can be represented in the form of a table of correspondence for the vectors B_i, N_j and D_k. Here for the vectors B_i and the vectors N_j belonging to the set (6.3) there will be pluses in all columns. For the vectors N_j belonging to a subset of (6.4) with one index i, the pluses are disposed in the same columns.

In what follows we will have to solve Problems 1 and 1′ for three types of sets $\mathbf{S}$, when all components q_{ij} of the vectors $Q_j \in \mathbf{S}$ are either real, i.e. $Q_j \in \mathbb{R}^n$; or rational, i.e. $Q_j \in \mathbb{Q}^n$; or integers, i.e. $Q_j \in \mathbb{Z}^n$.

7. About the computer program

For the case of a finite set of integer points $\mathbf{S} \in \mathbb{Z}^n$, the algorithm described in Sections 4–6 for the determination of all boundary subsets $\mathbf{S}_j^{(d)}$ and their normal cones $\mathbf{U}_j^{(d)}$ was implemented as a program for an IBM-compatible computer with MS-DOS environment [Soleev and Aranson 1994, Aranson 1996].

This program reads a file with the input data of the problem, which in a text form (ASCII code) contains: the dimension of the space n; the number of points s in the studied set $\mathbf{S}$; the number t of inequalities (6.1), which give the cone of the problem $\mathbf{K}$. Then in the data file follow the coordinates of the points Q_i of the set S and the coordinates of the vectors K_k in the inequalities (6.1). The values of the coordinates must be integers, lying in the range -32767 to 32767. Below, in Examples 7.1 and 7.2, the printouts of such files of input data are shown as Tables 1.3 and 1.6.

The results of the execution of the program are:

1) the dimension of the minimal linear subspace containing the set $\mathbf{S}$, and the vectors of the basis of its orthogonal complement (marked by the letter B);

2) the vectors normal to the hyperfaces (marked by the letter N); the coordinates of each such a vector are integers having the largest common divisor equal to 1;

3) the vectors of the intersections of the boundaries of the normal cones with the boundaries of the cone of the problem (marked by the letter D);

4) a table of correspondence, where the columns marked by the letter V, correspond to the vertices whose normal cones intersect the cone of the problem;

5) a table of the boundary subspaces, described by the numbers j of their points Q_j, together with lists of the skeletons of their normal cones, described by the numbers k of the vectors N_k, and the numbers m of the faces of a dimension larger by one, the intersection of which forms the given face.

These results are written by the program in a file, in which firstly the input data of the problem are copied. Printouts of the files with the results are shown in Examples 7.1 and 7.2 below. In Chapters 4 and 5 this program will be applied to concrete problems described by ordinary differential equations. The solution of systems of homogeneous linear inequalities by the Motzkin–Burger method was implemented as a complete collection of procedures in a separate file. These procedures for the solution of systems of homogeneous linear inequalities with integer coefficients can be used by other programs.

As an illustration the printouts of the input data and the results of the execution of the program for two examples are given. In the Example 7.1 the vertices of a four-dimensional octatope and three points lying in faces with dimensions 3, 2 and 1 respectively, and also a point inside the octatope were taken as points of the set **S**. The results of the execution of the program are a table of correspondence and a diagram of all faces of the four-dimensional octatope, and a table of boundary subsets. Example 7.2 is analogous to Example 7.1, but in it we take the negative orthant as the cone of the problem.

Example 7.1 (Tables 1.3–1.5). Table 1.3 gives the input data. Here the dimension of the problem is $n = 4$, the number of points $Q_j \in \mathbf{S}$ is $s = 12$, and for the cone of the problem **K** the number of vectors K_k in (6.1) is $t = 0$. These data are in the first line of table 1.3. In the remaining lines the vectors Q_j in the form of columns are given.

Table 1.3: Input data of Example 7.1.

$n = 4,\ s = 12,\ t = 0$

j	1	2	3	4	5	6	7	8	9	10	11	12
	–1	0	0	0	4	0	0	0	–4	0	0	0
Q_j	–1	–2	0	0	0	4	0	0	0	–4	0	0
	–1	–1	–2	0	0	0	4	0	0	0	–4	0
	–1	–1	–2	0	0	0	0	4	0	0	0	–4

Table 1.4: Table of correspondence of Example 7.1.

					j	1	2	3	4	5	6	7	8	9	10	11	12
i	t_i	B_i, N_i, D_i			T_j	d	d	d	d	V	V	V	V	V	V	V	V
1	N	–1	–1	–1	–1	+	+	+	–	–	–	–	–	+	+	+	+
2	N	1	–1	–1	–1	–	+	+	–	+	–	–	–	–	+	+	+
3	N	–1	1	–1	–1	–	–	+	–	–	+	–	–	+	–	+	+
4	N	1	1	–1	–1	–	–	+	–	+	+	–	–	–	–	+	+
5	N	–1	–1	1	–1	–	–	–	–	–	–	+	–	+	+	–	+
6	N	1	–1	1	–1	–	–	–	–	+	–	+	–	–	+	–	+
7	N	–1	1	1	–1	–	–	–	–	–	+	+	–	+	–	–	+
8	N	1	1	1	–1	–	–	–	–	+	+	+	–	–	–	–	+
9	N	–1	–1	–1	1	–	–	–	–	–	–	–	+	+	+	+	–
10	N	1	–1	–1	1	–	–	–	–	+	–	–	+	–	+	+	–
11	N	–1	1	–1	1	–	–	–	–	–	+	–	+	+	–	+	–
12	N	1	1	–1	1	–	–	–	–	+	+	–	+	–	–	+	–
13	N	–1	–1	1	1	–	–	–	–	–	–	+	+	+	+	–	–
14	N	1	–1	1	1	–	–	–	–	+	–	+	+	–	+	–	–
15	N	–1	1	1	1	–	–	–	–	–	+	+	+	+	–	–	–
16	N	1	1	1	1	–	–	–	–	+	+	+	+	–	–	–	–

Table 1.5: Boundary subsets of Example 7.1.

d	i	j	k	m
3	1	1, 2, 3, 9, 10, 11, 12	1	1
3	2	2, 3, 5, 10, 11, 12	2	1
3	3	3, 6, 9, 11, 12	3	1
3	4	3, 5, 6, 11, 12	4	1
3	5	7, 9, 10, 12	5	1
3	6	5, 7, 10, 12	6	1
3	7	6, 7, 9, 12	7	1
3	8	5, 6, 7, 12	8	1
3	9	9, 10, 11, 12	9	1
3	10	5, 8, 10, 11	10	1
3	11	6, 8, 9, 11	11	1
3	12	5, 6, 8, 11	12	1
3	13	7, 8, 9, 10	13	1
3	14	5, 7, 8, 10	14	1
3	15	6, 7, 8, 9	15	1
3	16	5, 6, 7, 8	16	1
2	1	2, 3, 10, 11, 12	1, 2	1, 2
2	2	3, 9, 11, 12	1, 3	1, 3
2	3	9, 10, 12	1, 5	1, 5
2	4	9, 10, 11	1, 9	1, 9
2	5	3, 5, 11, 12	2, 4	2, 4
2	6	5, 10, 12	2, 6	2, 6
2	7	5, 10, 11	2, 10	2, 10
2	8	3, 6, 11, 12	3, 4	3, 4
2	9	6, 9, 12	3, 7	3, 7
2	10	6, 9, 11	3, 11	3, 11
2	11	5, 6, 12	4, 8	4, 8
2	12	5, 6, 11	4, 12	4, 12
2	13	7, 10, 12	5, 6	5, 6
2	14	7, 9, 12	5, 7	5, 7
2	15	7, 9, 10	5, 13	5, 13
2	16	5, 7, 12	6, 8	6, 8
2	17	5, 7, 10	6, 14	6, 14
2	18	6, 7, 12	7, 8	7, 8
2	19	6, 7, 9	7, 15	7, 15
2	20	5, 6, 7	8, 16	8, 16
2	21	8, 10, 11	9, 10	9, 10
2	22	8, 9, 11	9, 11	9, 11
2	23	8, 9, 10	9, 13	9, 13
2	24	5, 8, 11	10, 12	0, 12
2	25	5, 8, 10	10, 14	0, 14
2	25	5, 8, 10	10, 14	0, 14
2	26	6, 8, 11	11, 12	1, 12
2	27	6, 8, 9	11, 15	1, 15

Table 1.5: Boundary subsets of Example 7.1 (continuation).

d	i	j	k	m
2	28	5, 6, 8	12, 16	2, 16
2	29	7, 8, 10	13, 14	3, 14
2	30	7, 8, 9	13, 15	3, 15
2	31	5, 7, 8	14, 16	4, 16
2	32	6, 7, 8	15, 16	5, 16
1	1	3, 11, 12	1, 2, 3, 4	1, 2, 5, 8
1	2	10, 12	1, 2, 5, 6	1, 3, 6, 13
1	3	10, 11	1, 2, 9, 10	1, 4, 7, 21
1	4	9, 12	1, 3, 5, 7	2, 3, 9, 14
1	5	9, 11	1, 3, 9, 11	2, 4, 10, 22
1	6	9, 10	1, 5, 9, 13	3, 4, 15, 23
1	7	5, 12	2, 4, 6, 8	5, 6, 11, 16
1	8	5, 11	2, 4, 10, 12	5, 7, 12, 24
1	9	5, 10	2, 6, 10, 14	6, 7, 17, 25
1	10	6, 12	3, 4, 7, 8	8, 9, 11, 18
1	11	6, 11	3, 4, 11, 12	8, 10, 12, 26
1	12	6, 9	3, 7, 11, 15	9, 10, 19, 27
1	13	5, 6	4, 8, 12, 16	11, 12, 20, 28
1	14	7, 12	5, 6, 7, 8	13, 14, 16, 18
1	15	7, 10	5, 6, 13, 14	13, 15, 17, 29
1	16	7, 9	5, 7, 13, 15	14, 15, 19, 30
1	17	5, 7	6, 8, 14, 16	16, 17, 20, 31
1	18	6, 7	7, 8, 15, 16	18, 19, 20, 32
1	19	8, 11	9, 10, 11, 12	21, 22, 24, 26
1	20	8, 10	9, 10, 13, 14	21, 23, 25, 29
1	21	8, 9	9, 11, 13, 15	22, 23, 27, 30
1	22	5, 8	10, 12, 14, 16	24, 25, 28, 31
1	23	6, 8	11, 12, 15, 16	26, 27, 28, 32
1	24	7, 8	13, 14, 15, 16	29, 30, 31, 32
0	1	12	1, 2, 3, 5, 4, 6, 7, 8	1, 2, 4, 7, 10, 14
0	2	11	1, 2, 3, 9, 4, 10, 11, 12	1, 3, 5, 8, 11, 19
0	3	10	1, 2, 5, 9, 6, 10, 13, 14	2, 3, 6, 9, 15, 20
0	4	9	1, 3, 5, 9, 7, 11, 13, 15	4, 5, 6, 12, 16, 21
0	5	5	2, 4, 6, 10, 8, 12, 14, 16	7, 8, 9, 13, 17, 22
0	6	6	3, 4, 7, 11, 8, 12, 15, 16	10, 11, 12, 13, 18, 23
0	7	7	5, 6, 7, 13, 8, 14, 15, 16	14, 15, 16, 17, 18, 24
0	8	8	9, 10, 11, 13, 12, 14, 15, 16	19, 20, 21, 22, 23, 24

In Tables 1.4, 1.5 the results of the computation are given. That is Table 1.4 is the table of correspondence, and Table 1.5, which is organized in the same way as Table 1.2, is the table of the boundary subsets. There were obtained 16 hyperplanes, 32 two-dimensional faces, 24 edges and 8 vertices. □

Example 7.2 (Tables 1.6–1.8). Table 1.6, unlike Table 1.3, has $t = 4$, and the vectors K_k from the inequalities (6.1) defining the cone of the problem $\mathbf{K}$ are given after the vectors Q_j. In Tables 1.7, 1.8 the results of the computation are given. That is Table 1.7 is the table of correspondence; Table 1.8 is organized in the same way as Tables 1.2 and 1.5, but in the fourth column are given only the numbers of those vectors N_k and D_k from the left part of Table 1.7, which constitute the skeleton of the intersection $\mathbf{U}_j^{(d)} \cap \mathbf{K}$. Besides only those faces $\mathbf{\Gamma}_j^{(d)}$ are shown for which that intersection is nonempty. There were obtained one hyperplane $\mathbf{\Gamma}_1^{(3)}$, four two-dimensional faces, six edges and four vertices. As it can be seen, the presence of a cone of the problem can significantly shorten the answer. □

Table 1.6: Input data of Example 7.2.

$n = 4,\ s = 12,\ t = 4$

j	1	2	3	4	5	6	7	8	9	10	11	12
	–1	0	0	0	4	0	0	0	–4	0	0	0
Q_j	–1	–2	0	0	0	4	0	0	0	–4	0	0
	–1	–1	–2	0	0	0	4	0	0	0	–4	0
	–1	–1	–2	0	0	0	0	4	0	0	0	–4

k	1	2	3	4
	–1	0	0	0
K_k	0	–1	0	0
	0	0	–1	0
	0	0	0	–1

Table 1.7: Table of correspondence of Example 7.2.

						j	1	2	3	4	5	6	7	8	9	10	11	12
i	t_i	B_i, N_i, D_i				T_j	d	d	d	d	V	V	V	V	V	V	V	V
1	N	–1	–1	–1	–1		+	+	+	–	–	–	–	–	+	+	+	+
2	D	–1	0	0	0		–	–	–	–	–	–	–	–	+	–	–	–
3	D	0	–1	0	0		–	–	–	–	–	–	–	–	–	+	–	–
4	D	–1	–1	0	0		–	–	–	–	–	–	–	–	+	+	–	–
5	D	0	0	–1	0		–	–	–	–	–	–	–	–	–	–	+	–
6	D	–1	0	–1	0		–	–	–	–	–	–	–	–	+	–	+	–
7	D	0	–1	–1	0		–	–	–	–	–	–	–	–	–	+	+	–
8	D	–1	–1	–1	0		–	–	–	–	–	–	–	–	+	+	+	–
9	D	0	0	0	–1		–	–	–	–	–	–	–	–	–	–	–	+
10	D	–1	0	0	–1		–	–	–	–	–	–	–	–	+	–	–	+
11	D	0	–1	0	–1		–	–	–	–	–	–	–	–	–	+	–	+
12	D	–1	–1	0	–1		–	–	–	–	–	–	–	–	+	+	–	+
13	D	0	0	–1	–1		–	–	+	–	–	–	–	–	–	–	+	+
14	D	–1	0	–1	–1		–	–	+	–	–	–	–	–	+	–	+	+
15	D	0	–1	–1	–1		–	+	+	–	–	–	–	–	–	+	+	+

Table 1.8: Boundary subsets of Example 7.2.

d	i	j	k	m
3	1	1, 2, 3, 9, 10, 11, 12	1	1
2	1	9, 10, 11	1, 8	1, 8
2	2	9, 10, 12	1, 12	1, 12
2	3	3, 9, 11, 12	1, 14	1, 14
2	4	2, 3, 10, 11, 12	1, 15	1, 15
1	1	9, 10	1, 4, 8, 12	1, 2
1	2	9, 11	1, 6, 8, 14	1, 3
1	3	10, 11	1, 7, 8, 15	1, 4
1	4	9, 12	1, 10, 12, 14	2, 3
1	5	10, 12	1, 11, 12, 15	2, 4
1	6	3, 11, 12	1, 13, 14, 15	3, 4
0	1	9	1, 2, 4, 6, 8, 10, 12, 14	1, 2, 4
0	2	10	1, 3, 4, 7, 8, 11, 12, 15	1, 3, 5
0	3	11	1, 5, 6, 7, 8, 13, 14, 15	2, 3, 4
0	4	12	1, 9, 10, 11, 12, 13, 14, 15	4, 5, 6

8. An infinite set S

We show here that Problem 1′ (see Section 6) with an infinite set **S** reduces to an analogous problem with a finite set.

Let in $\mathbb{R}^n$ be given a convex cone **C**. We call it an *integer* one if it has a skeleton of integer vectors, and a *forward* one if it does not contain a straight line. The *subset* **S**′ of the set **S** is called *dominant with respect to the forward cone* **C** if $\mathbf{S} \subset \mathbf{S}' + \mathbf{C}$. A dominant subset **S**′ is called the *minimal* one if none of its proper subsets is dominant for the set **S**, and we denote it as $\mathbf{S}/\mathbf{C}$. Let $\mathbf{S} \subset \mathbf{S}_1 \cup \mathbf{S}_2$; obviously, $\mathbf{S}/\mathbf{C} \subset \mathbf{S}_1/\mathbf{C} \cup \mathbf{S}_2/\mathbf{C}$. We recall the notations: $\mathbb{R}^n_+ = \{Q \geq 0\}$ and $\mathbb{Z}^n_+ = \mathbb{R}^n_+ \cap \mathbb{Z}^n$.

Lemma 8.1 *Let $R \in \mathbb{Z}^n$ and $\mathbf{S} \subset R + \mathbb{Z}^n_+$. Then the minimal dominant subset $\mathbf{S}/\mathbb{R}^n_+$ consists of a finite number of points.*

Proof. We will carry it out by induction on the dimension n. For $n = 1$ the assertion of the lemma is obvious, and the number of points equals one. Assume that the assertion of the lemma is true for the dimensions smaller than n. We prove it for the dimension n. Let $S = (s_1, \ldots, s_n) \in \mathbf{S}$. We divide the set $R + \mathbb{Z}^n_+$ into a finite number of parts in the following way:

$$R + \mathbb{Z}^n_+ = \mathbf{M}_0 \bigcup_{i=1}^{n} \bigcup_{j=r_i}^{s_i-1} \mathbf{M}^i_j,$$

where $\mathbf{M}_0 = S + \mathbb{Z}^n_+$ and $\mathbf{M}^i_j = (R + \mathbb{Z}^n_+) \cap \{Q : q_i = j\}$. We write $\mathbf{S}_0 = \mathbf{S} \cap \mathbf{M}_0$

and $\mathbf{S}_j^i = \mathbf{S} \cap \mathbf{M}_j^i$. Then, obviously,

$$\mathbf{S} = \mathbf{S}_0 \bigcup_{i=1}^{n} \bigcup_{j=r_i}^{s_i-1} \mathbf{S}_j^i. \tag{8.1}$$

The set $\mathbf{S}_0/\mathbb{R}_+^n$ has exactly one point S. Each of the sets $\mathbf{S}_j^i/\mathbb{R}_+^n$ has a finite number of points according to the inductive hypothesis, because $\mathbf{M}_j^i$ differs from $\mathbb{R}_+^{n-1}$ by a parallel translation along the vector $R + (j - r_i)E_i$, where E_i denotes the ith unit vector. Therefore, the totality of the points of all sets

$$\mathbf{S}_0/\mathbb{R}_+^n, \quad \mathbf{S}_j^i/\mathbb{R}_+^n, \quad j = r_i, \ldots, s_i - 1; \ i = 1, \ldots, n, \tag{8.2}$$

is finite. Since the set $\mathbf{S}/\mathbb{R}_+^n$ lies in the union of the sets (8.2), then it has a finite number of points. □

Corollary. *Let a finite set* $\mathbf{R} = \{R_1, \ldots, R_\tau\}$ *be in* $\mathbb{Z}^n$ *and* $\mathbf{S} \subset \mathbf{R} + \mathbb{R}_+^n$. *Then the dominant subset* $\mathbf{S}/\mathbb{R}_+^n$ *is finite.*

Example 8.1. Let $\mathbf{S} = \{Q\colon Q \in \mathbb{Z}_+^n,\ q_1 + \ldots + q_n \geq m,\ m \in \mathbb{Z}_+\}$. Then the dominant subset $\mathbf{S}/\mathbb{R}_+^n = \{Q\colon Q \in \mathbb{Z}^n,\ q_1 + \ldots + q_n = m\}$. □

Example 8.2. For $n = 2$ we consider the set $\mathbf{S}$ of integer points $Q = (q_1, q_2) \geq 0$ satisfying the inequality $2q_1 + 3q_2 \geq 12$. The dominant subset $\mathbf{S}/\mathbb{R}_+^2$ consists of five points: $(0,4)$, $(2,3)$, $(3,2)$, $(5,1)$, $(6,0)$, shown in Fig. 1.6. In these points $2q_1 + 3q_2$ equals 12 and 13. □

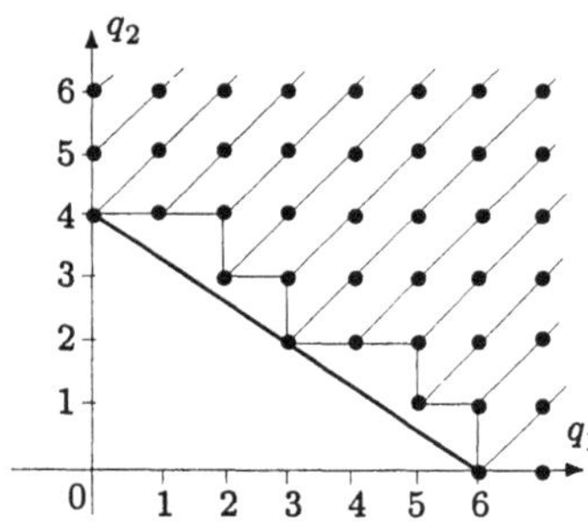

Figure 1.6: The dominant subsets $\mathbf{S}/\mathbb{R}_+^2$ and $\mathbf{\Gamma}/\mathbb{R}_+^2$ for Examples 8.2 and 8.3. The set $\mathbf{S}/\mathbb{R}_+^2 + \mathbb{R}_+^2$ is hatched.

Theorem 8.1 *Let a finite set* $\mathbf{R} \subset \mathbb{Z}^n$, $\mathbf{C}$ *be an integer forward closed cone in* $\mathbb{R}^n$ *and* $\mathbf{S} \subset (\mathbf{R} + \mathbf{C}) \cap \mathbb{Z}^n$. *Then the dominant set* $\mathbf{S}/\mathbf{C}$ *consists of a finite number of points.*

Proof. We denote $d = \dim \mathbf{C}$. Let the cone $\mathbf{C}^* \subset \mathbb{R}_*^n$ be the dual to the cone $\mathbf{C}$ (see Section 1), and the integer vectors $B_1, \ldots, B_{n-d}$ and $C_1, \ldots, C_\sigma$ form a basis of the linear subspace and the skeleton of the cone $\mathbf{C}^*$, i.e. the cone $\mathbf{C}$ is given by the equalities $\langle B_i, Q\rangle = 0$ and the inequalities $\langle C_j, Q\rangle \le 0$, $i = 1,\dots,n-d$; $j = 1,\dots,\sigma$. We form the matrix $\mathcal{C} = (B_1 \dots B_{n-d} C_1 \dots C_\sigma)$, and we make the change of coordinates $\tilde{Q} = -\mathcal{C}^* Q$, where $\mathcal{C}^*$ is the transposed matrix $\mathcal{C}$, $\tilde{Q} \in \mathbb{R}^\rho$ and $\rho = n - d + \sigma$. With this change, the space $\mathbb{R}^n$ turns into a linear subspace $\mathbf{L} \subset \mathbb{R}^\rho$, the set $\mathbf{R} \subset \mathbb{Z}^n$ turns into the set $\tilde{\mathbf{R}} \subset \mathbb{Z}^\rho \cap \mathbf{L}$, the cone $\mathbf{C}$ turns into $\mathbb{R}^\rho_+ \cap \mathbf{L}$, the set $\mathbf{S} \subset (\mathbf{R}+\mathbf{C}) \cap \mathbb{Z}^n$ turns into the set $\tilde{\mathbf{S}} \subset (\tilde{\mathbf{R}} + \mathbb{R}^\rho_+ \cap \mathbf{L}) \cap \mathbb{Z}^\rho$, and the set $\mathbf{S}/\mathbf{C}$ turns into the set $\tilde{\mathbf{S}}/(\mathbb{R}^\rho_+ \cap \mathbf{L})$. The change of coordinates is one-to-one for the points $\tilde{Q} \in \mathbf{L}$. Since $\tilde{\mathbf{S}} \subset \mathbf{L}$, then $\tilde{\mathbf{S}}/(\mathbb{R}^\rho_+ \cap \mathbf{L}) = \tilde{\mathbf{S}}/\mathbb{R}^\rho_+$. As it follows from Lemma 8.1, this dominant subset consists of a finite number of points. Therefore this is also the case for the set $\mathbf{S}/\mathbf{C}$. □

In the situation of Theorem 8.1, we can compute the dominant subset $\mathbf{S}/\mathbf{C}$ by means of the following **algorithm**. We take such a vector $K \in \mathbb{R}^n_*$, that $\langle K, Q\rangle > 0$ for all $Q \in \mathbf{C}\setminus\{0\}$. We write $\mathbf{S}_c = \mathbf{S} \cap \{Q\colon \langle K, Q\rangle = c\}$. Obviously, there are nonempty subsets $\mathbf{S}_c$ only for a discrete set $\{c_0 < c_1 < c_2 < \dots\}$ of values c, and each subset $\mathbf{S}_{c_k}$ is finite. All points $Q \in \mathbf{S}_{c_0}$ lie in $\mathbf{S}/\mathbf{C}$, i.e. $\mathbf{S}_{c_0}/\mathbf{C} = \mathbf{S}_{c_0}$. Let the set

$$(\mathbf{S}_{c_0} \cup \dots \cup \mathbf{S}_{c_k})/\mathbf{C} \tag{8.3}$$

consist of the points $Q_1, \dots, Q_l$. For each point $Q \in \mathbf{S}_{c_{k+1}}$ we consider the differences

$$Q - Q_j, \quad j = 1, \dots, l. \tag{8.4}$$

If at least one of them lies in the cone $\mathbf{C}$, then the point Q does not belong to the set $\mathbf{S}/\mathbf{C}$. If none of these differences lie in the cone $\mathbf{C}$, then the point Q lies in $\mathbf{S}/\mathbf{C}$ and also in

$$(\mathbf{S}_{c_0} \cup \dots \cup \mathbf{S}_{c_{k+1}})/\mathbf{C}. \tag{8.5}$$

In this way we find the set (8.5) and we can pass to $\mathbf{S}_{c_{k+2}}$. If it is known that the set $\mathbf{S}_{c_k}$ is dominant for the set

$$\mathbf{S}_{c_k} \cup \mathbf{S}_{c_{k+1}} \cup \dots,$$

then we can stop on the set $\mathbf{S}_{c_k}$, because the set (8.3) coincides with the set $\mathbf{S}/\mathbf{C}$ (see Example 8.1).

This algorithm is particularly simple for $\mathbf{C} = \mathbb{R}^n_+$. In that case $K = (1, 1, \dots, 1)$, and a point $Q \in \mathbf{S}_{c_{k+1}}$ belongs to $\mathbf{S}/\mathbf{C}$ only if each difference (8.4) contains a negative component.

The determination of the dominant subset is also suitable for a continuous set. Thus, for $\mathbf{\Gamma} = \mathrm{CNV}\,\mathbf{S}$ the dominant subset $\mathbf{\Gamma}/\mathbf{C}$ is that part of the boundary $\partial\mathbf{\Gamma}$ in which the faces $\mathbf{\Gamma}^{(d)}_j$ have normals lying inside the cone $\mathbf{C}^*$, dual to the cone $\mathbf{C}$. We designate as $\check{\mathbf{C}}^* = \mathbf{C}^* \setminus \partial\mathbf{C}^*$ the interior of the cone $\mathbf{C}^*$ and as $\partial_{\check{\mathbf{C}}^*}$ that part of the boundary $\partial\mathbf{\Gamma}$ of the set $\mathbf{\Gamma}$ which intersects with the supporting planes $\mathbf{H}_P$, where $P \in \check{\mathbf{C}}^*$. Obviously,

$$\mathbf{\Gamma}/\mathbf{C} = \partial_{\check{\mathbf{C}}^*}\mathbf{\Gamma} = \partial_{\check{\mathbf{C}}^*}(\mathbf{\Gamma}/\mathbf{C}). \tag{8.6}$$

On the other hand, we have

$$\partial_{\check{\mathbf{C}}^*}\mathbf{\Gamma} = \partial_{\check{\mathbf{C}}^*}\,\mathrm{CNV}\,(\mathbf{S}/\mathbf{C}). \tag{8.7}$$

Therefore,

$$(\mathrm{CNV}\,\mathbf{S})/\mathbf{C} = \partial_{\check{\mathbf{C}}^*}\,\mathrm{CNV}\,(\mathbf{S}/\mathbf{C}), \tag{8.8}$$

i.e. the extraction of the minimal dominant subset commutes with the construction of the convex hull.

Example 8.3. For the set $\mathbf{S}$ of Example 8.2 its convex hull $\mathbf{\Gamma} \stackrel{\text{def}}{=} \mathrm{CNV}\,\mathbf{S} = \{Q\colon q_1, q_2 \geq 0, 2q_1 + 3q_2 \geq 12\}$. The dominant subset $\mathbf{\Gamma}/\mathbb{R}^2_+$ is the segment $\{Q\colon q_1, q_2 \geq 0, 2q_1 + 3q_2 = 12\}$ shown in Fig 1.6 by the boldface line. □

Assume that in the situation of Theorem 8.1 all edges of the set $\mathbf{R} + \mathbf{C}$, which are parallel to the edges of the cone $\mathbf{C}$, are $\mathbf{E}_1, \ldots, \mathbf{E}_\sigma$, and that on each of them there is a point of the set $\mathbf{S}$. We denote these points as $G_j \subset \mathbf{S} \cap \mathbf{E}_j$, $j = 1, \ldots, \sigma$. Let the vector K be such that $\langle K, Q\rangle > 0$ for $Q \in \mathbf{C} \setminus \{0\}$ and $\varkappa = \max\langle K, G_j\rangle$ with $j = 1, \ldots, \sigma$. We consider the subset $\mathbf{S}' = \mathbf{S} \cap \{Q\colon \langle K, Q\rangle \leq \varkappa\}$ of the set $\mathbf{S}$. Obviously,

$$\partial_{\check{\mathbf{C}}^*}\,\mathrm{CNV}\,\mathbf{S}' = \partial_{\check{\mathbf{C}}^*}\,\mathrm{CNV}\,\mathbf{S}.$$

According to (8.7)

$$\partial_{\check{\mathbf{C}}^*}\mathbf{\Gamma} = \partial_{\check{\mathbf{C}}^*}\,\mathrm{CNV}\,(\mathbf{S}'/\mathbf{C}).$$

For the computation of the dominant part of the set $\mathbf{\Gamma}$, this property allows us to replace the infinite set $\mathbf{S}$ by its finite subset $\mathbf{S}'$. In particular, the procedure of successive extraction for increasing $\langle K, Q\rangle$ of the points of the dominant subset $\mathbf{S}/\mathbf{C}$ can be terminated, just as all points G_j are taken into account, i.e. with $c_k = \varkappa$. For instance, if $\mathbf{C} = \mathbb{R}^n_+$, $\mathbf{S} \subset \mathbb{Z}^n_+$ and on each coordinate axis q_j there is a point $g_j E_j \in \mathbf{S}$, $j = 1, \ldots, n$, then we can restrict ourselves to the subset $\mathbf{S}' = \mathbf{S} \cap \{Q\colon q_1 + \ldots + q_n \leq \varkappa\}$, where $\varkappa = \max g_j$ for $j = 1, \ldots, n$, and it is sufficient to compute its dominant subset $\mathbf{S}'/\mathbf{C}$.

The results of this section allow to solve Problem 1′ (see the beginning of Section 6) with an infinite set $\mathbf{S}$ in the case when $\mathbf{S} \subset (\mathbf{R} + \mathbf{K}^*) \cap \mathbb{Z}^n$, where $\mathbf{R} \subset \mathbb{Z}^n$ is a finite set, and the cone $\mathbf{K}^*$ is the dual to the cone of the problem $\mathbf{K}$. Since $\dim \mathbf{K} = n$, the cone $\mathbf{K}^*$ is forward. Firstly, according to the described algorithm we compute the dominant subset $\mathbf{S}/\mathbf{K}^*$ or $\mathbf{S}'/\mathbf{K}^*$, which is finite according to Theorem 8.1. Then we solve Problem 1′ for that subset as it was described in Sections 6, 7. Here according to the formula (8.8), for the set $\mathbf{S}$ we obtain all boundary subsets $\mathbf{S}_P$ with $P \in \check{\mathbf{K}}$ as intersections $\mathbf{S} \cap \mathbf{\Gamma}_j^{(d)}$, where $\mathbf{\Gamma}_j^{(d)} \subset \partial_{\check{\mathbf{K}}} \mathrm{CNV}\,(\mathbf{S}/\mathbf{K}^*)$.

We remark that for Problem 1′ with a finite set $\mathbf{S}$ we can also obtain a substantial simplification if at first we compute the subset $\mathbf{S}'$ or $\mathbf{S}'/\mathbf{K}^*$. The methods and algorithms of the present section will be used in Chapter 4, Section 3 and in Chapter 5.

9. Coherent boundary subsets

Let now several finite sets $\mathbf{S}_i$, $i = 1, \dots, m$ be given in $\mathbb{R}^n$. For a given vector $P \neq 0$ and for each $i = 1, \dots, m$ the set $\mathbf{S}_i$ has its own boundary subset $\mathbf{S}_{iP}$.

Problem 2. For each $P \neq 0$ find all boundary subsets $\mathbf{S}_{iP}$, $i = 1, \dots, m$.

This problem is solved in the following way. To each set $\mathbf{S}_i$, $i = 1, \dots, m$, there correspond the objects described earlier. Namely, the polyhedron $\mathbf{\Gamma}_i$, the collection of its faces $\left\{\mathbf{\Gamma}_{ij_i}^{(d_i)}\right\}$, their normal cones $\mathbf{U}_{ij_i}^{(d_i)}$, and also the boundary subsets $\mathbf{S}_{ij_i}^{(d_i)}$. An aggregate of boundary subsets $\mathbf{S}_{ij_i}^{(d_i)}$, $i = 1, \dots, m$, will be called *coherent* if there exists a vector $P \neq 0$ such that

$$\mathbf{S}_{ij_i}^{(d_i)} = \mathbf{S}_{iP}, \quad i = 1, \dots, m. \tag{9.1}$$

The set $\mathbf{\Pi}$ of all vectors P, to which there corresponds one coherent aggregate of boundary subsets (9.1) is called the *normal cone of the aggregate of subsets* (9.1). In other words, $\mathbf{\Pi}$ is the set of vectors P, which satisfy the system of linear relations

$$\begin{aligned} \mathbf{\Pi} = \{P\colon \langle P, Q_{j_1}^i \rangle &= \langle P, Q_{j_2}^i \rangle, \quad Q_{j_1}^i, Q_{j_2}^i \in \mathbf{S}_{iP}, \; j_1, j_2 = 1, \dots, s_i, \\ \langle P, Q_{j_1}^i \rangle &> \langle P, Q_l^i \rangle, \quad Q_l^i \in \mathbf{S}_i \setminus \mathbf{S}_{iP}, \; i = 1, \dots, m\}, \end{aligned} \tag{9.2}$$

where $\mathbf{S}_i = \{Q_1^i, \dots, Q_{s_i}^i\}$, $i = 1, \dots, m$. Obviously,

$$\mathbf{\Pi} = \mathbf{U}_{1j_1}^{(d_1)} \cap \ldots \cap \mathbf{U}_{mj_m}^{(d_m)} \stackrel{\mathrm{def}}{=} \mathbf{\Pi}_{j_1, \dots, j_m}^{(d_1, \dots, d_m)}. \tag{9.3}$$

The *tangent cone* $\mathbf{T}_{j_1, \dots, j_m}^{(d_1, \dots, d_m)}$ *of the aggregate of boundary subsets* (9.1) is the cone dual to the cone (9.3), i.e. the conical hull of the union of the tangent cones $\mathbf{T}_{1j_1}^{(d_1)} \cup \ldots \cup \mathbf{T}_{mj_m}^{(d_m)}$.

We demonstrate one of the methods to find the normal cones (9.3) of the aggregate of boundary subsets (9.1). We assume that for the set $\mathbf{S}_1$ the dual structural diagram for cones $\mathbf{U}_{1j}^{(d)}$ has been constructed. Each system of linear inequalities, determining the cone $\overline{\mathbf{U}}(Q_j^2)$, $j = 1, \dots, s_2$, is to be solved in each normal cone $\mathbf{U}_{1i}^{(0)}$, given by the skeleton. Instead of $\mathbf{U}_{1i}^{(0)}$ in the last but one line of the dual structural diagram, we write the intersection $\overline{\mathbf{U}}(Q_j^2) \cap \mathbf{U}_{1i}^{(0)}$ (for each fixed j separately). Then, on the basis of the following lines and the dashes of the dual structural diagram for $\mathbf{S}_1$, we restore the whole dual structural diagram of intersections for each $j = 1, \ldots, s_2$. From the diagram we extract all nonempty cones of intersection $\mathbf{U}_{2\gamma}^{(d_2')}$. In the next step, each system of linear inequalities, determining the cone $\overline{\mathbf{U}}(Q_j^3)$, $j = 1, \ldots, s_3$, is to be solved in each nonempty cone of intersection $\overline{\mathbf{U}}(Q_j^2) \cap \mathbf{U}_{1i}^{(0)}$ with the already found skeleton. Exactly as in the previous step, we restore the structural diagrams of intersections $\mathbf{U}_{3\delta}^{(d_3')}$. Thus, following this process, on the mth step we obtain all normal cones (9.3) and their coherent aggregates of boundary subsets.

In what follows, we will need those coherent boundary subsets among which there is no vertex, i.e. we have

Problem 2″. For the aggregate of sets $\mathbf{S}_1, \ldots, \mathbf{S}_m$, extract all cones (9.3) in which all $d_i > 0$.

Example 9.1. Assume that $n = 3$, the set $\mathbf{S}_1$ is (3.1), the set $\mathbf{S}_2$ is

$$Q_1^2 = (0,1,1), \quad Q_2^2 = (2,0,1), \quad Q_3^2 = (3,1,0), \quad Q_4^2 = (2,1,1) \tag{9.4}$$

and the cone of the problem is $\mathbf{K} = \{P < 0\}$. On Fig. 1.7 the tetrahedron $\mathbf{\Gamma}_2$ is shown.

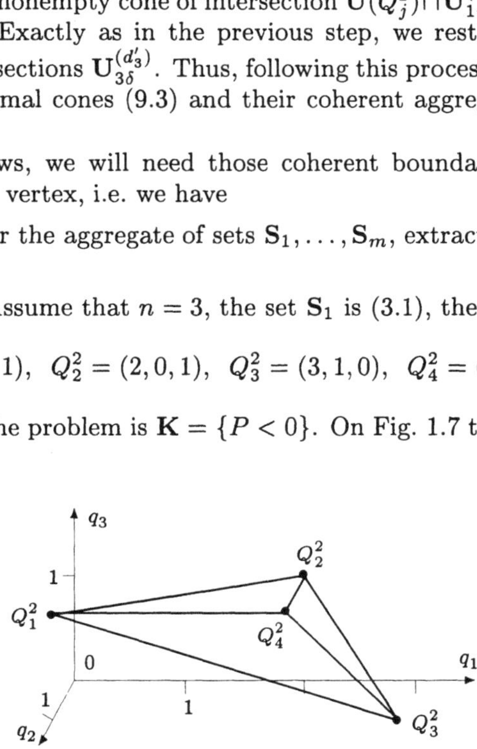

Figure 1.7: The tetrahedron $\mathbf{\Gamma}_2$ of Example 9.1.

It is immediately follows from Figs. 1.7 and 1.2 that for $\mathbf{\Gamma}_2$ the supporting planes with $P < 0$ pass only through the elements of the triangle $\mathbf{\Gamma}_{21}^{(2)}$ with vertices Q_1^2, Q_2^2, Q_3^2, and for $\mathbf{\Gamma}$, through the elements of the triangles $\mathbf{\Gamma}_1^{(2)}$, $\mathbf{\Gamma}_2^{(2)}$, $\mathbf{\Gamma}_3^{(2)}$. We put the linear relations (2.1) on the plane $\mathbf{H}_- = \{P\colon p_1 + p_2 + p_3 = -1\}$ with coordinates p_2, p_3 for those cones of the faces $\mathbf{U}_{ij}^{(2)}$, and for those cones of the edges $\mathbf{U}_{ij}^{(1)}$ which intersect the cone $\mathbf{K}$ of the problem. Since $\mathbf{\Gamma}_1 = \mathbf{\Gamma}$, we will omit the index $i = 1$. For the faces $\mathbf{\Gamma}_1^{(2)}$, $\mathbf{\Gamma}_2^{(2)}$, $\mathbf{\Gamma}_3^{(2)}$ and $\mathbf{\Gamma}_{21}^{(2)}$

we obtain the points

$$\begin{aligned}
\mathbf{H}_- \cap \mathbf{U}_1^{(2)} &= \{4p_2 = 2p_3 = -1\},\\
\mathbf{H}_- \cap \mathbf{U}_2^{(2)} &= \{2p_2 = 4p_3 = -1\},\\
\mathbf{H}_- \cap \mathbf{U}_3^{(2)} &= \{4p_2 = 4p_3 = -1\},\\
\mathbf{H}_- \cap \mathbf{U}_{21}^{(2)} &= \{3p_2 = 2p_3 = -1\}.
\end{aligned}$$

For the edges $\mathbf{\Gamma}_1^{(1)}$, $\mathbf{\Gamma}_2^{(1)}$, $\mathbf{\Gamma}_3^{(1)}$, $\mathbf{\Gamma}_4^{(1)}$, $\mathbf{\Gamma}_5^{(1)}$, $\mathbf{\Gamma}_6^{(1)}$ we obtain the segments and half-lines

$$\begin{aligned}
\mathbf{H}_- \cap \mathbf{U}_1^{(1)} &= \{4p_2 + 4p_3 = -3,\ 4p_2 < -14p_3 < -1\},\\
\mathbf{H}_- \cap \mathbf{U}_2^{(1)} &= \{4p_2 = -1 > 4p_3 > -2\},\\
\mathbf{H}_- \cap \mathbf{U}_3^{(1)} &= \{4p_3 = -1 > 4p_2 > -2\},\\
\mathbf{H}_- \cap \mathbf{U}_4^{(1)} &= \{2p_2 + p_3 = -1 < 4p_2\},\\
\mathbf{H}_- \cap \mathbf{U}_5^{(1)} &= \{p_2 + 2p_3 = -1 < 4p_3\},\\
\mathbf{H}_- \cap \mathbf{U}_6^{(1)} &= \{p_2 = p_3 > -1/4\},
\end{aligned}$$

shown in Fig. 1.3. For the edges $\mathbf{\Gamma}_{21}^{(1)} \supset \{Q_1^2, Q_2^2\}$, $\mathbf{\Gamma}_{22}^{(1)} \supset \{Q_1^2, Q_3^2\}$, $\mathbf{\Gamma}_{23}^{(1)} \supset \{Q_2^2, Q_3^2\}$ we obtain the half-lines

$$\begin{aligned}
\mathbf{H}_- \cap \mathbf{U}_{21}^{(1)} &= \{3p_2 + 2p_3 = -2 < 4p_3\},\\
\mathbf{H}_- \cap \mathbf{U}_{22}^{(1)} &= \{3p_2 + 4p_3 = -3 > 6p_3\},\\
\mathbf{H}_- \cap \mathbf{U}_{23}^{(1)} &= \{2p_3 = -1 > 3p_2\}.
\end{aligned}$$

Plotting these lines in the plane $\mathbf{H}_-$, we obtain a picture of the section $\mathbf{H}_-$ of the space $\mathbb{R}_*^3$ (Fig. 1.8). Bold and thin lines in Fig. 1.8 represent the sections of the normal cones $\mathbf{U}_j^{(1)}$ and $\mathbf{U}_{2j}^{(1)}$ of the edges of the polyhedra $\mathbf{\Gamma}$ and $\mathbf{\Gamma}_2$ respectively. The boundary of the section $\mathbf{K} \cap \mathbf{H}_-$ is represented there by hatching. From Fig. 1.8 it can be seen that in $\mathbf{K} \cap \mathbf{H}_-$ there are only two intersections of the normal cones of the edges and faces:

$$\begin{aligned}
\mathbf{H}_- \cap \mathbf{\Pi}_1 &\stackrel{\text{def}}{=} \mathbf{H}_- \cap \mathbf{U}_2^{(2)} \cap \mathbf{U}_{21}^{(1)} = \{2p_2 = 4p_3 = -1\},\\
\mathbf{H}_- \cap \mathbf{\Pi}_2 &\stackrel{\text{def}}{=} \mathbf{H}_- \cap \mathbf{U}_4^{(1)} \cap \mathbf{U}_{22}^{(1)} = \{5p_2 = 5p_3/3 = -1\}.
\end{aligned} \tag{9.5}$$

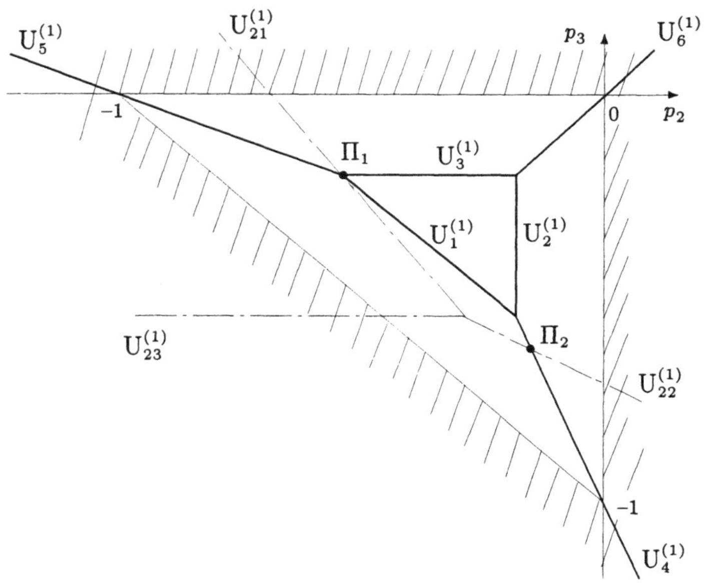

Figure 1.8: The intersection of the normal cones for the two sets $\mathbf{S}$ and $\mathbf{S}_2$ of Example 9.1 with the plane $p_1 + p_2 + p_3 = -1$ for the cone of the problem $p_1, p_2, p_3 < 0$. The dot-dash lines correspond to $\mathbf{S}_2$.

Since in $\mathbf{H}_-$ we have $p_1 = -1 - p_2 - p_3$, and in the first case $p_2 = -1/2$, $p_3 = -1/4$, then in it $p_1 = -1/4$ and $\mathbf{\Pi}_1 = \{P = -\lambda(1, 2, 1)\}$ with $\lambda > 0$; in the same way, in the second case $p_1 = -1/5$, i.e. $\mathbf{\Pi}_2 = \{P = -\lambda(1, 1, 3)\}$, where $\lambda > 0$. The pair $\mathbf{\Pi}_1$ and $\mathbf{\Pi}_2$ is the solution to Problem 2″ in the cone $\mathbf{K}$. □

10. Comparison with the Bugaev–Sintsov method

To find solutions to Problem 2″ with $m = n - 1$, Bugaev [1888] offered the "principle of the largest and the smallest exponent". He demonstrated that for $n = 2$ his method is analogous to the method of the Newton's polygon (see Section 3 and [Bruno 1979a, Chapter I]), and he described his method in detail for $n = 3$.

In our notations and with the elaborations of Sintsov [1898], Bugaev's method looks like this. To find the vectors $P = -(r_1, \ldots, r_{n-1}, 1) < 0$ lying in the required cones (9.3), firstly from each set $\mathbf{S}_i$ we extract the subset of its extremal points

$$\mathbf{S}_i^* = \{Q = (q_1, \ldots, q_{n-1}, q_n^*) \colon q_n^* = \min q_n \text{ with } \quad Q = (q_1, \ldots, q_{n-1}, q_n) \in \mathbf{S}_i\}.$$

Let the points $Q \in \mathbf{S}_i^*$ be numbered as follows: $\mathbf{S}_i^* = \{Q_1^i, \ldots, Q_{t_i}^i\}$. To each pair of points Q_j^i and Q_k^i from $\mathbf{S}_i^*$ we put in correspondence the equation

$$\langle R, Q_j^i \rangle = \langle R, Q_k^i \rangle \tag{10.1}$$

for a vector $R = (r_1, \ldots, r_{n-1}, 1) = -P$. We consider a system of such equations for $i = 1, \ldots, n-1$, where $j = j(i)$ and $k = k(i)$. Let R be a solution to that system, then for each i the following relations are to be verified

$$\langle R, Q_j^i \rangle \le \langle R, Q_s^i \rangle, \qquad 0 < s < t_i, \quad s \neq j, k. \tag{10.2}$$

If they all hold, then R is the required value $-P$. If at least one of them does not hold, then such R is not suitable. After the required R has been found, for each i we extract those $Q \in \mathbf{S}_i$ for which the scalar product $\langle R, Q \rangle$ is minimal, i.e. we extract the boundary subsets $\mathbf{S}_{iP}$.

In the case $n = 3$ Sintsov [1898] proposed a geometrical process to find vectors R. In the plane r_1, r_2, the solutions to the equation (10.1) form straight line $\mathbf{R}_{jk}^i$. The inequalities (10.2) on each straight line $\mathbf{R}_{jk}^i$ isolate the sets $\mathbf{L}_{jk}^i$. The point $R^* = (r_1^*, r_2^*, 1)$, corresponding to a solution to the relations (10.1) and (10.2), must lie both in a $\mathbf{L}_{jk}^1$ and in a $\mathbf{L}_{lm}^2$, i.e. it is the point of the intersection of the set $\{\mathbf{L}_{jk}^1\}$ with the set $\{\mathbf{L}_{lm}^2\}$. Hence, we must represent the sets $\{\mathbf{L}_{jk}^1\}$ and $\{\mathbf{L}_{lm}^2\}$ on the plane r_1, r_2, and find all points of their intersection. Sintsov proposed to plot all straight lines $\mathbf{R}_{jk}^i$ on the plane r_1, r_2 with its own color for each i, and to determine all points of intersections of straight lines with different colors. Then for each such point of intersection of the straight lines $\mathbf{R}_{jk}^1$ and $\mathbf{R}_{lm}^2$ it has to be ascertained whether it lies in the corresponding sets $\mathbf{L}_{jk}^1$ and $\mathbf{L}_{lm}^2$, i.e. whether the inequalities (10.2) hold in it, that one can easily see from the disposition of the point with respect to the other straight lines $\mathbf{R}_{jg}^1$ and $\mathbf{R}_{lh}^2$.

This construction is easily interpreted from our viewpoint, because our $P = -R$, and the objects in the hyperplane with coordinates $r_1, \ldots, r_{n-1}$ correspond to objects in the hyperplane $p_i = -r_i$, $i = 1, \ldots, n-1$, $p_n = -1$. We consider these objects more explicitly. From the points $Q \in \mathbf{S}_i$ only extreme points from $\mathbf{S}_i^*$ (and not all of them) fall on $\partial_{\{P<0\}}\mathbf{\Gamma}_i$. Let i be fixed; we take two points Q_j^i, $Q_k^i \in \mathbf{S}_i^*$. Then $\mathbf{L}_{jk}^i = -\overline{\mathbf{U}}(\{Q_j^i, Q_k^i\}) \cap \{p_n = -1\}$. According to Theorem 5.1, $\overline{\mathbf{U}}\left(\{Q_j^i, Q_k^i\}\right) = \overline{\mathbf{U}}_{il}^{(d)}$ for some $d > 0$ and l. Moreover, all $\overline{\mathbf{U}}_{il}^{(1)}$ are represented in that way. All remaining $\overline{\mathbf{U}}_{il}^{(d)}$ with $d > 1$ belong to the cones $\overline{\mathbf{U}}_{il}^{(1)}$, and they are intersections of these cones. Therefore the points of intersection of the sets $\mathbf{L}_{jk}^i$, $i = 1, \ldots, n-1$, give the intersections of the cones $\mathbf{U}_{il_i}^{(d_i)}$ $(i = 1, \ldots, n-1)$ with the hyperplane $p_n = -1$ for $d_i \ge 1$.

The Bugaev–Sintsov method has the following deficiencies: a) it is applicable only to finite sets; b) the case where the system of equations (10.1) for $i = 1, \dots, n-1$ has non-unique solution was not studied; c) the volume of the computation is too great, because the item-by-item examination of all possible pairs of points Q^i_j, $Q^i_k \in \mathbf{S}^*_i$ is considered, and the boundary subsets $\mathbf{S}^{(d)}_{is} = \mathbf{\Gamma}^{(d)}_{is} \cap \mathbf{S}_i$ are not extracted, although it can be done with the figures of the polyhedra $\mathbf{\Gamma}_i$.

In Fig. 1.9 the result of the Sintsov procedure for the sets (3.1) and (9.4) is shown in the plane $p_1 + p_2 + p_3 = -1$ with coordinates p_2, p_3. It is clear from the comparison of Figs. 1.8 and 1.9 that the computations of the coherent boundary subsets are simplified, if they are carried out with the application of the polyhedra $\mathbf{\Gamma}_i$. Sintsov pointed out that the extension of his method to a larger number of variables would lead to drawings in a multi-dimensional space.

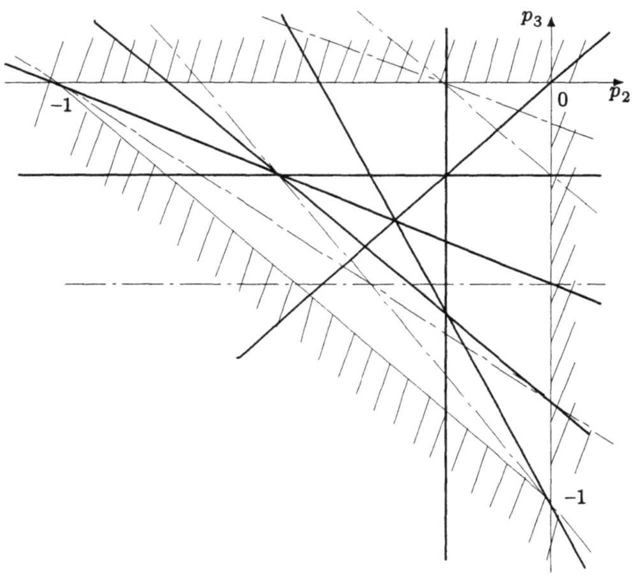

Figure 1.9: The result of the Sintsov procedure in the plane $p_1 + p_2 + p_3 = -1$ for the two sets $\mathbf{S}$ and $\mathbf{S}_2$ of Example 9.1 with the cone of the problem $p_1, p_2, p_3 < 0$. The dot-dash lines correspond to $\mathbf{S}_2$.

11. Linear transformations

Here vectors are considered as matrices-columns. We consider linear transformations of coordinates in $\mathbb{R}^n_*$

$$P' = \alpha P \tag{11.1}$$

with a nonsingular real matrix $\alpha = (\alpha_{ij})$, $\alpha_{ij} \in \mathbb{R}$, $\det \alpha \neq 0$. The scalar product $\langle P, Q\rangle$ for $P \in \mathbb{R}^n_*$ and $Q \in \mathbb{R}^n$ is preserved under the transformation (11.1) if simultaneously in $\mathbb{R}^n$ the linear transformation is made:

$$Q' = \alpha^{*-1} Q, \tag{11.2}$$

where α^* is the transposed matrix α. Then from the property of the scalar product $\langle \beta P, Q\rangle = \langle P, \beta^* Q\rangle$ we have

$$\langle P', Q'\rangle = \langle \alpha P, \alpha^{*-1} Q\rangle = \langle P, \alpha^* \alpha^{*-1} Q\rangle = \langle P, Q\rangle. \tag{11.3}$$

Thus, the scalar product (11.3) is preserved with the transformations (11.1) and (11.2) in $\mathbb{R}^n_*$ and $\mathbb{R}^n$ respectively, i.e. these spaces are *conjugate* (or *dual*) to each other.

The affine transformations (11.1) and (11.2) are one-to-one mappings of the spaces $\mathbb{R}^n_*$ and $\mathbb{R}^n$ into themselves. Here linear subspaces and manifolds are mapped into linear subspaces and manifolds with the dimension preserved. It is also true for the conical and convex hulls. Let a set $\mathbf{S}$ be given in $\mathbb{R}^n$, and the sets $\mathbf{\Gamma}$, $\mathbf{U}$, $\mathbf{\Gamma}_j^{(d)}$, $\mathbf{S}_j^{(d)}$ and $\mathbf{U}_j^{(d)}$, constructed in the previous sections, correspond to it. Let $\mathbf{S}'$ be the set obtained from the set $\mathbf{S}$ after the transformation (11.2); to it there correspond the sets $\mathbf{\Gamma}'$, $\mathbf{U}'$, $\mathbf{\Gamma}_j^{(d)\prime}$, $\mathbf{S}_j^{(d)\prime}$, $\mathbf{U}_j^{(d)\prime}$. Then $\mathbf{\Gamma}' = \alpha^{*-1}\mathbf{\Gamma}$, $\mathbf{\Gamma}_j^{(d)\prime} = \alpha^{*-1}\mathbf{\Gamma}_j^{(d)}$, $\mathbf{S}_j^{(d)\prime} = \alpha^{*-1}\mathbf{S}_j^{(d)}$; $\mathbf{U}' = \alpha\mathbf{U}$, $\mathbf{U}_j^{(d)\prime} = \alpha\mathbf{U}_j^{(d)}$, i.e. all constructions of the previous sections are invariant under the transformations (11.1), (11.2).

Problem 3. Let in $\mathbb{R}^n$ be given a linear subspace $\mathbf{L}^{(d)}$ of dimension $d < n$. It is required to make a linear transformation (11.2) such that the linear subspace $\alpha^{*-1}\mathbf{L}^{(d)}$ was a coordinate one.

We mention two methods of solution of Problem 3.

Method 1. Assume that $\mathbf{L}^{(d)}$ is the inner linear hull of the vectors $Q_1, \dots, Q_d$, i.e. these vectors form a basis of the subspace $\mathbf{L}^{(d)}$. We supplement it by the vectors $R_{d+1}, \dots, R_n$ to form a basis in $\mathbb{R}^n$. Then the matrix

$$\alpha^* = (Q_1 \dots Q_d \, R_{d+1} \dots R_n),$$

where the vectors Q_j and R_k are columns, gives a solution to Problem 3. Indeed, in that case $\alpha^* E_j = Q_j$, where E_j is the jth unit vector. Therefore

$$E_j = \alpha^{*-1} Q_j, \quad j = 1, \dots, d.$$

We can simplify the form of the matrix α in the following way. Since the vectors $Q_1, \dots, Q_d$ are linearly independent, then in the matrix $\mathcal{Q} = (Q_1 \dots Q_d)$ there exists a non-zero minor of the maximal order. Let it be the upper minor; we denote its matrix as C. Then

$$\mathcal{Q}C^{-1} = \begin{pmatrix} I_d \\ A \end{pmatrix}, \tag{11.4}$$

where I_d is the unit matrix of dimension $d \times d$ and A is a matrix of dimension $(n-d) \times d$. The columns of the matrix (11.4) also form a basis of the subspace $\mathbf{L}^{(d)}$. Therefore we can take

$$\alpha^* = \begin{pmatrix} I_d & 0 \\ A & I_{n-d} \end{pmatrix}, \quad \text{i.e. } \alpha = \begin{pmatrix} I_d & A^* \\ 0 & I_{n-d} \end{pmatrix}.$$

Method 2. Let the vectors $P_{d+1}, \dots, P_n \in \mathbb{R}^n_*$ form a basis in the orthogonal complement to the subspace $\mathbf{L}^{(d)}$, i.e. they are linearly independent and

$$\mathbf{L}^{(d)} = \{Q\colon \langle P_k, Q\rangle = 0,\ k = d+1, \dots, n\}.$$

We supplement them by the vectors $S_1, \dots, S_d$ to form a basis in $\mathbb{R}^n_*$ and we set

$$\alpha^{-1} = (S_1 \dots S_d \, P_{d+1} \dots P_n).$$

Then $\alpha^{-1}E_k = P_k$ and $E_k = \alpha P_k$, $k = d+1, \dots, n$, i.e. the vectors $E_{d+1}, \dots, E_n$ are orthogonal to the subspace $\alpha^{*-1}\mathbf{L}^{(d)}$.

We can simplify the form of the matrix α in the following way. Since the vectors $P_{d+1}, \dots, P_n$ are linearly independent, then in the matrix $\mathcal{P} = (P_{d+1} \dots P_n)$ there exists a non-zero minor of the maximal order. Let it be the lower minor; we denote its matrix as D. Then

$$\mathcal{P}D^{-1} = \begin{pmatrix} B \\ I_{n-d} \end{pmatrix}, \tag{11.5}$$

where B is a matrix of dimension $d \times (n-d)$. The columns of the matrix (11.5) also form a basis of the orthogonal complement to $\mathbf{L}^{(d)}$. Therefore we can take

$$\alpha^{-1} = \begin{pmatrix} I_d & B \\ 0 & I_{n-d} \end{pmatrix}, \quad \text{i.e. } \alpha = \begin{pmatrix} I_d & -B \\ 0 & I_{n-d} \end{pmatrix}.$$

A matrix α with all elements α_{ij} being integers and with $\det \alpha = \pm 1$ is called the *unimodular* one. For an unimodular matrix α, the inverse matrix

α^{-1} is also unimodular. The fundamental property of the transformations (11.1) and (11.2) with unimodular matrices α is that they are one-to-one transformations on the integer lattices $\mathbb{Z}_*^n$ and $\mathbb{Z}^n$ respectively.

Let in Problem 3 the subspace $\mathbf{L}^{(d)}$ be spanned on integer vectors $Q_1, \ldots, Q_d$; then we can find the unimodular matrix α by two methods, analogous to those considered before.

Method 1. We form a $(n \times n)$ matrix $\beta = (\mathcal{Q}\, 0)$. As it is known (see [Gantmacher 1967]) every square matrix β can be transformed into a diagonal matrix δ by means of elementary operations with its columns and lines, i.e. the square matrix β with integer elements can be represented as $\beta = \gamma\delta\gamma_1$, where γ and γ_1 are unimodular matrices, and δ is a diagonal matrix with integer elements. We can take the unimodular matrix γ as the matrix α^*.

Method 2. If the integer vectors $P_{d+1}, \ldots, P_n$ form a basis of the orthogonal complement to $\mathbf{L}^{(d)}$, then we assemble the matrix $\tilde{\beta} = (0\, \mathcal{P})$ and by means of elementary operations we represent it in the form $\tilde{\beta} = \tilde{\gamma}\tilde{\delta}\tilde{\gamma}_1$, where $\tilde{\gamma}$ and $\tilde{\gamma}_1$ are unimodular matrices, and $\tilde{\delta}$ is a diagonal matrix with integer elements. We can take the unimodular matrix $\tilde{\gamma}$ as the matrix α^{-1} .

For $n = 2$ the unimodular matrix α can be computed with the help of the algorithm of continued fractions (see [Khintchin 1961, Bruno 1979a]), and for $n \geq 3$ and $d = 1$ or $d = n - 1$ with the help of a generalization of continued fractions [Bruno and Parusnikov 1997].

Now in $\mathbb{R}^n$ we consider a parallel translation along the vector $T \in \mathbb{R}^n$, i.e. $Q' = Q + T$. After that transformation, all objects in $\mathbb{R}^n$ are displaced along the vector T; if in addition $P' = P$, then the objects in $\mathbb{R}_*^n$ are not changed and $\langle P', Q' \rangle = \langle P, Q \rangle + \langle P, T \rangle$. Let in $\mathbb{R}^n$ be given a finite set

$$\mathbf{S} = \{Q_1, \ldots, Q_s\}, \quad d = \dim \operatorname{Cnv} \mathbf{S}. \tag{11.6}$$

There exist a vector T and a matrix α such that the set

$$\alpha^{*-1}(\mathbf{S} + T) \tag{11.7}$$

lies in a coordinate subspace of dimension d. We can take any vector $-Q_j$ as the vector T, and then use the solution to Problem 3. If the set $\mathbf{S}$ is integer, i.e. $\mathbf{S} \subset \mathbb{Z}^n$, then we can take an integer vector T and an unimodular matrix α.

Problem 4. For the set (11.6), find a vector T and a matrix α such that the set (11.7) was in the non-negative orthant of a coordinate subspace of dimension d, i.e.

$$\operatorname{Con}\{E_1, \ldots, E_d\} \supset \alpha^{*-1}(Q_j + T) \geq 0, \quad j = 1, \ldots, s. \tag{11.8}$$

Solution. Let the vector T_1 and the matrix α be such that all points $Q_j' = \alpha^{*-1}(Q_j + T_1)$, $j = 1, \ldots, s$, lie in the coordinate subspace spanned on $E_1, \ldots, E_d$. Let $r_k = \min q'_{kj}$ with $j = 1, \ldots, s$, $k = 1, \ldots, d$. We set $R = (r_1, \ldots, r_d, 0, \ldots, 0)$; then $Q'_j - R \geq 0$, $j = 1, \ldots, s$. Therefore, the inequalities (11.8) are satisfied with the already found matrix α and with the vector $T = T_1 - \alpha^* R$.

Problem 5. Given a finite set $\tilde{\mathbf{S}} = \mathbf{S} \cup \{Q_{s+1}, \ldots, Q_t\}$ and its boundary subset (11.6), find a vector $\tilde{T}$ and a matrix $\tilde{\alpha}$ such that the set $\tilde{\alpha}^{*-1}(\tilde{\mathbf{S}} + \tilde{T})$ was in the non-negative orthant, i.e.

$$\tilde{\alpha}^{*-1}(Q_j + \tilde{T}) \geq 0, \quad j = 1, \ldots, t, \tag{11.9}$$

and the set $\tilde{\alpha}^{*-1}(\mathbf{S} + \tilde{T}) \subset \operatorname{Con}\{E_1, \ldots, E_d\}$.

Solution. As for the solution of Problem 4, we find a vector T and a matrix α satisfying the inequalities (11.8). Let $\mathbf{U}(\tilde{\mathbf{S}}; \mathbf{S})$ be the normal cone of the boundary subset $\mathbf{S}$ of the set $\tilde{\mathbf{S}}$. After the transformation $Q' = \alpha^{*-1}(Q + T)$, the set $\tilde{\mathbf{S}}$ changes into $\tilde{\mathbf{S}}'$, and $\mathbf{S}$ into the boundary subset $\mathbf{S}'$. Here $\mathbf{U}' \stackrel{\text{def}}{=} \mathbf{U}(\tilde{\mathbf{S}}'; \mathbf{S}') = \alpha \mathbf{U}(\tilde{\mathbf{S}}; \mathbf{S})$ is the transformed normal cone. Let $R' = (r'_1, \ldots, r'_n) \in \mathbf{U}'$. Obviously,

$$r'_1 = \ldots = r'_d = 0, \quad \langle R', Q'_k \rangle < 0, \quad k = s+1, \ldots, t. \tag{11.10}$$

In the hyperplane $\langle R', Q' \rangle = 0$ we choose a basis $V_1, \ldots, V_{n-1}$ with $V_j = E_j$, $j = 1, \ldots, d$. Then each point Q' of that hyperplane has the unique expansion $Q' = \sum_{j=1}^{n-1} w_j V_j$. In the hyperplane $\mathbf{H}' = \{Q' \colon \langle R', Q' \rangle = -1\}$ we take a point Q'_0 as the origin. Each point $Q' \in \mathbf{H}'$ has the unique expansion

$$Q' = Q'_0 + \sum_{j=1}^{n-1} w_j V_j. \tag{11.11}$$

Now through each point $Q'_k \in \tilde{\mathbf{S}}' \setminus \mathbf{S}'$ we draw a ray $\lambda Q'_k$, $\lambda \geq 0$. According to (11.10), the ray $\lambda Q'_k$ intersects the hyperplane $\mathbf{H}'$ in a point with

$$\lambda = \lambda_k \stackrel{\text{def}}{=} -\langle R', Q'_k \rangle^{-1} > 0, \quad k = s+1, \ldots, t,$$

i.e. that point is $\lambda_k Q'_k$. The set $\tilde{\mathbf{S}}'$ lies in the conical hull of the vectors $E_1, \ldots, E_d, \lambda_{s+1} Q'_{s+1}, \ldots, \lambda_t Q'_t$. We expand each point $\lambda_k Q'_k$ according to formula (11.11). We obtain

$$\lambda_k Q'_k = Q'_0 + \sum_{j=1}^{n-1} w_j^{(k)} V_j, \quad k = s+1, \ldots, t.$$

We set

$$\tau_j \leq \min_{s < k \leq t} w_j^{(k)}, \quad j = 1, \ldots, n-1, \tag{11.12}$$

and

$$V_n = Q_0' - \sum_{j=1}^{n-1} \tau_j V_j.$$

The set $\tilde{\mathbf{S}}'$ lies in the conical hull of the vectors $V_1, \dots, V_n$. Therefore it is now sufficient to make the linear transformation

$$Q'' = \beta^{*-1} Q',$$

which changes the vectors V_j into unit vectors, i.e.

$$\beta^* = (V_1 \dots V_n). \tag{11.13}$$

The property (11.9) is satisfied if we set $\tilde{\alpha} = \beta\alpha$ and $\tilde{T} = \beta^* T$.

Remark 11.1. If in Problem 5 the set $\tilde{\mathbf{S}}$ is integer, then there exists a solution to that problem with an integer vector $\tilde{T}$ and an unimodular matrix $\tilde{\alpha}$.

Indeed, in that case we can take the unimodular matrix α, and the vectors $R', V_1, \dots, V_{n-1}$ and Q_0' such integer ones that the greatest common divisor of the minors of the maximal order of the matrix $(V_1 \dots V_{n-1})$ equals unity. In addition, in (11.12) we can take integer τ_j. Then the matrix (11.13) is unimodular.

Now in $\mathbb{R}^n$ we consider a coherent aggregate of finite subsets

$$\mathbf{S}_{1j_1}^{(d_1)}, \dots, \mathbf{S}_{mj_m}^{(d_m)} \tag{11.14}$$

with a normal cone $\mathbf{\Pi}$. The co-dimension of the cone $\mathbf{\Pi}$, i.e. $d = n - \dim \mathbf{\Pi}$, is called the *dimension* of the aggregate (11.14). For each $k = 1, \dots, m$ we denote as $d(k)$ the dimension of the aggregate consisting of the first k boundary subsets of the aggregate (11.14), i.e. from

$$\mathbf{S}_{1j_1}^{(d_1)}, \dots, \mathbf{S}_{kj_k}^{(d_k)}.$$

Obviously, $d(k)$ is a monotonous non-decreasing function and $d(n) = d$.

Theorem 11.1 *There exist a matrix α and vectors $T_1, \dots, T_m$, such that:*

1) the subsets $\alpha^{*-1}\left(\mathbf{S}_{ij_i}^{(d_i)} + T_i\right)$, $i = 1, \dots, k$, lie in the coordinate subset spanned on $E_1, \dots, E_{d(k)}$ $(k = 1, \dots, m)$;

2) all sets $\alpha^{*-1}(\mathbf{S}_i + T_i)$, $i = 1, \dots, m$, lie in the non-negative orthant $Q' \geq 0$;

3) if all points of the set $\mathbf{S}_i$ are integers, then the matrix α is unimodular and the vectors $T_1, \dots, T_m$ are also integer.

The proof follows easily from the discussion in the present section.

Chapter 2

Singularities of algebraic equations

1. Implicit function

In the following, $X = (x_1, \ldots, x_n)$ is a real or complex vector, $Q = (q_1, \ldots, q_n) \in \mathbb{R}^n$ and $X^Q = x_1^{q_1} \ldots x_n^{q_n}$ is a monomial. We consider a function $f(X)$, analytic in the point $X = 0$. It expands into the Maclaurin power series

$$f(X) = \sum f_Q X^Q, \tag{1.1}$$

where $Q \in \mathbb{Z}_+^n$ and $f_Q \in \mathbb{C}$ or $f_Q \in \mathbb{R}$, converging in a neighborhood $\mathcal{U} = \{X : |x_i| < \epsilon, i = 1, \ldots, n\}$ of the point $X = 0$. We assume $f(0) = 0$, i.e. in the expansion (1.1) the coefficient f_0 equals zero. We consider the equation

$$f(X) = 0. \tag{1.2}$$

The totality of the solutions to the equation (1.2) in the neighborhood $\mathcal{U}$ forms an *analytical set* [Fuks 1962], or more accurately, an *analytical hypersurface*, passing through the point $X = 0$. The linear terms in the expansion (1.1) are

$$f_{E_1} x_1 + f_{E_2} x_2 + \ldots + f_{E_n} x_n,$$

where E_i is the ith unit vector; here

$$f_{E_i} = \left.\frac{\partial f}{\partial x_i}\right|_{X=0}, \quad i = 1, \ldots, n.$$

If there is at least one coefficient f_{E_i} not equal to zero, then the point $X = 0$ is a *simple point* of the hypersurface (1.2). In the opposite case, it is called a *critical* or *singular* one.

We begin with the resolution of the equation (1.2) in the vicinity of a simple point $X = 0$, assuming for definiteness that

$$a \stackrel{\text{def}}{=} f_{E_1} \neq 0. \tag{1.3}$$

We denote $X'' = (x_2, \ldots, x_n)$, $||Q|| = q_1 + \ldots + q_n$.

Theorem 1.1 *Assume that the function $f(X)$ is analytic at the origin, $f(0) = 0$ and that* (1.3) *is satisfied. Then the equation* (1.2) *has an analytical solution*

$$x_1 = \varphi(X'') = \sum \varphi_{R''} X''^{R''}, \tag{1.4}$$

where $R'' \in \mathbb{Z}_+^{n-1}$. Besides

$$\varphi_{R''} = \beta_{R''}(\{f_Q\})/a^{2k-1}, \tag{1.5}$$

where the $\beta_{R''}$ are polynomials of degree $2k-1$ of the coefficients f_Q with $||Q|| \leq k \stackrel{\text{def}}{=} ||R''||$.

Proof. For the first assertion, see [Goursat 1927, § 184; Bruno 1979a, Chapter I, Subsection 2.1]. We prove the second assertion. Substituting the expansion (1.4) into the expansion (1.1) and equating to zero the coefficients at $X''^{R''}$, we obtain equalities

$$a\varphi_{R''} + \gamma_{R''}(\{\varphi_{S''}\}) = 0, \tag{1.5$'$}$$

where $||S''|| \leq ||R''||$, and $\gamma_{R''}$ are finite sums of terms of the form

$$f_Q \varphi_{R''_1} \cdots \varphi_{R''_l} \tag{1.5$''$}$$

with $Q \geq 0$, $R''_i \geq 0$, $||Q|| \geq 2$ and

$$R''_1 + \ldots + R''_l = R'', \quad l = ||Q|| \geq 2. \tag{1.5$'''$}$$

Now, by induction on $k = ||R'||$, we prove the second assertion of the theorem. For $||R''|| = 1$, it obviously follows from Formula (1.5$'$). Let an integer $k > 1$ be fixed and the second assertion be true for all $||R''|| < k$. We prove it for $||R''|| = k$. By the inductive supposition, the term (1.5$''$) according to (1.5$'''$) includes a^{-1} in the power

$$2R''_1 - 1 + \ldots + 2R''_l - 1 \leq 2R'' - l \leq 2R'' - 2,$$

and it is the polynomial of the same power in other coefficients f_P (excluding f_Q). Hence, according to (1.5$'$) and (1.5$''$), for $||R''|| = k$ the second assertion of the theorem including Formula (1.5) is true. □

The first assertion of Theorem 1.1 is the Cauchy's theorem of implicit function. The polynomials $\beta_{R''}$ can be written in explicit form for small $||R''||$. For instance, for $n = 2$ we have

$$\beta_1 = -f_{10}, \quad \beta_2 = -f_{20}a^2 + f_{11}f_{10}a - f_{02}f_{10}^2.$$

Now we consider the series (1.1), in which the coefficients f_Q are functions of the coordinates $Y = (y_1, \dots, y_m)$ (real or complex), defined and unique in a domain $\mathcal{D}$. Let in that domain

$$|a| > \varepsilon > 0, \quad \varepsilon = \text{const.} \tag{1.6}$$

Then Theorem 1.1 holds for all $Y \in \mathcal{D}$. Let $\xi_1(Y), \dots, \xi_l(Y)$ be base functions of Y. We will consider the cases where all coefficients f_Q are rational functions of $\xi_1, \dots, \xi_l$, i.e. they are ratios of polynomials in them. According to (1.5), in these cases the coefficients $\varphi_{R''}$ of the series (1.4) are also rational functions of $\xi_1, \dots, \xi_l$, which can be explicitly computed for all exponents R'' with the norm $||R''||$ bounded by a very large number. Here the specific character of functions ξ_i does not matter; only property (1.6) is necessary. Usually the theorems of implicit function (like Theorem 1.1) have a local character (for small $|x_i|$). But with the mentioned dependence on Y, Theorem 1.1 is local with respect to X and global with respect to Y.

Following [Goursat 1927, §§ 161–164], we consider a numerical series

$$\sum a_Q, \tag{1.7}$$

in which the indices $Q = (q_1, \dots, q_n)$ run through a n-dimensional integer lattice $\mathbb{Z}^n$. Let the indices Q be represented by points in $\mathbb{R}^n$. We imagine in $\mathbb{R}^n$ an infinite increasing sequence of domains

$$\Omega_1 \subset \Omega_2 \subset \Omega_3 \subset \dots \subset \Omega_k \subset \dots,$$

having as the limit for $k \to \infty$ the whole space $\mathbb{R}^n$. Let

$$s_k = \sum a_Q \quad \text{over } Q \in \Omega_k,$$

i.e. the sum of all terms of the series (1.7) whose indices are in the domain Ω_k. If for k increasing indefinitely the sum s_k tends to a limit s and that limit does not depend on the choice of the sequence of domains $\{\Omega_k\}$, then the series (1.7) *converges to the sum* s. If for a sequence of domains $\{\Omega_k\}$ the corresponding partial sums of the series

$$\sum |a_Q| \tag{1.8}$$

tend to a limit, then the *series* (1.7) is *absolutely convergent*, and the sums of the series (1.7) and (1.8) do not depend on the choice of the sequence of domains $\{\Omega_k\}$.

Everything said about the series (1.7) is applicable to a multiple power series of the form (1.1). Assume that in the expansion (1.1) all similar terms have been collected; the set $\mathbf{S} = \{Q\colon\ f_Q \neq 0\}$ is called the *support of the expansion* (1.1), and it is designated as supp$f$ or $\mathbf{S}(f)$. Let in $\mathbb{R}^n$ be given a convex cone $\mathbf{T}$. We will say that the expansion (1.1) belongs to *class* $\mathcal{C}(\mathbf{T})$, if supp$f \subset \mathbf{T}$.

Assume that the cone $\mathbf{T}$ is a forward one and that vectors $T_1, \dots, T_l$ form its skeleton. If the series (1.1) is absolutely convergent for $X = X_0$, i.e. the series

$$\sum |f_Q||X_0|^Q$$

converges, then according to [Bruno 1979a, Chapter I, Subsection 3.4] the series (1.1) is absolutely convergent for all X that satisfy the inequalities

$$|X|^{T_i} \le |X_0|^{T_i}, \quad i = 1, \dots, l.$$

Therefore a series (1.1) of class $\mathcal{C}(\mathbf{T})$ will be called *convergent* if it is absolutely convergent in domains of the form

$$\mathcal{U}(\mathbf{T}, \varepsilon) = \{X\colon\ |X|^{T_i} \le \varepsilon\},$$

where $\varepsilon > 0$. For $n = 2$ the structure of such domains is considered in detail in [Bruno 1979a, Chapter I].

Remark 1.1. Let $\mathbf{C}$ be a polyhedral convex n-dimensional cone in $\mathbb{R}^n_*$, which is given by the inequalities $\langle P, T_i\rangle < 0$, $i = 1, \dots, l$. Then the cone $\mathbf{T}$ dual to $\mathbf{C}$ is the conical hull of the vectors $T_1, \dots, T_l$. It can be supposed that they form the skeleton of the cone $\mathbf{T}$. We put the set $\mathcal{U}(\mathbf{T}, \varepsilon)$ in correspondence to the cone $\mathbf{C} \subset \mathbb{R}^n_*$.

Theorem 1.2 [Bruno 1979a, Chapter I, Subsection 3.5] *The sum f of a series* (1.1) *of class $\mathcal{C}(\mathbf{T})$ is an analytic function in all inner points of the set of convergence, possibly excluding the coordinate subspaces.*

We consider now the cone $\mathbf{T}$, with the skeleton $E_1, T_2, \dots, T_l$, where the vectors $T_2, \dots, T_l$ have first coordinates equal to zero. Then Theorem 1.1 can be generalized in the following way.

Theorem 1.3 *Assume that $T_1 = E_1$, the vectors $T_2, \dots, T_l$ have the first coordinate equal to zero, and vectors $T_1, \dots, T_l$ form the skeleton of the cone $\mathbf{T}$, and $T_2'', \dots, T_l''$ form the skeleton of the cone $\mathbf{T}''$. Let the vector $K = (1, k_2, \dots, k_n) \in \mathbb{R}^n_*$ be such that $\langle K, T_i\rangle > 0$ for $i = 2, \dots, l$ and $\langle K'', Q''\rangle \ge 1$ for all $Q'' \in \mathbf{T}'' \cap \mathbb{Z}^{n-1}$, $Q'' \ne 0$. Assume that the series* (1.1) *belongs to class $\mathcal{C}(\mathbf{T})$ and is absolutely convergent in a set $\mathcal{U}(\mathbf{T}, \varepsilon)$. Assume that $f_0 = 0$ and that* (1.3) *is satisfied. Then the equation* (1.2) *has the solution* (1.4), *where the series φ belongs to class $\mathcal{C}(\mathbf{T}'')$ and absolutely converges in a set $\mathcal{U}(\mathbf{T}'', \varepsilon'')$. Here the formula* (1.5) *holds, where $\beta_{R''}$ are specific polynomials of power no larger than $2k - 1$ of the coefficients f_Q with $\langle K, Q\rangle \le k \stackrel{\text{def}}{=} \langle K'', R''\rangle$.*

Proof. It repeats almost completely the proof of Theorem 1.1. We have only to remark that $a\varphi_{R''}$ is the sum of terms of the form

$$f_Q \varphi_{R_1''} \cdots \varphi_{R_s''}, \tag{1.9}$$

where

$$Q'' + R_1'' + \ldots + R_s'' = R'', \quad Q = (q_1, Q''), \quad s \geq 1. \tag{1.10}$$

Here $\langle K'', Q'' \rangle \geq 0$, $\langle K'', R_j'' \rangle \geq 1$, and if $\langle K'', Q'' \rangle = 0$, then $s \geq 2$. Therefore, in the sum (1.10) there are at least two terms, Q'' and R_1'' or R_1'' and R_2'', for each of which the scalar product with the vector K'' is not less than one. Hence in the sum (1.10) the inequalities $\langle K'', R_j'' \rangle < k$ are satisfied for $j = 1, \ldots, s$. Now let us show by induction on k that the support of the series (1.4) lies in the cone $\mathbf{T}''$. Indeed, in the equality (1.10) the point Q'' belongs to the cone $\mathbf{T}''$ by hypothesis; therefore, if the vector R'' does not lie in the cone $\mathbf{T}''$, then according to the equality (1.10) at least one vector R_j'' does not lie in the cone $\mathbf{T}''$. But then by inductive assumption in the product (1.9) the corresponding factor $\varphi_{R_j''}$ equals zero, i.e. all terms (1.9) vanish. □

Remark 1.2. All assertions made after the proof of Theorem 1.1 concerning the dependence of the coefficients f_Q and $\varphi_{R''}$ on the additional coordinates remain in force for Theorem 1.3 as well.

2. Newton polyhedron

In the present section, we determine the truncations of polynomials, and we study the properties of the truncations, which are used for the truncation of a system of algebraic equations (see [Bruno 1965, 1979a]).

In what follows we will consider functions of τ only for $\tau \to +\infty$ in the real case and for $\tau \to \infty$ in the complex case. The real case will be considered below in detail. Similar reasoning can be used in the complex case, if τ is replaced by $|\tau|$ and the modula of the monomials are compared. We recall that for $\tau \to +\infty$ we have $\tau^{p_1} < \tau^{p_2}$ if $p_1 < p_2$,

$$\tau^p \to \begin{cases} 0, & \text{if } p < 0, \\ 1, & \text{if } p = 0, \\ +\infty, & \text{if } p > 0, \end{cases}$$

and $\tau^p > 0$ for $\tau > 0$ and any real p.

We call the vector $P = (p_1, \ldots, p_n) \in \mathbb{R}_*^n$ the *(vector) order of the vector function*

$$(b_1 \tau^{p_1}(1 + o(1)), \ldots, b_n \tau^{p_n}(1 + o(1))),$$

where $o(1)$ is a function tending to zero, and all $b_i \neq 0$.

Lemma 2.1 *Let*

$$x_i = b_i \tau^{p_i}(1 + o(1)), \quad i = 1, \dots, n, \quad P = (p_1, \dots, p_n) \neq 0, \tag{2.1}$$

$$x_i = b'_i \tau'^{p'_i}(1 + o(1)), \quad i = 1, \dots, n, \quad P' = (p'_1, \dots, p'_n) \neq 0, \tag{2.2}$$

be different parametrizations of the same curve in $\mathbb{R}^n$ *or* $\mathbb{C}^n$.

Then there exists $\lambda > 0$ *such that* $p_i = \lambda p'_i$, $i = 1, \dots, n$, *or in the vector notation*

$$P = \lambda P'.$$

Proof. Since (2.1) gives a one-to-one mapping of the half-line $a < \tau < +\infty$ on a *curve*, and (2.2) maps the half-line $a' < \tau' < +\infty$ one-to-one on the same curve, then there exists a one-to-one mapping $\tau' = \xi(\tau)$ such that $\tau' \to +\infty$ for $\tau \to +\infty$, and the following relations are satisfied:

$$b_i \tau^{p_i}(1 + o(1)) = b'_i [\xi(\tau)]^{p'_i}(1 + o(1)), \quad i = 1, \dots, n. \tag{2.3}$$

We find $\xi(\tau)$, analyzing one of these equalities. It was assumed that $P' \neq 0$, therefore, one of the p'_i differs from zero. Let $p'_1 \neq 0$; then from (2.3) for $i = 1$ we obtain

$$\xi(\tau) = [b_1/b'_1]^{1/p'_1} \cdot \tau^{p_1/p'_1}(1 + o(1)).$$

Since $\xi(\tau) \to \infty$ for $\tau \to \infty$, then $p_1/p'_1 = \lambda > 0$. After the substitution of the found expression for $\xi(\tau)$ in the equalities (2.3), they should turn into identities:

$$b_i \tau^{p_i}(1 + o(1)) \equiv b'_i [b_1/b'_1]^{p'_i/p'_1} \cdot \tau^{(p_1/p'_1)p'_i} \cdot (1 + o(1)), \quad i = 1, \dots, n,$$

from which we conclude that

$$p_i = (p_1/p'_1)p'_i = \lambda p'_i, \quad b_i = \varkappa^{p'_i} b'_i, \quad i = 1, \dots, n,$$

where $\varkappa = [b_1/b'_1]^{1/p'_1}$. □

Thus, if there is a parametrization for a curve, having order P, then the order P' of any other parametrization can differ from P only by a positive scalar factor. In other words, the order of a curve is determined exactly up to an arbitrary positive factor.

We call the vector $Q = (q_1, \dots, q_n)$, where the q_i are real, the *(vector) power* or *exponent* of the monomial

$$X^Q \stackrel{\text{def}}{=} x_1^{q_1} x_2^{q_2} \dots x_n^{q_n}. \tag{2.4}$$

If we are interested only in real values X, then we consider only the values of x_i for which $x_i^{q_i}$ are real, which is always the case for $x_i > 0$. In what follows, not stating it explicitly, for real X we will imply the two main cases:

1) the case where all considered vector powers have integer components. Then by X it will be understood an arbitrary point of the space $\mathbb{R}^n$;

2) the case where there are no restrictions on the vector powers. Then by X it will be understood a point from $\mathbb{R}^n_+ \stackrel{\text{def}}{=} \{X \geq 0\}$.

If in the product (2.4) we set

$$x_i = \tau^{p_i}, \quad i = 1, \ldots, n, \tag{2.5}$$

then

$$X^Q = x_1^{q_1} \ldots x_n^{q_n} = \tau^{p_1 q_1 + \ldots + p_n q_n} = \tau^{\langle P,Q \rangle},$$

i.e. we obtain a power function of τ, the exponent of which is the scalar product of the order P of the curve (2.5) and the power Q of the product (2.4).

Let $\mathbf{S}$ be a discrete set in $\mathbb{R}^n$. We consider a sum of products of the type (2.4) with real or complex coefficients

$$f(X) = \sum f_Q X^Q \quad \text{over} \quad Q \in \mathbf{S}, \tag{2.6}$$

assuming that similar terms are already collected, i.e. to each vector power Q corresponds one and only one term $f_Q X^Q$ with $f_Q \neq 0$, i.e. $\mathbf{S}$ is the support of the sum (2.6).

If we make the substitution (2.5) into $f(X)$, we obtain

$$f(\tau^{p_1}, \ldots, \tau^{p_n}) = \sum f_Q \tau^{\langle P,Q \rangle} \quad \text{over} \quad Q \in \mathbf{S}.$$

We extract from this sum all terms which have the maximal power in τ, i.e. all $Q' \in \mathbf{S}$ for which

$$\langle P, Q' \rangle = \sup \langle P, Q \rangle \stackrel{\text{def}}{=} c_P \quad \text{over} \quad Q \in \mathbf{S}. \tag{2.7}$$

According to Chapter 1, Section 1, those points $Q \in \mathbf{S}$ for which the scalar product $\langle P, Q \rangle$ attains the maximal value c_P form a boundary subset $\mathbf{S}_P$ of the set $\mathbf{S}$. In addition, to the set $\mathbf{S} = \operatorname{supp} f$ there corresponds the outer convex hull $\mathbf{\Gamma}(f) \stackrel{\text{def}}{=} \operatorname{CNV} \mathbf{S}$, which is called the *Newton polyhedron* of the sum f. The boundary $\partial \mathbf{\Gamma}$ consists of faces $\mathbf{\Gamma}_j^{(d)}$; to each face $\mathbf{\Gamma}_j^{(d)}$ there corresponds the boundary subset $\mathbf{S}_j^{(d)}$ and the tangent cone $\mathbf{T}_j^{(d)}$ in $\mathbb{R}^n$, and also the normal cone $\mathbf{U}_j^{(d)}$ in $\mathbb{R}^n_*$. If $P \in \mathbf{U}_j^{(d)}$, then $\mathbf{S}_P = \mathbf{S}_j^{(d)}$. To each boundary subset $\mathbf{S}_j^{(d)}$ of the set $\mathbf{S}$ there corresponds its subsum

$$\hat{f}_j^{(d)}(X) = \sum f_Q X^Q \quad \text{over} \quad Q \in \mathbf{S}_j^{(d)} \tag{2.8}$$

of the sum (2.6), which we call the *truncation of the sum* (2.6) *with respect to the order* P, and we denote it also as $\hat{f}_P(X)$. Here the normal and the tangent cones are related also to the truncation.

Example 2.1. Let

$$f(x_1, x_2, x_3) = x_1^2 + x_2^2 + x_3^2 + x_1 x_2 x_3. \tag{2.9}$$

Then

$$\begin{aligned} \widehat{f}_P &= x_1^2 + x_2^2 + x_3^2, \quad if \quad P = (-1,-1,-1); \\ \widehat{f}_P &= x_1 x_2 x_3, \quad if \quad P = (1,1,1); \\ \widehat{f}_P &= x_1^2 + x_2^2 + x_1 x_2 x_3, \quad if \quad P = (1,1,0), \end{aligned} \tag{2.10}$$

and so on. We remark that the truncations are different for the two orders, $P = (1,1,1)$ and $P = (-1,-1,-1)$, which differ only in sign. This is not surprising, since the vector $P = (1,1,1)$ corresponds to the substitution

$$x_1 = b_1\tau(1+o(1)), \quad x_2 = b_2\tau(1+o(1)), \quad x_3 = b_3\tau(1+o(1)),$$

and since $\tau \to \infty$ everywhere, then here $x_1, x_2, x_3 \to \infty$. The vector $P = (-1,-1,-1)$ corresponds to the substitution

$$x_1 = b_1\tau^{-1}(1+o(1)), \quad x_2 = b_2\tau^{-1}(1+o(1)), \quad x_3 = b_3\tau^{-1}(1+o(1)),$$

where $x_1, x_2, x_3 \to 0$. Thus, for $P = (1,1,1)$ the function $f(X)$ is considered in the vicinity of the point $X^0 = (\infty, \infty, \infty)$, and for $P = (-1,-1,-1)$ the function $f(X)$ is considered in the vicinity of the point $X^0 = (0,0,0)$. □

The truncation of the function $f(X)$ with respect to the order P gives the dominant term under substitution of the form (2.1), that is precisely expressed in the following **properties of the truncation**:

1) $\widehat{f}_P(X)$ is homogeneous with respect to order P, i.e.

$$\widehat{f}_P(b_1\tau^{p_1}, \ldots, b_n\tau^{p_n}) = f_P(b_1, \ldots, b_n)\tau^c;$$

2) $\widehat{f}_P(b_1\tau^{p_1}(1+o(1)), \ldots, b_n\tau^{p_n}(1+o(1))) = \widehat{f}_P(b_1, \ldots, b_n)\tau^c + o(1)\tau^c$; it is possible that $\widehat{f}_P(b_1, \ldots, b_n) = 0$;

3) $f(b_1\tau^{p_1}(1+o(1)), \ldots, b_n\tau^{p_n}(1+o(1))) = \widehat{f}_P(b_1, \ldots, b_n)\tau^c + o(1)\tau^c$.

The proofs, if built successively, are obvious. We remark that the property 3) expresses the essence of the truncation: the extraction of the part $\widehat{f}$, which determines the asymptotic behavior of the whole function f on a curve of order P.

Theorem 2.1 *If the curve* (2.1) *is a solution to the equation* $f(X) = 0$, *then its first approximation*

$$x_i = b_i\tau^{p_i}, \quad i = 1, \ldots, n, \tag{2.11}$$

is a solution to the corresponding truncated equation $\widehat{f}_P(X) = 0$.

The proof follows trivially from the properties of the truncation. From this we obtain that the vector coefficient $B = (b_1, \ldots, b_n)$ must be a root of the equation $\widehat{f}_P(B) = 0$ for the solution (2.1) of the equation $f(X) = 0$. This suggests a method of search for asymptotic (or local) expansions for solutions to the equation $f(X) = 0$. In order to reduce an arbitrary truncation to a standard form, we will use power transformations (see next Section). The *dimension of the truncation* $\widehat{f}_j^{(d)}$ equals the dimension d of the corresponding face $\mathbf{\Gamma}_j^{(d)}$.

Example 2.2 (continuation of Examples 3.1 and 5.1 of Chapter 1). Let the polynomial

$$f(x_1, x_2, x_3) = a_1 x_1 x_2 x_3 + a_2 x_1^4 + a_3 x_2^4 + a_4 x_3^4 + a_5 x_1^2 x_3^2$$

be given. The support of the polynomial is $\mathbf{S}(f) = \{Q_1 = (1,1,1), Q_2 = (4,0,0), Q_3 = (0,4,0), Q_4 = (0,0,4), Q_5 = (2,0,2)\}$. The Newton polyhedron $\mathbf{\Gamma}(f)$ is a tetrahedron. To each of its vertices, edges, faces there corresponds its normal cone. □

3. Power transformations

Let $\alpha = (\alpha_{ij})$ be a square matrix of dimension n with real elements α_{ij} and $\det \alpha \neq 0$. The transformation

$$y_i = x_1^{\alpha_{i1}} \ldots x_n^{\alpha_{in}}, \quad i = 1, \ldots, n, \tag{3.1}$$

is called a *power transformation* with matrix α. The inverse transformation

$$x_i = y_1^{\beta_{i1}} \ldots y_n^{\beta_{in}}, \quad i = 1, \ldots, n, \tag{3.2}$$

is also a power transformation with matrix $\beta = (\beta_{ij}) = \alpha^{-1}$. If we introduce the vectors $\log X = (\log x_1, \ldots \log x_n)$ and $\log Y$, then the power transformations (3.1) and (3.2) are linear transformations of these vectors: $\log Y = \alpha \log X$ and $\log X = \beta \log Y$.

We note some **properties of the power transformation** (3.1).

1. Under the transformation (3.1), $X^Q = Y^{\tilde{Q}}$, where

$$\tilde{Q} = \beta^* Q \tag{3.3}$$

and β^* is the transposed matrix β. Indeed,

$$X^Q = \exp\langle \log X, Q\rangle = \exp\langle \beta \log Y, Q\rangle = \exp\langle \log Y, \beta^* Q\rangle = Y^{\beta^* Q}.$$

Thus, under the power transformation (3.1), the vector exponents undergo linear transformations. This is also true for the sum

$$f(X) \stackrel{\text{def}}{=} \sum f_Q X^Q \quad \text{over } Q \in \mathbf{S}, \tag{3.4}$$

which under the transformation (3.2) is changed into the sum

$$f(X) = g(Y) \stackrel{\text{def}}{=} \sum g_{\tilde{Q}} Y^{\tilde{Q}} \quad \text{over } \tilde{Q} \in \tilde{\mathbf{S}}, \tag{3.5}$$

where $\tilde{Q} = \beta^* Q$, $\tilde{\mathbf{S}} = \beta^* \mathbf{S}$, $g_{\tilde{Q}} = f_Q$. The geometrical structures of the sets $\mathbf{S}$ and $\tilde{\mathbf{S}}$ are also linked by the linear transformation (3.3). Therefore under the transformation (3.1) a truncation is changed into a truncation, i.e. the operation of taking a truncation commutes with the power transformation.

2. Under the transformation (3.1), the curve

$$x_i = b_i \tau^{p_i}(1 + o(1)), \quad i = 1, \ldots, n, \quad \tau \to \infty,$$

with the vector order $P \stackrel{\text{def}}{=} (p_1, \ldots, p_n)$ is changed into the curve

$$y_i = \tilde{b}_i \tau^{\tilde{p}_i}(1 + o(1)), \quad i = 1, \ldots, n,$$

with the vector order

$$\tilde{P} \stackrel{\text{def}}{=} (\tilde{p}_1, \ldots, \tilde{p}_n) = \alpha P. \tag{3.6}$$

Indeed, $\log X = (P + o(1)) \log \tau$, $\log Y = (\tilde{P} + o(1)) \log \tau$ and $\log Y = \alpha \log X$, i.e. $(\tilde{P} + o(1)) \log \tau = (\alpha P + o(1)) \log \tau$. Therefore, in the space $\mathbb{R}^n_*$ of the orders P, the power transformation (3.1) induces the linear transformation (3.6). This transformation is dual to the transformation (3.3). So the spaces $\mathbb{R}^n$ and $\mathbb{R}^n_*$ are dual, and the scalar product $\langle P, Q \rangle$ is preserved under the power transformation (3.1). In particular, the cones of truncations $\mathbf{U}_k^{(d)}$ of the sum (3.4) are changed into the cones of truncations $\tilde{\mathbf{U}}_k{}^{(d)}$ of the sum (3.5). Generally speaking, all geometrical objects in the dual spaces $\mathbb{R}^n$ and $\mathbb{R}^n_*$ undergo the linear transformations (3.3) and (3.6) and keep all their linear properties unchanged.

3. The transformation (3.1) is a one-to-one mapping of the set $\{X : 0 < |x_i| < \infty,\ i = 1, \ldots, n\}$ on the set $\{Y : 0 < |y_i| < \infty,\ i = 1, \ldots, n\}$ for the complex X and Y if and only if the matrix α is unimodular, i.e. all α_{ij} are integers and $\det \alpha = \pm 1$. It is true for real X and Y if and only if all elements of matrices α and α^{-1} are rational numbers with odd denominators. In this chapter we will use mostly power transformations with unimodular matrices α.

4. Assume that in addition to (3.1), there is a power transformation

$$z_i = y_1^{\gamma_{i1}} y_2^{\gamma_{i2}} \ldots y_n^{\gamma_{in}}, \quad i = 1, \ldots, n,$$

with a matrix $\gamma = (\gamma_{ij})$. Then the coordinates Z are related to the coordinates X by the power transformation with matrix $\gamma\alpha$, i.e. the power transformations form a group, and power transformations with unimodular matrices form its subgroup.

Let $f(X)$ be an arbitrary Laurent polynomial, and d be the dimension of its Newton polyhedron $\mathbf{\Gamma} = \mathbf{\Gamma}(f)$; we will call d the *dimension of the polynomial* $f(X)$. In $\mathbb{R}^n_*$ we consider the maximal linear subspace $\mathbf{N}(f)$ normal to $\mathbf{\Gamma}(f)$. Obviously, $\dim \mathbf{\Gamma} + \dim \mathbf{N} = n$.

Theorem 3.1 *Let the dimension of the sum* (3.4) *be $d < n$. Then there exists a matrix α and a vector $T \in \mathbb{R}^n$ such that under the power transformation* (3.1)

$$X^T f(X) = g(y_1, \ldots, y_d).$$

Proof. According to the results of Chapter 1, Section 11, there exists a matrix α and a vector T such that the set (11.7) of Chapter 1 lies in coordinate subspace spanned on $E_1, \ldots, E_d$. But that means that $X^T f(X)$ changes under the power transformation (3.1) into the sum

$$g(Y) \stackrel{\text{def}}{=} \sum g_{\tilde{Q}} Y^{\tilde{Q}} \quad \text{over} \quad \tilde{Q} \in \alpha^{*-1}(\mathbf{S} + T), \tag{3.7}$$

in which for all vectors $\tilde{Q}$ the components $\tilde{q}_{d+1}, \ldots, \tilde{q}_n$ equal zero. Therefore, in the sum (3.7) the coordinates $y_{d+1}, \ldots, y_n$ appear only in power zero, i.e. they are absent. □

Theorem 3.2 *Let $\widehat{f}_j^{(d)}(X)$ be a truncation of the finite sum* (3.4). *There exists a matrix α and a vector $T \in \mathbb{R}^n$ such that under the power transformation* (3.1)

$$X^T \widehat{f}_j^{(d)}(X) = \widehat{g}(y_1, \ldots, y_d), \tag{3.8}$$

$$X^T f(X) = g(Y) \stackrel{\text{def}}{=} \sum g_{\tilde{Q}} Y^{\tilde{Q}} \quad \textit{over} \quad \tilde{Q} \in \tilde{\mathbf{S}}. \tag{3.9}$$

Here $g(Y)$ is a polynomial, i.e. in $\tilde{\mathbf{S}}$ all $\tilde{Q} \geq 0$.

The proof follows from the solution of Problem 5 in Chapter 1, Section 11.

Corollary 3.1 *For the truncation $\widehat{f}_j^{(d)}(X)$ of the Laurent polynomial* (3.4), *there exists a unimodular matrix α and an integer vector T such that the equalities* (3.8) *and* (3.9) *are satisfied. Here $\widehat{g}$ is a polynomial in $y_1, \ldots, y_d$, and g is a polynomial in $y_{d+1}, \ldots, y_n$.*

4. Asymptotic solution of an algebraic equation

We consider the equation

$$f(X) \stackrel{\text{def}}{=} \sum_{Q \in \mathbf{S}} f_Q X^Q = 0 \tag{4.1}$$

with a finite support $\mathbf{S} = \mathrm{supp} f \subset \mathbb{Z}^n$, i.e. $f(X)$ is a Laurent polynomial.

Finding those solutions to Equation (4.1), which have at least one coordinate identically equal to zero or infinity, will be considered as a solved problem, since it reduces to several problems analogous to the initial one, but with smaller dimensions. Therefore we will look for those solutions to Equation (4.1) in which there are no coordinates identically equal to zero or infinity. For such solutions, Equation (4.1) can be multiplied by any monomial, that induces a parallel translation of the set $\mathbf{S}$ in $\mathbb{R}^n$.

Equation (4.1) defines in the space of the coordinates X an algebraic hypersurface $\mathcal{F}$. We are interested in the parts of the hypersurface, where at least one coordinate x_i can tend to zero or infinity, i.e. the parts on which exists a curve of the form (2.1), satisfying Equation (4.1), with $P \neq 0$ and all $b_i \neq 0$. According to Theorem 2.1, the first approximation (2.11) to the curve (2.1) satisfies the corresponding truncated equation

$$\hat{f}_P(X) \stackrel{\mathrm{def}}{=} \sum_{Q \in \mathbf{S}_P} f_Q X^Q = 0. \tag{4.2}$$

According to the results of Chapter 1, Section 2, we have $\mathbf{S}_P = \mathbf{S}_j^{(d)}$ for $P \in \mathbf{U}_j^{(d)}$. The union of all $\mathbf{U}_j^{(d)}$ fills $\mathbb{R}_*^n \setminus \{0\}$, i.e. every vector $P \neq 0$ gets into a normal cone $\mathbf{U}_j^{(d)}$. Therefore we have to arrange a list of all boundary subsets $\mathbf{S}_j^{(d)}$ and identify their normal cones $\mathbf{U}_j^{(d)}$, i.e. to solve Problem 1 from Chapter 1, Section 2 for the set $\mathbf{S} = \mathrm{supp} f$.

We remark now that the truncation of zero dimension $\hat{f}_j^{(0)}$ is the monomial $f_Q X^Q$. The equation $\hat{f}_j^{(0)}(X) \stackrel{\mathrm{def}}{=} f_Q X^Q = 0$ has only those solutions X in which one of the coordinates x_i is identically zero. According to Theorem 2.1, the first approximation (2.11) to the curve (2.1) with all $b_i \neq 0$ must satisfy this equation. But it is impossible, since on the curve (2.11)

$$\hat{f}_j^{(0)}(X) \equiv f_Q B^Q \tau^{\langle P, Q \rangle} \neq 0.$$

Therefore we must exclude all vertices of the polyhedron $\mathbf{\Gamma}$ from the list of boundary subsets, and exclude all their normal cones $\mathbf{U}_j^{(0)}$ from the space $\mathbb{R}_*^n$.

Now we will look for the corresponding part of the hypersurface $\mathcal{F}$ for each boundary subset $\mathbf{S}_j^{(d)}$ with $d > 0$. For this we apply to the pair $f(X)$, $\hat{f}_j^{(d)}(X)$ the transformations mentioned in Corollary to Theorem 3.2, i.e. the multiplication by the monomial X^T and the power transformation (3.1). We obtain a pair: a Laurent polynomial and its truncation

$$g(Y) \stackrel{\mathrm{def}}{=} \sum g_{\tilde{Q}} Y^{\tilde{Q}} \quad \text{over} \quad \tilde{Q} \in \tilde{\mathbf{S}} \stackrel{\mathrm{def}}{=} \alpha^{*-1}(\mathbf{S} + T), \tag{4.3}$$

$$\hat{g}_j^{(d)}(y_1, \ldots, y_d) \overset{\text{def}}{=} \sum g_{\tilde{Q}} Y^{\tilde{Q}} \quad \text{over } \tilde{Q} \in \tilde{\mathbf{S}}_j^{(d)} \overset{\text{def}}{=} \alpha^{*-1}(\mathbf{S}_j^{(d)} + T), \tag{4.4}$$

where $\hat{g}$ is an ordinary polynomial and g is an ordinary polynomial in y_{d+1}, $\ldots$, y_n. Here $\tilde{P} = \alpha P \in \tilde{\mathbf{U}}_j^{(d)} \overset{\text{def}}{=} \alpha \mathbf{U}_j^{(d)}$, and the normal cone $\tilde{\mathbf{U}}_j^{(d)}$ lies in the coordinate subspace $\tilde{p}_1 = \ldots = \tilde{p}_d = 0$. The tangent cone $\tilde{\mathbf{T}}_j^{(d)}$ contains restrictions only on $\tilde{q}_{d+1}$, $\ldots$, $\tilde{q}_n$ and lies in the cone $\tilde{q}_{d+1}, \ldots, \tilde{q}_n \geq 0$.

The truncated equation

$$\hat{g}(Y) \overset{\text{def}}{=} \hat{g}_j^{(d)}(y_1, \ldots, y_d) = 0 \tag{4.5}$$

determines in the d-dimensional space a surface $\hat{\mathcal{G}}$ of a dimension not greater than $d-1$. Now we have to find the whole surface (4.5). We isolate at first the subset $\hat{\mathcal{G}}_c$ of all its critical points Y, for which

$$\hat{g}(Y) = \frac{\partial \hat{g}}{\partial y_1} = \ldots = \frac{\partial \hat{g}}{\partial y_d} = 0. \tag{4.6}$$

The remaining part of the surface $\hat{\mathcal{G}} \setminus \hat{\mathcal{G}}_c$ can be split into a finite number of parts $\hat{\mathcal{G}}_1$, $\ldots$, $\hat{\mathcal{G}}_k$, in each of which at least one of the partial derivatives $\partial \hat{g}/\partial y_i$ does not equal to zero and the coordinate y_i can be expressed as function of the remaining coordinates y_1, $\ldots$, y_d. Let $\hat{\mathcal{G}}_1$ be such a part of the surface $\hat{\mathcal{G}} \setminus \hat{\mathcal{G}}_c$ on which $\partial \hat{g}/\partial y_1 \neq 0$ and $y_1 = \psi_1(y_2, \ldots, y_d)$. In the Laurent polynomial $g(Y)$ we make the change of coordinates $y_1 = \psi_1 + z_1$, and we obtain the polynomial

$$h(z_1, \psi_1, y_2, \ldots, y_d, y_{d+1}, \ldots, y_n) \overset{\text{def}}{=} g(Y),$$

which we write as a polynomial $\check{h}$ of $\check{Z} = (z_1, y_{d+1}, \ldots, y_n)$:

$$\check{h} = \sum \check{h}_{\check{Q}} \check{Z}^{\check{Q}}, \tag{4.7}$$

with coefficients $\check{h}_{\check{Q}}$, depending on ψ_1, y_2, $\ldots$, y_d. In the "point" $z_1 = y_{d+1} = \ldots = y_n = 0$ we have $\partial \check{h}/\partial z_1 = (\partial \hat{g}/\partial y_1)(\psi_1, y_2, \ldots, y_d) \overset{\text{def}}{=} a \neq 0$, and the exponents $\check{Q}$ of the expansion (4.7) lie in the tangent cone $\check{\mathbf{T}}$, equal to the projection of the cone $\tilde{\mathbf{T}}$ on the subspace of the coordinates $\tilde{q}_1$, $\tilde{q}_{d+1}$, $\ldots$, $\tilde{q}_n$. Now Theorem 1.3 of implicit function can be applied to the equation $\check{h}(\check{Z}) = 0$, and we obtain its solutions in the form of the series

$$z_1 = \varphi(y_{d+1}, \ldots, y_n) \overset{\text{def}}{=} \sum \varphi_{R''} Y''^{R''},$$

where

$$Y'' = (y_{d+1}, \ldots, y_n), \quad R'' = (r_{d+1}, \ldots, r_n),$$

and the coefficients $\varphi_{R''}$ are polynomials in ψ_1, y_2, ..., y_d, divided by powers of

$$a \stackrel{\text{def}}{=} \frac{\partial \hat{g}}{\partial y_1}(\psi_1, y_2, \ldots, y_d).$$

Remembering that $y_1 = \psi_1 + z_1$ and going back from Y to the coordinates X with the help of the inverse of the power transformation (3.1), we obtain x_1, ..., x_n as functions of the parameters y_2, ..., y_n. These functions can be represented as power series of y_{d+1}, ..., y_n, the coefficients of which are rational functions of $\psi_1(y_2, \ldots, y_d)$, y_2, ..., y_d.

The question of the computability of these expansions in concrete cases rests on the question of the computability of the function ψ_1. In simple cases the function $\psi_1((y_2, \ldots, y_d)$ can be elementary, but in more complex cases it cannot be written explicitly, and it can only be computed approximately. In fact, the computation of this function can be postponed to the last moment. Initially one computes symbolically a sufficient number of coefficients of the mentioned series as rational functions of the symbols ψ_1, y_2, ..., y_d. Substituting in these coefficients sufficiently accurate values of the function ψ_1, we obtain their approximate values with an arbitrary accuracy.

In this way, near all parts of the surface $\hat{\mathcal{G}} \setminus \hat{\mathcal{G}}_c$ we can obtain asymptotic expansions for parts of the hypersurface $\mathcal{G}$, determined by the equation

$$g(Y) = 0. \tag{4.8}$$

There remains to study the hypersurface $\mathcal{G}$ near the set of critical points $\hat{\mathcal{G}}_c$. This set is defined by the system of algebraic equations (4.6), and it consists of a finite number of components $\hat{\mathcal{G}}_{cj}$ of different dimensions. Firstly, we consider a component of zero dimension

$$y_i = y_i^0, \quad i = 1, \ldots, d.$$

After the substitution

$$y_i = y_i^0 + z_i, \quad i = 1, \ldots, d,$$

in Equation (4.8), we obtain the equation

$$h(z_1, \ldots, z_d, y_{d+1}, \ldots, y_n) \stackrel{\text{def}}{=} g(Y) = 0, \tag{4.9}$$

whose solutions must be studied in the vicinity of the point $z_1 = \ldots = z_d = y_{d+1} = \ldots = y_n = 0$. Here h is a polynomial in its arguments. This is the very problem with which we began, but now there is a cone of the problem $\mathbf{K} = \{P\colon\ p_1, \ldots, p_d < 0, (p_{d+1}, \ldots, p_n) \in \tilde{\mathbf{U}}_j^{(d)}\}$. To solve it we must again apply the technique of the Newton polyhedron and of power transformations, but now the singularity in the coordinates z_1, ..., z_d, y_{d+1}, ..., y_n is simpler than it was in the coordinates X, since it is already partially resolved. Indeed, we must keep only those subsets of the set of obtained solutions to Equation (4.9) which lie in the set $\mathcal{U}(\mathbf{T}, \epsilon)$ corresponding to the cone $\mathbf{K}$ in agreement with Remark 1.1, for a sufficiently small $\varepsilon > 0$. Continuing this process, after a finite number of such steps we can hope to obtain a situation without singularity, which is covered by the theorem of implicit function.

Assume now that the component $\hat{\mathcal{G}}_{cj}$ of the set $\hat{\mathcal{G}}_c$ has a positive dimension e, $0 < e < d$. Splitting this component into parts, as it was done with the surface $\hat{\mathcal{G}} \setminus \hat{\mathcal{G}}_c$, and writing down explicit representations for each part, we can continue the resolving of the singularity near these parts, but we will not do it here, in order not to complicate the presentation.

Executing the described computations for all boundary subsets $\mathbf{S}_j^{(d)}$ with $d > 0$, we obtain asymptotic expansions in the parameters for the different parts of the hypersurface $\mathcal{F}$. Now we must match those expansions whose domains of application overlap in $\mathcal{F}$. These expansions belong either to one boundary subset $\mathbf{S}_j^{(d)}$ with $d > 0$, or to neighboring boundary subsets $\mathbf{S}_j^{(d)}$ and $\mathbf{S}_k^{(l)}$, having a non-empty intersection. But here we will not deal with the matching of asymptotic expansions.

Frequently it is necessary to study the hypersurface $\mathcal{F}$ not everywhere, but near a point X^0 or linear manifold $\mathcal{M}$. Then by a linear change of coordinates the point becomes the origin of coordinates, and the linear manifold becomes a coordinate subspace, and then we study only their vicinities, which is equivalent to the presence of the cone of the problem. In that case we have to solve Problem 1′ of Chapter 1, that decreases the volume of computations, but their character remains the same as it has been described above.

Example 4.1. Assume $n = 3$ and

$$f \stackrel{\text{def}}{=} x_1^2 - x_2^2 - x_3^2 + f_3(X), \tag{4.10}$$

where $X = (x_1, x_2, x_3)$, and f_3 is a polynomial containing no terms of degree less than three. We must solve the equation $f = 0$ in the vicinity of the point $X = 0$. The support supp$f$ consists of the points $Q_1 = (2,0,0)$, $Q_2 = (0,2,0)$, $Q_3 = (0,0,2)$ and points $Q = (q_1, q_2, q_3) \in \mathbb{Z}_+^3$ with $q_1 + q_2 + q_3 \geq 3$. Therefore the Newton polyhedron $\mathbf{\Gamma}(f)$ has a face $\mathbf{\Gamma}_1^{(2)}$ with vertices Q_1, Q_2, Q_3, and with the normal $N = -(1,1,1)$. The corresponding truncation is

$$\hat{f}_1^{(2)}(X) \stackrel{\text{def}}{=} x_1^2 - x_2^2 - x_3^2.$$

The power transformation

$$y_1 = x_1 x_3^{-1},$$

$$y_2 = x_2 x_3^{-1},$$

$$y_3 = x_3$$

with the inverse

$$x_1 = y_1 y_3,$$
$$x_2 = y_2 y_3,$$
$$x_3 = y_3$$

brings the polynomial (4.10) into the form

$$f = y_3^2(y_1^2 - y_2^2 - 1 + y_3 \tilde{f}(Y)),$$

where $\tilde{f}$ is a polynomial in $Y = (y_1, y_2, y_3)$. After reduction by y_3^2 we obtain

$$g(Y) \stackrel{\text{def}}{=} y_3^{-2} f = y_1^2 - y_2^2 - 1 + \sum_{k=1}^{m} \tilde{g}_k(y_1, y_2) y_3^k, \tag{4.11}$$

$$\hat{g} = y_1^2 - y_2^2 - 1,$$

where $\tilde{g}_k$ are polynomials in y_1, y_2 of degree not greater than $k+2$. The equation $\hat{g} = 0$ defines a hyperbola, which has no critical points. This equation has two solutions:

$$y_1 = \pm\sqrt{y_2^2 + 1}.$$

Let ψ_1 be one of these solutions. On it $\partial \hat{g}/\partial y_1 = 2y_1 = 2\psi_1$. In (4.11) we substitute $y_1 = \psi_1 + z_1$, then

$$g(Y) = h(z_1, \psi_1, y_2, y_3) \stackrel{\text{def}}{=} z_1^2 + 2\psi_1 z_1 + \sum_{k=0}^{m_1} \sum_{l=1}^{m_2} h_{kl}(\psi_1, y_2) z_1^k y_3^l,$$

where h_{kl} are polynomials in their arguments of degree no greater than $l-k+2$. In the point $z_1 = y_3 = 0$ we apply to the equation $h = 0$ Theorem 3.1 of implicit function. Here $a = 2\psi_1$. Therefore the solution to this equation is represented by a series of the form

$$z_1 = \sum_{k=1}^{\infty} \beta_k(\psi_1, y_2) y_3^k / a^{k+1}, \tag{4.12}$$

where β_k are polynomials in their arguments, and it can be shown that their degrees do not exceed $2k+2$. Going back to the coordinates X, we obtain the solution (4.12) in the form

$$x_1 = x_3 \psi_1 + \sum_{k=1}^{\infty} \beta_k(\psi_1, x_2 x_3^{-1}) x_3^{k+1} / (2\psi_1)^{k+1}. \tag{4.13}$$

We define $\omega = x_3 \psi_1$, i.e. $\omega = \pm\sqrt{x_2^2 + x_3^2}$. Since $\beta_k(\omega x_3^{-1}, x_2 x_3^{-1})$ are polynomials of degree $2k+2$, i.e. $\beta_k x_3^{2k+2}$ are homogeneous polynomials in

ω, x_2, x_3 of degree $2k+2$, which we designate as $2^{k+1}c_k$, then the expansion (4.13) has the form

$$x_1 = \omega + \sum_{k=1}^{\infty} c_k(\omega, x_2, x_3)/\omega^{k+1}. \tag{4.14}$$

This expansion describes two surfaces, corresponding to the two signs: $\omega = \pm\sqrt{x_2^2 + x_3^2}$. An expansion similar to (4.14) for solutions to Equation (4.10) was obtained in [Bruno 1996a] with the help of another approach. □

Remark 4.1. To find solutions to the system (4.6), one can use either the elimination method [Lefshets 1954, 1957] or the method of Gröbner bases [Becker and Weispfenning 1993].

Remark 4.2. The procedure described in the present section for the computation of asymptotic expansions for branches of solutions to equation (4.1) is also applicable in the case of an infinite support supp$f$ and a suitable cone of the problem according to Chapter 1, Section 8.

Remark 4.3. It is not quite clear whether each face $\mathbf{\Gamma}_j^{(d)}$ with $d > 0$ should be considered separately according to the mentioned process or it is sufficient to consider only the hyperfaces $\mathbf{\Gamma}_j^{(n-1)}$, attributing to each hyperface the cone which is the union of all cones $\mathbf{U}_k^{(d)}$ with $\mathbf{\Gamma}_k^{(d)} \subset \mathbf{\Gamma}_j^{(n-1)}$.

5. Implicit functions

We consider now a more general situation than in Sections 1–4. Let m ($m < n$) functions

$$f_1(X), \ldots, f_m(X) \tag{5.1}$$

be analytic at the origin $X = 0$ and expanded in power series

$$f_i(X) = \sum f_{iQ} X^Q, \quad i = 1, \ldots, m, \tag{5.2}$$

where $Q \in \mathbb{Z}_+^n$ and all $f_{i0} = 0$. We split vectors of dimension n in two subvectors of dimensions m and $n-m$ respectively; for instance, $X = (x_1, \ldots, x_n) = (X', X'')$, where $X' = (x_1, \ldots, x_m)$ and $X'' = (x_{m+1}, \ldots, x_n)$. Let

$$a \stackrel{\text{def}}{=} \det(f_{iE_j}), \quad i, j = 1, \ldots, m, \tag{5.3}$$

i.e.

$$a = \det \frac{\partial(f_1, \ldots, f_m)}{\partial X'} \quad \text{for} \quad X = 0.$$

Theorem 5.1 *Assume that the functions* (5.1) *are analytic at the origin* $X = 0$, *that all* $f_i(0) = 0$ *and that* $a \neq 0$ *in* (5.3). *Then the system of equations*

$$f_i(X) = 0, \quad i = 1, \ldots, m, \tag{5.4}$$

has a unique analytical solution

$$x_i = \varphi_i(X'') \stackrel{\text{def}}{=} \sum \varphi_{iR''} X''^{R''}, \quad i = 1, \dots, m, \tag{5.5}$$

where $R'' \in \mathbb{Z}_+^{n-m}$. *Here the coefficients are*

$$\varphi_{iR''} = \beta_{iR''}(\{f_{jQ}\})/a^{2k-1}, \tag{5.6}$$

where the $\beta_{iR''}$ *are polynomials of the coefficients* f_{jQ} *with* $\|Q\| \le k \stackrel{\text{def}}{=} \|R''\|$.

The proof is analogous to the proof of Theorem 1.1.

Remark 5.1. As in Section 1, in Theorem 5.1 one can assume that the coefficients f_{iQ} are global functions of Y only if the condition $a(Y) \neq 0$ is satisfied. Then the coefficients $\varphi_{iR''}$ are also global functions of Y.

We consider now generalized power series, as in Theorems 1.2 and 1.3.

Theorem 5.2 *Assume that* $T_i = E_i$, $i = 1, \dots, m$, *that the parts* T_j' *of the vectors* $T_{m+1}, \dots, T_l$ *are equal to zero, that vectors* $T_1, \dots, T_l$ *form the skeleton of the cone* $\mathbf{T} \subset \mathbb{R}^n$, *and that vectors* $T''_{m+1}, \dots, T''_l$ *form the skeleton of the cone* $\mathbf{T}'' \subset \mathbb{R}^{n-m}$. *Let the vector* $K = (1, \dots, 1, k_{m+1}, \dots, k_n)$ *be such that* $\langle K, T_j \rangle > 0$ *for* $j = m+1, \dots, l$, *and* $\langle K'', Q'' \rangle \ge 1$ *for all* $Q'' \in \mathbf{T}'' \cap \mathbb{Z}^{n-m} \setminus \{0\}$. *Assume that the series* (5.1) *belong to class* $\mathcal{C}(\mathbf{T})$ *and converge absolutely in a set* $\mathcal{U}(\mathbf{T}, \varepsilon)$. *Let all* $f_{i0} = 0$ *and* $a \neq 0$ *in* (5.3).

Then the system of equations (5.4) *has the solution* (5.5), *where the series* φ_i *belong to class* $\mathcal{C}(\mathbf{T}'')$ *and converge absolutely in a set* $\mathcal{U}(\mathbf{T}'', \varepsilon'')$. *Here formula* (5.6) *is correct, where* $\beta_{iR''}$ *are some polynomials of order no larger than* $2k-1$ *of the coefficients* f_{jQ} *with* $\langle K, Q \rangle \le k \stackrel{\text{def}}{=} \langle K'', R'' \rangle$.

The proof is analogous to the proofs of Theorems 1.3 and 5.1. Remark 5.1 remains in force for Theorem 5.2.

6. Truncated systems of equations

We consider now the aggregate of sums

$$f_i(X) \stackrel{\text{def}}{=} \sum f_{iQ} X^Q \text{ over } Q \in \mathbf{S}_i, \quad i = 1, \dots, m. \tag{6.1}$$

We assume that similar terms in each sum f_i are collected. To each of these sums f_i ($i = 1, \dots, m$) we put in correspondence the support $\mathbf{S}_i$, the polyhedron $\mathbf{\Gamma}_i$, its faces $\mathbf{\Gamma}_{ik_i}^{(d_i)}$, the truncations $\hat{f}_{ik_i}^{(d_i)}(X)$, the normal cones $\mathbf{U}_{ik_i}^{(d_i)}$, and other objects defined before. Here the first subscript i indicates that the object belongs to f_i. The aggregate of the truncated sums

$$\hat{f}_{iP}(X) \stackrel{\text{def}}{=} \sum f_{iQ} X^Q \text{ over } Q \in \mathbf{S}_{iP}, \quad i = 1, \dots, m, \tag{6.2}$$

is called *truncation of the aggregate* (6.1) with respect to the order $P = (p_1, \ldots, p_n)$, if for each $i = 1, \ldots, m$ the sum $\widehat{f}_{iP}(X)$ is the truncation of the sum f_i with respect to the order P.

The set $\mathbf{\Pi}$ of the vectors P in $\mathbb{R}^n_*$ is called the *cone of truncation* of the aggregate (6.2), if for each $P \in \mathbf{\Pi}$ the truncation with respect to the order P of the aggregate (6.1) is the aggregate (6.2), and conversely, if (6.2) is the truncation of the aggregate (6.1) with respect to the order P, then $P \in \mathbf{\Pi}$. The cone $\mathbf{\Pi}$ coincides with the normal cone $\mathbf{\Pi}$ of some aggregate of boundary subsets $\mathbf{S}^{(d_i)}_{ik_i}$, determined in Chapter 1, Section 9. Let d_i be the dimension of the truncation $\widehat{f}_{iP}$ in the sense of Section 3. The *dimension of the truncation of the aggregate* (6.2) is the codimension of the cone of truncation of that aggregate, i.e.

$$d = n - \dim \mathbf{\Pi} \leq d_1 + d_2 + \ldots + d_m.$$

Together with the aggregates of sums (6.1) and (6.2), we consider the system of equations

$$f_i(X) = 0, \quad i = 1, \ldots, m, \tag{6.3}$$

and the *truncated system of equations*

$$\widehat{f}_{iP}(X) = 0, \quad i = 1, \ldots, m. \tag{6.4}$$

Theorem 6.1 *Assume that the curve* (2.1) *is a solution to the full system of equations* (6.3), *and that the vector P lies in the cone $\mathbf{\Pi}$ of the truncated system* (6.4). *Then the first approximation to the curve* (2.1)

$$x_i = b_i \tau^{p_i}, \quad i = 1, \ldots, n, \tag{6.5}$$

is a solution to the truncated system (6.4).

Proof. We use the properties of the truncation from Section 2. According to property 3)

$$f(b_1\tau^{p_1}(1+o(1)), \ldots, b_n\tau^{p_n}(1+o(1))) = \widehat{f}(b_1, \ldots, b_n)\tau^c + o(1)\tau^c$$

and according to property 1)

$$\widehat{f}(b_1\tau^{p_1}, \ldots, b_n\tau^{p_n}) = \widehat{f}(b_1, \ldots, b_n)\tau^c.$$

Therefore (6.5) is a solution to the truncated system (6.4). □

A similar assertion is in [Bernshtein 1975].

We assume that in $\mathbb{R}^n_*$ an open convex cone $\mathbf{K}$ is given, and that from the sets $\mathbf{S}_i = \mathbf{S}(f_i)$ it is necessary to extract only those boundary subsets $\mathbf{S}^{(d_i)}_{ik_i}$ for which the normal cones $\mathbf{U}^{(d_i)}_{ik_i}$ intersect the cone $\mathbf{K}$. Here $\mathbf{K}$ is the *cone of the problem*. When a vicinity of a point X^0 is studied, the cone of the problem is determined by that point. For instance, the cone of the problem is $\mathbf{K} = \{P\colon\ p_1 < 0, \ldots, p_n < 0\}$ for the study of the vicinity of the point $X^0 = 0$, and

$$\mathbf{K} = \{P\colon\ p_1 < 0, \ldots, p_l < 0, p_{l+1} > 0, \ldots, p_n > 0\}.$$

if the point X^0 is such that $x_i^0 = 0$, $i = 1, \ldots, l$ and $x_j^0 = \infty$, $j = l+1, \ldots, n$.

For $m = 1$ all definitions and constructions of this section coincide with the corresponding definitions and constructions of Section 2.

Example 6.1. Assume that we have the system of equations

$$\begin{aligned} f(x_1, x_2, x_3) &\stackrel{\text{def}}{=} a_1 x_1 x_2 x_3 + a_2 x_1^4 + a_3 x_2^4 + a_4 x_3^4 + a_5 x_1^2 x_3^2 = 0, \\ f_2(x_1, x_2, x_3) &\stackrel{\text{def}}{=} a_{21} x_2 x_3 + a_{22} x_1^2 x_3 + a_{23} x_1^3 x_2 + a_{24} x_1^2 x_2 x_3 = 0. \end{aligned} \tag{6.6}$$

The Newton polyhedron $\mathbf{\Gamma} = \mathbf{\Gamma}(f)$ and its accessories were constructed in Chapter 1, Examples 3.1 and 5.1 (see also Chapter 2, Example 2.2), and the Newton polyhedron $\mathbf{\Gamma}_2 = \mathbf{\Gamma}(f_2)$ in Chapter 1, Example 9.1. Also for the cone of the problem $\mathbf{K} = \{P < 0\}$ there were found two intersections of normal cones $\mathbf{U}_{1j}^{(d_1)} \cap \mathbf{U}_{2k}^{(d_2)}$ with d_1, $d_2 > 0$. They are $\mathbf{\Pi}_1 = \{P = -\lambda(1,2,1)\} = \mathbf{U}_2^{(2)} \cap \mathbf{U}_{21}^{(1)}$ and $\mathbf{\Pi}_2 = \{P = -\lambda(1,1,3)\} = \mathbf{U}_4^{(1)} \cap \mathbf{U}_{22}^{(1)}$, where $\lambda > 0$. To the cone $\mathbf{\Pi}_1$ there corresponds the truncated system

$$\begin{aligned} \widehat{f}(x_1, x_2, x_3) &\stackrel{\text{def}}{=} a_1 x_1 x_2 x_3 + a_2 x_1^4 + a_4 x_3^4 + a_5 x_1^2 x_3^2 = 0, \\ \widehat{f}_2(x_1, x_2, x_3) &\stackrel{\text{def}}{=} a_{21} x_2 x_3 + a_{22} x_1^2 x_3 = 0; \end{aligned} \tag{6.7}$$

and to the cone $\mathbf{\Pi}_2$, the truncated system

$$\widehat{f}(x_1, x_2, x_3) \stackrel{\text{def}}{=} a_2 x_1^4 + a_3 x_2^4 = 0, \tag{6.8}$$

$$\widehat{f}_2(x_1, x_2, x_3) \stackrel{\text{def}}{=} a_{21} x_2 x_3 + a_{23} x_1^3 x_2 = 0. \ \square \tag{6.9}$$

We assume that the expansions

$$x_k = g_k(Z) = \sum g_{kR'} Z^{R'} \quad \text{over } R' \in \mathbf{S}'_k, \quad k = 1, \ldots, n, \tag{6.10}$$

where $Z \in \mathbb{C}^l$, $R' \in \mathbb{R}^l$, represent a l-parameter solution to the system of equations (6.3), and $X \to X^0$ for $Z \to Z^0$. All constructions described above, related to the extraction of truncations, are valid for the expansions (6.10). Let the vector $P' \in \mathbb{R}^l_*$ be fixed, and the truncation of the aggregate of the functions $\{g_k\}$ with respect to order P' be

$$\widehat{g}_k(Z), \quad k = 1, \ldots, n,$$

i.e. with the substitution

$$z_i = \tau^{p'_i}(b'_i + o(1)), \quad i = 1, \ldots, l, \tag{6.11}$$

we obtain

$$\widehat{g}_k = \tau^{\langle P', R'_k \rangle}(\text{const} + o(1)), \quad R'_k \in \mathbf{S}'^{(d_k)}_{kj_k}, \quad k = 1, \ldots, n.$$

We denote $p_k = \langle P', R'_k \rangle$, $k = 1, \ldots, n$. Then the vector $P = (p_1, \ldots, p_n)$ is the order of the solution (6.10), corresponding to the vector P' of the substitution (6.11), i.e. to the order P' of the parameters Z.

Theorem 6.2 [Bruno and Soleev 1994b] *Assume that:*

a) the expansions (6.10) *give a l-parameter solution to the system* (6.3)*;*

b) the truncation of that solution with respect to the order P' is

$$x_k = \widehat{g}_k(Z), \quad k = 1, \ldots, n, \tag{6.12}$$

and on the curves (6.11) *it has the order P;*

c) (6.4) *is the truncation of the system of equations* (6.3) *with respect to the order P.*

Then the truncation (6.12) *of the solution* (6.10) *is a solution to the truncated system of equations* (6.4).

Proof. Assume that (6.10) represents a l-parameter solution to the system of equations (6.3), and that $X \to X^0$ for $Z \to Z^0$. To each sum $g_i(Z)$, $i = 1, \ldots, n$, there corresponds the Newton polyhedron $\mathbf{\Gamma}'_i$, the collection of its faces $\{\mathbf{\Gamma}'^{(d_i)}_{ij_i}\}$, its normal cones $\mathbf{U}'^{(d_i)}_{ij_i}$, and also the boundary subsets $\mathbf{S}'^{(d_i)}_{ij_i}$ and the truncations $\widehat{g}^{(d_i)}_{ij_i}$. We consider the intersection

$$\mathbf{U}'^{(d_1)}_{1j_1} \cap \ldots \cap \mathbf{U}'^{(d_n)}_{nj_n} \stackrel{\text{def}}{=} \mathbf{\Pi}'^{(d_1, \ldots, d_n)}_{j_1, \ldots, j_n}. \tag{6.13}$$

We assume that the cone (6.13) is non-empty and the aggregate of truncations

$$\widehat{g}_k(Z) \stackrel{\text{def}}{=} \sum g_{kR'} Z^{R'} \ \text{ over } R'_k \in \mathbf{S}'^{(d_k)}_{kj_k}, \quad k = 1, \ldots, n$$

corresponds to it. Then for $Z \to Z^0$ the aggregare of truncations (6.12) is the first approximation to the aggregare of solutions (6.10). Let the vector $P' = (p'_1, \ldots, p'_l)$ be in the cone (6.13), i.e. the truncation (6.12) is a truncation of the aggregare (6.10) with respect to the order P'. Substituting (6.11) into (6.10), we have

$$\begin{gathered} x_k = g_k = \tau^{\langle P', R'_k \rangle}(\widehat{g}_k(b'_1, \ldots, b'_l) + o(1)), \\ R'_k \in \mathbf{S}'^{(d_k)}_{kj_k}, \quad k = 1, \ldots, n. \end{gathered} \tag{6.14}$$

We denote

$$p_k = \langle P', R'_k \rangle, \quad b_k = \widehat{g}_k(b'_1, \ldots, b'_l), \quad k = 1, \ldots, n; \tag{6.15}$$

Then equalities (6.14) are of the form

$$x_k = g_k(Z(\tau)) \stackrel{\text{def}}{=} \tau^{p_k}(b_k + o(1)), \quad k = 1, \ldots, n. \tag{6.16}$$

Here the vector $P = (p_1, \ldots, p_n)$ is the order of the solution (6.10), corresponding to the vector P' of the substitution (6.11), and in (6.15) all $b_k \not\equiv 0$, since $\widehat{g}_k \not\equiv 0$, and, varying b'_i, it is always possible to obtain that all $b_k \neq 0$.

According to condition c) of the theorem, the system (6.4) is the truncation of the system of equations (6.3) with respect to the order P. Therefore we take advantage of the properties of the truncations from Section 2. According to Property 3), after the substitution (6.16), we have

$$f_i(\tau^{p_1}(b_1 + o(1)), \ldots, \tau^{p_n}(b_n + o(1))) = \widehat{f}_i(b_1, \ldots, b_n)\tau^{c_i} + o(1),$$
$$c_i = \langle P, Q \rangle, \quad Q \in \mathbf{S}^{(d_i)}_{ij_i}, \quad i = 1, \ldots, m,$$

and from Property 1) of truncations (see Section 2)

$$\widehat{f}_i(b_1\tau^{p_1}, \ldots, b_n\tau^{p_n}) = \widehat{f}_i(b_1, \ldots, b_n)\tau^{c_i}, \quad i = 1, \ldots, m.$$

The truncation (6.12) with respect to the order P has the form

$$x_k = b_k\tau^{p_k}, \quad k = 1, \ldots, n$$

after the substitution $z_i = b'_i\tau^{P'_i}$, $i = 1, \ldots, l$ (compare with (6.11)). □

In other words, the first approximation to a solution is a solution to the corresponding first approximation to the system of equations. The opposite assertion is not true in general, i.e. a solution to the first approximation (truncation) of the system of equations is not necessarily the first approximation to a solution of this system. For a generalization of Theorem 6.2 to other classes of solutions, see [Bruno 1996a].

Remark 6.1. Only the asymptotic coordinates x_k which tend either to zero or to infinity are essential for the extraction of truncations. If in (6.1) the functions f_i depend also on neutral coordinates Y in such a way that the coefficients f_{iQ} are functions of Y, which remain bounded away from 0 and ∞, then the whole construction described above for the extraction of truncations is also applicable in that case. With that the coefficients $g_{kR'}$ in (6.10) are also bounded functions of Y.

Remark 6.2. It can happen that the truncated aggregate (6.2) is degenerate, i.e. there exists more functional dependencies between the $\widehat{f}_{ij_i}^{(d_i)}(X)$ than between the $f_i(X)$. Then by means of an operation over the f_i, from the aggregare (6.1) we can pass to an equivalent aggregare, which would have another aggregare of truncations. For instance, if $f_1 = \varphi_3 + \varphi_9$, $f_2 = \varphi_3$, where φ_k are homogeneous polynomials of degree k and $X^0 = 0$, then for $P = (-1, \dots, -1)$ we have $\widehat{f}_{1P} = \widehat{f}_{2P} = \varphi_3$, but for the aggregare $f_3 = f_1 - f_2$, f_2, we have $\widehat{f}_{3P} = \varphi_9$, $\widehat{f}_{2P} = \varphi_3$. The procedure of modification of the given equations in order to improve the truncated system is presented in Section 8 for some cases.

7. Linear transformations of power exponents

We consider now transformations of the system of equations (6.3), which in the space $\mathbb{R}^n$ of the power exponents Q correspond to parallel translations (its own for each f_i) and to affine transformations (one for all f_i). These transformations make the geometry of the power exponents (or the power geometry) more rich and give a method for the solution of truncated systems.

Finding those solutions to the system (6.3) in which one of the coordinates is identically equal to zero can be considered as a solved problem, since it reduces to a problem analogous to the original one, but with a smaller dimension. Therefore we will look for solutions to the system (6.3) in which no coordinate is identically equal to zero. For such solutions, in each of the equations of the system (6.3) we can execute a reduction by any product of powers of the coordinates. If in the ith equation we make a reduction by X^{-T_i}, then in it the vector power exponents become equal to $\tilde{Q} = Q + T_i$. i.e. the set $\mathbf{S}_i$ has a parallel shift by the vector T_i.

Let $f(X)$ be a sum (2.6) and d be the dimension of its Newton polyhedron $\mathbf{\Gamma} = \mathbf{\Gamma}(f)$; in Section 3 this was called the dimension of the sum $f(X)$. In $\mathbb{R}^n_*$ we consider the maximal linear space $\mathbf{N}(f)$ normal to $\mathbf{\Gamma}(f)$. Obviously, $\dim \mathbf{\Gamma} + \dim \mathbf{N} = n$. Similarly, for the aggregate of sums $f_i(X)$, $i = 1, \dots, m$, we consider the polyhedra $\mathbf{\Gamma}(f_i)$ and their normal spaces $\mathbf{N}(f_i)$. We denote

$$\mathbf{N} = \mathbf{N}(f_1) \cap \dots \cap \mathbf{N}(f_m)$$

and $d = n - \dim \mathbf{N}$. The quantity d is called the *dimension of* the mentioned *aggregate* of sums (see Section 6). This definition of the dimension coincides with the definition given earlier for the truncated system, and always $d < n$ for them.

We assume that for the system of equations (6.1), (6.3) there is a truncated system of equations (6.2), (6.4).

Theorem 7.1 *If the dimension of the system* (6.4) *equals d, then there exists a matrix α such that the power transformation* (3.1) *with the matrix α and* *appropriate reductions reduces the system* (6.4) *to a system of m equations for d variables. If in the expansions* (6.2) *all exponents Q are integers, then there exists an unimodular matrix α.*

The proof is similar to the proof of Theorem 3.1, only for each i we choose its own vector T_i such that

$$X^{T_i}\hat{f}_i(X) = \hat{g}_i(y_1,\dots,y_d), \quad i = 1,\dots,m.$$

We denote the dimension of the subsystem of the first i equations of the system (6.4) as $d(i)$. Obviously, $d(i) \le d(i+1)$ and $d(m) = d$. From Theorem 7.1 there follows

Theorem 7.2 *There exists a power transformation* (3.1) *and reductions such that the system* (6.4) *of dimension d reduces to a system of m equations for d variables, having a quasitriangular form: the subsystem of its first i equations depends on $d(i)$ variables* ($i = 1,\dots,m$). *If in* (6.2) *all exponents Q are integers, then there exists a transformation* (3.1) *with an unimodular matrix α.*

Indeed, according to Theorem 7.1 we reduce the whole system (6.4) to a system of d variables. Then, also making power transformations for these d variables, we reduce the subsystem of its first $i = m-1$ equations to a subsystem of $d(m-1)$ variables, and so on with i decreasing from m to 1.

Corollary 7.1 *We assume that in the polynomial system* (6.4) $d(i) = i$ *for* $i < n$, *and if* $m > n-1$, *then* $d(j) = n-1$ *for* $n \le j \le m$. *Then there exists an unimodular matrix α such that the power transformation* (3.1) *with appropriate reductions reduces the system* (6.4) *to a triangular form*

$$\begin{aligned} &\widehat{g}_i(y_1,\dots,y_i) = 0, \quad i = 1,\dots,\min(n-1,m), \\ &\widehat{g}_j(y_1,\dots,y_{n-1}) = 0, \quad j = n,\dots,m. \end{aligned} \tag{7.1}$$

Corollary 7.2 *We consider a system* (6.4) *in which the dimension of each polynomial $\widehat{f}_i$ equal to one. Then the power transformation with reductions transforms the system to the form* (7.1), *where each of the functions $\widehat{g}_i$ is a polynomial in one monomial Y^{R_i}:*

$$\widehat{g}_i = h_i(Y^{R_i}), \quad i = 1,\dots,m,$$

with $R_i = (r_{i1},\dots,r_{ii},0,\dots,0)$ *for* $i < n$. *If we have found all roots* $z_i^{(1)}, \dots, z_i^{(l_i)}$ *of each equation*

$$h_i(z_i) = 0, \quad i = 1,\dots,m,$$

then the system (7.1) *reduces to* $l_1 l_2 \dots l_m$ *systems*

$$Y^{R_i} = z_i^{(j_i)}, \quad i = 1,\dots,m,$$

where $j_i = 1, \ldots, l_i$.

Assume that in the situation of Corollary 7.2 S_i is the unit vector of the linear hull of the support $\mathbf{S}(\widehat{f}_i)$, i.e. the difference between two neighboring integer points from the segment $\mathbf{\Gamma}(\widehat{f}_i)$, $i = 1, \ldots, m$. Obviously, $S_i \in \mathbb{Z}^n$. Then $r_{11} \ldots r_{ii} = \pm$LCD (largest common divisor) of minors of the maximal order of the matrix

$$(S_1 \ldots S_i), \quad i = 1, \ldots, \min(m, n-1);$$

i.e. $r_{11} = 1$ always. If all $r_{ii} = \pm 1$, then by additional power transformations one can obtain $r_{ij} = \delta_{ij}$ (Kronecker symbol). Then $Y^{R_i} = y_i$ for $i = 1, \ldots, \min(m, n-1)$.

By a change of the order of enumeration of the sums $\widehat{f}_i$ in the system (6.4), we can obtain an additional simplification in the transformed system

$$\widehat{g}_i(y_1, \ldots, y_d) \stackrel{\text{def}}{=} X^{T_i} \widehat{f}_i(x_1, \ldots, x_n) = 0, \quad i = 1, \ldots, m, \tag{7.2}$$

obtained according to Theorem 7.2. The necessary transpositions for small dimensions ($n \leq 6$) can be seen immediately from the transformed system (7.2). In general, the choice of the best enumeration requires the examination of different variants. We will not analyze this here.

Theorem 7.3 *Assume that* (6.2) *is a truncation of the aggregate of sums* (6.1) *and that $d(i)$ is the dimension of the aggregate of the first i sums in* (6.2). *There exists a matrix α and vectors T_1, ..., T_m such that as a result of the power transformation* (3.1) *and reductions the aggregate* (6.1) *transforms into the aggregate of sums*

$$g_i(Y) \stackrel{\text{def}}{=} \sum g_{i\tilde{Q}} Y^{\tilde{Q}} = X^{T_i} f_i(X), \quad i = 1, \ldots, m,$$

where $\tilde{Q} \geq 0$, and its truncation (6.2) *transforms into truncation*

$$\widehat{g}_i(y_1, \ldots, y_{d(i)}) \stackrel{\text{def}}{=} X^{T_i} \widehat{f}_i(X), \quad i = 1, \ldots, m.$$

If supports of all sums $f_i(X)$ are integer, then the matrix α can be taken as unimodular, and the vectors T_1, ..., T_m as integer.

The proof follows from Chapter 1, Theorem 11.1.

Remark 7.1. In the situation of Theorem 7.3 for Laurent polynomials (6.1), there exist a matrix α and vectors T_1, ..., T_m such that all $\hat{g}_i$ are ordinary polynomials in y_1, ..., y_d, and all g_i are polynomials in y_{d+1}, ..., y_n and Laurent polynomials in y_1, ..., y_d. Such matrices α are easier to find than the matrices α of Theorem 7.3.

Example 7.1. Let us simplify the truncated systems (6.7) and (6.8), (6.9) of Example 6.1. We have for the system (6.7) $Q_1 - Q_2 = (-3, 1, 1)$, $Q_2^2 - Q_1^2 =$

$(2,-1,0)$. We compose from these vectors the first two lines of the matrix α and we select the third line so as to obtain $\det\alpha = 1$. We obtain the unimodular matrix

$$\alpha = \begin{pmatrix} -3 & 1 & 1 \\ 2 & -1 & 0 \\ -1 & 1 & 0 \end{pmatrix}, \quad \alpha^{-1} = \begin{pmatrix} 0 & 1 & 1 \\ 0 & 1 & 2 \\ 1 & 2 & 1 \end{pmatrix}.$$

The power transformation with the matrix α and its inverse are

$$\begin{aligned} y_1 &= x_1^{-3}x_2x_3, & x_1 &= y_2y_3, \\ y_2 &= x_1^2x_2^{-1}, & x_2 &= y_2y_3^2, \\ y_3 &= x_1^{-1}x_2, & x_3 &= y_1y_2^2y_3. \end{aligned} \tag{7.3}$$

After the power transformation (7.3) and the reduction of the first equation of the truncated system (6.7) by $y_2^4y_3^4$, and the second one by $y_1y_2^3y_3^3$, we obtain the system

$$\begin{aligned} \widehat{g}(y_1,y_2) &\stackrel{\text{def}}{=} a_1y_1 + a_2 + a_4y_1^4y_2^4 + a_5y_1^2y_2^2 = 0, \\ \widehat{g}_2(y_2) &\stackrel{\text{def}}{=} a_{21} + a_{22}y_2 = 0. \end{aligned} \tag{7.4}$$

Here according to Remark 7.1, the full system (6.6) becomes

$$\begin{aligned} g(y_1,y_2,y_3) &\stackrel{\text{def}}{=} a_1y_1 + a_2 + a_3y_3^4 + a_4y_1^4y_2^4 + a_5y_1^2y_2^2 = 0, \\ y_1g_2(y_1,y_2,y_3) &\stackrel{\text{def}}{=} a_{21} + a_{22}y_2 + a_{23}y_1^{-1}y_2y_2^2 + a_{24}y_2^2y_3^2 = 0. \end{aligned} \tag{7.5}$$

According to the formulae (9.5), (3.2), (3.1) and the Example 9.1 of Chapter 1, to the truncated system (6.8), (6.9) there correspond $\mathbf{S}_4^{(1)} = \{Q_2, Q_3\}$ and $\mathbf{S}_{22}^{(1)} = \{Q_1^2, Q_3^2\}$. We form the differences: $Q_3 - Q_2 = 4(-1,1,0)$, $Q_1^2 - Q_3^2 = (-3,0,1)$. We write the vectors $(-1,1,0)$ and $(-3,0,1)$ in the first two lines of the matrix α, and we take a unit vector as the third line. We obtain the unimodular matrix

$$\alpha = \begin{pmatrix} -1 & 1 & 0 \\ -3 & 0 & 1 \\ 1 & 0 & 0 \end{pmatrix}, \quad \alpha^{-1} = \begin{pmatrix} 0 & 0 & 1 \\ 1 & 0 & 1 \\ 0 & 1 & 3 \end{pmatrix}.$$

The power transformation with this matrix α and its inverse are

$$\begin{aligned} y_1 &= x_1^{-1}x_2, & x_1 &= y_3, \\ y_2 &= x_1^{-3}x_3, & x_2 &= y_1y_3, \\ y_3 &= x_1, & x_3 &= y_2y_3^3. \end{aligned}$$

After this power transformation and the reduction of the equation (6.8) by x_1^4, and the equation (6.9) by $x_1^3x_2$, we obtain the truncated system

$$\begin{aligned} \widehat{g}(y_1) &\stackrel{\text{def}}{=} a_2 + a_3y_1^4 = 0, \\ \widehat{g}_2(y_2) &\stackrel{\text{def}}{=} a_{21}y_2 + a_{23} = 0. \end{aligned} \tag{7.6}$$

8. Asymptotic solution of a system of equations

We consider the *fundamental problem*. Let the system of equations

$$f_i(X) \stackrel{\text{def}}{=} \sum_{Q\in\mathbf{S}_i} f_{iQ}X^Q = 0, \quad i = 1,\ldots,m, \tag{8.1}$$

where the $f_i(X)$ are Laurent polynomials, and the convex cone $\mathbf{K}$ in $\mathbb{R}_*^n$ be given. It is required to find all such multiparameter solutions to the system (8.1), which lie in the set $\mathcal{U}(\mathbf{K}^*,\varepsilon)$, where $\mathbf{K}^*$ is the cone dual to the cone $\mathbf{K}$, and $\varepsilon > 0$ is sufficiently small, and through which one can draw a curve of the form (2.1) with $P \in \mathbf{K}$ ($P \neq 0$).

A fundamental problem is called the *reduced* one if we are looking for branches in which none of the coordinates is equal identically to zero (or infinity). Obviously, a fundamental problem falls apart into a n-dimensional reduced problem and a finite number of reduced problems in which some coordinates x_i are set to be zeros (or infinities), i.e. constants, and the cone of the problem is the intersection of the cone of the fundamental problem $\mathbf{K}$ with the set

$$\{P\colon\ p_i = -\infty,\ \text{if } x_i \equiv 0; p_j = +\infty,\ \text{if } x_j = \infty\}.$$

To solve the reduced problem with the cone of the problem $\mathbf{K}$, for each f_i we form the Newton polyhedron $\mathbf{\Gamma}_i$, and we single out its faces $\mathbf{\Gamma}_{ik}^{(d)}$ with normal cones $\mathbf{U}_{ik}^{(d)}$. Here it is sufficient to single out all faces $\mathbf{\Gamma}_{ik}^{(d)}$, for which the intersection $\mathbf{U}_{ik}^{(d)} \cap \mathbf{K}$ is non-empty. We introduce $\mathbf{K}_{ik}^{(d)} = \mathbf{K} \cap \mathbf{U}_{ik}^{(d)}$, and we consider all possible non-empty intersections

$$\mathbf{K}_{1k_1}^{(d_1)} \cap \ldots \cap \mathbf{K}_{mk_m}^{(d_m)} \stackrel{\text{def}}{=} \mathbf{\Pi}_\lambda, \quad \lambda = 1,\ldots,l. \tag{8.2}$$

Let $\mathbf{\Pi}_\lambda$ be one such intersection, and let the truncated system

$$\widehat{f}_i(X) \stackrel{\text{def}}{=} \sum_{Q \in \mathbf{S}_{ik_i}^{(d_i)}} f_{iQ} X^Q = 0, \quad i = 1, \ldots, m. \tag{8.3}$$

correspond to it.

If there is a $d_i = 0$, then $\widehat{f}_i = aX^Q$. The solutions to the equation $aX^Q = 0$ have one of the coordinates identically equal to zero (or infinity), they cannot be solutions to the reduced problem. Thus it is necessary to consider only truncated systems (8.3) in which all $d_i > 0$. According to the truncated system (8.3) we make a power transformation and the reductions mentioned in Theorem 7.1. Then the system (8.1) transforms into the system

$$g_i(y_1, \ldots, y_n) \stackrel{\text{def}}{=} X^{T_i} f_i(x_1, \ldots, x_n) = 0, \quad i = 1, \ldots, m, \tag{8.4}$$

and the truncated system (8.3) transforms into the system

$$\widehat{g}_i(y_1, \ldots, y_d) \stackrel{\text{def}}{=} X^{T_i} \widehat{f}_i(x_1, \ldots, x_n) = 0, \quad i = 1, \ldots, m, \tag{8.5}$$

where d is the dimension of the truncated system (8.3). We assume that the $\widehat{g}_i$ are ordinary polynomials in $y_1, \ldots, y_d$, and that in the g_i the coordinates $y_{d+1}, \ldots, y_n$ appear with non-negative powers. It is always possible to achieve this according to Remark 7.1. Here the cone of the truncation (8.5) is

$$\tilde{\mathbf{\Pi}}_\lambda = \alpha \mathbf{\Pi}_\lambda \subset \{P : p_1 = \ldots = p_d = 0\}.$$

Now we have to find those solutions

$$y_i = c_i \tau^{p_i}(1 + o(1)), \quad i = 1, \ldots, n,$$

to the full system (8.4) which have the vector order $P \in \tilde{\mathbf{\Pi}}_\lambda$. According to Theorem 6.1, in such solutions the values $y_i = c_i$, $i = 1, \ldots, d$, satisfy the system (8.5). Here all $c_1, \ldots, c_d$ are different from zero and infinity. The tangent cone $\tilde{\mathbf{T}}$ of the truncated system (8.5) contains restrictions only on $q_{d+1}, \ldots, q_n$, and it lies in the cone $q_{d+1}, \ldots, q_n \geq 0$.

If $d < m$, then in the generic case the truncated system (8.5) has no solutions. If it has an isolated linear manifold of solutions, then in the system (8.4) we have to make a linear change of coordinates, which transfers that manifold into a coordinate subspace, and then to solve it in the vicinity of that subspace, using the Newton polyhedra if necessary. If the system (8.5) has a continuous set of solutions, which is not a linear manifold, then such a *case* is called the *degenerate* one; we will consider it later .

We consider now the system (8.5) with

$$m \leq d. \tag{8.6}$$

In the d-dimensional space with coordinates $y_1, \ldots, y_d$, the truncated system of equations (8.5) determines an algebraic manifold $\hat{\mathcal{G}}$. We have to find now the whole manifold $\hat{\mathcal{G}}$. We extract firstly the subset $\hat{\mathcal{G}}_c$ of all critical points $y_1, \ldots, y_d$, in which the equalities (8.5) hold, and the matrix

$$\frac{\partial(\hat{g}_1, \ldots, \hat{g}_m)}{\partial(y_1, \ldots, y_d)} \tag{8.7}$$

of dimension $d \times m$ has a rank less then m. The set $\hat{\mathcal{G}} \setminus \hat{\mathcal{G}}_c$ can be split into a finite number of parts $\hat{\mathcal{G}}_1, \ldots, \hat{\mathcal{G}}_l$, in each of which there is at least one non-zero minor of order m

$$\det \frac{\partial(\hat{g}_1, \ldots, \hat{g}_m)}{\partial(y_{i_1}, \ldots, y_{i_m})}$$

of the matrix (8.7), and the corresponding coordinates $y_{i_1}, \ldots, y_{i_m}$ can be expressed as functions of the remaining coordinates among $y_1, \ldots, y_d$. Let $\hat{\mathcal{G}}_1$ be that part of the set $\hat{\mathcal{G}} \setminus \hat{\mathcal{G}}_c$ on which

$$\det(\partial(\hat{g}_1, \ldots, \hat{g}_m)/\partial(y_1, \ldots, y_m)) \neq 0$$

and $y_i = \psi_i(y_{m+1}, \ldots, y_d)$, $i = 1, \ldots, m$. In the whole system (8.4) we make the change of coordinates

$$y_i = \psi_i + z_i, \quad i = 1, \ldots, m, \tag{8.8}$$

and we obtain the functions

$$h_i(z_1, \ldots, z_m, \psi_1, \ldots, \psi_m, y_{m+1}, \ldots, y_n) \stackrel{\text{def}}{=} g_i(Y), \quad i = 1, \ldots, m,$$

which we write as polynomials in $\check{Z} = (z_1, \ldots, z_m, y_{d+1}, \ldots, y_n)$:

$$\check{h}_i = \sum \check{h}_{i\check{Q}} \check{Z}^{\check{Q}},, \quad i = 1, \ldots, m, \tag{8.9}$$

with coefficients $\check{h}_{i\check{Q}}$ depending on $\psi_1, \ldots, \psi_m, y_{m+1}, \ldots, y_d$. In the "point" $\check{Z} = 0$ we have

$$\det \frac{\partial(\check{h}_1, \ldots, \check{h}_m)}{\partial(z_1, \ldots, z_m)} = \det \left\{ \left. \frac{\partial(g_1, \ldots, g_m)}{\partial(y_1, \ldots, y_m)} \right|_{y_i = \psi_i} \right\} \stackrel{\text{def}}{=} a \neq 0, \tag{8.10}$$

and the exponents $\check{Q}$ of the expansions (8.9) lie in the tangent cone $\check{\mathbf{T}}$, equal to the projection of the cone $\tilde{\mathbf{T}}$ in the subspace of the coordinates $q_1, \ldots, q_m$, $q_{d+1}, \ldots, q_n$. Now to the system of equations

$$h_i(\check{Z}) = 0, \quad i = 1, \ldots, m,$$

one can apply the Implicit Function Theorem 5.2 and obtain its solution in the form of a series

$$z_i = \phi(y_{d+1}, \dots, y_n) \stackrel{\text{def}}{=} \sum \varphi_{iR''} Y''^{R''}, \quad i = 1, \dots, m, \tag{8.11}$$

where $Y'' = (y_{d+1}, \dots, y_n)$, $R'' = (r_{d+1}, \dots, r_n)$, and the coefficients $\varphi_{iR''}$ are polynomials in $\psi_1, \dots, \psi_m, y_{m+1}, \dots, y_d$, divided by powers of a from (8.10). Remembering the change (8.8) and going back from Y to X with the help of the power transformation, we obtain a parametric representation $X = X(y_{m+1}, \dots, y_n)$ for some solutions to the initial system of equations (8.1). In a similar way we obtain the solutions related to the other parts $\hat{\mathcal{G}}_j$ of the set $\hat{\mathcal{G}} \setminus \hat{\mathcal{G}}_c$.

The set of the critical points $\hat{\mathcal{G}}_c$ of the manifold $\hat{\mathcal{G}}$ is an algebraic manifold itself. It consists of some connected components $\hat{\mathcal{G}}_{c1}, \dots, \hat{\mathcal{G}}_{ck}$. If $\hat{\mathcal{G}}_{cj}$ is an isolated linear manifold, then in the full system (8.4) we make a linear change of coordinates, transforming that manifold into a coordinate subspace, and we study the obtained system in the vicinity of that subspace. This is again a fundamental problem, but with a narrower set of solutions. If $\hat{\mathcal{G}}_{cj}$ is not a linear manifold, then this *case* is also called a *degenerate* one.

At present the unified strategy for the study of degenerate cases is not yet elaborated. In [Bruno and Soleev 1991a, § 4] two such strategies are presented. The first consists in obtaining by means of manipulations of the polynomials $g_1, \dots, g_m$ a new polynomial g_{m+1}, which is a function of $g_1, \dots, g_m$, and which has a truncation $\hat{g}_{m+1}$, and this truncation is not a function of the truncations $\hat{g}_1, \dots, \hat{g}_m$. If such a polynomial cannot be obtained, then the polynomials $g_1, \dots, g_m$ are functionally dependent, and one can find that dependence. The second strategy consists in the introduction of new coordinates $y_{n+i} = \alpha_i$, $i = 1, \dots, s$, where $\alpha_1, \dots, \alpha_s$ form a polynomial basis of the ideal of the manifold $\hat{\mathcal{G}}_{cj}$, and in the writing of the functions $g_1, \dots, g_m$ as polynomials in $y_{n+i}, \dots, y_{n+s}$.

So, we have described here one step of the procedure for the solution of the problem posed in the present section. As a result of that step, the primary fundamental problem falls apart into a finite number of secondary fundamental problems, each of which in a sense is simpler than the initial problem. Some of them have a unique solution-branch, or they have no one such solution. For such problems the procedure of extraction of the branches can be considered as completed.

For the remaining secondary fundamental problems the procedure must be continued. After a finite number of such steps one can separate all branches with multiplicity one; however the number of steps is not known a priori (see [Kukles and Grus 1958]).

The most studied case is $m = n - 1$, when all polynomials $f_1, \dots, f_{n-1}$ are functionally independent. The system (8.1) defines an algebraic curve and we have to find its branches [Bruno and Soleev 1991a]. If $d = m = n - 1$, then the classical theorem by Bezout asserts that the number of complex solutions to the system of truncated equations (8.5) can be estimated through their degrees (it equals the product of these degrees). A more exact number of solutions to the system of equations was obtained by Bernshtein [1975] (compare [Kushnirenko 1975a, 1976; Khovanskii 1978b]). It is equal to the mixed Minkovski volume $V(\mathbf{\Gamma}_1, \ldots, \mathbf{\Gamma}_m)$ of the Newton polyhedra $\mathbf{\Gamma}_1, \ldots, \mathbf{\Gamma}_m$ of the corresponding polynomials (see [Khovanskii 1988]).

We call the *complexity of the truncated system* (8.3) the $(n-1)$-dimensional mixed Minkovski volume of the corresponding translated faces $\mathbf{\Gamma}_{ik}^{(d_i)} + T_i$, $i = 1, \ldots, m$, where $-T_i \in \mathbf{\Gamma}_{ik}^{(d_i)}$, and we call the *complexity of the reduced problem* the sum of the complexities of all truncated systems whose normal cones intersect the cone of the problem. Then in the generic case, the number of complex branches of the reduced problem is equal to its complexity.

There are formulae for the number of branches of solutions to the system (8.1) in a small neighborhood of the isolated critical point $X = 0$ (see [Grin′ 1971; Sather 1973; Fucuda et al. 1986; Szafraniec 1992]). We shall show one of them. We designate as b the number of branches of the solutions to the system (8.1) in a small neighborhood of the isolated critical point $X = 0$, and we put

$$\omega \overset{\text{def}}{=} x_1^2 + \ldots + x_n^2, \quad \delta \overset{\text{def}}{=} \det[\partial(F, \omega)/\partial X],$$

where $F \overset{\text{def}}{=} (f_1, \ldots, f_{n-1})$. Obviously, $\delta(0) = 0$. Let $H \overset{\text{def}}{=} (F, \delta)$.

Theorem 8.1 [Szafraniec 1992] *If the origin of the coordinates is an isolated critical point of the system* (8.1) *with* $m = n-1$, *then for the system* $H(X) = 0$ *it is also an isolated singular point and* $b = 2 \deg H$.

Szafraniec [1992] asserts that this formula was programmed on a computer, and that the program uses a diagram of the initial exponents of the system (8.1).

Example 8.1 (continuation of Examples 6.1 and 7.1). With the values $a_1 = 40$, $a_2 = -a_3 = 25$, $a_4 = -1$, $a_5 = 16$, $a_{2i} = -1$, $i = 1, 2, 3, 4$, we consider the system (6.6). Then each of the truncated systems (7.4) and (7.6) has two real and two complex simple solutions. Thus, the system (7.4) has two real solutions:

$$Y'^0_1 = (-1, -1),$$

$$Y'^0_2 = (5, -1);$$

and two complex ones:

$$Y'^0_{3,4} = (-2 \pm i, -1). \tag{8.12}$$

Since all roots are simple and $d = n - 1$, then the branches are isolated, i.e. the implicit function Theorem 5.2 is applicable. Substituting in the system

(7.5) $Y' = Y'^0 + Z'$ and computing first terms of the expansions $Z'(y_3)$, we obtain

$$z_1 = -14y_3^2/3 + O(y_3^4), \quad z_2 = -2y_3^2 + O(y_3^4) \text{ near } Y_1'^0,$$

$$z_1 = -68y_3^2/15 + O(y_3^4), \quad z_2 = -4y_3^2/5 + O(y_3^4) \text{ near } Y_2'^0,$$

$$z_1 = (13 \mp 6i)y_3^2/5 + O(y_3^4), \quad z_2 = -(2 \pm i)y_3^2/5 + O(y_3^4) \text{ near } Y_{3,4}'^0.$$

Going back to the initial coordinates according to (7.3), we find two real branches $\mathcal{F}_j$:

$$x_1 = -y_3 + b_j' y_3^3 + O(y_3^5),$$

$$x_2 = -y_3^2 + b_j' y_3^4 + O(y_3^6),$$

$$x_3 = b_j y_3 + c_j y_3^3 + O(y_3^5), \quad j = 1, 2,$$

where

$$b_1' = -2, \quad b_1 = -1, \quad c_1 = -2/3; \quad b_2' = -4/5, \quad b_2 = 5, c_2 = 52/15.$$

The complex roots (8.12) of the system (7.4) give two complex branches $\mathcal{F}_{5,6}$:

$$x_1 = -y_3 - \frac{1}{5}(7 \pm i)y_3^3 + O(y_3^5),$$

$$x_2 = -y_3^2 - \frac{1}{5}(7 \pm i)y_3^4 + O(y_3^6),$$

$$x_3 = (-2 \pm i)y_3 + \frac{1}{5}(3 \mp 21i)y_3^3 + O(y_3^5).$$

The truncated system (7.6) has two real simple solutions: $y_{1_{1,2}}^0 = \pm 1$, $y_2^0 = -1$; and two complex simple solutions: $y_{1_{3,4}}^0 = \pm i$, $y_2^0 = -1$. For these solutions we obtain respectively two real branches $\mathcal{F}_{3,4}$:

$$x_2 = \pm x_1 - \frac{7}{5}x_1^2 + o(x_1^2), \quad x_3 = -x_1^3 \pm x_1^4 + o(x_1^4);$$

and two complex branches $\mathcal{F}_{7,8}$:

$$x_2 = \pm i x_1 - \frac{7}{5}x_1^2 + o(x_1^2), \quad x_3 = -x_1^3 \pm i x_1^4 + o(x_1^4).$$

The role of the parameter τ for the discovered branches $\mathcal{F}_{1,2,5,6}$ and $\mathcal{F}_{3,4,7,8}$ is played by y_3^{-1} and x_1^{-1} respectively.

In Fig. 2.1 it is shown the approximate disposition of the real half-branches $\mathcal{F}_1$ and $\mathcal{F}_2$ for $y_3 > 0$, $\mathcal{F}_3$ and $\mathcal{F}_4$ for $x_1 > 0$ in a small neighborhood of the point $X = 0$; the dashed lines represent curves with $x_2 < 0$. □

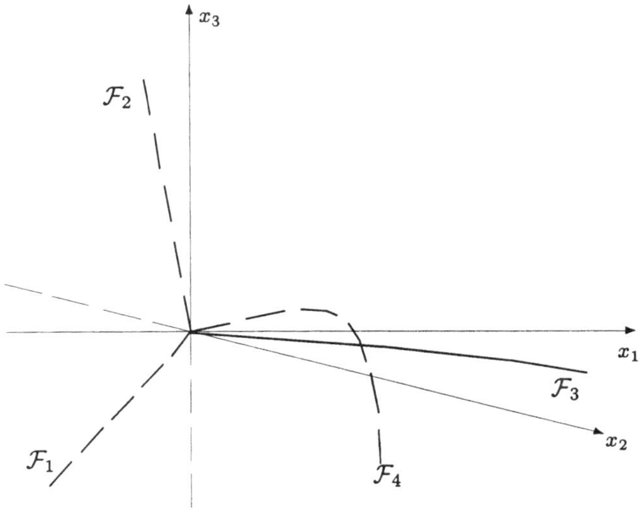

Figure 2.1: The disposition of the half-branches $\mathcal{F}_1$–$\mathcal{F}_4$ of the curve of Example 8.1 near the point $X = 0$.

We consider now the situation with an infinite support, analyzed in Chapter 1, Section 8. If $\mathbf{S}$ is the support of a Laurent series f and $\mathbf{K}$ is the cone of the problem, then we call the convex hull of the set $\partial_{\mathbf{K}}\mathbf{\Gamma}$ the *dominant Newton polyhedron* for f.

We consider now the fundamental problem, where the supports $\mathbf{S}_i = \text{supp} f_i$ are infinite, integer valued, and lie in sets of the form $\mathbf{C}^* + \{(Q_1^i, \dots, Q_{l_i}^i)\}$ (different for different i), and the cone of the problem $\mathbf{K}$ lies inside the integer cone $\mathbf{C}$. For the solution of such a fundamental problem one applies the procedure described earlier in the case of finite supports $\text{supp} f_i$. The difference only is that now for substitutions of the form $y_j = y_j^0 + z_j$ ($y_j^0 \neq 0$), $j = 1, \dots, d$, one obtains series in non-negative powers of z_j (instead of polynomials). In particular, if the functions f_i, $i = 1, \dots, m$, are analytic in the vicinity of the point $X = 0$, then they are expanded in Maclaurin series

$$f_i = \sum f_{iQ} X^Q, \quad Q \geq 0.$$

The system (8.1) determines an analytic set (see [Hartshorne 1977]). Here $\mathbf{C} = \{P\colon\ P \leq 0\}$, $\mathbf{K} = \{P\colon\ P < 0\}$ and $Q \geq 0$ for all $\mathbf{S}_i$.

After the extraction of the truncated system and a power transformation, the initial cones $\mathbf{C}$ and $\mathbf{K}$ are transformed into some cones $\tilde{\mathbf{C}}$ and $\tilde{\mathbf{K}}$ of a more general nature. That is why the fundamental problem at the very beginning was formulated in a form invariant with respect to power transformations.

About the replacement of analytic equations by algebraic ones, see also [MacMillan 1912a].

9. Positional functions of mechanisms

In this section, the singularities of positional functions of robotics (see [Wang and Waeldron 1987, Kieffer and Litvin 1991]) are divided in two types. The first type comprises singularities as functions of the control coordinates. The second type are the singularities near which the coordinates of position and control do not form a n-dimensional manifold, where n is the number of degrees of freedom. On the examples of two plane mechanisms it is demonstrated that the singularities of the second type take place only in the exceptional mechanisms, whose parameters satisfy certain conditions. The results of this section were published in [Bruno and Soleev 1994a, 1995a].

1. Classification. Various mechanisms used in robotics are described by systems of algebraic equations of the form

$$f_i(X,Y)=0, \quad i=1,\dots,m, \tag{9.1}$$

where $X=(x_1 \dots, x_m)$ are the position coordinates of the mechanism, $Y=(y_1,\dots,y_n)$ are the control coordinates and the f_i are polynomials. Here the number of equations m coincides with the number of position coordinates X, and the number n is called the *number of degrees of freedom of the mechanism.*

To each value of the control vector coordinates $Y=Y^0$ there corresponds, in general, a finite number of values of vector $X(Y^0)$ satisfying the system (9.1). The many-valued vector function $X(Y)$ of these values is called the *positional function of the mechanism.* Assume that the point $X=X^0$, $Y=Y^0$ satisfies the system of equations (9.1). We are interested in the behavior of the positional function in the vicinity of that point. We consider the Jacobian of the system (9.1)

$$J(X,Y) \stackrel{\text{def}}{=} \det(\partial f_i/\partial x_j), \quad i,j=1,\dots,m,$$

computed in the point X, Y. If

$$J(X^0,Y^0)\neq 0, \tag{9.2}$$

then according to Theorem 5.1 the solutions to the system (9.1) are represented in the form

$$x_i - x_i^0 = \varphi_i(Y-Y^0), \quad i=1,\dots,m,$$

where the φ_i are power series absolutely convergent for sufficiently small $\sum\limits_{k=1}^{n} |y_k - y_k^0|$. We call such a point X^0, Y^0 a *simple* one for the system (9.1) (and the corresponding position of the mechanism is also *simple*). In its vicinity the positional function is a single-valued analytic function of the controll coordinates. A point X^0, Y^0, where the property (9.2) is not satisfied, is called the *singular* one. It corresponds to the *singular position of the mechanism.* In the vicinity of a singular point, the positional function $X(Y)$ is not a single-valued analytic function of the controlling coordinates Y.

We denote as $Z = (X, Y)$ the vector of dimension $m + n$ formed by the vectors X and Y as follows

$$z_i = x_i, \quad i = 1, \dots, m,$$

$$z_{m+j} = y_j, \quad j = 1, \dots, n.$$

We consider the Jacobian matrix of the system (9.1) of dimension $m \times (m + n)$:

$$M(Z) \stackrel{\text{def}}{=} (\partial f_i / \partial z_j), \quad i = 1, \dots, m, \quad j = 1, \dots, m + n. \tag{9.3}$$

Let

$$1 \leq i_1 < i_2 < \dots < i_n \leq m + n$$

be a set of n different ordered integers, which we denote as $I = (i_1, \dots, i_n)$. We denote as M_I the minor of order m of the matrix (9.3) obtained by crossing out the columns with indices $i_1, \dots, i_n$.

If for a set I

$$M_I(X^0, Y^0) \neq 0, \tag{9.4}$$

then according to Theorem 5.1 the solutions to the system (9.1) are represented in the form

$$U - U^0 = \Phi(V - V^0), \tag{9.5}$$

where the vector V contains all z_i whose indices i belong to the set I, and U are the remaining coordinates z_i. Here $\Phi = (\varphi_1, \dots, \varphi_m)$ is a convergent vector power series, i.e. a single-valued analytic vector-function. A singular point X^0, Y^0 is called the *singular point of the first type* if the inequality (9.4) is satisfied in it at least for one set of indices I, and it is called the *singular point of the second type* in the opposite case, i.e. if in it all minors of dimension m of the matrix M are equal to zero.

In the vicinity of a singular point of the first type, the n-dimensional solution to the system (9.1) is unique, and it is represented in the form (9.5). In the vicinity of a singular point of the second type, such a representation is not possible, but we can obtain solutions for small $Z - Z_0$ according to Section 8.

We remark now that singular points of the first type are determined by a system of $m + 1$ equations for $m + n$ unknowns. The majority of singular points belong to the first type, because singular points of the second type are determined by a system of $m + (m+n)!/(m!n!)$ equations. In fact not all equations of the form $M_I = 0$ are independent, and therefore the number of independent equations determining singular points of the second type is less than $m + (m+n)!/(m!n!)$; but in any case it is larger than $m+n$. That means that singular points of the second type are possible only in exceptional cases, when the coefficients of the polynomials f_i in the system (9.1) satisfy some algebraic equations. These equations are obtained if from the system of equations

$$\begin{aligned} &f_i = 0, \quad i = 1, \dots, m, \\ &M_I = 0 \quad \text{for all sets} \quad I = (i_1, \dots, i_n) \end{aligned}$$

we eliminate the coordinates X and Y. If we avoid such exceptional cases, selecting the parameters of the mechanism (i.e. the coefficients of the polynomials f_i), then we obtain mechanisms with singular points of the first type only, near which the positional function can be completely studied with the help of the Cauchy's theorem of implicit functions, as mentioned above.

In studies [Wang and Waeldron 1987; Kieffer and Litvin 1991; Lunev and Misyurin 1993, 1994] and others, this property was not taken into account, and for the study of the positional function near a singular point of the first type instead of the theorem of implicit functions there were used more complicated methods intended for the analysis of complicated singularities.

So, before analyzing the singularities of the solutions to the system (9.1), we must isolate all its singular points and identify the cases where singular points of the second type are possible. This is explained below in the analysis of two mechanisms.

Remark 9.1. In actual mechanisms, the division of the variables Z in positional coordinates X and control coordinates Y can be done in many ways. This division is essential for the determination of the singular points of the first type.

2. Plane four-link. We consider a plane four-linked mechanism with hinges at the points O, A, B, C (Fig. 2.2). The lengths of its links, $OA = l_1$, $AB = l_2$, $BC = l_3$ and of the base $OC = a$ are positive constants.

Let the points O, A, B, C in the plane x, y have respectively coordinates $(0,0)$, (x,y), (x_1, y_1), $(a, 0)$. Then the equations of the links of this mechanism are

$$\begin{aligned} f_1 &\stackrel{\text{def}}{=} x^2 + y^2 - l_1^2 = 0, \\ f_2 &\stackrel{\text{def}}{=} (x_1 - x)^2 + (y_1 - y)^2 - l_2^2 = 0, \\ f_3 &\stackrel{\text{def}}{=} (x_1 - a)^2 + y_1^2 - l_3^2 = 0. \end{aligned} \tag{9.6}$$

Here $m = 3$, $n = 1$. As a control coordinate we can take any one of the four coordinates x, y, x_1, y_1.

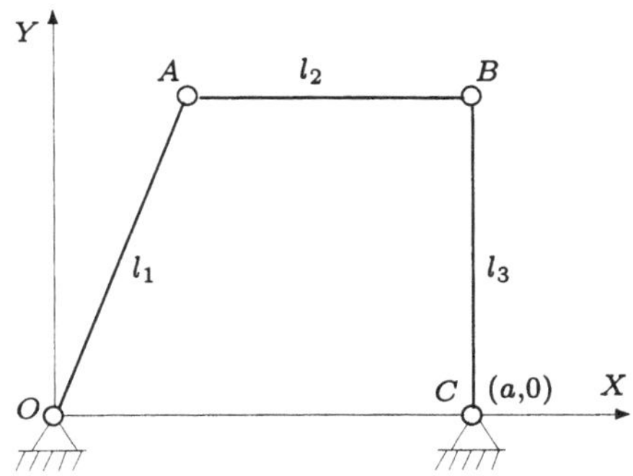

Figure 2.2: A plane four-link.

Theorem 9.1 *The four-link mechanism has a singularity of the second type only if its parameters satisfy at least one of the eight equations*

$$\sigma_1 l_1 + \sigma_2 l_2 + \sigma_3 l_3 = a, \quad \text{where} \quad \sigma_i = \pm 1. \tag{9.7}$$

Here the singularities of the second type arise only for a horizontal position of the mechanism, i.e. when $y = y_1 = 0$.

Proof. The Jacobian matrix of the system (9.6) is

$$M = 2\begin{pmatrix} x & y & 0 & 0 \\ x - x_1 & y - y_1 & x_1 - x & y_1 - y \\ 0 & 0 & x_1 - a & y_1 \end{pmatrix}.$$

Its minors of dimension 3 are

$$M_1 = 8yD_2, \quad M_2 = 8xD_2, \quad M_3 = 8yD_1, \quad M_4 = 8(x_1 - a)D_1,$$

where

$$D_1 = \begin{vmatrix} x & y \\ x - x_1 & y - y_1 \end{vmatrix}, \quad D_2 = \begin{vmatrix} x_1 - x & y_1 - y \\ x_1 - a & y_1 \end{vmatrix}.$$

Since

$$x^2 + y^2 = l_1^2 \neq 0, \quad (x_1 - x)^2 + y_1^2 = l_3^2 \neq 0,$$

then all four minors M_i are equal to zero only if

$$D_1 = D_2 = 0.$$

Therefore singular points of the second type satisfy a system of five equations for four variables. $D_1 = 0$ only if the three points O, A, B are in the same straight line. Similarly, $D_2 = 0$ only if the three points A, B, C are in the same straight line. Therefore $D_1 = D_2 = 0$ only if all four points O, A, B, C are in the same straight line. Since the points O and C lie in the x axis, then the points A and B must also lie in it. This is possible only if the lengths of the links satisfy one of the relations (9.7). □

Let y_1 be the controlling coordinate. Then the Jacobian is

$$J = M_4 = 8(x_1 - a)D_1.$$

The singular points of the first type are those for which either $x_1 = a$, or $D_1 = 0$, i.e. the links AB and BC lie in the same straight line.

The second proof of Theorem 9.1. From the condition $D_1 = 0$ follows that the points O, A, B lie in the same straight line. Here let l be the distance between the point B and the point O, i.e.

$$l = l_1 \pm l_2. \tag{9.8}$$

Then the positional equations of the point B are

$$x_1^2 + y_1^2 = l^2, \quad (x_1 - a)^2 + y_1^2 = l_3^2. \tag{9.9}$$

Their difference gives

$$2ax_1 - a^2 = l^2 - l_3^2, \quad \text{i.e.} \quad x_1 = \frac{l^2 - l_3^2 + a^2}{2a}.$$

Substituting the last value into the first equation (9.9), we obtain a quadratic equation for y_1

$$g \stackrel{\text{def}}{=} y_1^2 + \frac{(l^2 - l_3^2 + a^2)^2}{4a^2} - l^2 = 0.$$

The singular solutions to this equation are those for which $\partial g/\partial y_1 = 0$, i.e. $y_1 = 0$. Such a solution is possible only if the discriminant

$$d \stackrel{\text{def}}{=} (l^2 - l_3^2 + a^2)^2 - 4a^2 l^2 = 0.$$

But, factorizing it, we obtain

$$\begin{aligned} d &= (l^2 - l_3^2 + a^2 - 2al)(l^2 - l_3^2 + a^2 + 2al) = \\ &= [(l-a)^2 - l_3^2][(l+a)^2 - l_3^2] = \\ &= (l - a - l_3)(l - a + l_3)(l + a - l_3)(l + a + l_3). \end{aligned}$$

Taking into account (9.8), we find that $d = 0$ only if one of the equalities (9.7) is satisfied. □

We remark that the equality

$$-l_1 - l_2 - l_3 = a$$

is not possible, since all $l_i > 0$ and $a > 0$. The equalities

$$\begin{aligned} l_1 + l_2 + l_3 &= a, \\ -l_1 - l_2 + l_3 &= a, \\ -l_1 + l_2 - l_3 &= a, \\ l_1 - l_2 - l_3 &= a \end{aligned} \tag{9.10}$$

do not realize in functioning mechanisms, because the length of one link equals the sum of the lengths of the three other links. Thus, in functioning mechanisms, only three of the eight equalities (9.7) can be realized:

$$\begin{aligned} l_1 + l_2 - l_3 &= a, \\ l_1 - l_2 + l_3 &= a, \\ -l_1 + l_2 + l_3 &= a \end{aligned} \tag{9.11}$$

(see Fig. 2.3). The first and the third of them are known in the theory of mechanisms as the conditions of the rotation [Dimentberg 1950].

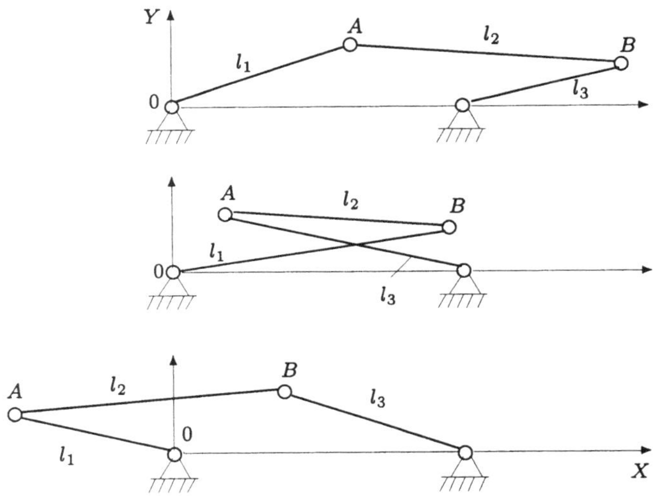

Figure 2.3: The realizable singular cases (9.11) of the four-link.

Theorem 9.2 *The singular solutions of the second type are isolated among all solutions to the system* (9.6) *in cases* (9.10), *and they lie in one-dimensional sets of solutions in cases* (9.11).

Proof. Assume that (9.7) is satisfied. We consider the solutions to the system (9.6) near the point

$$x^0 = \sigma_1 l_1, \quad y^0 = 0, \quad x_1^0 = a - \sigma_3 l_3, \quad y_1^0 = 0,$$

i.e. we put

$$x = x^0 + z, \qquad x_1 = x_1^0 + z_1,$$

where z, y, z_1, y_1 are small deviations from the singular point x^0, y^0, x_1^0, y_1^0. Substituting these values in the equations (9.6), opening there parentheses and collecting similar terms, we obtain the system

$$\begin{aligned} &2\sigma_1 l_1 z + z^2 + y^2 = 0, \\ &2\sigma_2 l_2 (z_1 - z) + (z_1 - z)^2 + (y_1 - y)^2 = 0, \\ &-2\sigma_3 l_3 z_1 + z_1^2 + y_1^2 = 0. \end{aligned} \tag{9.12}$$

Here the equality (9.7) was used in the second equation. According to Theorem 1.1 of implicit function, applied separately to the first and to the third equations (9.12), we obtain

$$z = -\frac{y^2}{2\sigma_1 l_1} + O(y^4), \quad z_1 = \frac{y_1^2}{2\sigma_3 l_3} + O(y_1^4).$$

Substituting these values in the second equation (9.12), we obtain

$$\sigma_2 l_2 \left(\frac{y_1^2}{\sigma_3 l_3} + \frac{y^2}{\sigma_1 l_1} \right) + (y_1 - y)^2 + O(y^4, y_1^4, y^2 y_1^2) = 0. \tag{9.13}$$

The character of the small solutions to these equations is determined by the quadratic form

$$\psi \stackrel{\text{def}}{=} \sigma_2 l_2 \left(\frac{y_1^2}{\sigma_3 l_3} + \frac{y^2}{\sigma_1 l_1} \right) + (y_1 - y)^2,$$

which is the first approximation to the left part of the equation (9.13). Namely, if the discriminant D of that form is positive, then the form ψ is positively or negatively definite, and the equation $\psi = 0$ has only the isolated solution $y = y_1 = 0$. But then also the full equation (9.13) has only the isolated solution $y = y_1 = 0$. If $D < 0$, the equation $\psi = 0$ has two branches of solutions, intersecting at the point $y = y_1 = 0$, and the same is true for the solutions to the equation (9.13). Computations with (9.7) taken into account give

$$\begin{aligned} D = \left(1 + \frac{\sigma_2 l_2}{\sigma_1 l_1}\right)\left(1 + \frac{\sigma_2 l_2}{\sigma_3 l_3}\right) - 1 = \frac{\sigma_2 l_2}{\sigma_1 l_1} + \frac{\sigma_2 l_2}{\sigma_3 l_3} + \frac{\sigma_2^2 l_2^2}{\sigma_1 l_1 \sigma_3 l_3} = \\ = \frac{\sigma_2 l_2}{\sigma_1 l_1 \sigma_3 l_3}(\sigma_3 l_3 + \sigma_1 l_1 + \sigma_2 l_2) = a \frac{\sigma_2 l_2}{\sigma_1 l_1 \sigma_3 l_3}. \end{aligned}$$

Since the parameters a, l_1, l_2, l_3 are positive, the sign of D is the same as in the product $\sigma_1\sigma_2\sigma_3$.

So, if $\sigma_1\sigma_2\sigma_3 > 0$, then the solution $y = y_1 = 0$ is isolated; if $\sigma_1\sigma_2\sigma_3 < 0$, then it lies in a one-dimensional set of solutions. That is what distinguishes the cases (9.10) and (9.11). □

Theorem 9.2 confirms that the cases (9.10) cannot be realized in mechanisms.

Remark 9.2. If the first and the third equalities (9.11) are satisfied simultaneously, then $l_1 = l_3$ and $l_2 = a$, i.e. the four-link is a parallelogram. Such a parallelogram is used in the links of the big wheels of locomotives. It is remarkable that the functioning mechanism has the degenerate values of the parameters, for which the positional function has a singular point of the second type. There are other examples of functioning mechanisms with such degenerate values of the parameters.

3. A plane mechanism with sliders. Fig. 2.4 shows a plane mechanism with one degree of freedom, comprising a structural group of the third class [Lunev and Misyurin 1993]. This mechanism has five rotating joints (the points O, A, B, C, D) and two translating ones (the points D and C). The point C moves along the x axis and has one coordinate x_2; the point D moves along the straight line L with the equation $ax+by+d = 0$ and has coordinates x_3, y_3. The rotating joints A, B have coordinates x_0, y_0 and x_1, y_1 respectively.

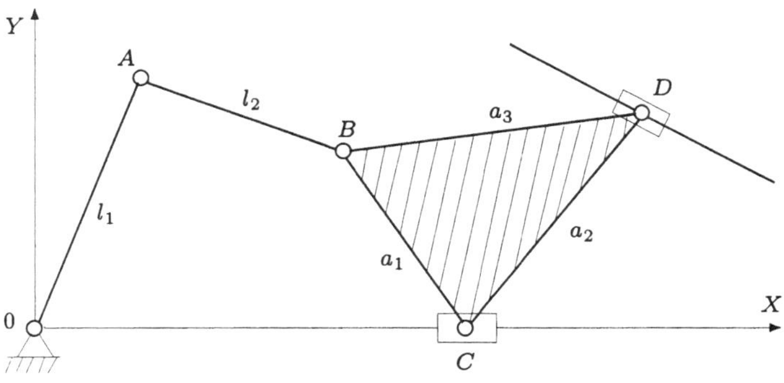

Figure 2.4: A plane mechanism with sliders.

We introduce the notation

$$u_{ij} = x_i - x_j, \quad v_{ij} = y_i - y_j, \quad i,j = 0,\ 1,\ 2,\ 3,$$

l_1, l_2, a_1, a_2, a_3 are the constant lengths of the segments OA, AB, BC, CD, BD respectively, and $\varkappa = a_3/a_1$.

The coordinates of the vertex D of the triangle BCD can be expressed through the coordinates of the side BC and the angle α between the sides BC and BD in the following way:

$$u_{31} = u_{21}\varkappa\cos\alpha - y_1\varkappa\sin\alpha, \quad v_{31} = -u_{21}\varkappa\sin\alpha - y_1\varkappa\cos\alpha.$$

With the notation we obtain the following system of equations of constraints of the mechanism:

$$\begin{aligned}
&x_0^2 + y_0^2 = l_1^2,\\
&u_{10}^2 + v_{10}^2 = l_2^2,\\
&u_{21}^2 + y_1^2 = a_1^2,\\
&u_{31} = u_{21}\varkappa\cos\alpha - y_1\varkappa\sin\alpha,\\
&v_{31} = -u_{21}\varkappa\sin\alpha - y_1\varkappa\cos\alpha,\\
&ax_3 + by_3 + d = 0.
\end{aligned} \tag{9.14}$$

This is a system of six equations for the seven variables x_0, y_0, x_1, y_1, x_2, x_3, y_3. Three of the equations are linear. We will find all singular solutions of the second type to this system. First of all, from the last three (linear) equations of the system (9.14) we eliminate the variables x_3 and y_3. We obtain one equation $\xi x_1 + \eta u_{21} + \zeta - y_1 = 0$, where

$$\begin{aligned}
\xi = a/\delta, \quad \eta = (a\varkappa\cos\alpha - b\varkappa\sin\alpha)/\delta, \quad \zeta = d/\delta,\\
\delta = a\varkappa\sin\alpha + b\varkappa\cos\alpha - b, \quad \varkappa = a_3/a_1, \quad \alpha = \angle CBD.
\end{aligned} \tag{9.15}$$

Now we consider the system of four equations for five variables

$$\begin{aligned}
&x_0^2 + y_0^2 = l_1^2,\\
&u_{10}^2 + v_{10}^2 = l_2^2,\\
&u_{21}^2 + y_1^2 = a_1^2,\\
&2(\xi x_1 + \eta u_{21} + \zeta - y_1) = 0.
\end{aligned} \tag{9.16}$$

Its Jacobian matrix is

$$M = 2\begin{pmatrix} x_0 & y_0 & 0 & 0 & 0\\ u_{01} & v_{01} & u_{10} & v_{10} & 0\\ 0 & 0 & u_{12} & y_1 & u_{21}\\ 0 & 0 & \xi-\eta & -1 & \eta \end{pmatrix}.$$

The minors of the maximal order of that matrix are

$$\begin{aligned}
&M_1 = 2^4 y_0 F, \qquad M_2 = 2^4 x_0 F, \qquad M_3 = 2^4 D_0 D_1,\\
&M_4 = 2^4 D_0 D_2, \qquad M_5 = 2^4 D_0 D_3,
\end{aligned}$$

where

$$F \stackrel{\text{def}}{=} \begin{vmatrix} u_{10} & v_{10} & 0 \\ 0 & y_1 & u_{21} \\ \xi & -1 & \eta \end{vmatrix} = \eta y_1 u_{10} + \xi u_{21} v_{10} + u_{21} u_{10}, \quad D_0 \stackrel{\text{def}}{=} \begin{vmatrix} x_0 & y_0 \\ u_{01} & v_{01} \end{vmatrix},$$

the D_i are the minors of the matrix

$$N \stackrel{\text{def}}{=} \begin{pmatrix} u_{12} & y_1 & u_{21} \\ \xi - \eta & -1 & \eta \end{pmatrix},$$

obtained by crossing out its ith column ($i = 1, 2, 3$). We find the conditions under which $M_1 = \ldots = M_5 = 0$.

If $D_0 \neq 0$, then the rank of the matrix N must be equal to unity, which is possible only in two cases:

1) $y_1 = x_1 - x_2 = 0$, i.e. $a_1 = 0$ and the points B and C coincide, and the triangle BCD degenerates into a segment BD;

2) $\xi = 0$, i.e. $a = 0$ and the straight line L is horizontal; in that case $y_3 = \text{const}$, $y_1 = \text{const}$, $u_{21} = \text{const}$, and besides the equality $y_1\eta + u_{21} = 0$ must be satisfied.

In both cases the parameters of the mechanism satisfy some relation.

We will assume that $a \neq 0$ and $a_1 \neq 0$; then the equalities

$$M_3 = M_4 = M_5 = 0$$

are possible only for $D_0 = 0$. Therefore, the singular points of the second type are determined by a system of six equations (the system (9.16) and the equations $F = 0$, $D_0 = 0$) for five variables x_0, y_0, x_1, y_1, x_2. It is hard to eliminate all variables from this system and to obtain an equation for its parameters. Therefore we choose another way, analogous to the second proof of Theorem 9.1.

The equality $D_0 = 0$ means that the points O, A and B lie in the same straight line. We consider that case. Then the system (9.16) has the form

$$\begin{aligned} &x_1^2 + y_1^2 = l^2, \quad l = l_1 \pm l_2, \\ &u_{21}^2 + y_1^2 = a_1^2, \\ &\xi x_1 + \eta u_{21} + \zeta - y_1 = 0. \end{aligned} \tag{9.17}$$

This is a system of three equations for three variables x_1, y_1, x_2. We eliminate from it x_1 and y_1. Subtracting the first equation (9.17) from the second, we obtain

$$-2x_1x_2 + x_2^2 = a_1^2 - l^2,$$

i.e.

$$x_1 = \frac{x_2^2 + l^2 - a_1^2}{2x_2}.$$

Now from the third equation (9.17) we obtain

$$y_1 = (\xi - \eta)x_1 + \eta x_2 + \zeta = \frac{(\xi + \eta)x_2^2 + \nu}{2x_2} + \zeta, \quad \nu \stackrel{\text{def}}{=} (\xi - \eta)(l^2 - a_1^2).$$

Substituting these values of x_1 and y_1 in the first equation (9.17), we obtain

$$g(x_2) \stackrel{\text{def}}{=} (x_2^2 + l^2 - a_1^2)^2 + ((\xi + \eta)x_2^2 + 2\zeta x_2 + \nu)^2 - 4l^2 x_2^2 = 0. \tag{9.18}$$

Since x_1 and y_1 are uniquely expressed through x_2, then the singular solutions to the system (9.17) of the second type correspond to singular solutions of the second type to the equation $g(x_2) = 0$. Its singular roots are those for which $dg/dx_2 = 0$, i.e. the multiple roots. They exist only if the discriminant D' of the polynomial $g(x_2)$ of the fourth degree equals zero. The equation $D' = 0$ gives the condition on the parameters of the system (i.e. of the mechanism), which is the only one that makes the singular solutions of the second type possible.

According to (9.18), $g(x_2) = g_4 x_2^4 + 4g_3 x_2^3 + 6g_2 x_2^2 + 4g_1 x_2 + g_0$, where

$$\begin{aligned} g_0 &= (l^2 - a_1^2)^2((\xi - \eta)^2 + 1), \\ g_1 &= (\xi - \eta)(l^2 - a_1^2)\zeta, \\ g_2 &= ((\xi^2 - \eta^2)(l^2 - a_1^2) - (l^2 + a_1^2) + 2\zeta^2)/3, \\ g_3 &= (\xi + \eta)\zeta, \\ g_4 &= (\xi + \eta)^2 + 1. \end{aligned} \tag{9.19}$$

According to [Lekkerkerker 1969, § 42.2] the discriminant of this polynomial is

$$D' = I^3 - 27J^2, \tag{9.20}$$

where

$$I = g_4 g_0 - 4g_3 g_1 + 3g_2^2, \quad J = \begin{vmatrix} g_4 & g_3 & g_2 \\ g_3 & g_2 & g_1 \\ g_2 & g_1 & g_0 \end{vmatrix}. \tag{9.21}$$

Thus, we have proved

Theorem 9.3 *Assume that $a \neq 0$ and $a_1 \neq 0$ in the mechanism of Fig. 2.4. Then singular positions of the second type are possible in it only if its parameters a, b, d, l_1, l_2, a_1, a_3, α satisfy the equation $D' = 0$ according to the formulae* (9.15) – (9.21).

10. Historical and bibliographical remarks

In this chapter we have considered the system of equations

$$f_i(x_1,\dots,x_n)=0, \quad i=1,\dots,m, \tag{10.1}$$

where $x_1,\dots,x_n$ are real or complex variables and $f_i(x_i,\dots,x_n)$ are respectively real or complex functions, analytic in the vicinity of a point $X^0=(x_1^0,\dots,x_n^0)$, which is a root of the system (10.1), i.e. all $f_i(X^0)=0$.

Assume that $m<n$ and that X^0 is a simple root of the system (10.1), i.e. in that point the rank of the matrix $A=(\partial f_i/\partial x_j)$ equals m. Then according to the theorem of implicit functions, in a small vicinity of the point X^0 the system (10.1) can be solved for m coordinates, and its solutions form a single manifold, which passes through X^0. For instance, if the left minor of order m of the matrix A differs from zero, then the solutions to the system (10.1) can be found in the form

$$x_i=\varphi_i(x_{m+1},\dots,x_n), \quad i=1,\dots,m.$$

If on the other hand X^0 is a critical (or singular) point (i.e. the rank of the matrix A is smaller than m in it), then through that point can pass several branches, which are formed by the solutions to the system (10.1). For each of them, there can exist its own local *uniformization*

$$x_i-x_i^0=\varphi_i(\tau_1,\dots,\tau_{n-m}), \quad i=1,\dots,n, (10.2) \tag{10.2}$$

where φ_i expand in power series in $\tau_1,\dots,\tau_{n-m}$. It is easy to demonstrate that different branches of the set (10.1) correspond to different uniformizations (10.2). Therefore we consider the branch (10.2) as separated if an initial piece of the expansion is computed for it, and if it is proved that there is no other branch with the same initial piece of the expansion.

Before we give a short review of the methods of finding the branches and of resolution of the singularity near the singular point $X^0=0$, let us quote Moiseev [1983, p. 295]: "More than three hundred years ago, Newton elaborated a method subsequently known as the "Newton diagram", which allows to find all solutions to the equation

$$F(X,Y)=0$$

under the conditions that $F(0,0)=0$, and that the function be an analytic one of its variables. The method of the Newton diagram is at present the only means to build effective numerical methods for the determination of all solutions to this problem. Numerical realizations of Newton's method are well developed only for the scalar case, where the variables X and Y are scalar. The case where the dimension of the variable X is large already leads to huge computational difficulties. If the dimension of the variable Y is also larger than 1, then the methods of numerical realization of Newton's ideas are unknown. Thus, the elaboration of numerical methods of post-bifurcational analysis now is one of the most important problems of computational mathematics, the solution of which determines the fate of numerous applied studies". In the present chapter, we presented a method of solution of the problem mentioned in the citation.

Thus, for the case $n = 2$ Newton [1711] proposed in 1670 a geometrical approach (see also [Chebotarev 1943; Bruno 1973a, 1979a]), allowing to obtain small solutions to the equation

$$f(x,y) \stackrel{\text{def}}{=} \sum_{p,q} a_{p,q} x^p y^q = 0 \tag{10.3}$$

in the form of several "Puiseux series"

$$y = \sum_{j=1} b_j x^{p_j}, \quad 0 < p_1 < p_2 < \dots, \tag{10.4}$$

with rational p_j. For this purpose, the series $f(x,y)$ from (10.3) is associated with the set $\mathbf{S}$ of the points (p,q) in the plane p,q, where $a_{pq} \neq 0$, and then the boundary of the convex hull of the set $\mathbf{S}$ (*the Newton's open polygon*) is isolated. The sum of the terms

$$\sum{}' a_{pq} x^p y^q \stackrel{\text{def}}{=} \widehat{f}(x,y),$$

whose points (p,q) lie in one edge of the Newton's broken line, gives the "truncated equation"

$$\widehat{f}(x,y) = 0,$$

from which the first term of the series (10.4) is found. After the substitution

$$y = b_1 x^{p_1} + z$$

the second term of the expansion (10.4) is found in a similar way, and so on. But the expansions (10.4) do not give the local uniformization of the branches: one branch (10.2) corresponds to several different expansions (10.4).

Therefore Weierstrass [1902] (see also [Bruno 1979a; Bruno and Soleev 1991a]) proposed a procedure of resolution of a singularity, which preserves the integer nature of the exponents and leads to the local uniformization of the branches. The procedure is analogous to the sigma-process (see [Walker 1950]), and it can be executed with the help of power transformations.

In fact Newton [1711] found only one extreme edge of the Newton's open polygon. The whole open polygon was for the first time considered by Puiseux [1850].

There were proposed several generalizations of these methods for systems (10.1) with $n \geq 3$.

The *method of successive elimination* of the coordinates from the system (10.1) with $m = n - 1$ and its reduction to one equation (10.3) was proposed by Lefshets ([1954; 1957, Chapter VIII, § 8]), and it was elaborated in detail by Trenogin [1960], Aizengendler [1966], Vainberg and Aizengendler [1966], Vainberg and Trenogin [1969]. That method gives systems of equations of the form

$$g_i(x_1, \ldots, x_n) = 0, \quad i = 1, \ldots, k,$$

with $k < n - 1$ for $(n - l)$-dimensional components of solutions to the system (10.1), and it gives the expansions in fractional powers in one of the coordinates for one-dimensional components. The method of elimination is convenient for theoretical substantiation in problems of multidimensional nonlinear branching. Thus, with that method one can prove the existence of uniformizations (10.2) for all one-dimensional components of solutions to the system (10.1). But that method, notwithstanding its theoretical elegance and its mathematical strictness, is not always convenient for practical applications, because it requires cumbersome computations with power series.

Comparatively recently the *method of Gröbner's bases* was proposed as an alternative to the method of elimination. The essence of the method consists in the following. For a given initial system of algebraic equations one computes another (graduated) system of equations, which is called *Gröbner's bases.* Here both systems of polynomials generate the same ideal. That bases for the first time was proposed by Hironaka [1964], who called it the standard bases. An algorithm for its computation was elaborated by Buchberger [1970]. He called it Gröbner's bases to honor his scientific supervisor. At present, the development of various algorithms of that method is under way [Becker and Weispfenning 1993]. But in the meanwhile they require rather cumbersome computations. Besides, the simplified system of equations forming the Gröbner's bases is to be solved somehow.

The *method of Newton's open polygons*, proposed by MacMillan [1912a, b], consists in the following. Each series f_i is written in the form

$$f_i = \sum_{p,q} f_{iq}^{(p)} x_n^q, \quad i = 1, \ldots, m = n - 1,$$

where $f_{iq}^{(p)}$ is a homogeneous polynomial in $x_1, \ldots, x_{n-1}$ of degree p. In the plane p, q to each series f_i one puts in correspondence its set $\widetilde{\mathbf{S}}_i$ of the points (p, q), where $f_{iq}^{(p)} \not\equiv 0$, and its Newton's open polygon. Then the edges and the vertices of these open polygons lying in supporting straight lines with the same slope are singled out. Using the points of the sets lying in the isolated edges and vertices, we build the truncated system

$$\widehat{f}_1 = \ldots = \widehat{f}_{n-1} = 0, \tag{10.5}$$

whose solution gives the first terms of homogeneous expansions

$$x_i = \sum_{j=1}^{\infty} b_{ij} x_n^{p_j}, \quad i = 1, \ldots, n-1, \tag{10.6}$$

where the p_j are rational numbers, $0 < p_j < p_{j+1}$, and b_j = const. MacMillan points out that not all solutions to the system (10.1) have the homogeneous form (10.6), and not all of them can be found with the help of Newton's open polygons. Therefore he offers additional methods for finding the solutions in which $x_1, \ldots, x_{n-1}$ have different orders of smallness in x_n.

Evidently unaware of the work of MacMillan, Grawes [1955] proposed to consider the system (10.1) in the form of one vector equation

$$F \stackrel{\text{def}}{=} \sum_{p,q} F_q^{(p)} x^q = 0$$

with one Newton's open polygon on the plane p, q. A shortcoming of that approach is that according to Grawes, in the truncated vector equation $\widehat{F} = 0$, some components of the vector $\widehat{F}$ can be identically zero, while according to MacMillan, in the truncated system (10.5) each $\widehat{f}_i$ contains some terms $f_{iq}^{(p)} x^q$ and is not identically zero. Therefore in the Grawes method the sets of solutions to the truncated vector equations can have a large dimension, which makes difficult even the finding of branches with homogeneous expansions (10.6).

Grawes' construction with one Newton's open polygon was developed by Botachev [1976, 1985]. He formally follows from the system of equations (10.1) to the vector equation

$$F(Y, x_n) \stackrel{\text{def}}{=} (f_1(Y, x_n), \ldots, f_m(Y, x_n)) = 0, \quad Y = (x_1, \ldots, x_{n-1}). \tag{10.7}$$

That passage to vector form of expression allows one not to distinguish formally the two-dimensional and many-dimensional cases. So, Botachev applies the method of Newton's diagram to the vector equation (10.7).

The author proposed in [1962] a geometry of the power exponents for the study of the solutions to a system of ordinary differential equations in the vicinity of a singular point. In particular, there was proposed a polyhedron, which was called in [Gindikin 1973] the "Newton polyhedron". In the book [Bruno 1979a], on the basis of that approach there was proposed a local method, the application of which to the equation (10.3) is analyzed in detail in Chapter 1, Sections 2, 3. Its application to the system (10.1) with $m = n - 1$

is briefly discussed in Chapter 4, Section 2. That approach was developed in [Soleev 1983; Bruno and Soleev 1990, 1991a, b, 1992]. The present chapter develops that local method of resolution of the singularity of the system (10.1), which is based upon the application of the geometry of power exponents, including the Newton polyhedron.

The *method of the Newton polyhedra* consists in the following. Each

$$f_i(X) \stackrel{\text{def}}{=} \sum f_{iQ} X^Q \quad \text{over} \quad Q \in \mathbf{S}_i \quad i = 1, \dots, m, \tag{10.8}$$

where $Q = (q_1, \dots, q_n)$, $X^Q = x_1^{q_1} \dots x_n^{q_n}$, f_{iQ} are numerical coefficients, is put in correspondence in $\mathbb{R}^n$ with the *support*

$$\mathbf{S}_i \stackrel{\text{def}}{=} \operatorname{supp} f_i \stackrel{\text{def}}{=} \{Q\colon\ f_{iQ} \neq 0\}$$

and with its convex hull $\mathbf{\Gamma}(f_i)$, which is called the *Newton polyhedron* of the sum $f_i(X)$. Using the faces of the polyhedra $\mathbf{\Gamma}(f_1), \dots, \mathbf{\Gamma}(f_m)$ we compose a "truncated system"

$$\widehat{f}_1(X) = \dots = \widehat{f}_m(X) = 0 \tag{10.9}$$

and the corresponding set $\mathcal{U}(\varepsilon)$ in such a way that the first approximation to the solution (10.2) in the set $\mathcal{U}(\varepsilon)$ is the solution to the system (10.9). Here the truncated system (10.9) is quasi-homogeneous, and in its set $\mathcal{U}(\varepsilon)$, the power transformation

$$y_i = X^{S_j}, \quad j = 1, \dots, n,$$

reduces the truncated system (10.9) to a system with a smaller number of variables.

Bernshtein [1975] applied the method of the Newton polyhedra to the computation of the number of branches of the algebraic curve of a special kind, but he did not consider the problem of the computation of the asymptotic expansions studied firstly in [Bruno 1976].

Khovanskii [1983], with the help of the Newton polyhedra, studied the global properties of polynomials, which are invariant under power transformations. Therefore instead of $\mathbb{C}^n$ he considers them in torus manifolds $(\mathbb{C}^*)^n$, where $\mathbb{C}^* = \mathbb{C} \setminus \{0\}$. He also uses power transformations for the resolution of singularities (see also [Koushnerenko 1976; Khovanskii 1978a, b, 1981, 1988, 1992; Pukhlikov and Khovanskii 1992; Gelfond and Khovanskii 1996]).

The works of Bugaev [1888] and Sintsov [1898] are close to the method of the Newton polyhedra. Bugaev considered the case of finite sums (10.8), and for the extraction of the truncated systems (10.9) he proposed to examine all possible variants. Sintsov proposed a geometrical method, which corresponds to the construction of the normal cones of the faces of the Newton polyhedra

$\mathbf{\Gamma}(f_i)$ (for more details see Chapter 1, Section 10 and [Bruno and Soleev 1991a]).

Apart from these general methods, many special methods were proposed for the solution of particular problems of branching (see the review [Sather 1973]). For instance, the Nazarov method [Zabreiko and Kats 1971]. The review by Botachev [1985] misrepresents the essence of the methods of analysis of multi-dimensional branching as well as the history of their development. The methods enumerated above (apart from the method of Weierstrass for $n = 2$ and the method of the Newton polyhedra for an arbitrary n) give the solutions to the system (10.1) in the form of series in fractional powers of some coordinates (of one coordinate for a curve), i.e. the methods do not give the local uniformization of the branches. The approach with methods of the Lie Group Analysis, see in [Loginov and Trenogin 1975; Loginov 1985].

Newton polyhedra were used in many works: for the estimation of the number of solutions to a system of n equations in n unknowns [Bernshtein 1975; Bernshtein et al. 1976; Kushnirenko 1975a, b, 1976], for the computation of asymptotics of oscillating integrals [Varchenko 1976; Vasil′ev 1977; Varchenko and Khovanskii 1985], for the computation of the topological characteristics of algebraic manifolds [Danilov and Khovanskii 1986] (compare with [Zariski 1940; Milnor 1968; Shafarevich 1972; Hartshorne 1977; Abyhankar 1982; Bierstone and Milman 1989]), and also for other problems (see the reviews in [Bruno 1979a, Chapter IV; Khovanskii 1983; Arnol′d et al. 1984; Bruno 1995b, c] and the papers [Saito 1983; Kazarnovskii 1984, 1997; Bruno and Soleev 1990, 1991a; Yossin 1991]). These works usually considered problems of the generic case with given Newton polyhedra, and the studies had a rather theoretical character. An attempt to solve concrete problems required the creation of an algorithm for the computation of the Newton polyhedra [Soleev 1982; Soleev and Aranson 1994; Aranson 1996]. Such an algorithm was presented in Chapter 1.

Unlike the mentioned papers by Bernshtein, Khovanskii and others, the present chapter has not assumed that the system (10.1) belonged to the generic case. In order to simplify the presentation, we have only assumed that the system (10.1) did not belong to an exceptional set.

Chapter 3

Asymptotics of solutions to a system of ordinary differential equations

1. Local theorems of existence

Let us consider a real autonomous system of ordinary differential equations

$$dx_i/dt \stackrel{\text{def}}{=} \dot{x}_i = \varphi_i(x_1, \ldots, x_n), \quad i = 1, \ldots, n, \tag{1.1}$$

where φ_i are sufficiently smooth functions. Most of what is to follow is valid for the complex system (1.1) with analytic φ_i as well, but here we restrict ourselves to the real systems. In the vector notation $X = (x_1, \ldots, x_n)$, $\Phi = (\varphi_1, \ldots, \varphi_n)$ System (1.1) has the form

$$\dot{X} = \Phi(X). \tag{1.2}$$

A *solution* to the system is the vector-function $X(t)$; in the phase space of coordinates X it forms the *integral curve* $X = \Xi(\tau)$, where τ is a parameter in the curve. The integral curve $X = \Xi(\tau)$ satisfies either System (1.2) with the substitution of variable $\tau = \tau(t)$, or System $\Phi(X) = 0$. In the first case the integral curve is called a *nontrivial* one, and in the second case it is called a *trivial* one.

Two curves are the *asymptotics* of each other in a neighborhood of the point X^0, if there exist such parametrizations $x_i = \xi_i(\tau)$ for one of them and $x_i = \eta_i(\tau)$ for another that $\xi_i \to x_i^0$ and $\xi_i = \eta_i(1 + o(1))$, $i = 1, \ldots, n$, when $\tau \to \infty$. The property of the curves to be asymptotics does not depend on parameters. It is an expression of the fact that the curves tend to each other faster then to the point X^0.

Let us consider System (1.2) in a neighborhood of the point $X = X^0$. If $\Phi(X^0) \neq 0$, then X^0 is a *simple point* of System (1.2). If $\Phi(X^0) = 0$, then X^0 is a *stationary* (or singular) point.

Theorem 1.1 [Petrovskii 1964] *Through the simple point X^0 there passes the unique integral curve of System* (1.2).

It is the Cauchy theorem of existence and uniqueness of a solution. Let X^0 be a stationary point. By the parallel translation $X = X^0 + \tilde{X}$ it can be placed in the origin $\tilde{X} = 0$. Near the stationary point $X = 0$ System (1.2) has the form

$$\dot{X} = AX + \tilde{\Phi}(X), \tag{1.3}$$

where A is a square n-matrix and $\tilde{\Phi} = O(|X|^2)$. Let $\lambda_1, \ldots, \lambda_n$ be eigenvalues of the matrix A. If $\Lambda \stackrel{\text{def}}{=} (\lambda_1, \ldots, \lambda_n) \neq 0$, then the stationary point is called the *elementary* one. If $\Lambda = 0$, then it is called the *nonelementary* one.

Let $0 \leq d < n$. We consider System (1.1) of the following form:

$$\begin{aligned} \dot{x}_i &= \varphi_i(X), \qquad i = 1, \ldots, d; \\ \dot{x}_j &= x_j f_j(X), \quad j = d+1, \ldots, n, \end{aligned} \tag{1.4}$$

where φ_i and f_j are regular functions of X. This system has $n - d$ invariant hyperplanes

$$x_j = 0, \quad j = d+1, \ldots, n. \tag{1.5}$$

Their intersection

$$x_{d+1} = \ldots = x_n = 0 \tag{1.6}$$

is also an invariant subspace. We split each n-vector in two subvectors: d-vector and $(n-d)$-vector. For example, $X = (X', X'')$, where $X' = (x_1, \ldots, x_d)$, $X'' = (x_{d+1}, \ldots, x_n)$. We introduce the vector $\log X = (\log x_1, \ldots, \log x_n)$. Now System (1.4) can be written in the form

$$\begin{aligned} \dot{X}' &= \Phi'(X), \\ d\log X''/dt &= F''(X). \end{aligned}$$

If we put $X'' = 0$ in right parts of that system, then we obtain the system

$$\begin{aligned} \dot{X}' &= \Phi'(X', 0) \stackrel{\text{def}}{=} \hat{\Phi}'(X'), \\ d\log X''/dt &= F''(X', 0) \stackrel{\text{def}}{=} \hat{F}''(X'). \end{aligned} \tag{1.7}$$

Let $X^0 = (X'^0, 0)$ be a point in subspace (1.6). If $\Phi'(X^0) \stackrel{\text{def}}{=} \hat{\Phi}'(X'^0) \neq 0$, then X^0 is a simple point of System (1.4), and according to Theorem 1.1 there passes through it the unique integral curve of System (1.4), that lies entirely in the invariant subspace (1.6). If $\Phi'(X^0) \stackrel{\text{def}}{=} \hat{\Phi}'(X'^0) = 0$, then X^0 is the stationary point both of the complete system (1.4), and of System (1.7). If by the substitution of variables $X = X^0 + \tilde{X}$ we transfer the point in the origin $\tilde{X} = 0$ and omit the tilde, then System (1.4) in the form (1.3) takes the form

$$\begin{aligned} \dot{X}' &= A_{11}X' + A_{12}X'' + \tilde{\Phi}'(X), \\ d\log X''/dt &= \Lambda'' + \tilde{F}''(X), \end{aligned} \tag{1.8}$$

where $\Lambda'' = F''(0)$, i.e. the matrix A in notation (1.3) has the block form

$$A = \begin{pmatrix} A_{11} & A_{12} \\ 0 & A_{22} \end{pmatrix},$$

where A_{22} is the diagonal matrix $\operatorname{diag}\{F''(0)\}$. We note that the values $\lambda_j \overset{\text{def}}{=} f_j(0)$, $j = d+1, \dots, n$ are real eigenvalues of the matrix A. The integral curve $X = \Xi(\tau)$ of System (1.8), that goes into the stationary point $X = 0$ when $\tau \to +\infty$, and that has all components of the vector $\Xi''(\tau)$ different from zero, will be called the *h-asymptotical curve*.

The *order of a scalar function* $\xi(\tau)$ defined when $\tau \in (\tau_0, +\infty)$ is the number

$$p = \lim \frac{\log|\xi(\tau)|}{\log \tau} \quad \text{when } \tau \to +\infty,$$

if it exists. It is easy to see that for a power function $\xi(\tau) = b\tau^\mu$ the order is $p = \mu$, and it is the same for the function $b\tau^\mu(1 + o(1))$. The *order of a vector function* $\Xi(\tau) = (\xi_1(\tau), \dots, \xi_n(\tau))$ defined when $\tau \in (\tau_0, +\infty)$ is the vector $P = (p_1, \dots, p_n)$, composed of orders of components $\xi_i(\tau)$ of the vector $\Xi(\tau)$. Under the substitution of parameter τ to τ' the vector order P changes into $P' = \lambda P$, where the factor $\lambda > 0$ (compare Lemma 2.1, Chapter 2). In Section 2, Chapter 2 the order was defined only for the functions with the exponential asymptotics, but now it is defined for the broader class of functions and in particular for logarithms and multiple logarithms. Indeed, $p(\log \tau) = 0$.

Theorem 1.2 *Let* $\Lambda'' \neq 0$ *in System* (1.8) *and* $X = \Xi(\tau)$ *be its nontrivial h-asymptotical integral curve, which has the part* $\Xi''(\tau)$ *with the vector order* $P'' \neq 0$. *Then*

$$\Lambda'' = \varkappa P'', \quad \varkappa \in \mathbb{R} \setminus \{0\}. \tag{1.9}$$

Proof. Since $\Lambda'' \neq 0$, then one of the numbers $\lambda_{d+1}, \dots, \lambda_n$ is different from zero. Let it be $\lambda_n \neq 0$ for definiteness. When $\tau \to +\infty$ by definition

$$\lim \frac{\log|\xi_j(\tau)|}{\log|\xi_n(\tau)|} = \frac{\lim(\log|\xi_j|/\log\tau)}{\lim(\log|\xi_n|/\log\tau)} = \frac{p_j}{p_n}, \quad j = d+1, \dots, n.$$

On the other hand, since $\Xi(\tau)$ is a nontrivial integral curve of System (1.8), then

$$d[\log|\xi_j|]/dt = \lambda_j + \tilde{f}_j(\Xi) \to \lambda_j.$$

Since all $\xi_j \to 0$ when $\tau \to +\infty$, then the numerator and denominator of the ratio $\log|\xi_j|/\log|\xi_n|$ tend to ∞. But their derivatives with respect to t have finite limits. Hence the L'Hôpital rule is applicable, by which

$$\lim \frac{\log|\xi_j|}{\log|\xi_n|} = \frac{\lim d[\log|\xi_j|]/dt}{\lim d[\log|\xi_n|]/dt}, \quad j = d+1, \dots, n,$$

i.e.

$$\frac{p_j}{p_n} = \frac{\lambda_j}{\lambda_n}, \quad j = d+1, \dots, n.$$

But this is equivalent to the equality (1.9). □

Corollary 1.1 *System* (1.8) *with* $\Lambda'' \neq 0$ *has nontrivial h-asymptotical integral curves, only if* $\Lambda'' \geq 0$ *or* $\Lambda'' \leq 0$.

Indeed, since h-asymptotical integral curve tends to zero, then it has all the orders $p_i \leq 0$. In particular, $P'' \leq 0$, and according to Theorem 1.2 all non-zero components of the vector Λ'' must have the same sign.

We consider the **four cases of the vector** $\Lambda'' = (\lambda_{d+1}, \dots, \lambda_n)$.

H1. All numbers $\lambda_{d+1}, \dots, \lambda_n$ are different from zero and have the same sign, i.e. $\Lambda'' > 0$ or $\Lambda'' < 0$.

H2. There are two numbers $\lambda_j \neq 0$ and $\lambda_k \neq 0$ of different signs.

N3. There are $\lambda_j = 0$ and $\lambda_k \neq 0$; moreover all $\lambda_k \neq 0$ are of the same sign, i.e. $\Lambda'' \neq 0$, and either $\Lambda'' \geq 0$ and $\Lambda'' \not> 0$ or $\Lambda'' \leq 0$ and $\Lambda'' \not< 0$.

H4. $\Lambda'' = 0$.

Theorem 1.3 *In Case* H1 *System* (1.8) *has at least* $(n-d-1)$*-parameter family of nontrivial h-asymptotical integral curves* $X = \Xi(\tau)$ *with*

$$\xi_j = b_j \tau^{p_j}(1 + o(1)), \quad b_j \neq 0, \quad p_j = -\lambda_j/\lambda_n, \quad j = d+1, \dots, n, \tag{1.10}$$

where $b_n = \pm 1$, *and* $b_{d+1}, \dots, b_{n-1}$ *can be taken as parameters of the family.*

Proof. It is given for analytic System (1.4), (1.3), assuming that reader is familiar with [Bruno 1971, §10] and [Bruno 1979a, Chapter III, §1]. Let $\Lambda'' > 0$ for definiteness. It can always be obtained by the change of sign of the time t. Let $\lambda_1, \dots, \lambda_d$ be eigenvalues of the matrix A_{11} in (1.8). We enumerate them in such a way that

$$\text{Re}\,\lambda_i \leq 0, \quad i = 1, \dots, l; \quad \text{Re}\,\lambda_j > 0, \quad j = l+1, \dots, d.$$

We make a linear transformation of coordinates $X' \to Y'$, reducing the matrix A_{11} to the Jordan form. According to the Dulak theorem (see also [Bruno 1971, §10]) the System (1.8) has an invariant analytic manifold

$$y_i = \psi_i(y_{l+1}, \dots, y_d, x_{d+1}, \dots, x_n), \quad i = 1, \dots, l. \tag{1.11}$$

The Taylor series for ψ_i do not contain constant and linear terms. On this manifold a system is induced, which according to [Bruno 1971; 1979a, Chapter III, §1] by the analytic change of coordinates

$$\begin{aligned} y_j &= z_j + \zeta_j(z_{l+1}, \ldots, z_n), \quad j = l+1, \ldots, d, \\ x_k &= z_k(1 + h_k(z_{l+1}, \ldots, z_n)), \quad k = d+1, \ldots, n, \end{aligned} \tag{1.12}$$

is reduced to the resonant normal form

$$\begin{aligned} \dot z_j &= \lambda_j z_j + \omega_j(z_{l+1}, \ldots, z_n), \quad j = l+1, \ldots, d, \\ \dot z_k &= z_k(\lambda_k + \rho_k(z_{l+1}, \ldots, z_n)), \quad k = d+1, \ldots, n, \end{aligned} \tag{1.13}$$

where ω_j and ρ_k are polynomials, besides ρ_k contain only the terms const $\cdot$ $z_{l+1}^{q_{l+1}} \ldots z_n^{q_n}$, which have

$$q_{l+1}\lambda_{l+1} + \ldots + q_n\lambda_n = 0, \tag{1.14}$$

integer $q_{l+1}, \ldots, q_n \geq 0$ and $q_{l+1} + \ldots + q_n \geq 1$. Since all Re $\lambda_{l+1}, \ldots,$ Re $\lambda_n >$ 0, then Equation (1.14) has no solutions among the mentioned $q_{l+1}, \ldots, q_n$, i.e. all $\rho_k \equiv 0$. Therefore in the normal form (1.13) the subsystem

$$\dot z_k = z_k \lambda_k, \quad k = d+1, \ldots, n$$

is isolated. Hence System (1.13) has solutions of the form

$$z_j = z_j(t), \quad j = l+1, \ldots, d; \qquad z_k = c_k \exp(\lambda_k t), \quad k = d+1, \ldots, n,$$

where c_k are arbitrary constants. These solutions tend to zero when $t \to -\infty$. To them there correspond the integral curves

$$\begin{aligned} z_j &= \eta_j(\tau), \quad j = l+1, \ldots, d, \\ z_k &= b_k \tau^{p_k}, \quad p_k = -\lambda_k/\lambda_n, \quad k = d+1, \ldots, n, \end{aligned} \tag{1.15}$$

where $b_1 = \pm 1$, and $b_{d+1}, \ldots, b_{n-1}$ are arbitrary constants. The integral curves (1.15) tend to zero when $\tau \overset{\text{def}}{=} |z_n|^{-1} \to +\infty$. Returning to coordinates Y', X'' by means of formulae (1.12) and (1.11), and then to coordinates X by means of the inverted linear transformation, the integral curves (1.15) would change into the integral curves (1.10) with the same values of p_j and b_j. □

There is a similar proof for the smooth system (1.4), (1.3), which can be carried out using the results by Samovol [1982].

We note some **properties of h-asymptotical integral curves** of System (1.8), which correspond to Cases H1–H4.

1. In Cases H1–H3 System (1.8) has no h-asymptotical trivial integral curves, because they lie in the hyperplanes (1.5), for which $\lambda_j \neq 0$.

2. We revise the number of h-asymptotical integral curves in Case H1. Let $\lambda_1, \ldots, \lambda_d$ be eigenvalues of the matrix A_{11} in (1.8). If all Re $\lambda_1, \ldots,$ Re λ_d have other sing then λ_n, then curves (1.15) exhaust all h-asymptotical integral curves of System (1.8). But if k numbers among Re $\lambda_1, \ldots,$ Re λ_d have the same sign as λ_n, then System (1.8) has $(k+n-d-1)$-parameter family of h-asymptotical integral curves, i.e. each curve (1.15) corresponds to the k-parameter family of h-asymptotical integral curves, which are disposed in the manifold (1.11). Obviously, $k = d - l$ and $0 < k \leq d$.

3. According to Corollary to Theorem 1.2, in Case H2 System (1.8) has no h-asymptotical integral curves.

4. In Case H3 System (1.8) may have h-asymptotical integral curves with the logarithmical asymptotics. Here System (1.7) has no h-asymptotical integral curves. For example, System

$$\dot{x}_1 = x_1^2, \quad \dot{x}_2 = x_2 \tag{1.16}$$

with $n = 2$ and $d = 0$ have integral curves

$$x_1 = (\log \tau + c)^{-1}, \quad x_2 = \tau^{-1}, \tag{1.17}$$

going into the stationary point $x_1 = x_2 = 0$ with the order $P = (0, -1)$ proportional to the vector $(\lambda_1, \lambda_2) = (0, 1)$. For System (1.16) System (1.7) is $\dot{x}_1 = 0$, $\dot{x}_2 = x_2$. Its integral curves $x_1 = \text{const} \neq 0$, $x_2 = \tau^{-1}$ are not h-asymptotical, although the curves (1.17) are h-asymptotical ones (compare below Section 6).

5. To isolate h-asymptotical integral curves in Cases H3 and H4 the more elaborate analysis is necessary, that takes into account non-linear in X'' terms in right hand sides of System (1.8), which are absent in System (1.7). In Case H4 with $\Lambda' \stackrel{\text{def}}{=} (\lambda_1, \ldots, \lambda_d) \neq 0$ and in Case H3 it is necessary for the purpose to compute the resonant normal form of System (1.8) [Bruno 1979a]. In Case H4 with $\Lambda' = 0$ it is necessary to build the Newton polyhedron, as it is described below in Section 4.

2. The power transformation

A power function x^μ with real exponent μ is single-valued in the following three cases:

1) $\mu \in \mathbb{R}$, $x \in \mathbb{R}_+$, i.e. x is real and $x > 0$;
2) the number μ is rational with odd denominator, $x \in \mathbb{R}$;
3) $\mu \in \mathbb{Z}$, $x \in \mathbb{C}$.

Further we will suppose that the domain of definition of all power functions we are dealing with lies in the mentioned above domains of uniqueness depending upon the form of the corresponding exponents. If a coordinate x_i appears with different exponents, then its domain of definition corresponds to the worst of them.

We consider an autonomous system of differential equations

$$dx_i/dt \stackrel{\text{def}}{=} \dot{x}_i = \varphi_i(X), \quad i = 1, \dots, n, \tag{2.1}$$

where $X = (x_1, \dots, x_n)$ and φ_i are finite or infinite sums of monomials $\varphi_{iQ} X^Q = \varphi_{iQ} x_1^{q_1} \dots x_n^{q_n}$, $\varphi_{iQ} = \text{const}$. We write System (2.1) in the form

$$d \log x_i/dt \stackrel{\text{def}}{=} (\log^{\cdot} x_i) = f_i(X) \stackrel{\text{def}}{=} \sum f_{iQ} X^Q, \quad i = 1, \dots, n, \tag{2.2}$$

where $f_i = \varphi_i x_i^{-1}$, and in the vector form

$$d \log X/dt = F(X) \stackrel{\text{def}}{=} \sum F_Q X^Q \text{ over } Q \in \mathbf{S}, \tag{2.3}$$

where $\log X = (\log x_1, \dots, \log x_n)$, $F(X) = (f_1, \dots, f_n)$, $F_Q = (f_{1Q}, \dots, f_{nQ}) = \text{const}$. The set $\mathbf{S} = \{Q : F_Q \neq 0\}$ is called the *support of System* (2.3), and it is denoted as supp F or $\mathbf{S}(F)$.

Let $\alpha = (\alpha_{ij})$ be a nonsingular square matrix of dimension n with real elements α_{ij}. The substitution of variables

$$y_i = x_1^{\alpha_{i1}} x_2^{\alpha_{i2}} \dots x_n^{\alpha_{in}}, \quad i = 1, \dots, n, \tag{2.4}$$

is called *the power transformation.* It is linear in logarithms:

$$\log Y = \alpha \log X, \quad \log X = \alpha^{-1} \log Y. \tag{2.5}$$

We note **the properties of power transformation** (2.4).

1. It is a one-to-one mapping of the orthant $X > 0$, i.e. all $0 < x_i < \infty$, in the orthant $Y > 0$.

2. It is defined for all $x_i \in \mathbb{R} \setminus \{0\}$ if and only if all numbers α_{ij} are rational with odd denominators.

3. The transformation (2.4) is a one-to-one mapping of the set $\{X \in \mathbb{R}^n,\ 0 < |x_i| < \infty,\ i = 1, \dots, n\}$ into the set $\{Y \in \mathbb{R}^n,\ 0 < |y_i| < \infty,\ i = 1, \dots, n\}$ if and only if all numbers α_{ij} are rational with odd denominators and the nominator of the rational number $\det \alpha$ is odd. These properties are obviously satisfied if the matrix α is a *unimodular* one (i.e. α_{ij} are integers and $\det \alpha = \pm 1$). In this case it is more convenient to consider the space with coordinates x_i as the direct product of n "circles" with points $x_i = 0$ and $x_i = \infty$ (i.e. as the "toric manifold").

4. The transformation (2.4) is a one-to-one mapping of the set $\{X \in \mathbb{C}^n, 0 < |x_i| < \infty, i = 1, \ldots, n\}$ in the set $\{Y \in \mathbb{C}^n, 0 < |y_i| < \infty, i = 1, \ldots, n\}$ if and only if the matrix α is a unimodular one. In this case it is more convenient to consider the space with coordinates X as the direct product of n complex spheres.

5. In all points $\{X\colon 0 < |x_i| < \infty, i = 1, \ldots, n\}$ where the transformation (2.4) is defined it is an analytical one.

6. The power transformations form a group that is isomorphic to the group of nonsingular real matrices. In particular, the power transformation that is the inverse to the transformation (2.4) have the matrix α^{-1}.

Lemma 2.1 *Let the power transformation* (2.4) *change System* (2.3) *into System*

$$d \log Y/dt = G(Y) \stackrel{\text{def}}{=} \sum G_R Y^R. \tag{2.6}$$

Then

$$R = \alpha^{*-1} Q \quad \textit{and} \quad G_R = \alpha F_Q, \tag{2.7}$$

where α^* is the transposed matrix α.

Proof. According to (2.4) and (2.5) the following equalities are valid

$$X^Q = \exp\langle Q, \log X\rangle = \exp\langle Q, \alpha^{-1} \log Y\rangle = \exp\langle \alpha^{*-1} Q, \log Y\rangle = Y^{\alpha^{*-1}Q}.$$

Now, according to (2.4) and (2.3) we have

$$d \log Y/dt = \alpha d \log X/dt = \alpha \sum F_Q X^Q = \sum \alpha F_Q Y^{\alpha^{*-1}Q}.$$

Comparing this formula with (2.6) we obtain (2.7). □

Lemma 2.1 shows that under the power transformation (2.4) the *vector exponents* Q and the *vector coefficients* F_Q undergo linear transformations as vectors of the dual spaces $\mathbb{R}^n$ and $\mathbb{R}^n_*$ respectively, since their scalar products are preserved:

$$\langle G_R, R_1\rangle = \langle \alpha F_Q, \alpha^{*-1} Q_1\rangle = \langle F_Q, Q_1\rangle.$$

In particular, under the power transformation (2.4) the support $\mathbf{S}$ of System (2.3) undergoes linear transformation $\mathbf{S}(G) = \alpha^{*-1}\mathbf{S}(F)$.

Let $T = (t_1, \ldots, t_n) \in \mathbb{R}^n$. *The power change of time*

$$dt_1 = X^T dt \tag{2.8}$$

transforms System (2.3) into the system

$$d \log X/dt_1 = X^{-T} F(X) = \sum F_Q X^{Q-T} \text{ over } Q \in \mathbf{S}, \tag{2.9}$$

i.e. all vector exponents Q of System (2.3) are translated in parallel by the vector $-T$.

Let $\mathbf{\Gamma} = \mathbf{\Gamma}(F)$ be the convex hull of the support $\mathbf{S}(F)$ of System (2.3). The number $d(F) = \dim(\mathbf{\Gamma}(F))$ is called the *dimension of System* (2.3). If the support $\mathbf{S}$ of System (2.3) consists of a finite number of points $Q_1, \dots, Q_s$, then its dimension d is equal to the number of linearly independent vectors among the remainders $Q_j - Q_s$, $j = 1, \dots, s-1$. The definition of dimension by the remainders of vector exponents is explained by the fact that in System (2.3) we consider only those solutions $X(t)$ which have no coordinates identically equal to zero. For such solutions, in System (2.3) we can make the change of time (2.8).

If $d = 0$, then System (2.3) has the form $d \log X/dt = F_Q X^Q$. After the power change of time $dt_1 = X^Q dt$, i.e. (2.8) with $T = Q$, we obtain the system $d \log X/dt_1 = F_Q$, i.e. $x_i = c_i \exp(f_{iQ} t_1)$, $i = 1, \dots, n$, and $dt_1 = \exp(\langle F_Q, Q\rangle t_1)\, dt$. If $\varkappa \stackrel{\text{def}}{=} \langle F_Q, Q\rangle = 0$, then $t_1 = t$ and $x_i = c_i \exp(f_{iQ} t)$. If $\varkappa \neq 0$, then $\exp(-\varkappa t_1) = -\varkappa t$ and $x_i = c_i(-\varkappa t)^{-f_{iQ}/\varkappa}$.

We will simplify System (2.3) using linear transformations of exponents Q and affine transformations of coefficients F_Q.

Theorem 2.1 *If System* (2.3) *has the dimension* $d < n$, *then there exist the power transformation* (2.4) *and the power change of time* (2.8), *which reduce System* (2.3) *to the form*

$$d \log Y/dt_1 = G(y_1, \dots, y_d). \tag{2.10}$$

The proof follows from Lemma 2.1 and from the paragraph before the formulation of Problem 4 in Section 11, Chapter 1.

From System (2.10) there splits subsystem of the order d

$$dy_i/dt = y_i g_i(y_1, \dots, y_d), \quad i = 1, \dots, d. \tag{2.11}$$

The variables $y_{d+1}, \dots, y_n$ are obtained in quadratures along solutions to subsystem (2.11).

Corollary 2.1 *The systems of dimension* 1 *are solved in quadratures.*

Indeed, in this case System (2.10) has the form

$$dy_i/dt = y_i g_i(y_1), \quad i = 1, \dots, n,$$

that implies

$$y_i = y_i^0 \exp \int \frac{g_i(y_1)}{g_1(y_1)} \frac{dy_1}{y_1}, \quad i = 2, \dots, n.$$

The inverse transformation from Y to X gives x_i as functions of y_1.

If the set $\mathbf{S}$ consists of a finite number of points $Q_1, \dots, Q_s$, then the most simple construction of the matrix α of Theorem 2.1 is pointed out in Method 1 of solution of Problem 3 in Section 11, Chapter 1. Namely, out of vectors $Q_j - Q_s$, $j = 1, \dots, s-1$ some d linearly independent vectors

$U_1, \dots, U_d$ are selected, and they form d upper lines of the matrix α. Then the next lines $U_{d+1}, \dots, U_n$ of the matrix are chosen as linearly independent from the preceding ones.

Example 2.1. We consider the equation of Emden–Fowler [Rosov 1985] $\ddot{x} = at^\sigma x^\mu$, where $a = \pm 1$, real numbers σ and μ are constants. We assume $x = x_1$, $t = x_3$, and we write the equation in the form of the autonomous system

$$\begin{aligned} \dot{x}_1 &= x_2 = x_1(x_1^{-1}x_2), \\ \dot{x}_2 &= ax_1^\mu x_3^\sigma = x_2(ax_1^\mu x_2^{-1}x_3^\sigma), \\ \dot{x}_3 &= 1 = x_3(x_3^{-1}). \end{aligned} \tag{2.12}$$

Here $n = 3$, $Q_1 = (-1, 1, 0)$, $Q_2 = (\mu, -1, \sigma)$, $Q_3 = (0, 0, -1)$. We compose the remainders of vectors of the support:

$$\begin{aligned} U_1 &\stackrel{\text{def}}{=} Q_1 - Q_3 = (-1, 1, 1), \\ U_2 &\stackrel{\text{def}}{=} Q_2 - Q_3 = (\mu, -1, \sigma + 1). \end{aligned}$$

The values of 2-minors of the (2×3)-matrix formed by these vectors are $M_1 = \sigma + 2$, $M_2 = \sigma + \mu + 1$, $M_3 = 1 - \mu$. Here the minor M_i is obtained by crossing out the ith column of the matrix. All three minors vanish only if $\sigma = -2$ and $\mu = 1$. In this case the Emden–Fowler equation is explicitly integrable as the Euler equation. Further we will suppose that $(\sigma, \mu) \neq (-2, 1)$, then the vectors U_1 and U_2 are linearly independent and the dimension of System (2.12) equals two. We put $T = Q_3$, $U_3 = (0, 0, 1)$, i.e. we make the change of time $dt_1 = x_3^{-1}dt$ and the power transformation (2.4) with the matrix α:

$$\begin{aligned} y_1 &= x_1^{-1}x_2x_3, \\ y_2 &= x_1^\mu x_2^{-1}x_3^{\sigma+1}, \\ y_3 &= x_3 \end{aligned} \qquad \alpha = \begin{pmatrix} -1 & 1 & 1 \\ \mu & -1 & \sigma + 1 \\ 0 & 0 & 1 \end{pmatrix},$$

where $\det \alpha = 1 - \mu$. The inverse matrix and transformation are

$$\alpha^{-1} = \frac{1}{1-\mu} \begin{pmatrix} -1 & -1 & \sigma + 2 \\ -\mu & -1 & \sigma + \mu + 1 \\ 0 & 0 & 1 - \mu \end{pmatrix}, \quad \begin{aligned} x_1 &= (y_1^{-1}y_2^{-1}y_3^{\sigma+2})^{1/(1-\mu)}, \\ x_2 &= (y_1^{-\mu}y_2^{-1}y_3^{\sigma+\mu+1})^{1/(1-\mu)}, \\ x_3 &= y_3. \end{aligned} \tag{2.13}$$

System (2.12) transforms into the system

$$\begin{aligned} d\log y_1/dt_1 &= 1 - y_1 + ay_2 \stackrel{\text{def}}{=} g_1, \\ d\log y_2/dt_1 &= \sigma + 1 + \mu y_1 - ay_2 \stackrel{\text{def}}{=} g_2, \\ d\log y_3/dt_1 &= 1. \end{aligned} \tag{2.14}$$

In this way, the Emden–Fowler equation is reduced to the quadratic system in the plane formed by the first two equations of System (2.14). □

The second method to compute the matrix α is based upon the use of normals to the set $\mathbf{\Gamma}$ or $\mathbf{S}$ (see Section 11, Chapter 1).

Example 2.2. System (2.1), where φ_i are homogeneous polynomials of the order $l+1$, are written in the form (2.2), where $f_i(X) = x_i^{-1}\varphi_i(X)$ are homogeneous functions of the order l. Here the dimension of System (2.3) is $d = n-1$, the normal to the support is the vector $(1, \dots, 1)$, and the matrix

$$\alpha = \begin{pmatrix} 1 & 0 & \dots & -1 \\ 0 & 1 & \dots & -1 \\ \dots & \dots & \dots & \dots \\ 0 & 0 & \dots & 1 \end{pmatrix} \tag{2.15}$$

has the form corresponding to the normal according to Method 2 of solution of Problem 3 in Section 11, Chapter 1. The corresponding power transformation (2.4)

$$y_i = x_i x_n^{-1}, \quad i = 1, \dots, n-1, \quad y_n = x_n, \tag{2.16}$$

has the inverse transformation

$$x_i = y_i x_n = y_i y_n, \quad i = 1, \dots, n-1, \quad x_n = y_n.$$

Since

$$\log y_i = \log x_i - \log x_n, \quad i = 1, \dots, n-1,$$

then

$$\begin{aligned} d\log y_i/dt &= f_i(X) - f_n(X), \quad i = 1, \dots, n-1, \\ d\log y_n/dt &= f_n(X). \end{aligned} \tag{2.17}$$

But

$$f_i(X) = y_n^l f_i(y_1, \dots, y_{n-1}, 1), \quad i = 1, \dots, n.$$

If now in System (2.17) we make the power change of time $dt_1 = y_n^l dt = x_n^l dt$, then we obtain the system

$$\begin{aligned} d\log y_i/dt_1 &= f_i(y_1, \dots, y_{n-1}, 1) - f_n(y_1, \dots, y_{n-1}, 1), \quad i = 1, \dots, n, \\ d\log y_n/dt_1 &= f_n(y_1, \dots, y_{n-1}, 1), \end{aligned}$$

whose right hand sides depend only on $n-1$ variables. □

It may happen that in the power transformation (2.4) or in the change of time (2.8) there is a coordinate x_i whose exponents α_{ij} or t_i are worse then in the original system. For example, they are not rational with odd denominators. In this case in System (2.3) it is necessary to introduce a new parameter $\sigma_i = \operatorname{sgn} x_i$, and to make the power transformation and the change of time for $|x_i|$.

Theorem 2.2 *If System* (2.3) *has dimension $d < n$ and all vector exponents Q integer, then there exist such an integer vector T and an unimodular matrix α that the change of time* (2.8) *and the power transformation* (2.4) *reduce System* (2.3) *to the form* (2.10).

The proof repeats that of the Theorem 2.1. In Section 11, Chapter 1 two methods were indicated to compute the unimodular matrix α (as solution to Problem 3).

Lemma 2.2 *Let the power transformation* (2.4) *transform the curve $X = \Xi(\tau)$ with the vector order P into the curve $Y = \mathrm{H}(\tau)$ with the vector order $\tilde{P}$. Then $\tilde{P} = \alpha P$.*

Proof. Since $\log Y = \alpha \log X$, then $\log|\mathrm{H}(\tau)|/\log\tau = \alpha\log|\Xi(\tau)|/\log\tau$. If τ will tend to infinity, then in the left hand side of the equality the limit is $\tilde{P}$, and in the right hand side it is αP. Hence $\tilde{P} = \alpha P$. □

3. The generalized power transformations

Here we are dealing with the simplification of the system

$$d\log X/dt = \sum F_Q X^Q \text{ over } Q \in \mathbf{S} \tag{3.1}$$

with the finite support $\mathbf{S} = \{Q_1, \ldots, Q_s\}$. We will denote the vector coefficient F_{Q_j} as F_j, and we will write System (3.1) in the form

$$d\log X/dt = \sum_{j=1}^{s} F_j X^{Q_j}. \tag{3.2}$$

The linear subspace $\mathbf{F} \subset \mathbb{R}^n_*$ spanned on vectors $F_1, \ldots, F_s$ is called the *subspace of coefficients* of System (3.2). The number r is called the *rank* of System (3.2), if there is exactly r linearly independent vectors among the $F_1, \ldots, F_s$. Obviously, $1 \le r \le n$ and $r = \dim \mathbf{F}$.

Theorem 3.1 *If System* (3.2) *has the rank $r < n$, then there exists such a matrix α that the power transformation* (2.4) *reduces System* (3.2) *to the form*

$$\begin{aligned} d\log y_i/dt &= g_i(Y), \quad i = 1, \ldots, r, \\ d\log y_j/dt &= 0, \quad j = r+1, \ldots, n. \end{aligned} \tag{3.3}$$

The proof ia analogous to that of Theorem 2.1, if instead of $Q_j - T$ we write F_j, and instead of α we write α^{*-1}, since $G_j = \alpha F_j$. In particular, the methods of construction of the matrix α using the vectors $Q_j - T$, which were described in Section 11, Chap 1, are applicable for the construction of the matrix α^{*-1} using the vectors F_j.

Corollary 3.1 *System* (3.2) *of the rank* r *is reduced to the system with* r *variables and* $n - r$ *arbitrary constants.*

Indeed, from System (3.3) we obtain $y_j = y_j^0 = \text{const}\,,\ j = r+1, \ldots, n$.

Corollary 3.2 *System of the rank* 1 *can be solved in quadratures*

$$t - t_0 = \int \frac{dy_1}{y_1 g_1(y_1, y_2^0, \ldots, y_n^0)}.$$

Theorem 3.1 means that System (3.2) of the rank r has $n-r$ first integrals of the form $X^{U_j} = y_j^0 = \text{const}\,,\ j = r+1, \ldots, n$, besides vectors $U_j = (\alpha_{j1}, \ldots, \alpha_{jn})$ are linearly independent.

The function F of the right hand side of System (3.2) can be defined by two matrices of dimension $s \times n$: 1) by the matrix $\mathcal{F}$, whose jth line is F_j; 2) by the matrix $\mathcal{Q}$, whose jth line is Q_j. System (3.2) is also can be defined by these matrices. Under the power transformations of System (3.2) the matrix of the dimension $s \times s$

$$\mathcal{B} = \mathcal{Q}\mathcal{F}^* \tag{3.4}$$

is not changed, since its elements are $b_{ij} = \langle Q_i, F_j \rangle$. Further we will suppose that the change of time (2.8) with $T = Q_s$ is already done in System (3.2), i.e. the rank of the matrix $\mathcal{Q}$ is the dimension d of System (3.2) and $Q_s = 0$.

The power transformations (2.4) introduced in Section 2 preserve the number n of coordinate. Now we introduce the *generalized power transformations*

$$y_j = X^{R_j}, \quad j = 1, \ldots, m, \tag{3.5}$$

with an arbitrary number m. If $m < n$ and vectors R_j are linearly independent, then Transformation (3.5) can be completed to the ordinary power transformation (2.4). In other cases Transformation (3.5) can not be reduced to (2.4); it is analogous to the Veronese mapping, which is used in algebraic geometry [Shafarevich 1972].

Theorem 3.2 *The transformation*

$$y_j = X^{Q_j}, \quad j = 1, \ldots, s, \tag{3.6}$$

reduces the solution of System (3.2) *of dimension* d *to solution of the system*

$$d \log Y / dt = \mathcal{B} Y \tag{3.7}$$

and to $n-d$ quadratures. Here $Y = (y_1, \dots, y_s)$ and $y_s \equiv 1$.

Proof. According to (3.6) and to the definition of the matrix $\mathcal{F}$ System (3.2) can be written in the form

$$(\log\dot{X}) = \mathcal{F}^* Y. \tag{3.8}$$

Taking the logarithm of (3.6) we obtain

$$\log Y = \mathcal{Q} \log X. \tag{3.9}$$

Taking the derivative of this equation with respect to t and using (3.4) we obtain $(\log\dot{Y}) = \mathcal{Q}(\log\dot{X}) = \mathcal{Q}\mathcal{F}^* Y = \mathcal{B}Y$, i.e. System (3.7).

Let $Y = Y(t)$ be a solution to System (3.7). Since the rank of the matrix $\mathcal{Q}$ is d, then in the matrix $\mathcal{Q}$ there is a minor of the order d different from zero. Let it be the minor

$$\begin{vmatrix} q_{11} & \dots & q_{1d} \\ \dots & \dots & \dots \\ q_{d1} & \dots & q_{dd} \end{vmatrix} \neq 0$$

for definiteness. Then from first d equalities of the relation (3.9) the $\log x_1, \dots, \log x_d$ can be linearly expressed through $\log y_1, \dots, \log y_d$ and $\log x_{d+1}, \dots, \log x_n$. Consequently, coordinates $x_1, \dots, x_d$ express through the products of powers of coordinates $y_1, \dots, y_d, x_{d+1}, \dots, x_n$. From (3.8) we obtain

$$d \log x_j / dt = \sum_{k=1}^{s} f_{kj} y_k, \quad j = d+1, \dots, n, \tag{3.10}$$

$$x_j = x_j^0 \exp \int \sum_{k=1}^{s} f_{kj} y_k dt, \quad j = d+1, \dots, n.$$

Since $Q_s = 0$, then $y_s \equiv 1$. □

Example 3.1. Let $\mathcal{B} = 0$, i.e. all Q_j be orthogonal to all F_k. System (3.7) is $d \log Y/dt = 0$, i.e. $y_j = y_j^0 = \text{const}$, $j = 1, \dots, s$. From (3.8) we obtain

$$x_j = x_j^0 \exp \int \sum_{k=1}^{s} f_{kj} y_k^0 dt, \quad j = 1, \dots, n,$$

where $y_k^0 = (X^0)^{Q_k}$, $k = 1, \dots, s$. Consequently, $x_j = x_j^0 \exp(f_j(X^0)t)$, $j = 1, \dots, n$. □

Example 3.2. We consider the Blasius equation $\dddot{x} + x\ddot{x} = 0$ known in theory of boundary layer of viscous incompressible fluid [Van Dyke 1964]. We put

$x_1 = x$ and write this equation in the form of an autonomous system of equations:

$$\dot{x}_1 = x_2 = x_1(x_1^{-1}x_2),$$
$$\dot{x}_2 = x_3 = x_2(x_2^{-1}x_3),$$
$$\dot{x}_3 = -x_1x_3 = x_3(-x_1).$$

Here $n = 3$ and

$$\mathcal{Q} = \begin{pmatrix} -1 & 1 & 0 \\ 0 & -1 & 1 \\ 1 & 0 & 0 \end{pmatrix}, \quad \mathcal{F} = \begin{pmatrix} 1 & 0 & 0 \\ 0 & 1 & 0 \\ 0 & 0 & -1 \end{pmatrix}.$$

We put $T = Q_3 = (1, 0, 0)$, then the matrix $\mathcal{Q}'$ composed of the lines $Q_j - T$ is

$$\mathcal{Q}' = \begin{pmatrix} -2 & 1 & 0 \\ -1 & -1 & 1 \\ 0 & 0 & 0 \end{pmatrix}.$$

We make the change of time $dt_1 = x_1 dt$ and the generalized power transformation $y_1 = x_1^{-2}x_2$, $y_2 = x_1^{-1}x_2^{-1}x_3$, $y_3 = 1$. Then the considered system turns into the system of the form (3.7)

$$\begin{aligned} d\log y_1/dt_1 &= -2y_1 + y_2, \\ d\log y_2/dt_1 &= -y_1 - y_2 - 1. \end{aligned} \tag{3.11}$$

To simplify the transition to X from Y we suppose $y_4 = x_1$. The transition to y_1, y_2, y_4 from X is given by an ordinary power transformation with the unimodular matrix α, and the third equation of the system for y_1, y_2, y_4 is $d\log y_4/dt_1 = -1$. □

We recall that the dimension d of System (3.2) is the rank of the matrix $\mathcal{Q}$, and the rank r of this system is the rank of the matrix $\mathcal{F}$. If we denote the rank of the matrix $\mathcal{B}$ as b, then $0 \le b \le d$, $r \le n, s$.

Theorem 3.3 *Let* $d = \operatorname{rank}\mathcal{Q}$ *and* $b = \operatorname{rank}\mathcal{B}$. *By means of the transformation* (3.6) *and the power transformation*

$$\log z_i = \langle L_i, \log Y\rangle, \quad i = 1, \dots, s, \tag{3.12}$$

the solution of System (3.2) *is reduced to solution of the system of the form* (3.2) *with* b *variables and* $d - b$ *arbitrary constants and to* $n - d$ *integrations.*

Proof. Taking into account Theorem 3.2 it is sufficient to prove that System (3.7) obtained by the substitution (3.6) from (3.2) is reduced to the system with b variables and $d-b$ arbitrary constants. For this purpose, first of all we note that the variables X of System (3.2) correspond not to the whole space of variables Y, but only to a manifold of the dimension d. Indeed, let us isolate the linear subspace $\mathbf{L} = \{L: \mathcal{Q}^* L = 0\}$ in $\mathbb{R}^s$. Since $\operatorname{rank} \mathcal{Q} = d$, then $\dim \mathbf{L} = s - d$. For all $L = (l_1, \ldots, l_s) \in \mathbf{L}$ we have $l_1 Q_1 + \ldots + l_s Q_s = 0$. By virtue of (3.9)

$$\langle \log Y, L \rangle = \langle \mathcal{Q} \log X, L \rangle = \langle \log X, \mathcal{Q}^* L \rangle = 0 \tag{3.13}$$

if $L \in \mathbf{L}$, i.e. $Y^L = 1$.

Let us now isolate in $\mathbb{R}^s$ the linear subspace $\mathbf{L}_1 = \{L: \mathcal{B}^* L = 0\}$. Since the rank of the matrix $\mathcal{B}$ is $b \leq d$, then $\dim \mathbf{L}_1 = s - b \geq s - d$. Now we note that $\mathbf{L}_1 \supset \mathbf{L}$, i.e. if $\mathcal{Q}^* L = 0$, then $\mathcal{B}^* L = 0$. Indeed, by (3.4) we have $\mathcal{B}^* L = \mathcal{F} \mathcal{Q}^* L = 0$.

Let us select a basis $L_{d+1}, \ldots, L_s$ in $\mathbf{L}$ and complement it to the basis in $\mathbf{L}_1$ by vectors $L_{b+1}, \ldots, L_d$, and complement this basis in $\mathbf{L}_1$ to the basis in $\mathbb{R}^s$ by vectors $L_1, \ldots, L_b$. The power transformation (3.12) transforms System (3.7) into the system

$$d \log z_i / dt = \langle d \log Y / dt, L_i \rangle = \langle \mathcal{B} Y, L_i \rangle = \langle Y, \mathcal{B}^* L_i \rangle \stackrel{\text{def}}{=} g_i(Z), \quad i = 1, \ldots, s. \tag{3.14}$$

By virtue of the selection of vectors L_i we have $\mathcal{B}^* L_j = 0$ for $j > b$, i.e. $z_j = \text{const}$, and besides by virtue of (3.13) $\log z_k = 0$ for $k > d$, i.e. $z_k = 1$. Consequently, System (3.14) as the transformed system (3.7) is

$$\begin{aligned} &d \log z_i / dt = g_i(z_1, \ldots, z_d), \quad i = 1, \ldots, b, \\ &d \log z_j / dt = 0, \quad j = b+1, \ldots, d, \\ &z_k = 1, \quad k = d+1, \ldots, s. \end{aligned}$$

Hence we obtain

$$z_j = z_j^0 = \text{const}, \quad j = b+1, \ldots, d,$$

$$d \log z_i / dt = g_i(z_1, \ldots, z_b, z_{b+1}^0, \ldots, z_d^0), \quad i = 1, \ldots, b. \tag{3.15}$$

Thus, the solution of System (3.7) is reduced to the solution of System (3.15). According to the proof of Theorem 3.2 coordinates $x_1, \ldots, x_d$ are expressed through the products of powers of coordinates $y_1, \ldots, y_d, x_{d+1}, \ldots, x_n$, and coordinates $y_1, \ldots, y_s$ are expressed through the products of powers of coordinates $z_1, \ldots, z_s$. Since $z_{d+1} = \ldots = z_s = 1$, then coordinates $y_1, \ldots, y_s$ are expressed through the products of powers of coordinates $z_1, \ldots, z_d$. Consequently, in Equations (3.10) the right hand sides are functions of $z_1, \ldots, z_d$, and System (3.2) is reduced to the system

$$\begin{aligned} &d\log z_i/dt = g_i(z_1, \ldots, z_d), \quad i = 1, \ldots, b, \\ &d\log z_j/dt = 0, \quad j = b+1, \ldots, \\ &d\log x_k/dt = h_k(z_1, \ldots, z_d), \quad k = d+1, \ldots, n. \square \end{aligned}$$

Lemma 3.1 *If nontrivial integral curve $X = \Xi(\tau)$ of System* (3.2) *has the vector order P when $\tau \to \infty$, then $P \in \mathbf{F}$.*

Proof. Since $X = \Xi(\tau)$ is a nontrivial integral curve, then it corresponds to the solution $X(t)$ to System (3.2). Let r be the rank of System (3.2). If $r = n$, then $\mathbf{F} = \mathbb{R}^n_*$, $P \in \mathbf{F}$.

If $r < n$, then there exist $n-r$ linearly independent vectors $R_1, \ldots, R_{n-r} \in \mathbb{R}^n$, which have

$$\langle F_i, R_j \rangle = 0 \quad \text{for} \quad i = 1, \ldots, s, \quad j = 1, \ldots, n-r.$$

But this means that

$$\langle F(X), R_j \rangle = 0, \quad j = 1, \ldots, n-r,$$

and for solutions to System (3.2)

$$\langle d\log |X|/dt, R_j \rangle = 0, \quad j = 1, \ldots, n-r.$$

If we integrate over t the last equalities, we obtain

$$\langle \log |X|, R_j \rangle = c_j = \text{const}, \quad j = 1, \ldots, n-r,$$

i.e. $\langle \log |\Xi(\tau)|, R_j \rangle = c_j$, $j = 1, \ldots, n-r$ or

$$\left\langle \frac{\log |\Xi(\tau)|}{\log \tau}, R_j \right\rangle = \frac{c_j}{\log \tau}, \quad j = 1, \ldots, n-r.$$

If we let τ tend to infinity, we obtain as the limit the equalities

$$\langle P, R_j \rangle = 0, \quad j = 1, \ldots, n-r.$$

But this means that $P \in \mathbf{F}$. $\square$

Example 3.3. Let us demonstrate that in Lemma 3.1 the nontriviality of the integral curve is essential. For the system

$$d\log |x_1|/dt = x_1 - x_2^2, \quad d\log |x_2|/dt = 0$$

$\mathbf{F} = \lambda(1,0)$, $\lambda \in \mathbb{R}$. But there is a trivial integral curve $x_1 = \tau^2$, $x_2 = \tau$. Its order is $P = (2,1)$, i.e. $P \notin \mathbf{F}$. $\square$

4. Truncated systems

We consider the system

$$d\log X/dt = F(X) \stackrel{\text{def}}{=} \sum F_Q X^Q \quad \text{over } Q \in \mathbf{S}. \tag{4.1}$$

We suppose that in the right hand side the similar terms are already collected and $\mathbf{S} = \{Q: F_Q \neq 0\}$. The exterior convex hull of the support $\mathbf{S}$, i.e. CNV $\mathbf{S}$, is called the *Newton polyhedron* of System (4.1), and it is denoted as $\mathbf{\Gamma}(F)$. Its boundary $\partial\mathbf{\Gamma}$ consists of faces $\mathbf{\Gamma}_j^{(d)}$ of various dimensions d. To the faces there correspond to boundary subsets $\mathbf{S}_j^{(d)}$ and tangent cones $\mathbf{T}_j^{(d)}$ in $\mathbb{R}^n$, and also the normal cones $\mathbf{U}_j^{(d)}$ in $\mathbb{R}_*^n$ (see Chapter 1). Each face $\mathbf{\Gamma}_j^{(d)}$ is put in correspondence with the *truncated system*

$$d\log X/dt = \hat{F}_j^{(d)}(X) \stackrel{\text{def}}{=} \sum F_Q X^Q \quad \text{over } Q \in \mathbf{S}_j^{(d)}. \tag{4.2}$$

We denote the components of the truncation $\hat{F}$ as $\check{f}_1, \ldots, \check{f}_n$, since not every $\check{f}_i$ is necessarily a truncation $\hat{f}_i$ of the function f_i, and some of $\check{f}_i$ may be identical to zero. The number d is called the *dimension* of Truncation (4.2). Obviously, $0 \leq d < n$. The linear hull of vectors F_Q for $Q \in \mathbf{S}_j^{(d)}$, i.e. of vector coefficients of Truncation (4.2), is called the *subspace of coefficients of the truncated System* (4.2), and it is denoted as $\mathbf{F}_j^{(d)}$. Thus, to the truncated System (4.2) there relate the following geometrical objects: the boundary subset $\mathbf{S}_j^{(d)}$; the face $\mathbf{\Gamma}_j^{(d)}$ and the tangent cone $\mathbf{T}_j^{(d)}$ in $\mathbb{R}^n$, and also the normal cone $\mathbf{U}_j^{(d)}$, the set of vector coefficients $\{F_Q\}$ and the subspace of coefficients $\mathbf{F}_j^{(d)}$ in $\mathbb{R}_*^n$.

Let the power transformation (2.4) and the power change of time (2.8) transform System (4.1) into the system

$$d\log Y/dt_1 = G(Y) \stackrel{\text{def}}{=} \sum G_R Y^R \quad \text{over } R \in \tilde{\mathbf{S}} \stackrel{\text{def}}{=} \mathbf{S}(G), \tag{4.3}$$

and its Truncation (4.2) into the system

$$d\log Y/dt_1 = \hat{G}(Y) \stackrel{\text{def}}{=} \sum G_R Y^R \quad \text{over } R \in \tilde{\mathbf{S}}_j^{(d)}. \tag{4.4}$$

Here according to Lemma 2.1 the relations (2.7) are satisfied, and System (4.4) is the truncation of System (4.3) with the normal cone $\tilde{\mathbf{U}}_j^{(d)} = \alpha\mathbf{U}_j^{(d)}$, the vector coefficients $\{G_R\} = \alpha\{F_Q\}$, the subspace of coefficients $\mathbf{G}_j^{(d)} = \alpha\mathbf{F}_j^{(d)}$, the support $\tilde{\mathbf{S}}_j^{(d)} = \alpha^{*-1}\mathbf{S}_j^{(d)}$, the tangent cone $\tilde{\mathbf{T}}_j^{(d)} = \alpha^{*-1}\mathbf{T}_j^{(d)}$. According to Lemma 2.2, the curve $X = \Xi(\tau)$ of the order P is transformed into the curve $Y = \mathrm{H}(\tau)$ of the order $\tilde{P} = \alpha P$. In particular, the curves

$$x_i = b_i \tau^{p_i}(1 + o(1)), \quad b_i \neq 0, \quad i = 1, \ldots, n, \tag{4.5}$$

$$x_i = b_i \tau^{p_i}, \quad b_i \neq 0, \quad i = 1, \ldots, n, \tag{4.6}$$

are transformed respectively into the curves

$$y_i = \tilde{b}_i \tau^{\tilde{p}_i}(1 + o(1)), \quad b_i \neq 0, \quad i = 1, \ldots, n, \tag{4.7}$$

$$y_i = \tilde{b}_i \tau^{\tilde{p}_i}, \quad b_i \neq 0, \quad i = 1, \ldots, n, \tag{4.8}$$

where

$$\tilde{P} = \alpha P, \quad \log \tilde{B} = \alpha \log B. \tag{4.9}$$

If a curve $X = \Xi(\tau)$ is an integral one for System (4.1) (or (4.2)), then its image $Y = \mathrm{H}(\tau)$ is the integral curve for System (4.3) (or (4.4)).

Theorem 4.1 *For System* (4.1) *and for its Truncation* (4.2) *of the dimension d there exist the matrix α and the vector T such that the power transformation* (2.4) *and the power change of time* (2.8) *reduce them to such a System* (4.3) *and to such a truncation* (4.4) *of* (4.3) *that*

1) $\hat{G}_j^{(d)}(Y) \equiv \hat{G}_j^{(d)}(y_1, \ldots, y_d)$, *i.e. it does not depend on* $y_{d+1}, \ldots, y_n$;

2) *the support* $\tilde{\mathbf{S}} \stackrel{\text{def}}{=} \mathbf{S}(G) \geq 0$, *i.e. all* $R_j \geq 0$ *in* (4.3);

3) *if* $\mathbf{S}(F) \subset \mathbb{Z}^n$, *then the matrix α can be chosen as unimodular one, and the vector T as integer one, and so* $\mathbf{S}(G) \subset \mathbb{Z}^n$.

The proof follows from Lemma 2.1 of this Chapter and from the solution of Problem 5 in Section 11, Chapter 1.

We describe now more explicitly System (4.3) and its Truncation (4.4), which are given by Theorem 4.1. We split each n-dimensional vector into two subvectors: d–dimensional and $(n-d)$–dimensional ones. For example, $Y = (Y', Y'')$, where $Y' = (y_1, \ldots, y_d)$, $Y'' = (y_{d+1}, \ldots, y_n)$. The cone $\mathbf{U}'' \stackrel{\text{def}}{=} \tilde{\mathbf{U}}_j^{(d)}$ of Truncation (4.4) lies in the subspace $\tilde{P}' = 0$, and it contains the negative orthant $\mathbf{U}'' \supset \{\tilde{P}'' < 0\}$. Conversely, the tangent cone $\tilde{\mathbf{T}}_j^{(d)}$ is isolated only by inequalities on R''. Let $\mathbf{T}''$ be its projection into the subspace $R' = 0$, then $\mathbf{T}'' \subset \mathbb{R}_+^{n-d} \stackrel{\text{def}}{=} \{R'' \geq 0\}$. The cones $\mathbf{U}''$ and $\mathbf{T}''$ are dual to each other. System (4.3) has the form

$$\begin{aligned} dy_j/dt_1 = \psi_i(Y) &\stackrel{\text{def}}{=} \sum \psi_{iR''}(Y')Y''^{R''}, \quad i = 1, \ldots, d, \\ d \log y_j/dt_1 = g_j(Y) &\stackrel{\text{def}}{=} \sum g_{jR''}(Y')Y''^{R''}, \quad j = d+1, \ldots, n, \end{aligned} \tag{4.11}$$

where exponents R'' lie in the forward cone $\mathbf{T}''$. System (4.11) is defined in the set $\mathcal{U}(\mathbf{T}'', \varepsilon)$. The truncated System (4.4) is obtained from System (4.11) if we substitute $Y'' = 0$ in its right hand sides, and it has the form

$$\begin{aligned} dy_j/dt_1 &= \psi_{i0}(Y'), \quad i = 1, \dots, d, \\ d \log y_j/dt_1 &= g_{j0}(Y'), \quad j = d+1, \dots, n, \end{aligned} \tag{4.12}$$

where $\psi_{i0} \stackrel{\text{def}}{=} y_i \breve{g}_i$ and $g_{j0} \stackrel{\text{def}}{=} \breve{g}_j$.

To study solutions to System (4.11) for small Y'' it is necessary to study solutions to its truncated System (4.12), and first of all to its first subsystem. When $Y'' \to 0$ the limit sets for solutions to the complete system (4.11) can be only the limit sets for solutions to the first Subsystem (4.12). They are so called α- and ω-limit sets [Nemytskii and Stepanov 1947]. These are stationary points, limit cycles, invariant tori and the more complicated objects, which include the mentioned ones along with the connecting them separatrix invariant sets. In the complete system (4.11) the neighborhood of a stationary point, the neighborhood of a periodic solution or the neighborhood of an invariant torus of the first subsystem of Truncation (4.12) can be studied introducing the local coordinates and reducing the system to the normal form [Bruno 1979a]. But the search of nonlocal limit sets which are different from the stationary points is a transcendental task having up to-day no sufficiently general algorithm of solution. So, we will dwell upon only the stationary points $Y'^0 = (y_1^0, \dots, y_d^0)$ of the first subsystem (4.12), which have all coordinates y_i^0 different from zero or infinity. A stationary point with a zero (or infinite) coordinate is a stationary point of other truncation of the considered system having the dimension less then d, and it is not considered here. To each stationary point Y'^0 there corresponds the $(n - d - 1)$-parameter family of nontrivial power integral curves of the truncated System (4.12):

$$y_i = y_i^0, \quad i = 1, \dots, d, \quad y_j = b_j \tau^{p_j}, \quad j = d+1, \dots, n, \tag{4.13}$$

where $p_j = -\lambda_j/\lambda_n$, $\lambda_j = g_{j0}(Y'^0)$, $b_{d+1}, \dots, b_{n-1}$ are arbitrary constants and $b_n = \pm 1$. The question is: are there integral curves of the complete system (4.11), which are asymptotically close to curves (4.13)? To answer this question we remark that the pair of Systems (4.11) and (4.12) is analogous to the pair of Systems (1.4) and (1.7). The only difference is that Truncation (1.7) of System (1.4) has the fixed normal cone

$$\{P \colon P' = 0,\ P'' < 0\} \tag{4.14}$$

and the fixed tangent cone $\{Q : Q'' \geq 0\}$, but the normal cone $\mathbf{U}''$ and the tangent cone $\mathbf{T}''$ of Truncation (4.12) are arbitrary dual cones with the properties $\mathbf{U}'' \supset \{P'' < 0\}$ and $\mathbf{T}'' \subset \{Q'' \geq 0\}$.

Let Y'^0 be a stationary point of the first subsystem (4.12). The integral curve $Y = \mathrm{H}(\tau)$ of System (4.11) is called the *g-asymptotical* one, if:

a) $\mathrm{H}'(\tau) \to Y'^0$ when $\tau \to +\infty$;

b)$\mathrm{H}''(\tau) \subset \mathcal{U}(\mathbf{T}'', \varepsilon)$ for all $\varepsilon > 0$ when $\tau \in (\tau_0(\varepsilon), +\infty)$;

c) all components $\eta_j(\tau)$ of the vector $\mathrm{H}''(\tau)$ are different from zero;

d) the vector $\mathrm{H}''(\tau)$ has the vector order $P'' \in \overline{\mathbf{U}}''$.

The h-asymptotical integral curves defined in Section 1 are the special case of g-asymptotical ones, when the tangent cone $\mathbf{T}'' = \mathbb{R}_+^{n-d} \stackrel{\text{def}}{=} \{Q'' \geq 0\}$, and the normal cone $\mathbf{U}''$ is (4.14). Indeed, the properties a) and c) in both definitions coincide; in the mentioned special case the property b) means that $\Xi''(\tau) \to 0$ when $\tau \to +\infty$, and the property d) is automatically satisfied for integral curves of System (1.4) with $\Xi''(\tau) \to 0$.

We will denote as before $\lambda_j = \breve{g}_j(Y'^0)$, $j = d+1, \ldots, n$. We form the vector $\Lambda'' = (\lambda_{d+1}, \ldots, \lambda_n)$.

Theorem 4.2 *Let* $\Lambda'' \neq 0$ *in System* (4.11) *and* $Y = \mathrm{H}(\tau)$ *be its nontrivial g-asymptotical integral curve, whose part* $\mathrm{H}''(\tau)$ *has the vector order* $P'' \neq 0$. *Then the equality* (1.9) *holds.*

Proof. Since the curve $Y = \mathrm{H}(\tau)$ is g-asymptotical, then $P'' \in \overline{\mathbf{U}}''$ according to the property d) of the definition. We shall distinguish two cases: $P'' \in \mathbf{U}''$ and $P'' \in \partial\overline{\mathbf{U}}''$ and shall use some well known geometrical facts (see for example [Chernikov 1968, Pshenichnyi 1980]).

If the vector P is inside the normal cone $\mathbf{U}''$, i.e. $P'' \in \mathbf{U}''$, then in the normal cone $\mathbf{U}''$ there are $n-d$ linearly independent vectors $B''_{d+1}, \ldots, B''_n$ such that $P'' = \mu_{d+1}B''_{d+1} + \ldots + \mu_n B_n$, where all $\mu_i > 0$, i.e. the vector P'' lies inside the conic hull of vectors $B''_{d+1}, \ldots, B''_n$, which is a $(n-d)$-dimensional *simplicial cone* with the skeleton from these vectors (see the end of Section 1, Chapter 1). We compose the matrix $\beta = -(B''_{d+1} \ldots B''_n)$, we put $\gamma = \beta^{-1}$, and we make the power transformation $\log Z'' = \gamma \log Y''$. After that System (4.11) turns into a system with respect to Y' and Z'' of the same form as (4.11), the vector P'' turns into the vector $\tilde{P}'' \stackrel{\text{def}}{=} \gamma P'' = -(\mu_{d+1}E_{d+1} + \ldots + \mu_n E_n) < 0$, and the vector Λ'' turns into the vector $\tilde{\Lambda}'' \stackrel{\text{def}}{=} \gamma\Lambda''$, i.e. we obtain a system of the form (1.8), which has the integral curve

$$Y' = \mathrm{H}'(\tau), \quad Z'' = \tilde{\mathrm{H}}''(\tau) \tag{4.15}$$

obtained from the curve $Y = \mathrm{H}(\tau)$ after the power transformation. Here the vector order of the part $\tilde{\mathrm{H}}''(\tau)$ equals $\tilde{P}'' = \alpha P'' < 0$, i.e. the part tends to zero. Since $\mathrm{H}(\tau)$ is a g-asymptotical integral curve, then $\mathrm{H}'(\tau) \to Y'^0$, and all components of the vector $\tilde{\mathrm{H}}''(\tau)$ are different from zero. Thus, the integral curve (4.15) is an h-asymptotical one, and Theorem 1.2 is applicable to it. Theorem 1.2 gives $\tilde{\Lambda}'' = \varkappa\tilde{P}''$, i.e. $\alpha\Lambda'' = \varkappa\alpha P''$, which is equivalent to Equality (1.9).

If vector P'' lies in the boundary of the normal cone $\mathbf{U}''$, i.e. $P'' \in \partial\overline{\mathbf{U}}''$, then we take as vectors $B_{d+1}, \ldots, B_n$ the skeleton of such a closed simplicial cone that lies entirely in $\overline{\mathbf{U}}''$ and contains the vector P''. The further discussion is the same as in the previous case, when $P \in \mathbf{U}''$. Only now $\tilde{P}'' \leq 0$. But $\tilde{\mathrm{H}}''(\tau) \to 0$ according to the property b) of g-asymptotical curves. Indeed, $\tilde{\mathrm{H}}''(\tau) \subset \mathcal{U}(\tilde{\mathbf{T}}'', \varepsilon)$ with every $\varepsilon > 0$ for $\tau \in (\tau_0(\varepsilon), +\infty)$, where $\tilde{\mathbf{T}}'' = \alpha^{*-1}\mathbf{T}'' \subset \mathbb{R}_+^{n-d}$, i.e. the curve $Z'' = \tilde{\mathrm{H}}''(\tau)$ gets into arbitrary neighborhood $|z_j| < \varepsilon$, $j = d+1, \ldots, n$ for sufficiently large τ. Consequently, $\tilde{\mathrm{H}}'' \to 0$ when $\tau \to +\infty$. Hence the curve (4.15) tends to the stationary point, and it is an h-asymptotical one. Theorem 1.2 is applicable to the curve, which completes the proof. □

We point out **four cases of disposition of vectors** $\pm\Lambda''$ with respect to the cone $\mathbf{U}''$.

G1. One of the vectors $\pm\Lambda''$ lies in the normal cone $\mathbf{U}''$; we denote it as $\sigma\Lambda''$.

G2. None of the vectors $\pm\Lambda''$ lies in the closer $\overline{\mathbf{U}}''$ of the normal cone.

G3. One of the vectors $\pm\Lambda''$ lies in the boundary of the normal cone $\mathbf{U}''$.

G4. $\Lambda'' = 0$.

For $\mathbf{U}'' = \{P'' < 0\}$ cases G1–G4 turn into cases H1–H4 respectively.

Theorem 4.3 *In Case* G1 *System* (4.11) *has at least* $(n-d-1)$*-parameter family of g-asymptotical integral curves with the power asymptotics* (4.13).

Proof is analogous to that of Theorem 4.2 for the case $P'' \in \mathbf{U}''$. Only instead of the vector P'' the vector $\sigma\Lambda''$ should be taken. The power transformation $\log Z'' = \gamma \log Y''$ gives a system with the vector $\sigma\tilde{\Lambda}'' = \gamma\sigma\Lambda'' < 0$, i.e. we obtain the system for Y', Z'' of the form (1.8), which relates to Case H1. According to Theorem 1.3 the system has $(n-d-1)$-parameter family of h-asymptotical integral curves. Under the inverse power transformation they turn into g-asymptotical integral curves of System (4.11), that proves the theorem. □

Remark 4.1. Corollary 1.1 and properties 1–5 of h-asymptotical integral curves at the end of Section 1 remain in force for g-asymptotical integral curves of System (4.11) if Cases H1–H4 are replaced by Cases G1–G4 respectively.

In this way, the procedure of search for asymptotics of integral curves of System (4.1) may look like this. For every truncated System (4.2) its own power transformation (2.4) is made as well as change of time (2.8), which are described in Theorem 4.1. In the obtained truncated System (4.12) all stationary points Y'^0 of its first subsystem having different from zero or infinity components $y_1^0, \ldots, y_d^0$ are found. In every such point the values $\lambda_j = \check{f}_j(y_1^0, \ldots, y_d^0)$, $j = d+1, \ldots, n$ are computed, and Theorems 4.3, 4.2 are applied, which in Case G1 give power asymptotics (as integral curves of the truncated system), and in Case G2 indicate their absence. In Cases G3 and G4 it is necessary to investigate further the neighborhood of the stationary point $(Y'^0, 0)$. In Case G1 the asymptotics can also be made more precise. But if we formulate Cases G1–G4 invariantly with respect to the power transformations, then we can obtain an algorithm allowing to find power asymptotics without the power transformations and to isolate cases of the type G3 and G4, which require the power transformations to complete analysis. It is the subject of the next section.

If in System (4.12) the equation $\hat{G}'(Y') = 0$ has a continuous set of solutions Y'^0, then in that set the vector $\hat{G}''(Y'^0)$ may vary continuously. By virtue of equality $\hat{G}''(Y'^0) = \varkappa P$ this leads to the existence of integral curves with continuously varying vector order.

Example 4.1. The system

$$\begin{aligned} d\log y_1/dt_1 &= 0, \\ d\log y_2/dt_1 &= y_1, \\ d\log y_3/dt_1 &= 1 \end{aligned} \tag{4.16}$$

has solutions

$$\begin{aligned} y_1 &= c, \\ y_2 &= \exp ct_1, \\ y_3 &= \exp t_1, \end{aligned}$$

where c is an arbitrary constant. For these solutions the vector order is $P = (0, c, 1) = G$. Here $n = 3$, $d = 1$, $G'(Y') = 0$ and $G'' = (y_1, 1)$. The system $G'(Y') = 0$ has a continuous set of solutions $y_1 = c$, to which there correspond the continuous set of values of the vector $G''(Y') = (c, 1) = P''$. The power transformation

$$\begin{aligned} x_1 &= y_1 y_2, \\ x_2 &= y_2, \\ x_3 &= y_3, \end{aligned}$$

and the change of time $dt_1 = y_2 dt$ transform System (4.16) into the system

$$\begin{aligned} d\log x_1/dt &= x_1, \\ d\log x_2/dt &= x_1, \\ d\log x_3/dt &= x_2, \end{aligned}$$

that turns into the system of the example at the end of [Bruno 1973b], if we put $x = x_2$, $y = x_1$, $z = x_3$. □

5. The power asymptotics

Lemma 5.1 *Let the truncated System* (4.2) *have the integral curve* (4.6), *and also the vector* P *be normal to the face* $\mathbf{\Gamma}_j^{(d)}$ *corresponding to Truncation* (4.2). *Then*

$$\hat{F}(B) = \varkappa P, \quad \varkappa \in \mathbb{R}. \tag{5.1}$$

Proof. Since the vector P is normal to the face $\mathbf{\Gamma}_j^{(d)}$, then according to the property of truncations

$$\hat{F}(b_1\tau^{p_1}, \ldots, b_n\tau^{p_n}) = \hat{F}(B)\tau^c. \tag{5.2}$$

If (4.6) is a trivial integral curve, then $\hat{F}(b_1\tau^{p_1}, \ldots, b_n\tau^{p_n}) = 0$ and $\hat{F}(B) = 0$ according to (5.2), i.e. the formula (5.1) is valid with $\varkappa = 0$.

If (4.6) is a nontrivial integral curve, then since

$$d\log(b_i\tau^{p_i})/dt = p_i d\log\tau/dt, \quad i = 1, \ldots, n$$

according to (5.2) System (4.2) takes the form of the identity $Pd\log\tau/dt \equiv \hat{F}(B)\tau^c$ on this curve. Now $d\log\tau/dt \neq 0$, hence $\tau^{-c}d\log\tau/dt = \text{const} \neq 0$. This gives the formula (5.1) with $\varkappa \neq 0$. □

Corollary 5.1 *In the situation of Lemma 5.1*

$$\langle\hat{F}(B), R_i\rangle = 0, \quad i = 1, \ldots, d, \tag{5.3}$$

where vectors $R_1, \ldots, R_d$ *form the basis of* d*-dimensional linear subspace that is parallel to the face* $\mathbf{\Gamma}_j^{(d)}$.

We remark that the condition of the vector P being normal to the face $\mathbf{\Gamma}_j^{(d)}$ is also written in the form

$$\langle P, R_i\rangle = 0, \quad i = 1, \ldots, d. \tag{5.4}$$

Lemma 5.2 *If for the truncated System* (4.2) *the vectors* P *and* B *satisfy Equalities* (5.1) *and* (5.4), *then* (4.6) *is an integral curve of System* (4.2).

Proof. On the curve (4.6)

$$d\log X/dt = Pd\log\tau/dt \tag{5.5}$$

and $\hat{F}(X) = \hat{F}(B)\tau^c$ by virtue of (5.4). If $\varkappa = 0$ in (5.1), then $\hat{F}(B) = 0$, and (4.6) is a trivial integral curve of System (4.2). If in (5.1) $\varkappa \neq 0$, then on the curve (4.6) System (4.2) is reduced to the single equation

$$d\log\tau/dt = \varkappa\tau^c, \tag{5.5'}$$

connecting τ and t. Consequently, this curve is an integral one for System (4.2). □

Lemma 5.3 *If for the truncated System* (4.2) *the vectors P and B satisfy Equality* (5.1) *with $\varkappa \neq 0$ and Equality* (5.3), *then* (4.6) *is an integral curve of System* (4.2).

Proof. In consequence of Equality (5.3) the vector $\hat{F}(B)$ is normal to the face $\mathbf{\Gamma}_j^{(d)}$. In consequence of Equality (5.1) with $\varkappa \neq 0$ the vector P is also normal to the face $\mathbf{\Gamma}_j^{(d)}$. Hence on the curve (4.6) Equalities (5.2) and (5.5) are valid, i.e. System (4.2) is reduced to the single equation (5.5′) on this curve. Consequently, the curve (4.6) is an integral one for System (4.2). □

Remark 5.1. If on the curve (4.6) Equality (5.1) with $\varkappa \neq 0$ is valid, then one can obtain $|\varkappa| = 1$ by the substitution of parameter $\tau = \tilde{\tau}^{|\varkappa|}$, i.e. $\varkappa = -1$ or $\varkappa = 1$.

Theorem 5.1 *If complete System* (4.1) *has an integral curve* (4.5) *with $P \in \mathbf{U}_j^{(d)}$, then the curve* (4.6) *is an integral one for the corresponding truncated System* (4.2), *i.e. Equality* (5.1) *is valid.*

Conversely, if truncated System (4.2) *with the normal cone $\mathbf{U}_j^{(d)}$ has a nontrivial integral curve* (4.6) *with $P \in \mathbf{U}_j^{(d)}$, then the complete System* (4.1) *has a nontrivial integral curve of the form* (4.5) (*may be not a single one*).

Proof. We begin with the direct assertion. If $\hat{F}(B) = 0$, then by virtue of (5.2) the curve (4.6) is a trivial integral curve of the truncated System (4.2). Equality (5.1) is valid.

If $\hat{F}(B) \neq 0$, then after the power transformation of Theorem 4.1 the pair of Systems (4.1) and (4.2) turns into the pair of Systems (4.11) and (4.12), the curves (4.5) and (4.6) turn into the curves (4.7) and (4.8), the vector $\hat{F}(B)$ turns into the vector $\hat{G}(\tilde{B})$, and the normal cone $\mathbf{U}_j^{(d)}$ turns into the cone $\mathbf{U}'' \stackrel{\text{def}}{=} \tilde{\mathbf{U}}_j^{(d)}$. All the relations between vectors in $\mathbb{R}^n$ and $\mathbb{R}_*^n$ are preserved. Here $\tilde{P}' = 0$, i.e. when $\tau \to +\infty$ the part Y' of the integral curve (4.7) must tend to the constant value $\tilde{B}' = Y'^0$, that has all components different from zero or infinity. Such a value Y'^0 may be only a stationary point of the first subsystem of System (4.12). In this point $\hat{G}''(\tilde{B}') = \Lambda''$, Solution (4.7) is a g-asymptotical one, and $\tilde{P} \in \mathbf{U}''$. According to Theorem 4.2 Equality $\hat{G}(\tilde{B}') = \varkappa\tilde{P}$ is satisfied, that means the fulfillment of Equality (5.1). Since vector P is normal to the face $\mathbf{\Gamma}_j^{(d)}$, then according to Lemma 5.2 the curve (4.6) is an integral one for the truncated System (4.2).

Conversely, if the integral curve (4.6) of the truncated System (4.2) is a nontrivial one, then $\hat{F}(B) \neq 0$, and according to Lemma 5.1 the relation (5.1) with $\varkappa \neq 0$ is satisfied. Consequently, one of the values $\pm\hat{F}(B)$ lies in the normal cone $\mathbf{U}_j^{(d)}$. The transformation of Theorem 4.1 turns System

(4.1) into System (4.3), which has the stationary point $Y'^0 = \tilde{B}'$, since the truncated System (4.4) has such a stationary point as the limit of the integral curve (4.8). System (4.4) in the neighborhood of the point Y'^0 is related to Case G1, since one of the values $\pm\hat{G}''(B) \stackrel{\text{def}}{=} \pm\hat{G}''(B')$ lies in the cone $\tilde{\mathbf{U}}_j^{(d)}$. According to Theorem 4.3, in the neighborhood of point Y'^0 System (4.3) has the family of solutions (4.7), in which $\tilde{b}_{d+1}, \ldots, \tilde{b}_n$ are arbitrary constants. In particular, there is a solution (4.7) with the same values of these constants as in Solution (4.8). Returning to variables X this Solution (4.7) turns into Solution (4.5), that completes the proof. □

Let the vector B with all nonzero components $b_i \neq 0$ satisfy System (5.3) corresponding to the truncated System (4.2). We denote $\Lambda = \hat{F}_j^{(d)}(B)$. We will distinguish **four cases F**:

F1. $\Lambda \neq 0$ and at least one of vectors $\pm\Lambda$ lies in the normal cone $\mathbf{U}_j^{(d)}$ of Truncation (4.2); let it be the vector $\sigma\Lambda$.

F2. $\Lambda \neq 0$ and no one of vectors $\pm\Lambda$ lies in the closure $\overline{\mathbf{U}}_j^{(d)}$ of the normal cone $\mathbf{U}_j^{(d)}$.

F3. $\Lambda \neq 0$ and at least one of vectors $\pm\Lambda$ lies in the boundary of the cone $\mathbf{U}_j^{(d)}$; we denote it as $\sigma\Lambda$.

F4. $\Lambda = 0$.

Now, the procedure of computation of power asymptotics (4.6) of solutions to System (4.1) may be organized in the following way. Firstly, we compute the Newton polyhedron of System (4.1), i.e. all its faces $\mathbf{\Gamma}_j^{(d)}$ and their normal cone $\mathbf{U}_j^{(d)}$ are computed. To each face $\mathbf{\Gamma}_j^{(d)}$ we put in correspondence its truncated System (4.2). For it, System (5.3) is composed, that consists of d equations in n variables $b_1, \ldots, b_n$. For every solution $B = B^0$ to System (5.3) the vector $\Lambda \stackrel{\text{def}}{=} \hat{F}(B^0)$ is computed. Then we decide to which Case F1–F4 the vector Λ belongs. If it belongs to Case F1, then vectors $B = B^0$ and $P = \sigma\Lambda$ give the required asymptotics (4.6), which correspond to Solution (4.5) to the complete System (4.1) according to Theorem 5.1. If vector Λ belongs to Case F2, then this value B^0 does not give power asymptotics. If vector Λ belongs to Case F3, then this pair $B = B^0$ and $P = \sigma\Lambda$ must be memorized, since it may correspond to a solution to the complete System (4.1) with the vector order P and with the logarithmic asymptotics. If $\Lambda = 0$, then this value B^0 is also must be memorized for further analysis.

In fact, the bulk of computations by this procedure can be significantly reduced, if we use the criteria given below. We recall that the truncated System (4.2) has its subspace of coefficients $\mathbf{F}_j^{(d)}$. According to Lemma 3.1 for solutions to the truncated System (4.2) the vector order $P \in \mathbf{F}_j^{(d)}$. Since we are interested only in those solutions, which have $P \in \mathbf{U}_j^{(d)}$ or $P \in \partial\overline{\mathbf{U}}_j^{(d)}$, then the following criteria are valid.

Criterion 1. If $\mathbf{U}_j^{(d)} \cap \mathbf{F}_j^{(d)} = \varnothing$, then Truncation (4.2) has no nontrivial integral curves whose orders lie in the normal cone $\mathbf{U}_j^{(d)}$ of Truncation (4.2).

Criterion 2. If $\partial\overline{\mathbf{U}}_j^{(d)} \cap \mathbf{F}_j^{(d)} = \varnothing$, then Truncation (4.2) has no nontrivial solutions whose orders P lie in $\partial\overline{\mathbf{U}}_j^{(d)}$.

Criterion 3. If System $\hat{F}(B) = 0$ has no solutions with all $b_i \neq 0$, then the truncated System (4.2) has no trivial solutions.

We indicate how to use Criterion 1. Let $\mathbf{S} = \{Q_1, \dots, Q_s\}$, and let the boundary subset $\mathbf{S}_j^{(d)}$ consist of points $Q_1, \dots, Q_l$. Then the subspace $\mathbf{F}_j^{(d)}$ is the linear hull of vectors $F_1, \dots, F_l$, if we denote $F_k \stackrel{\text{def}}{=} F_{Q_k}$. If $P \in \mathbf{F}_j^{(d)}$, then $P = \nu_1 F_1 + \dots + \nu_l F_l$, where $\nu_i \in \mathbb{R}$, and

$$\langle P, Q_i \rangle = \sum_{j=1}^{l} \nu_j \langle F_j, Q_i \rangle, \quad i = 1, \dots, s.$$

We write the condition $P \in \mathbf{U}_j^{(d)}$ in the form

$$\begin{gathered} \sum_{j=1}^{l} \nu_j \langle F_j, Q_1 \rangle = \dots = \sum_{j=1}^{l} \nu_j \langle F_j, Q_l \rangle, \\ \sum_{j=1}^{l} \nu_j \langle F_j, Q_i \rangle < \sum_{j=1}^{l} \nu_j \langle F_j, Q_1 \rangle, \quad i = l+1, \dots, s. \end{gathered} \tag{5.6}$$

Since the scalar products $\langle F_j, Q_i \rangle \stackrel{\text{def}}{=} b_{ij}$ are uniquely determined by System (4.1) and by Truncation (4.2), then relations (5.6) are a system of linear equations and inequalities in $\nu_1, \dots, \nu_l$. If this system has no solutions, then $\mathbf{U}_j^{(d)} \cap \mathbf{F}_j^{(d)} = \varnothing$. In the same way Criterion 2 can be used. To use these criteria it is convenient to compute the matrix $\mathcal{B}$ with elements $b_{ij} = \langle F_j, Q_i \rangle$ for System (4.1) (see Section 3).

If according to the notation (5.5) of Chapter 1 the boundary subset $\mathbf{S}_j^{(d)}$ (i.e. Truncation (4.2)) is defined by the collection of indices $[j_1, \dots, j_l]$, then the vector $P \in \mathbf{U}_j^{(d)} \cap \mathbf{F}_j^{(d)}$ has the form

$$P = \sum_{k=1}^{l} \nu_{j_k} F_{j_k}, \tag{5.6$'$}$$

where $\nu_{j_1}, \dots, \nu_{j_l}$ satisfy the relations

$$\begin{gathered} \sum_k \nu_{j_k} b_{j_1 j_k} = \dots = \sum_k \nu_{j_k} b_{j_l j_k}, \quad k = 1, \dots, l, \\ \sum_k \nu_{j_k} b_{i j_k} < \sum_k \nu_{j_k} b_{j_1 j_k}, \quad 1 \le i \le s, \quad i \neq j_1, \dots, j_l. \end{gathered} \tag{5.7}$$

The system of linear equations in (5.7) comprises $l-1$ homogeneous equations in l variables, hence it always has a solution. Even though System (5.7) has solutions (5.6′), it may happen that instead of System (5.3) it is more convenient to solve the system $\hat{F}(B) = \varkappa P$. It is particularly true for the cases when all solutions (5.6′) to System (5.7) are collinear, i.e. their set $\mathbf{U}_j^{(d)} \cap \mathbf{F}_j^{(d)}$ is one-dimensional.

Example 5.1. For $x_3 \to 0$ or $x_3 \to \infty$ we find the power asymptotics of integral curves of System (2.12) with $a = 1$. Here $n = s = 3$,

$$F = (x_1^{-1}x,\ x_1^{\mu}x_2^{-1}x_3^{\sigma},\ x_3^{-1}).$$

Matrices $\mathcal{Q}$, $\mathcal{F} = E$, $\mathcal{B} = \mathcal{Q}\mathcal{F}^* = \mathcal{Q}$ are

$$\mathcal{Q} = \begin{pmatrix} -1 & 1 & 0 \\ \mu & -1 & \sigma \\ 0 & 0 & -1 \end{pmatrix} = \mathcal{B}, \quad \mathcal{F} = E. \tag{5.8}$$

The Newton polyhedron $\mathbf{\Gamma}$ here is a triangle. It has three vertices Q_1, Q_2, Q_3, three edges, and it is also a face of itself, since it is two-dimensional when $n = 3$. There are altogether 7 faces. According to Criterion 3 no one of truncations has trivial solutions, since a solution B to any equation $f_i(B) = 0$ has at least one zero or infinite component. Therefore Criterion 3 need not be checked here. We check Criterions 1 and 2. Here a truncation will be given by the collection of indices $[j_1, \dots, j_l]$ of those points Q_j which belong to the corresponding boundary subset (see Section 5, Chapter 1).

[1]. $\mathbf{F} = \nu_1 F_1$, $\mathbf{U} = \{P: \langle P, Q_i\rangle < \langle P, Q_1\rangle,\ i = 2, 3\}$, i.e.

$$\begin{aligned} \mathbf{F} \cap \mathbf{U} &= \{P = \nu_1 F_1 : \nu_1\langle F_1, Q_i\rangle < \nu_1\langle F_1, Q_1\rangle,\ i = 2,3\} = \\ &= \{P = \nu_1 F_1 : \nu_1 \mu < -\nu_1,\ 0 < -\nu_1\}, \end{aligned}$$

whence we obtain inequalities $\nu_1 < 0$ and $\nu_1(\mu + 1) < 0$. This system has solutions $\nu_1 < 0$ only when $\mu + 1 > 0$.

For Criterion 2 we obtain two systems:

$$\nu_1 = 0, \quad \nu_1(\mu + 1) < 0;$$

$$\nu_1 < 0, \quad \nu_1(\mu + 1) = 0.$$

The first of them has no solutions, and the second has solutions $\nu_1 < 0$ when $\mu + 1 = 0$.

This truncation does not satisfy us, because it has $P = (\nu_1, 0, 0)$, but we are interested only in solutions which have $x_3 \to 0$ or $x_3 \to \infty$, i.e. $p_3 \neq 0$.

Further we are going to write relations (5.7) through the elements $b_{ij} \stackrel{\text{def}}{=} \langle F_j, Q_i \rangle$ of the matrix $\mathcal{B}$ in (5.8) finding such conditions on ν_1, ν_2, ν_3 that for the chosen truncation

$$\mathbf{F} \cap \mathbf{U} = \{P: P = \nu_1 F_1 + \nu_2 F_2 + \nu_3 F_3\} = \{P = (\nu_1, \nu_2, \nu_3)\}. \tag{5.9}$$

Here we are interested only in truncations that have $p_3 \stackrel{\text{def}}{=} \nu_3 \neq 0$, i.e. the term $F_3 X^{Q_3}$ belongs to the truncation. That means that in the collection of indices j the index 3 must be present. There are altogether $1 + 2 + 1 = 4$ of such boundary subsets. Therefore we discard the truncation [2].

[3]. Relations (5.7) for intersection (5.9) are

$$\nu_3 b_{13} < \nu_3 b_{33}, \quad \nu_3 b_{23} < \nu_3 b_{33},$$

i.e. according to (5.8) $0 < -\nu_3$, $nu_3 \sigma < -\nu_3$. This system has solutions $\nu_3 < 0$ only if $\sigma + 1 > 0$.

For Criterion 2 we obtain two boundary systems:

$$0 = -\nu_3, \quad \nu_3 \sigma < -\nu_3;$$

$$0 < -\nu_3, \quad \nu_3 \sigma = -\nu_3.$$

The first system has no solutions, and the second has solutions

$$\nu_3 < 0 \ \text{ only when } \ \sigma + 1 = 0. \tag{5.10}$$

We discard the truncation [1,2].

[1,3]. Relations (5.7) for intersection (5.9) are

$$\nu_1 b_{11} + \nu_3 b_{13} = \nu_1 b_{31} + \nu_3 b_{33} > \nu_1 b_{21} + \nu_2 b_{23},$$

i.e.

$$-\nu_1 = -\nu_3 > \nu_1 \mu + \nu_3 \sigma \ \text{ or } \ \nu_1 = \nu_3, \ \ 0 > \nu_3(\mu + \sigma + 1).$$

The boundary system for Criterion 2 is

$$-\nu_1 = -\nu_3 = \nu_1 \mu + \nu_3 \sigma.$$

It has solutions

$$\nu_1 = \nu_3 \neq 0 \ \text{ only when } \ \mu + \sigma + 1 = 0. \tag{5.11}$$

Further instead of elements b_{ij} we are going to write their numerical values directly.

[2,3]. Relations (5.7) for intersection (5.9) are $-\nu_2 + \nu_3 \sigma = -\nu_3 > \nu_2$, i.e. $\nu_2 = \nu_3(\sigma + 1)$ and $\nu_3(\sigma + 2) < 0$. The boundary system has solutions

$$\nu_2 = \nu_3(\sigma + 1) \ \text{ only when } \ \sigma + 2 = 0. \tag{5.12}$$

[1,2,3]. Relations (5.7) consist only of equations

$$-\nu_1 + \nu_2 = \nu_1\mu - \nu_2 + \nu_3\sigma = -\nu_3,$$

i.e.

$$\nu_2 = \nu_1(\mu + \sigma + 1)/(\sigma + 2), \quad \nu_3 = \nu_1(1 - \mu)/(\sigma + 2),$$

if $\sigma + 2 \neq 0$.

Now we note that in all cases the intersection (5.9) is one-dimensional. Hence instead of System (5.3) the system

$$\hat{F}(B) = \varkappa P, \quad \varkappa = \pm 1 \tag{5.13}$$

may be solved, where vectors B and P are normalized in such a way that the third equation of System (5.13) be satisfied, i.e. $\hat{f}_3(B) = \varkappa p_3$. For example,

$$b_3 = \varkappa\nu_3, \quad p_3 = \nu_3 = \pm 1. \tag{5.14}$$

[3]. $\hat{F} = (0, 0, x_3^{-1})$, $P = (0, 0, -1)$ when $\sigma + 1 > 0$. Under the normalization (5.14) System (5.13) reduces here to the equation $\varkappa = \varkappa$. We obtain power asymptotics $X = (b_1, b_2, \pm\tau^{-1})$, where b_1, b_2 are arbitrary constants. When $\sigma + 1 = 0$ the same vector B satisfies System (5.13), and the vector P is orthogonal to the edge corresponding to the collection [2,3], i.e. according to classification of Section 5 this is Case F3 with the intermediate truncation [2,3] (see Section 6).

[1,3]. $\hat{F} = (x_1^{-1}x_2, 0, x_3^{-1})$, $P = (\nu_3, 0, \nu_3)$, where $\nu_3(\mu + \sigma + 1) < 0$, i.e. $\nu_3 = -\operatorname{sgn}(\mu + \sigma + 1)$. System (5.13) reduces to the single equation $b_1^{-1}b_2 = \varkappa\nu_3$, i.e. $b_2 = \varkappa\nu_3 b_1$. The power asymptotics are $X = (b_1\tau^{\nu_3}, \pm b_1, \pm\tau^{\nu_3})$, where b_1 is an arbitrary constant. When $\mu + \sigma + 1 = 0$ the same vectors B satisfy System (5.13), and the vector P is orthogonal to the triangle $\mathbf{\Gamma}$, i.e. it is Case F3, where the intermediate truncation is the complete system.

[2,3]. $\hat{F} = (0, x_1^{\mu}x_2^{-1}x_3^{\sigma}, x_3^{-1})$, $P = (0, \nu_3(\sigma + 1), \nu_3)$ when $\nu_3(\sigma + 2) < 0$, i.e. $\nu_3 = -\operatorname{sgn}(\sigma + 2)$. System (5.13) reduces to the equation

$$b_1^{\mu}b_2^{-1}(\varkappa\nu_3)^{\sigma} = \varkappa\nu_3(\sigma + 1), \quad \text{i.e. } b_2 = \varkappa^{\sigma-1}\nu_3^{\sigma-1}b_1^{\mu}/(\sigma + 1).$$

The power asymptotics are

$$X = (b_1, (\pm 1)^{\sigma-1}b_1^{\mu}(\sigma + 1)^{-1}\tau^{\nu_3(\sigma+1)}, \pm\tau^{\nu_3}). \tag{5.15}$$

When $\sigma + 2 = 0$ the same vector B satisfies System (5.13), and the vector P is normal to the triangle $\mathbf{\Gamma}$, i.e. it is Case F3, where the intermediate truncation is the complete system.

[1,2,3]. $\hat{F} = F$, $P = ((\sigma + 2)(1 - \mu)^{-1}\nu_3, (\mu + \sigma + 1)(1 - \mu)^{-1}\nu_3, \nu_3)$. System (5.13) is

$$b_1^{-1}b_2 = \frac{\sigma + 2}{1 - \mu}\nu_3, \quad b_1^{\mu}b_2^{-1}(\varkappa\nu_3)^{\sigma} = \frac{\mu + \sigma + 1}{1 - \mu}\nu_3.$$

Its solution is

$$b_1 = \left[\frac{(\sigma+2)(\mu+\sigma+1)}{(1-\mu)^2(\varkappa\nu_3)^\sigma}\right]^{1/(\alpha-1)}, \quad b_2 = \frac{\sigma+2}{1-\mu} b_1 \nu_3,$$

when

$$\sigma + 2 \neq 0, \quad \mu + \sigma + 1 \neq 0, \quad 1 - \mu \neq 0.$$

The power asymptotics are

$$X = (b_1 \tau^{\sigma+2}, b_2 \tau^{\mu+\sigma+1}, \pm\tau^{1-\mu}). \tag{5.16}$$

Since the truncated system here coincides with the complete one, this asymptotics coincides with the exact solution to the original system. We may presume that in it $\tau > 0$.

Thus, we have found four power asymptotics of integral curves of System (2.12) with $x_3 \to 0$ or $x_3 \to \infty$ and have isolated three Cases F3: (5.10)–(5.12). It should be mentioned that not all values of parameters $\sigma, \mu, \varkappa$ imply that all written asymptotics are real. Therefore in each case the real asymptotics have to be specially selected among the found above depending upon the values of these parameters. □

6. Logarithmic asymptotics

We demonstrate how for integral curves of System (4.1) one can find all asymptotics with multiple logarithms of the form

$$x_i = b_i \tau^{p_i^0} (\log \tau)^{p_i^1} \dots (\log^{(k)} \tau)^{p_i^k}, \quad i = 1, \dots, n, \tag{6.1}$$

where $\log^{(k)} \tau$ means the k-multiple logarithm $\log(\log(\dots(\log \tau)\dots))$, $p_i^j \in \mathbb{R}$, all $b_i \neq 0$, $k > 0$, vectors $P^0 = (p_1^0, \dots, p_n^0) \neq 0$ and $P^k = (p_1^k, \dots, p_n^k) \neq 0$.

We demonstrate how it is done in Case F3 according to classification of Section 5. We consider at first the case with $d = 0$, i.e. the truncated System (4.2) corresponds to the vertex Q_1:

$$d \log X/dt = F_1 X^{Q_1} \stackrel{\text{def}}{=} \widehat{\widehat{F}}(X) \tag{6.2}$$

with the normal cone $\mathbf{U}$, and the vector σF_1 belongs to $\overline{\mathbf{U}} \backslash \mathbf{U} = \partial \overline{\mathbf{U}}$. Since the boundary $\partial \overline{\mathbf{U}}$ of the cone $\overline{\mathbf{U}}$ consists of other normal cones, then the vector σF_1 lies in one of them. We denote it as $\mathbf{U}_k^{(e)}$, where $e > 0$. It corresponds to the *intermediate truncated system*

$$d \log X/dt = \widehat{F}_k^{(e)}(X) \stackrel{\text{def}}{=} \widehat{F}(X). \tag{6.3}$$

We perform now the power transformation (2.4) and the change of time $dt_1 = X^{Q_1}dt$ in such a way that the point Q_1 turned into $R = 0$, the face $\mathbf{\Gamma}_k^{(e)}$ placed in the coordinate subspace of coordinates $r_1, \ldots, r_e$, and the whole support of System (4.1) turned into the set lying in the non-negative orthant $R \geq 0$. Then System (4.1) would turn into System (4.11), Truncation (6.3) into Truncation (4.12), and Truncation (6.2) into the system

$$d \log Y/dt_1 = G_1 \stackrel{\text{def}}{=} \alpha F_1. \tag{6.4}$$

Here d should be understood as e, Y' as vector $(y_1, \ldots, y_e)$, and Y'' as vector $(y_{e+1}, \ldots, y_n)$. The part G_1' of the vector G_1 equals zero; all components of the part G_1'' are different from zero, and they have the same sign. System (6.4) is the truncation of System (4.12) having the normal cone $\mathbf{U} \subset \mathbb{R}^n$. Let $\mathbf{U}'$ be its projection into the subspace $P'' = 0$. Now we are going to make the *logarithmic transformation*

$$z_j = \log |y_j|, \quad j = e+1, \ldots, n \tag{6.5}$$

in System (4.11) with Truncation (4.12). Then truncation (4.12) turns into the system

$$dy_i/dt_1 = \psi_{i0}(Y'), \quad i = 1, \ldots, e, \tag{6.6}$$

$$dz_j/dt_1 = g_{j0}(Y'), \quad j = e+1, \ldots, n, \tag{6.7}$$

and the transformed complete system (4.11) differs from System (6.6), (6.7) by exponentially small with z_j addends in right hand sides, since addends in the right hand sides of Equations (4.11) contain coordinates $y_{e+1}, \ldots, y_n$ with positive powers q, but $y_j^q = \exp(qz_j)$ and $z_j \to -\infty$ when $y_j \to 0$. Now we are going to study System (6.6), (6.7) with the help of the Newton polyhedron. For this we write it in the form

$$d \log Y'/dt_1 = H'(Y'), \tag{6.8}$$

$$d \log z_j/dt_1 = z_j^{-1} \breve{g}_j(y_1, \ldots, y_e), \quad j = e+1, \ldots, n. \tag{6.9}$$

The support of this system lies in $n - e + 1$ linear manifolds

$$\mathbf{M}_0 = \{Q: q_{e+1} = \ldots = q_n = 0\},$$

$$\mathbf{M}_j = \{Q: q_j = -1,\ q_k = 0,\ e < k \leq n,\ k \neq j\}, \quad j = e+1, \ldots, n,$$

and it has $Q' \geq 0$. The cone of the problem here is $\mathbf{K} = \{P' \in \mathbf{U}',\ P'' > 0\}$, since $z_j \to -\infty$ when $y_j \to 0$. Now for System (6.8), (6.9) we compose the Newton polyhedron, and we isolate all its truncations, whose normal cone intersect the cone of the problem $\mathbf{K}$. We recall that in (6.4) the part G_1' of the vector G_1 equals zero, and all components of the part G_1'' are different from zero. Consequently, the truncation of the function $z_j^{-1}\breve{g}_j(Y')$ with respect to the order $P \in \mathbf{K}$ is $z_j^{-1}\breve{g}_j(0) \stackrel{\text{def}}{=} z_j^{-1}g_{1j}$, $j = e+1, \dots, n$. The point $Q' = 0$ is absent in the support of Subsystem (6.8) (since $H'(0) = G_1' = 0$), therefore Subsystem (6.8) has several different truncations for $P \in \mathbf{K}$, but all of them are truncations of Subsystem (6.8) for $P' \in \mathbf{U}'$. Let one of them be

$$d\log Y'/dt_1 = \widehat{H'}(Y'). \tag{6.10}$$

It corresponds to a truncation of System (6.8), (6.9) consisting of Subsystem (6.10) and equations

$$d\log z_j/dt_1 = z_j^{-1}g_{1j}, \quad j = e+1, \dots, n \tag{6.11}$$

and corresponding to a vector order $P \in \mathbf{K}$, where the vector $P' \in \mathbf{U}'$ and P' corresponds to Truncation (6.10). Now we have to obtain power or other integral curves $Y' = \mathrm{H}'(\tau)$ of System (6.10); they correspond to $\log y_j = z_j = g_{1j}t_1$, $j = e+1, \dots, n$, as components of the integral curve of the truncated System (6.10), (6.11). If the integral curve $Y' = \mathrm{H}'(\tau)$ is a nontrivial one, and it relates to Case F1, then it corresponds to the integral curve of the complete System (4.11) having the same asymptotics. If we encounter Case F3 in the truncated System (6.10), then the procedure of computation of logarithmic asymptotics must be continued until we come to the power asymptotics. We note that these computations use only the intermediate truncated System (6.3).

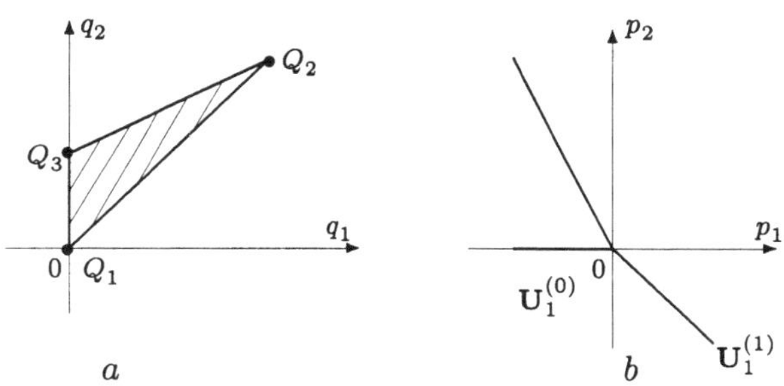

Figure 3.1: Newton's polygon (a) and the normal cones (b) for System (6.12).

Example 6.1. For $n = 2$ and a natural number k, we consider the system

$$\begin{aligned} \dot{x}_1 &= x_1 + a_1 x_1^{k+1} x_2^k, \\ \dot{x}_2 &= -x_2 + a_2 x_1^k x_2^{k+1} + x_2^2. \end{aligned} \tag{6.12}$$

Its support consists of three points: $Q_1 = 0$, $Q_2 = (k, k)$ and $Q_3 = (0, 1)$.

Fig. 3.1 shows the Newton polygon and the normal cones of faces. The normal cone of the vertex $\mathbf{\Gamma}_1^{(0)} = Q_1$ is $\mathbf{U}_1^{(0)} = \{P : p_2 < 0,\, p_1 < -p_2\}$. To the point Q_1 there corresponds the vector coefficient $F_1 = (1, -1)$ lying in the boundary of the cone $\overline{\mathbf{U}}_1^{(0)}$. This part of the boundary of the cone $\overline{\mathbf{U}}_1^{(0)}$ is the normal cone $\mathbf{U}_1^{(1)}$ of the edge $\mathbf{\Gamma}_1^{(1)} \supset \{Q_1, Q_2\}$. Thus, we have Case F3 with $d = 0$, $e = 1$, where the final truncated System (6.2) is

$$\dot{x}_1 = x_1, \quad \dot{x}_2 = -x_2,$$

the intermediate truncated System (6.3) is

$$\dot{x}_1 = x_1 + a_1 x_1^{k+1} x_2^k, \quad \dot{x}_2 = -x_2 + a_2 x_1^k x_2^{k+1}. \tag{6.13}$$

After the power transformation $y_1 = x_1 x_2$, $y_2 = x_2$ System (6.13) turns into the system

$$(\log y_1)^{\cdot} = a y_1^k, \quad (\log y_2)^{\cdot} = -1 + a_2 y_1^k, \tag{6.14}$$

where $a = a_1 + a_2$. After the logarithmic transformation $z_2 = \log y_2$ we obtain the system

$$(\log y_1)^{\cdot} = a y_1^k, \quad (\log z_2)^{\cdot} = z_2^{-1}(-1 + a_2 y_1). \tag{6.15}$$

Fig. 3.2 shows Newton polygons of Systems (6.14) and (6.15).

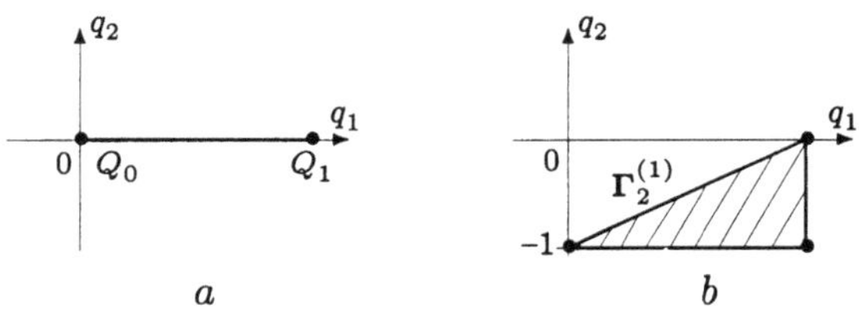

Figure 3.2: Newton's polygons for System (6.14) (a) and System (6.15) (b).

Fig. 3.2, b demonstrates that System (6.15) has a truncation corresponding to the slanting edge $\mathbf{\Gamma}_2^{(1)}$:

$$(\log y_1)^{\cdot} = a y_1^k, \quad (\log z_2)^{\cdot} = -z_2^{-1},$$

whence we obtain the nontrivial power integral curve $z_2 = (aky_1^k)^{-1}$. Since $z_2 \to -\infty$ when $y_2 \to 0$, then this curve is real only in the following cases:

$$\begin{aligned} &\text{when} \quad \operatorname{sgn} y_1 = \operatorname{sgn} a \quad \text{and} \quad k \quad \text{is} \quad \text{odd;} \\ &\text{when} \quad a < 0 \quad \text{and} \quad k \quad \text{is} \quad \text{even.} \end{aligned} \tag{6.16}$$

Returning to y_2 in cases (6.16) we obtain the logarithmic asymptotics of integral curves of System (6.14)

$$y_1 = (a^{-1}k^{-1}\log|\tau|)^{-1/k}, \quad y_2 = \pm\tau^{-1}, \quad \tau \to +\infty.$$

That gives the asymptotics of integral curves of System (6.12)

$$x_1 = \pm\tau(a^{-1}k^{-1}\log|\tau|)^{-1/k}, \quad x_2 = \pm\tau^{-1}, \quad \tau \to +\infty,$$

with the order $P = (1, -1) \in \mathbf{U}_1^{(1)}$. The asymptotics exist not always but only in cases (6.16) (compare with [Bruno 1979a, Chapter II, Subsection 1.11, c)]). □

We consider now Case F3 for the truncated System (4.2) with $d > 0$ and with the normal cone $\mathbf{U}_j^{(d)}$. The vector B with all nonzero components satisfies System (5.3), and one of the vectors $\pm\widehat{F}(B)$ (denoted as $\sigma\widehat{F}(B)$) lies in the border $\partial\overline{\mathbf{U}}_j^{(d)}$ of the cone $\overline{\mathbf{U}}_j^{(d)}$. That means that the vector $\sigma\widehat{F}(B)$ lies in a normal cone $\mathbf{U}_k^{(e)}$. It corresponds to the truncated System (6.3). Thus, we have the complete System (4.1)

$$d\log X/dt = F(X), \tag{6.16$'$}$$

its *intermediate truncation* (6.3)

$$d\log X/dt = \widehat{F}(X) \tag{6.17}$$

with the normal cone $\mathbf{U}_k^{(e)}$, and also an initial (or final) Truncation (4.2) with the normal cone $\mathbf{U}_j^{(d)}$, that we denote as

$$d\log X/dt = \widehat{\widehat{F}}(X), \tag{6.18}$$

since it is also a truncation of the truncated System (6.17). Here the cone $\mathbf{U}_k^{(e)}$ lies in the border of the cone $\overline{\mathbf{U}}_j^{(d)}$.

Now we split each n-vector in three parts: d-vector, $(e-d)$-vector and $(n-e)$-vector. For example, $Y = (Y', Y'', Y''')$, where $Y' = (y_1, \dots, y_d)$, $Y'' = (y_{d+1}, \dots, y_e)$, $Y''' = (y_{e+1}, \dots, y_n)$. Now we make the power transformation (2.4) and the change of time (2.8) in such a way that Systems (6.16′), (6.17), (6.18) were transformed into systems

$$d\log Y/dt_1 = G(Y), \tag{6.19}$$

$$d\log Y/dt_1 = \widehat{G}(Y', Y''), \tag{6.20}$$

$$d\log Y/dt_1 = \widehat{\widehat{G}}(Y') \tag{6.21}$$

respectively, where $\widehat{G}(Y', Y'') = G(Y', Y'', 0)$ and $\widehat{\widehat{G}}(Y') = \widehat{G}(Y', 0)$ with the normal cones $\mathbf{U}' = \alpha\mathbf{U}_k^{(e)} \subset \{P: P' = 0, P'' = 0\}$ and $\mathbf{U}'' = \alpha\mathbf{U}_j^{(d)} \subset \{P: P' = 0\}$ respectively. The support of System (6.19) lies in the non-negative orthant $Q \geq 0$. The vector $\tilde{B} = B^\alpha$. The vector $\Lambda = \widehat{\widehat{G}}(\tilde{B}')$ has $\Lambda' = 0$, $\Lambda'' = 0$, and the part Λ''' has all components different from zero and of the same sign. After the parallel translation

$$V' = Y' - \tilde{B}', \quad V'' = Y'', \quad V''' = Y'''$$

we obtain the complete System (6.19) and its truncations (6.20) and (6.21) in the form

$$d \log V/dt_1 = H(V), \tag{6.22}$$

$$d \log V/dt_1 = \widehat{H}(V', V''), \tag{6.23}$$

$$d \log V/dt_1 = \widehat{\widehat{H}}(V') \tag{6.24}$$

respectively. We have to investigate them near the point $V' = 0$, $V'' = 0$, $V''' = 0$, where $\widehat{H} = \Lambda$.

Let $\lambda_1, \ldots, \lambda_d$ be eigenvalues of the linear part of the subsystem $d \log V'/dt_1 = \widehat{\widehat{H}}'(V')$ of System (6.24).

If all $\lambda_1 = \ldots = \lambda_d = 0$, then we obtain the situation analogous to that we have studied in the case $d = 0$. Namely, System (6.23) has the form (4.12), and it can be studied with the help of the logarithmic transformation (6.5). The difference is only that now $V' \to 0$, and some of the components of the vector $V'(\tau)$ may turn to zero. Therefore, the cone of the problem with $P' < 0$ must be taken into account, while we isolate truncations of the vector function $\widehat{H}'(V', V'')$, $\widehat{H}''(V', V'')$.

Let now not all $\lambda_1, \ldots, \lambda_d$ be equal to zero. Let $\lambda_1, \ldots, \lambda_l \neq 0$ and $\lambda_{l+1} = \ldots = \lambda_d = 0$ for the sake of definiteness. Then the subsystem

$$\begin{aligned} d \log V'/dt_1 &= \widehat{H}'(V', V''), \\ d \log V''/dt_1 &= \widehat{H}''(V', V'') \end{aligned} \tag{6.25}$$

of System (6.23) has the *zero invariant manifold* $\widehat{M}$:

$$v_k = \zeta_k(v_{l+1}, \ldots, v_d, V''), \quad k = 1, \ldots, l \tag{6.26}$$

corresponding to zero eigenvalues. Generally speaking, it is not unique, but it has the unique Taylor series in analytical System (6.25). System (6.24) then has the invariant manifold $\widehat{\widehat{M}}$:

$$v_k = \zeta_k(v_{l+1}, \ldots, v_d, 0), \quad k = 1, \ldots, l.$$

If besides

$$\lambda \neq \langle \Lambda''', Q''' \rangle \quad \text{for} \quad Q''' \geq 0, \;\; Q''' \in \mathbb{Z}^{n-e}, \;\; k = 1, \ldots, l,$$

then System (6.22) has the invariant manifold M:

$$v_k = \tilde{\zeta}_k(v_{l+1}, \ldots, v_d, V'', V'''), \quad k = 1, \ldots, l,$$

where $\tilde{\zeta}_k \equiv \zeta_k$ for $V''' = 0$ [Bruno 1971, § 10]. As a result of the reduction of Systems (6.22), (6.23), (6.24) on manifolds M, $\widehat{M}$, $\widehat{\widehat{M}}$ respectively we obtain the systems

$$\begin{aligned} d \log W/dt_1 &= J(W), \\ d \log W/dt_1 &= \widehat{J}(W', V''), \\ d \log W/dt_1 &= \widehat{\widehat{J}}(W') \end{aligned} \tag{6.27}$$

respectively, where $W' = (v_{l+1}, \ldots, v_d)$, $W = (W', V'', V''')$. These are the same triple of systems as the triple of Systems (6.22)–(6.24) in the case $\lambda_1 = \ldots = \lambda_d = 0$, and the approach mentioned for the case is applicable to Systems (6.27). Although the invariant manifold is given by an infinite power series, the coefficients of the initial terms in Systems (6.27) depend only on the coefficients of the initial terms in the expansions (6.26), i.e. on the coefficients of the initial terms of System (6.25). The computations here are analogous to that of the coefficients of the normal forms [Bruno 1971, 1979a, 1995d].

Remark 6.1. In all cases the logarithmic asymptotics of integral curves are determined from the intermediate truncated System (6.20), more exactly by the system

$$\begin{aligned} d \log Y'/dt &= \widehat{G}'(Y', Y''), \\ d \log Y''/dt &= \widehat{G}''(Y', Y''), \\ d \log Y'''/dt &= \Lambda''' \end{aligned}$$

obtained from System (6.20). But System (6.20) corresponds to the intermediate truncated System (6.3). Consequently, to find the logarithmic asymptotics the power transformation should be done only for the corresponding truncated system.

Example 6.2. We consider the case with $n = 3$, $d = 1$, $e = 2$ for the system of the type (6.20) directly, that is obtained after the power transformation. Such is the system

$$\dot{y}_1 = y_1(y_1 - 1 - y_2), \quad \dot{y}_2 = y_2^2, \quad \dot{y}_3 = -y_3. \tag{6.28}$$

Here $Y' = y_1$, $Y'' = y_2$ and $Y''' = y_3$. The final truncated System (6.21) here is

$$\dot{y}_1 = y_1(y_1 - 1), \quad \dot{y}_2 = y_2^2, \quad \dot{y}_3 = -y_3. \tag{6.29}$$

Its subsystem $d\log Y'/dt = \widehat{\widehat{G}}'(Y)$ corresponding to the first equation of System (6.29) has the stationary point $Y'^0 = y_1 = 1$, where $\widehat{\widehat{G}}(Y'^0) = (0,0,1)$ and the eigenvalue $\lambda_1 = 1$. After the substitution $v_1 = y_1 - 1$ System (6.28) takes the form

$$\dot{v}_1 = (v_1 + 1)(v_1 - y_2), \quad \dot{y}_2 = y_2^2, \quad \dot{y}_3 = -y_3. \tag{6.30}$$

The subsystem of its first two equations has zero manifold $v_1 = y_2/(1 - y_2)$; on it System (6.30) induces the system $\dot{y}_2 = y_2^2$, $\dot{y}_3 = -y_3$, which turns into System (6.14) with $k = 1$, $a = 1$ and $a_2 = 0$ under the substitution $y_2, y_3 \to y_1, y_2$, i.e. it gives the logarithmic asymptotics

$$y_1 = 1, \quad y_2 = (\log|\tau|)^{-1}, \quad y_3 = \pm\tau^{-1}, \quad \tau \to +\infty. \square$$

Earlier we have analyzed the first step of the procedure, that reduces the search of the logarithmic asymptotics to the search of the power integral curves of the truncated system of the form (6.8), (6.9), i.e. of subsystems of System (6.8). But here again Cases F3 may appear, they must be subjected anew to the same step of the procedure, which leads to appearance of asymptotics with double logarithm. But with one step of the procedure we get a $(n - d - k)$-dimensional system from a n-dimensional one. Here $0 < d < n$, $0 < d < e < n$, $0 \le k \le e - d$, i.e. $0 < d + k \le e < n$. Therefore in order to get one autonomous equation one can make no more then $n - 1$ such steps in a system of the order n.

Thus, we have proved

Theorem 6.1 *The multiplicity k of logarithms in logarithmic asymptotics (6.1) of integral curves of a system of the order n is not greater then $n - 1$.*

Example 6.3. We demonstrate that it is possible to obtain logarithm of multiplicity $n - 1$ in the asymptotics (6.1) if $d = 0$ and d' is maximal in every step. The system

$$(\log x_i)^{\cdot} = \prod_{j=i+1}^{n} x_j, \quad i = 1, \ldots, n-1,$$
$$(\log x_n)^{\cdot} = 1$$

has an integral curve

$$x_{i+1} = \log x_i, \quad i = 1, \ldots, n-1,$$

i.e.

$$x_1 = \tau, \quad x_i = \log^{(i-1)}\tau, \quad i = 2, \ldots, n, \quad \tau \to +\infty. \square$$

Example 6.4. (the continuation of Example 5.1). We find logarithmic asymptotics in Cases F3 of Example 5.1, i.e. in Cases (5.10)–(5.12), when the final truncations are [3], [1,3] and [2,3], and the intermediate truncations are [2,3], [1,2,3] and [1,2,3] respectively. Since the study of these cases requires the power transformations to be made, it is desirable to choose them in such a way that they would be suitable for several of these cases at once. Here it is sufficient to use one transformation, that translates the point Q_3 into the origin, and the vectors $Q_1 - Q_3$ and $Q_2 - Q_3$ into the unit ones. It is just the transformation that was made in Example 2.1. Therefore it is possible now to use System (2.14).

We consider the case (5.10) with $\sigma + 1 = 0$. In this case $\widehat{\widehat{G}} = (1, 0, 1)$ and $\widehat{G}$ is obtained from G when $y_1 = 0$, i.e. the intermediate truncated system is

$$\begin{aligned} d\log y_1/dt_1 &= 1 + y_2, \\ d\log y_2/dt_1 &= -y_2, \\ d\log y_3/dt_1 &= 1. \end{aligned} \tag{6.31}$$

After the logarithmic transformation $z_1 = \log y_1$, $z_3 = \log y_3$ System (6.31) is

$$\begin{aligned} d\log z_1/dt_1 &= z_1^{-1}(1 + y_2), \\ d\log y_2/dt_1 &= -y_2, \\ d\log z_3/dt_1 &= z_3^{-1}. \end{aligned}$$

Its support consists of four points. To the vector $P = (1, -1, 1)$ there corresponds the truncation

$$\begin{aligned} d\log z_1/dt_1 &= z_1^{-1}, \\ d\log y_2/dt_1 &= -y_2, \\ d\log z_3/dt_1 &= z_3^{-1}, \end{aligned}$$

which has the power solution $z_1 = y_2^{-1} = z_3^{-1}$. To it there correspond the logarithmic asymptotics of integral curves of System (2.14) $Y = (y_3, (\log|y_3|)^{-1}, y_3)$ and of System (2.12)

$$x_1 = x_2 = (\log|x_3|)^{1/(\mu-1)} \tag{6.32}$$

by virtue of (2.13).

In Case (5.11) we have $\sigma + \mu + 1 = 0$, and the final truncation is [1,3], i.e. $\widehat{\widehat{G}}$ is obtained from G in (2.14) when $y_2 = 0$. Then the first equation of System (2.14) has the stationary point $y_1 = 1$ with the eigenvalue $\lambda_1 = -1$.

After the substitution $y_1 = 1 + v_1$ the system of the first two equations (2.14) takes the form

$$
\begin{aligned}
dv_1/dt_1 &= (1+v_1)(-v_1+y_2) = -v_1 + y_2 - v_1^2 + v_1 y_2,\\
d\log y_2/dt_1 &= \mu v_1 - y_2.
\end{aligned}
$$

It has the zero integral curve $v_1 = y_2 + \ldots$, on which it induces the equation $d\log y_2/dt_1 = (\mu - 1)y_2 + \ldots$ This equation together with the equation $d\log z_3/dt_1 = z_3^{-1}$ form the system having for $P = (-1, 1)$ the truncation

$$
\begin{aligned}
d\log y_2/dt_1 &= (\mu - 1)y_2,\\
d\log z_3/dt_1 &= z_3^{-1}
\end{aligned}
$$

with the power integral curve $y_2 = (1-\mu)^{-1} z_3^{-1}$, to which there correspond the asymptotics of integral curves of System (2.14)

$$
Y = (1, (1-\mu)^{-1}(\log|y_3|)^{-1}, y_3)
$$

and of System (2.12)

$$
\begin{aligned}
x_1 &= x_3[(1-\mu)\log|x_3|]^{1/(1-\mu)},\\
x_2 &= [(1-\mu)\log|x_3|]^{1/(1-\mu)}.
\end{aligned}
\tag{6.33}
$$

In Case (5.12) we have $\sigma + 2 = 0$, and the final truncation is [2,3], i.e. $\widehat{\widehat{G}}$ is obtained from G in (2.14) when $y_1 = 0$. Then the second equation of System (2.14) has the stationary point $y_2 = -1$ with the eigenvalue $\lambda_2 = 1$. After the substitution $y_2 = -1 + v_2$ the system of the first two equations (2.14) is

$$
\begin{aligned}
d\log y_1/dt_1 &= -y_1 + v_2,\\
dv_2/dt_1 &= (-1+v_2)(\mu y_1 - v_2).
\end{aligned}
$$

It has the zero invariant manifold, that is the curve $v_2 = \mu y_1 + O(y_1^2)$, on which it induces the equation $d\log y_1/dt_1 = (\mu-1)y_1 + O(y_1^2)$. Together with the third equation $d\log z_3/dt_1 = z_3^{-1}$ it gives the same asymptotics as in the previous case, i.e. the asimptotics

$$
Y = ((1-\mu)^{-1}(\log|y_3|)^{-1}, -1, y_3)
$$

for System (2.14) and the asymptotics

$$
\begin{aligned}
x_1 &= (-1)^{1/(1-\mu)}[(1-\mu)\log|x_3|]^{1/(1-\mu)},\\
x_2 &= (-1)^{1/(1-\mu)} x_3^{-1}[(1-\mu)\log|x_3|]^{\mu/(1-\mu)}
\end{aligned}
\tag{6.34}
$$

for System (2.12) by virtue of (2.13).

Thus, for integral curves of System (2.12) with $a = 1$ we have obtained three logarithmic asymptotics: (6.32), (6.33) and (6.34), when $\sigma + 1 = 0$, $\sigma + \mu + 1 = 0$ and $\sigma + 2 = 0$ respectively. The power asymptotics (5.15) turns into (6.32) when $\sigma + 1 = 0$, and power asymptotics (5.16) turns into (6.33) and (6.34) when $\sigma + \mu + 1 = 0$ and $\sigma + 2 = 0$ respectively. □

From discussion of this Section it follows:

Theorem 6.2 *The complete system* (6.16′) *has an integral curve with the logarithmic asymptotics* (6.1) *if and only if Truncation* (6.17) *of System* (6.16′) *with respect to the order* $P^0 = (p_1^0, \ldots, p_n^0)$ *has an integral curve with the same asymptotics* (6.1).

7. The simplex systems

If the number s of points of the support $\mathbf{S}$ in System (4.1) is greater by one then its dimension, i.e.

$$s = d + 1, \tag{7.1}$$

then all points Q_j are vertices, and the Newton polyhedron $\mathbf{\Gamma}$ is a d-dimensional simplex (see the end of Section 1, Chapter 1). Here the order of the system is $n \geq d$. We call such a System (4.1) the *simplex* one. It is an important class of systems. Thus, in all examples of this Chapter except Example 2.2 the simplex systems are considered. Let $Q_1, \ldots, Q_s$ be all the points of the support of the simplex System (4.1). We put $Q_i' = Q_i - Q_1$, $i = 1, \ldots, s$, and in System (4.1) we make the transformation

$$\begin{aligned} dt_1 &= X^{Q_1} dt, \quad y_i = X^{Q_i'}, \quad i = 1, \ldots, s, \\ y_j &= X^{R_j}, \quad j = s + 1, \ldots, n + 1, \end{aligned} \tag{7.2}$$

where vectors $R_{s+1}, \ldots, R_{n+1}$ complement the vectors $Q_2', \ldots, Q_s'$ to the basis in $\mathbb{R}^n$. Here $y_1 \equiv 1$. If $d = n$, i.e. $s = n+1$, then vectors R_j are not necessary, and the coordinates y_j are absent. System (4.1) takes the form

$$d \log y_i / dt_1 = \sum_{j=1}^{s} b_{ij} y_j, \quad i = 1, \ldots, n + 1, \tag{7.3}$$

where all $b_{1j} = 0$. That system has an independent subsystem of the first s equations, one of which (the first) is trivial. According to the notation of Section 3 we write System (7.3) in the vector form

$$d \log Y / dt_1 = \mathcal{B} Y', \tag{7.4}$$

where the rectangular $((n+1)\times s)$-matrix $\mathcal{B} = (b_{ij})$, $Y = (y_1, \dots, y_{n+1})$, $Y' = (y_1, \dots, y_s)$. For each $k = 2, \dots, s$ the transformation

$$\log z_i = \log y_i - \log y_k, \quad i = 1, \dots, n+1, \quad dt_k = y_k dt_1 \tag{7.5}$$

transforms System (7.4) into the system

$$d\log Z/dt_k = \mathcal{B}^{(k)} Z, \tag{7.6}$$

where $\mathcal{B}^{(k)} = (b_{ij}^{(k)}) = (b_{ij} - b_{kj})$, i.e. the ith line of the matrix $\mathcal{B}^{(k)}$ is the remainder of ith and kth lines of the matrix $\mathcal{B}$. It may be written in the form $\mathcal{B}^{(k)} = (I - I_k)\mathcal{B}$, where I is the unit $(n+1)$-matrix, and I_k has units in the kth column and other elements being zeros. In this way, we obtain $s-1$ more Systems (7.6) analogous to System (7.4), which are equivalent to System (7.4) and the original System (4.1). System (7.6) is obtained from System (4.1) directly, if at first we make the power change of time (2.8) with $T = Q_k$ and then the transformation (7.2), where $Q_i' = Q_i - Q_k$.

System (7.3), (7.4) has *the basic* independent *subsystem* of the first s equations

$$d\log Y'/dt_1 = \mathcal{B}'Y', \tag{7.7}$$

where $Y' = (y_1, \dots, y_s)$ and $\mathcal{B}'$ is the square s-matrix composed of the first s lines of the matrix $\mathcal{B}$. Here the first equation of System (7.7) is trivial: $d\log y_1/dt_1 = 0$, and coordinates $y_{s+1}, \dots, y_{n+1}$ are obtained from $Y'(t)$ by quadratures. The transformation (7.5) transforms System (7.7) into the basic subsystem

$$d\log Z'/dt_k = \mathcal{B}'^{(k)} Z'$$

of System (7.6). Here $\mathcal{B}'^{(k)} = (I - I_k)\mathcal{B}'$ and $z_k \equiv 1$. When the search of asymptotics of solutions to the original System (4.1) is under way all the stationary points of the basic subsystem (7.7) should be taken into account, including those where all $y_2, \dots, y_s$ are different from zero.

When $s = 1$ System (7.7) is trivial: $d\log y_1/dt = 0$, $y_1 \equiv 1$.

When $s = 2$ System (7.7) is the equation $d\log y_2/dt_1 = b_{21} + b_{22}y_2$, that is solved in quadratures:

$$\int \frac{dy_2}{y_2(b_{21} + b_{22}y_2)} = dt_1.$$

When $s = 3$ System (7.7) has the form

$$\dot{x}_i = x_i(a_i + b_i x_1 + c_i x_2), \quad i = 1, 2. \tag{7.8}$$

We introduce the coordinate $x_0 \equiv 1$. Then for $k = 1$ the transformation (7.5) is

$$\log y_0 = -\log x_1, \quad \log y_2 = \log x_2 - \log x_1, \quad dt_1 = x_1 dt.$$

It transforms System (7.8) into the system

$$\begin{aligned} d\log y_0/dt_1 &= -a_1 y_0 - b_1 - c_1 y_2, \\ d\log y_2/dt_1 &= (a_2 - a_1)y_0 + b_2 - b_1 + (c_2 - c_1)y_2. \end{aligned} \tag{7.9}$$

When $k = 2$ the transformation (7.5) is

$$\log z_0 = -\log x_2, \quad \log z_1 = \log x_1 - \log x_2, \quad dt_2 = x_2 dt.$$

It transforms System (7.8) into the system

$$\begin{aligned} d\log z_0/dt_2 &= -a_2 z_0 - b_2 z_1 - c_2, \\ d\log z_1/dt_2 &= (a_1 - a_2)z_0 + (b_1 - b_2)z_1 + c_1 - c_2. \end{aligned} \tag{7.10}$$

System (7.8) was studied in [Coppel 1966, Reyn 1987, Worz–Buseckros 1993]. It has a simple first integral if

$$\omega \overset{\text{def}}{=} b_1 a_2 (c_1 - c_2) + a_1 c_2 (b_2 - b_1) = 0. \tag{7.11}$$

From Systems (7.9) and (7.10), it is easy to see that this condition is invariant under the transformations of the form (7.5). Bautin [1954] had proved that System (7.8) had no limit cycles.

Example 7.1. We isolate the cases of integrability of the Emden–Fowler equation $\ddot{x} = at^\sigma x^\mu$ (see Examples 2.1, 5.1 and 6.4). According to Example 2.1 this equation is equivalent to System (2.14) with the basic subsystem of the first two equations. According to (7.8) and (7.11) for this subsystem $\omega = -a(\mu + 2\sigma + 3)$. Consequently, the Emden–Fowler equation is integrable explicitly under the condition $\mu + 2\sigma + 3 = 0$. Other cases of its integrability:

$$\sigma = 0; \quad \mu + \sigma + 3 = 0; \quad \mu = 2,\ \sigma = -15/7 \text{ and } \sigma = -20/7$$

(see [Berkovich 1998]). But in these cases the corresponding System (7.8) has no autonomous first integral. □

Example 7.2. According to (7.8) and (7.11), we have $\omega = 4 \neq 0$ for System (3.11), i.e. the Blasius equation has no simple first integral. □

Example 7.3. We find some asymptotics of solutions to the equation

$$\dddot{x} = x^3, \tag{7.12}$$

reducing it to the system of the form (7.8). Assuming $x_1 = x$ we write Equation (7.12) as an autonomous system

$$\begin{aligned} \dot{x}_1 &= x_2 = x_1(x_1^{-1}x_2), \\ \dot{x}_2 &= x_3 = x_2(x_2^{-1}x_3), \\ \dot{x}_3 &= x_1^3 = x_3^{-1}(x_1^3 x_3^{-1}). \end{aligned} \tag{7.13}$$

The support $\mathbf{S}$ of this system consists of three points:

$$Q_1 = (-1, 1, 0), \quad Q_2 = (0, -1, 1), \quad Q_3 = (3, 0, -1);$$

and the dimension of the system equals two, i.e. the system is a simplex one. We put $T = Q_1$,

$$U_1 = Q_2 - Q_1 = (1, -2, 1),$$
$$U_2 = Q_3 - Q_1 = (4, -1, -1).$$

In order to select the third vector U_3, we compute the vector product $[U_1, U_2]$, or the minors of the matrix

$$\begin{pmatrix} 1 & -2 & 1 \\ 4 & -1 & -1 \end{pmatrix},$$

with corresponding signs. We obtain $[U_1, U_2] = (3, 5, 7)$. Therefore the vector $U_3 = (1, 1, -1)$ gives the matrix $(U_1 U_2 U_3)$ with the determinant one, since it equals to $\langle [U_1, U_2], U_3 \rangle$. The power transformation (2.4) and the change of time (2.8), i.e.

$$\begin{aligned} y_1 &= x_1 x_2^{-2} x_3, & x_1 &= y_1^2 y_2^{-1} y_3^3, \\ y_2 &= x_1^4 x_2^{-1} x_3^{-1}, & x_2 &= y_1^3 y_2^{-2} y_3^5, \\ y_3 &= x_1 x_2 x_3^{-1}, & x_3 &= y_1^5 y_2^{-3} y_3^{-7}, \end{aligned} \tag{7.14}$$

$dt_1 = x_1^{-1} x_2 dt$, transform System (7.13) into the system

$$\begin{aligned} d \log y_1 / dt_1 &= 1 - 2y_1 + y_2, \\ d \log y_2 / dt_1 &= 4 - y_1 - y_2, \\ d \log y_3 / dt_1 &= 1 + y_1 - y_2. \end{aligned} \tag{7.15}$$

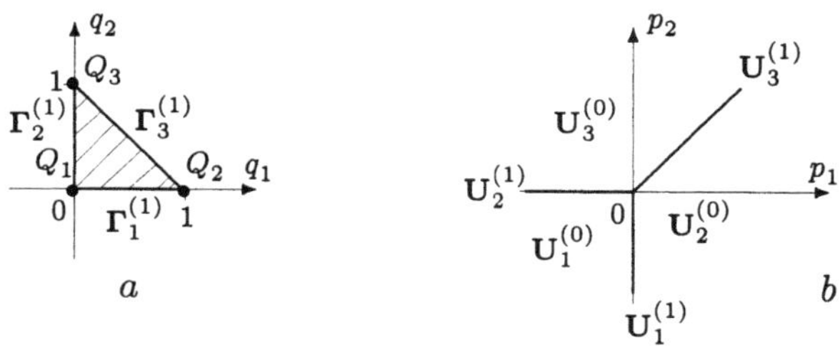

Figure 3.3: Newton polygon (a) and the normal cones (b) for System (7.15).

Fig. 3.3 shows the Newton polygon $\mathbf{\Gamma}$ of this system and projections of its normal cones in the plane p_1, p_2.

The subsystem of the first two equations of System (7.15) has four stationary points: (0,0), (1/2,0), (0,4), (5/3,7/3), and three stationary points in infinity: $y/x = 0$, $1/2$, ∞. There are some integral curves of the subsystem tending to each of them. To compute such integral curves we have to consider the neighborhood of the point in corresponding local coordinates. We consider the stationary point (5/3,7/3). At the point the right hand part of the third equation (7.15) equals 1/3. Therefore it corresponds to solutions

$$y_1 = 5/3, \quad y_2 = 7/3, \quad y_3 = c\exp(1/3)t_1,$$

where c is an arbitrary constant. According to (7.14) we obtain

$$x_1 = \frac{5^2}{3\cdot 7}c^3 \exp t_1, \quad x_2 = \frac{5^3}{3\cdot 7^2}c^5 \exp\frac{5}{3}t_1, \quad x_3 = \frac{5^5}{3^2 7^3}c^7 \exp\frac{7}{3}t_1.$$

Since

$$dt = x_1 x_2^{-1}\,dt_1 = \frac{7}{5}c^{-2}\exp -\frac{2}{3}t_1\,dt_1,$$

then

$$t = -\frac{3\cdot 7}{2\cdot 5}c^{-2}\exp -\frac{2}{3}t_1, \quad \exp t_1 = \left[-\frac{10}{21}c^2 t\right]^{-3/2},$$

where $t < 0$, and

$$x_1 = \frac{5^2}{3\cdot 7}\left[-\frac{10}{21}t\right]^{-3/2}, \quad x_2 = \frac{5^3}{3\cdot 7^2}\left[-\frac{10}{21}t\right]^{-5/2},$$
$$x_3 = \frac{5^5}{3^2 7^3}\left[-\frac{10}{21}t\right]^{-7/2}. \tag{7.16}$$

It is the exact solution to System (7.13) and to Equation (7.12).

In order to obtain more information about solutions to System (7.13), it is necessary to investigate in more detail the solutions to the basic subsystem formed by the first two equations (7.15), first of all near the stationary point $(5/3, 7/3)$. The matrix of the linear part of this subsystem is

$$\begin{pmatrix} -10/3 & 5/3 \\ -7/3 & -7/3 \end{pmatrix}$$

with the characteristic equation $\lambda^2 + (17/3)\lambda + 35/3 = 0$ and eigenvalues

$$\lambda_{1,2} = \frac{-17 \pm \sqrt{17^2 - 4\cdot 105}}{6} = \frac{-17 \pm \sqrt{-131}}{6}.$$

Consequently, this stationary point is a focus, and near the solution (7.16) the other solutions to Equation (7.12) oscillate somehow. □

8. A big example

We find the asymptotics of solutions to the equation

$$\dddot{x} = t^{\sigma} x^{\mu} \tag{8.1}$$

when $t \to \infty$ [Bruno 1965, 1973b], acting similarly to Examples 5.1 and 6.4. Here σ and μ are real numbers. We put $x_1 = x$, $x_2 = \dot{x}$, $x_3 = \ddot{x}$, $x_4 = t$ and write Equation (8.1) as the system

$$\begin{aligned}
\dot{x}_1 &= x_2 = x_1(x_1^{-1}x_2), \\
\dot{x}_2 &= x_3 = x_2(x_2^{-1}x_3), \\
\dot{x}_3 &= x_1^{\mu}x_4^{\sigma} = x_3(x_1^{\mu}x_3^{-1}x_4^{\sigma}), \\
\dot{x}_4 &= 1 = x_4(x_4^{-1}).
\end{aligned} \tag{8.2}$$

The support of this system and the collection of vector coefficients consist of four points each:

$$\begin{aligned}
Q_1 &= (-1, 1, 0, 0), & F_1 &= (1, 0, 0, 0), \\
Q_2 &= (0, -1, 1, 0), & F_2 &= (0, 1, 0, 0), \\
Q_3 &= (\mu, 0, -1, \sigma), & F_3 &= (0, 0, 1, 0), \\
Q_4 &= (0, 0, 0, -1), & F_4 &= (0, 0, 0, 1).
\end{aligned} \tag{8.3}$$

We compose the remainders:

$$\begin{aligned}
\tilde{Q}_1 &= Q_1 - Q_4 = (\ -1, \quad 1, \quad 0, \quad 1 \quad), \\
\tilde{Q}_2 &= Q_2 - Q_4 = (\quad 0, \ -1, \quad 1, \quad 1 \quad), \\
\tilde{Q}_3 &= Q_3 - Q_4 = (\quad \mu, \quad 0, \ -1, \quad \sigma + 1 \ ).
\end{aligned} \tag{8.4}$$

These vectors are linearly independent if the matrix of dimension 3×4 formed by the three vectors (8.4) has at least one non-zero minor of the order 3. The direct computations give the values of these minors

$$M_1 = \sigma + 3, \quad M_2 = -(\mu + \sigma + 2), \quad M_3 = 2\mu + \sigma + 1, \quad M_4 = \mu - 1, \tag{8.5}$$

where M_i is the determinant of the matrix obtained by crossing out the ith column. According to (8.5) all four minors turn to zero only when $\mu = 1$ and $\sigma = -3$. If $\mu = 1$ and $\sigma = -3$, Equation (8.1) is an Euler equation, and it is solved explicitly. Therefore we will suppose further that

$$(\mu, \sigma) \neq (1, -3). \tag{8.6}$$

Thus, when Condition (8.6) is satisfied, the three vectors (8.4) are linearly independent, i.e. the dimension of System (8.2) equals three. Since the number s of points of the support equals four, then the Newton polyhedron $\mathbf{\Gamma}$ is a tetrahedron, and each point Q_j is its vertex. Consequently, System (8.2) is a simplex one. Every collection of indices $1, 2, 3, 4$ generates its truncation. There are altogether $4 + 6 + 4 + 1 = 15$ truncations. Since we are interested only in solutions with $x_4 \to \infty$, then the cone of the problem

$$p_4 > 0. \tag{8.7}$$

Now, using Criteria 1–3, we discard those truncations, which do not contain the required solutions. Since the right hand side of each equation is a monomial in System (8.2), then no one of truncations has trivial solutions with components different from zero or infinity. Consequently, Criterion 3 need not be checked. To check Criteria 1 and 2 we compose the matrix $\mathcal{B} = \mathcal{Q}\mathcal{F}^* = (b_{ij})$ with elements $b_{ij} = \langle F_j, Q_i \rangle$ (see (3.4)). Since the matrix $\mathcal{F}$ is the unit one, then $\mathcal{B} = \mathcal{Q}$, i.e. the matrix $\mathcal{B}$ is formed by lines Q_j placed on the left in (8.3). Now for each truncation $\hat{F}_j^{(d)}(X)$ we are going to find such $P = \nu_1 F_1 + \nu_2 F_2 + \nu_3 F_3 + \nu_4 F_4$, which lie in the intersection $\mathbf{U}_j^{(d)} \cap \mathbf{F}_j^{(d)}$. Since $F_i = E_i$ are unit vectors, then $P = (\nu_1, \nu_2, \nu_3, \nu_4)$, and according to (8.7) we are interested only in those truncations which have

$$\nu_4 > 0, \tag{8.8}$$

i.e. the point Q_4 belongs to the corresponding boundary subset or number 4 belongs to the collection of indices $[j_1, \ldots, j_l]$ determining this boundary subset and the truncation. There are $1 + 3 + 3 + 1 = 8$ of such truncations. For each of them we compose relations of the form (5.7) on the ν_i, which determine the intersection $\mathbf{U}_j^{(d)} \cap \mathbf{F}_j^{(d)}$. We begin with truncations of smaller dimension.

[4]. $\mathbf{F} \cap \mathbf{U} = \{P\colon P = \nu_4 E_4,\ \nu_4\langle F_4, Q_i\rangle < \nu_4\langle F_4, Q_4\rangle,\ i = 1,2,3\}$, i.e. $\nu_4 b_{i4} < \nu_4 b_{44}$, $i = 1,2,3$ or $0 < -\nu_4$, $0 < -\nu_4$, $\nu_4\sigma < -\nu_4$. From here we obtain $\nu_4 < 0$ and $\sigma + 1 > 0$. Since the value ν_4 contradicts to the inequality (8.8), then this truncation does not suit us. The verification of Criterion 2 for $\nu_4 \neq 0$ leads to relations $\nu_4 < 0$ and $\sigma + 1 = 0$, which also contradict to the inequality (8.8).

[1,4]. $\nu_2 = \nu_3 = 0$, the relations (5.7) are

$$\begin{aligned} \nu_1 b_{11} + \nu_4 b_{14} &= \nu_1 b_{41} + \nu_4 b_{44}, \\ \nu_1 b_{21} + \nu_4 b_{24} &< \nu_1 b_{11} + \nu_4 b_{14}, \\ \nu_1 b_{31} + \nu_4 b_{34} &< \nu_1 b_{11} + \nu_4 b_{14}. \end{aligned}$$

Writing the values b_{ij} according to (8.3) we obtain $-\nu_1 = -\nu_4$, $0 < -\nu_1$, $\nu_1\mu + \nu_4\sigma < -\nu_1$, whence it follows $\nu_1 = \nu_4 < 0$ and $\sigma + \mu + 1 > 0$. Since $\nu_4 < 0$, then this truncation does not suit us.

Further, instead of the elements b_{ij}, we are going to write their numerical values directly.

[2,4]. $\nu_1 = \nu_3 = 0$, $-\nu_2 = -\nu_4$, $\nu_2 < -\nu_2$, $\sigma\nu_4 < -\nu_4$, i.e.

$$\nu_2 = \nu_4 < 0 \quad \text{and} \quad \sigma + 1 > 0.$$

This truncation does not suit us also.

[3,4]. $\nu_1 = \nu_2 = 0$, $-\nu_3 + \sigma\nu_4 = -\nu_4$, $0 < -\nu_4$, $\nu_3 < -\nu_4$, i.e.

$$\nu_4 < 0, \quad (\sigma + 1)\nu_4 = \nu_3 \quad \text{and} \quad \sigma + 2 > 0.$$

This truncation does not suit us also.

[1,2,4]. $\nu_3 = 0$, $-\nu_1 + \nu_2 = -\nu_2 = -\nu_4$, $\nu_1\mu + \nu_4\sigma < -\nu_4$, i.e.

$$\nu_4 = \nu_2 = \nu_1/2 \quad \text{and} \quad (\sigma + 2\mu + 1)\nu_4 < 0.$$

Since we are interested only in cases with $\nu_4 > 0$, then $\sigma + 2\mu + 1 < 0$. Criterion 2 is not satisfied when $\sigma + 2\mu + 1 = 0$.

[1,3,4]. $\nu_2 = 0$, $-\nu_1 = \nu_1\mu - \nu_3 + \nu_4\sigma = -\nu_4 > \nu_3$, i.e.

$$\nu_4 = \nu_1 = \nu_3/(\sigma + \mu + 1) \quad \text{and} \quad (\sigma + \mu + 2)\nu_4 < 0.$$

We have $\nu_4 > 0$ under the condition $\sigma + \mu + 2 < 0$. Criterion 2 is not satisfied when $\sigma + \mu + 2 = 0$.

[2,3,4]. $\nu_1 = 0$, $-\nu_2 + \nu_3 = -\nu_3 + \nu_4\sigma = -\nu_4 > \nu_2$, i.e.

$$\nu_4 = \nu_3/(\sigma + 1) = \nu_2/(\sigma + 2) \quad \text{and} \quad \sigma + 3 < 0 \quad \text{if} \quad \nu_4 > 0.$$

Criterion 2 is not satisfied when $\sigma + 3 = 0$.

[1,2,3,4]. $-\nu_1 + \nu_2 = -\nu_2 + \nu_3 = \nu_1\mu - \nu_3 + \nu_4\sigma = -\nu_4$, i.e.

$$\nu_1 : \nu_2 : \nu_3 : \nu_4 = (\sigma + 3) : (\sigma + \mu + 2) : (2\mu + \sigma + 1) : (1 - \mu).$$

If $1 - \mu \neq 0$, then this truncation (which coincides with the complete system) suits us, since there are no restrictions on ν_4.

Thus, using Criteria 1 and 2, we have selected four truncations with $p_4 > 0$ (the last ones) out of eight truncations with $p_4 \neq 0$, and we have found the conditions on the parameters, with which the truncations might have the required solutions.

Now, using these four truncations, we find asymptotics of such integral curves of System (8.2), on which $x_4 \to \infty$. We note that in each of the selected four truncations $\widehat{F}_j^{(d)}$ the intersection $\mathbf{U}_j^{(d)} \cap \mathbf{F}_j^{(d)}$ is one-dimensional.

Therefore the vector order P of the required solutions is determined to within the scalar factor. Since $x_4 \to \infty$, then in all cases the $\pm x_4$ may be taken as parameter τ. Thus, for each truncation we are going to find the coefficients $b_4 = \pm 1, b_1, b_2, b_3$ of its integral curve

$$x_1 = b_1\tau^{p_1}, \quad x_2 = b_2\tau^{p_2}, \quad x_3 = b_3\tau^{p_3}, \quad x_4 = \pm\tau, \tag{8.9}$$

where $P = (p_1, p_2, p_3, 1) \in \mathbf{U}_j^{(d)} \cap \mathbf{F}_j^{(d)}$, $B = (b_1, b_2, b_3, b_4)$ and equations

$$\widehat{F}(B) = \varkappa P, \quad \varkappa = \pm 1 \tag{8.10}$$

are satisfied. Since $\check{f}_4 = x_4^{-1}$, then the last of them means that $b_4 = \varkappa$.

[1,2,4]. $P = (2, 1, 0, 1)$, $\sigma + 2\mu + 1 < 0$,

$$\widehat{F} = (x_1^{-1}x_2, x_2^{-1}x_3, 0, x_4^{-1}).$$

The system of equations (8.10) is reduced to the two equations $b_1^{-1}b_2 = 2\varkappa$, $b_2^{-1}b_3 = \varkappa$, whence we obtain $b_3 = \varkappa b_2 = 2b_1$, where $b_1 = c \neq 0$ is an arbitrary constant. Thus, the power asymptotics here are

$$x_1 = cx_4^2, \quad x_2 = 2cx_4, \quad x_3 = 2c, \quad \sigma + 2\mu + 1 < 0. \tag{8.11}$$

When $\sigma + 2\mu + 1 = 0$ the obtained values B also satisfy System (8.10), and the vector P is normal to all vectors $\tilde{Q}_1, \tilde{Q}_2, \tilde{Q}_3$ in (8.4). Therefore it is Case F3, where the intermediate truncation is the complete system (8.2).

[1,3,4]. $P = (1, 0, \sigma + \mu + 1, 1)$, $\sigma + \mu + 2 < 0$,

$$\widehat{F} = (x_1^{-1}x_2, 0, x_1^{\mu}x_3^{-1}x_4^{\sigma}, x_4^{-1}).$$

For solutions (8.9) System (8.10) consists of two equations

$$b_1^{-1}b_2 = \varkappa, \quad b_1^{\mu}b_3^{-1} = \varkappa^{(\sigma-1)}(\sigma + \mu + 1).$$

Their solutions are

$$b_2 = \varkappa b_1, \quad b_3 = \varkappa^{(1-\sigma)}(\sigma + \mu + 1)^{-1}b_1^{\mu},$$

where $b_1 = c \neq 0$ is an arbitrary constant. Thus, we obtain asymptotics

$$x_1 = cx_4, \quad x_2 = c, \quad x_3 = (\sigma + \mu + 1)^{-1}cx_4^{\sigma+\mu+1}, \quad \sigma + \mu + 2 < 0. \tag{8.12}$$

When $\sigma + \mu + 2 = 0$ the obtained values B satisfy System (8.10), and the vector P is normal to all vectors $\tilde{Q}_1, \tilde{Q}_2, \tilde{Q}_3$ in (8.4). It is Case F3, where the intermediate truncation is System (8.2).

[2,3,4]. $P = (0, \sigma + 2, \sigma + 1, 1)$, $\sigma + 3 < 0$,

$$\widehat{F} = (0, x_2^{-1}x_3, x_1^{\mu}x_3^{-1}x_4^{\sigma}, x_4^{-1}).$$

For curves (8.9) System (8.10) consists of two equations

$$b_2^{-1} b_3 = \varkappa(\sigma+2), \quad b_1^{\mu} b_3^{-1} = \varkappa^{\sigma} = \varkappa(\sigma+1).$$

Its solutions are

$$b_2 = \varkappa^{\sigma}(\sigma+1)^{-1}(\sigma+2)^{-1} b_1^{\mu}, \quad b_3 = \varkappa^{\sigma-1}(\sigma+1)^{-1} b_1^{\mu},$$

where $b_1 = c \neq 0$ is an arbitrary constant. Thus, we obtain asymptotics

$$x_1 = c, \quad x_2 = (\sigma+1)^{-1}(\sigma+2)^{-1} c^{\mu} x_4^{\sigma+2}, \quad x_3 = (\sigma+1)^{-1} c^{\mu} x_4^{\sigma+1},$$
$$\sigma + 3 < 0. \tag{8.13}$$

When $\sigma + 3 = 0$ the obtained values B satisfy System (8.10), and vector P is normal to all vectors $\tilde{Q}_1, \tilde{Q}_2, \tilde{Q}_3$ in (8.4). It is Case F3, where the intermediate truncated system is System (8.2).

[1,2,3,4].

$$P = ((\sigma+3)(1-\mu)^{-1}, (\sigma+\mu+2)(1-\mu)^{-1}, (2\mu+\sigma+1)(1-\mu)^{-1}, 1),$$

$$\widehat{F} = F = (x_1^{-1} x_2, x_2^{-1} x_3, x_1^{\mu} x_3^{-1} x_4^{\sigma}, x_4^{-1}).$$

System (8.10) for curves (8.9) consists of three equations

$$\begin{aligned} b_1^{-1} b_2 &= \varkappa(\sigma+3)(1-\mu)^{-1}, \\ b_2^{-1} b_3 &= \varkappa(\sigma+\mu+2)(1-\mu)^{-1}, \\ b_1^{\mu} b_3^{-1} \varkappa^{\sigma} &= \varkappa(2\mu+\sigma+1)(1-\mu)^{-1}. \end{aligned}$$

Its solutions are

$$\begin{aligned} b_1 &= \left[\varkappa^{1-\sigma}(\sigma+2\mu+1)(\sigma+\mu+2)(\sigma+3)(1-\mu)^{-3})\right]^{1/(\mu-1)}, \\ b_2 &= \varkappa(\sigma+3)(1-\mu)^{-1} b_1, \\ b_3 &= (\sigma+\mu+2)(\sigma+3)(1-\mu)^{-2} b_1, \end{aligned} \tag{8.14}$$

$$\sigma + 2\mu + 1 \neq 0, \quad \sigma + \mu + 2 \neq 0, \quad \sigma + 3 \neq 0, \quad \mu - 1 \neq 0. \tag{8.15}$$

Here the curve (8.9) with coefficients (8.14) under the restrictions (8.15) is the exact integral curve of System (8.2).

We consider now the three found cases F3. In all of them the intermediate truncation is System (8.2) itself, i.e. $e = 3$, and the final truncations are three two-dimensional truncations, i.e. $d = 2$, the supports of which contain the common point Q_4. Therefore we make such a transformation that translates the point Q_4 into the origin and places the two-dimensional faces into two-dimensional coordinate subspaces. Then one power transformation would suffice for all three cases. Therefore we make the power transformation (compare with (8.4)) with the matrix α:

$$\begin{aligned} y_1 &= x_1^{-1}x_2x_4, \\ y_2 &= x_2^{-1}x_3x_4, \\ y_3 &= x_1^{\mu}x_3^{-1}x_4^{\sigma+1}, \\ y_4 &= x_4, \end{aligned} \qquad \alpha = \begin{pmatrix} -1 & 1 & 0 & 1 \\ 0 & -1 & 1 & 1 \\ \mu & 0 & -1 & \sigma+1 \\ 0 & 0 & 0 & 1 \end{pmatrix},$$

and we make the change of time $dt_1 = x_4^{-1}dt$. The determinant is $\det\alpha = \mu - 1$. We restrict ourselves with the case $\mu \neq 1$. The matrix $\beta = (\mu-1)\alpha^{-1}$ and the inverse power transformation are

$$\beta = \begin{pmatrix} 1 & 1 & 1 & -(\sigma+3) \\ \mu & 1 & 1 & -(\mu+\sigma+2) \\ \mu & \mu & 1 & -(2\mu+\sigma+1) \\ 0 & 0 & 0 & \mu-1 \end{pmatrix}, \qquad \begin{aligned} x_1 &= [y_1y_2y_3y_4^{-(\sigma+3)}]^{1/(\mu-1)}, \\ x_2 &= [y_1^{\mu}y_2y_3y_4^{-(\mu+\sigma+2)}]^{1/(\mu-1)}, \\ x_3 &= [y_1^{\mu}y_2^{\mu}y_3y_4^{-(2\mu+\sigma+1)}]^{1/(\mu-1)}, \\ x_4 &= y_4. \end{aligned} \tag{8.16}$$

System (8.2) is transformed into the system

$$\begin{aligned} d\log y_1/dt_1 &= 1 - y_1 + y_2 \stackrel{\text{def}}{=} g_1, \\ d\log y_2/dt_1 &= 1 - y_2 + y_3 \stackrel{\text{def}}{=} g_2, \\ d\log y_3/dt_1 &= \sigma + 1 + \mu y_1 - y_3 \stackrel{\text{def}}{=} g_3, \\ d\log y_4/dt_1 &= 1 \stackrel{\text{def}}{=} g_4. \end{aligned} \tag{8.17}$$

The vector P is transformed into the vector $\tilde{P} = \alpha P$. The points Q_1, Q_2, Q_3, Q_4 are transformed into the points $\tilde{Q}_1 = E_1$, $\tilde{Q}_2 = E_2$, $\tilde{Q}_3 = E_3$, $\tilde{Q}_4 = 0$ respectively.

We consider now the first Case F3 with $\sigma + 2\mu + 1 = 0$ and $\check{f}_3 \equiv 0$, where the final truncation $\widehat{\widehat{G}}$ is obtained from System (8.17) when $y_3 = 0$ in its right hand sides and the truncated system is

$$\begin{aligned} d\log y_1/dt_1 &= 1 - y_1 + y_2, \\ d\log y_2/dt_1 &= 1 - y_2, \\ d\log y_3/dt_1 &= -2\mu + \mu y_1, \\ d\log y_4/dt_1 &= 1. \end{aligned}$$

The subsystem of the first two equations has the stationary point $(y_1, y_2) = (2, 1)$ with eigenvalues $-2, -1$. When $y_3 = 0$ in this point $G = (0, 0, 0, 1)$. We make the substitution $y_1 = 2 + v_1$, $y_2 = 1 + v_2$ in System (8.17), then we obtain the system

$$\begin{aligned}
dv_1/dt_1 &= (2 + v_1)(-v_1 + v_2) = -2v_1 + 2v_2 - v_1^2 + v_1 v_2, \\
dv_1/dt_1 &= (1 + v_2)(-v_2 + v_3) = -v_2 + v_3 - v_2^2 + v_2 v_3, \\
d\log y_3/dt_1 &= \mu v_1 - y_3, \\
d\log y_4/dt_1 &= 1.
\end{aligned}$$

This system has the zero invariant curve

$$v_1 = \varphi_1(y_3) = \frac{1}{2} y_3 + \ldots, \qquad v_2 = \varphi_2(y_3) = \frac{1}{2} y_3 + \ldots,$$

and it induces on the curve the equation

$$dy_3/dt_1 = -y_3^2/2 + \ldots \tag{8.18}$$

A system on the zero curve induced by System (8.17) consists of Equation (8.18) and the equation

$$d\log z/dt_1 = z^{-1}, \tag{8.19}$$

where $z = \log y_4$. The corresponding truncated system

$$d\log y_3/dt_1 = -y_3^2/2, \quad d\log z/dt_1 = z^{-1}$$

has the power integral curve $z = y_3^{-2}$, i.e. $y_3 = 1/\sqrt{\log|y_4|}$, $y_4 \to \infty$. That gives the asymptotics of integral curves of System (8.17): $Y = (2, 1, (\log|\tau|)^{-1/2}, \pm\tau)$, which is transformed by the transformation (8.16) into

$$\begin{aligned}
x_1 &= 2^{1/(\mu-1)} x_4^2 (\log|x_4|)^{1/2(1-\mu)}, \\
x_2 &= 2^{\mu/(\mu-1)} x_4 (\log|x_4|)^{1/2(1-\mu)}, \\
x_3 &= 2^{\mu/(\mu-1)} (\log|x_4|)^{1/2(1-\mu)}.
\end{aligned} \tag{8.20}$$

We consider now Case F3 with $\sigma + \mu + 2 = 0$ and $\check{f}_2 \equiv 0$. The basic (nonlocal) coordinates now are y_1 and y_3, and the coordinate y_2 is small. From the first and the third equation (8.17) we obtain the system in the plane $y_2 = 0$

$$dy_1/dt_1 = y_1(1 - y_1), \quad dy_3/dt_1 = y_3(\sigma + 1 + \mu y_1 - y_3).$$

It has the stationary point $(y_1^0, y_3^0) = (1, -1)$ with eigenvalues $-1, 1$. When $y_2 = 0$ in this point $G = (0, 0, 0, 1)$. We make the substitution $y_1 = 1 + v_1$, $y_3 = -1 + v_3$ in System (8.17), then we obtain the system

$$\begin{aligned} dv_1/dt_1 &= (1+v_1)(-v_1+y_2) = -v_1 + y_2 - v_1^2 - v_1 y_2, \\ d\log y_2/dt_1 &= -y_2 + v_3, \\ dv_3/dt_1 &= (-1+v_3)(\mu v_1 - v_3) = -\mu v_1 + v_3 + \mu v_1 v_3 - v_3^2. \end{aligned}$$

This system has the zero invariant curve

$$v_1 = \varphi_1(y_2) = y_2 + \ldots, \quad v_3 = \varphi_3(y_2) = \mu y_2 + \ldots,$$

and it induces on the curve the equation $dy_2/dt_1 = (\mu - 1)y_2^2 + \ldots$ Together with Equation (8.19) it gives the truncated system

$$d\log y_2/dt_1 = (\mu - 1)y_2, \quad d\log z/dt_1 = z^{-1}$$

with the power integral curve $z = 1/(1-\mu)y_2$, which gives the asymptotics of integral curves of System (8.17)

$$Y = (1, [(1-\mu)\log|y_4|]^{-1}, -1, y_4),$$

and of System (8.2)

$$\begin{aligned} x_1 &= x_4[(\mu - 1)\log|x_4|]^{1/(1-\mu)}, \\ x_2 &= [(\mu - 1)\log|x_4|]^{1/(1-\mu)}, \\ x_3 &= x_4^{-1}\{-[(1-\mu)\log|x_4|]^{-\mu}\}^{1/(\mu-1)} \end{aligned} \tag{8.21}$$

by means of transformation (8.16).

We consider now Case F3 with $\sigma + 3 = 0$ and $\check{f}_1 \equiv 0$. In the plane $y_1 = 0$ the subsystem of the second and the third equations of System (8.17) has the stationary point $(y_2^0, y_3^0) = (-1, -2)$ with eigenvalues $1, 2$. In its neighborhood System (8.17) has the zero invariant curve

$$y_2 = -1 + \mu y_1 + \ldots, \quad y_3 = -2 + \mu y_1 + \ldots,$$

on which System (8.17) takes the form

$$d\log y_1/dt_1 = (\mu - 1)y_1 + \ldots, \quad d\log z/dt_1 = z^{-1},$$

which leads to asymptotics analogous to the previous case:

$$Y = ([(1-\mu)\log|y_4|]^{-1}, -1, -2, y_4),$$

i.e. according to (8.16)

$$\begin{aligned} x_1 &= [2(1-\mu)^{-1}(\log|x_4|)^{-1}]^{1/(\mu-1)}, \\ x_2 &= x_4^{-1}\{2[(1-\mu)\log|x_4|]^{-\mu}\}^{1/(\mu-1)}, \\ x_3 &= x_4^{-2}\{-2[(\mu-1)\log|x_4|]^{-\mu}\}^{1/(\mu-1)}. \end{aligned} \tag{8.22}$$

Thus, for solutions to System (8.2) with $x_4 \to \infty$ we have obtained the four power asymptotics of the form (8.9), namely, (8.11), (8.12), (8.13) and (8.14) with restrictions (8.15). If one of the first three inequalities (8.15) turns to equality, then instead of the asymptotics (8.9), (8.14) there is the logarithmic asymptotics (8.20) or (8.21), or (8.22) respectively. In addition we note that the integral curve (8.9), (8.14) corresponds to the stationary point

$$(y_1, y_2, y_3) = \left(\frac{\sigma+3}{1-\mu}, \frac{\sigma+\mu+2}{1-\mu}, \frac{\sigma+2\mu+1}{1-\mu}\right)$$

of the subsystem of the first three equations of System (8.17).

Exercise. Verify that Solution (8.9) with coefficients (8.14) and $b_4 = \varkappa$ found for Equation (8.1) in Case [1,2,3,4] coincides with Solution (7.16) to Equation (7.12) when $\sigma = 0$, $\mu = 3$.

9. Remarks

If System (4.1) depends on small or large parameters $\mu_1, \ldots, \mu_m$ (i.e. $\mu_j \to 0$ or $\mu_j \to \infty$), then these parameters may be included in coordinates x_j: if $\mu_j = x_j$, then $dx_j/dt = 0$ in System (4.1). Therefore the algorithms of this Chapter are applicable also for these systems, only the exponents of small and large parameters must be taken into account. A certain specificity is in the power transformation. Let $\mathbf{M} = (\mu_1, \ldots, \mu_m)$ be small and large parameters, $X = (x_1, \ldots, x_m)$ be independent variables. Then the power transformation has the block-triangular form

$$\begin{aligned} \log \mathrm{N} &= \alpha_{11} \log \mathrm{M}, \\ \log Y &= \alpha_{21} \log \mathrm{M} + \alpha_{22} \log X, \end{aligned}$$

where α_{11} and α_{22} are nonsingular square matrices of the dimensions m and n respectively, and α_{21} is the rectangular matrix of the dimension $n \times m$. The power change of time has the form $dt_1 = \mathbf{M}^S X^T dt$. Under such power transformations and change of time all the theorems of Chapter 3 remain valid also for systems with asymptotic parameters.

The study in Chapter 2 was given in some respects in the two levels: the first one is in the neighborhood of an ordinary simple point, the second one is in the neighborhood of a generalized simple point, where an arbitrary tangent cone is present. Chapter 3 has three such levels: the first one is the neighborhood of an ordinary elementary stationary point (Section 1, h-asymptotics and Cases H1–H4); the second one is the neighborhood of a generalized elementary stationary point (Section 4, g-asymptotics and Cases G1–G4); the third one concerns solutions to an arbitrary truncated system (Section 5, f-asymptotics and Cases F1–F4). The presence of the third level allows to find power asymptotics directly using the truncated system and without making the power transformation, which simplifies significantly the algorithm of their search.

The truncated systems are useful for an approximate location of all solutions to the original system of ODE, and not only of its asymptotic solutions tending to zero or infinity. It was demonstrated in the book [Bruno 1979a], where to each truncation in coordinate space X there had been put in correspondence the set $\mathcal{U}(\varepsilon)$, in which this truncation was the first approximation to the original system of ODE.

The material of Chapter 3 (except Sections 6, 7) is the presentation of old papers by the author [Bruno 1962, 1965, 1973b], which till now are not understood. Thus, it was written in the rewiew [Coleman 1966] of the paper [Bruno 1965]: "The methods and results are not very clear". Although many scientists have been (and are now) interested in the questions of search and the study of asymptotic solutions to a system of ODE. The history of papers [Bruno 1962, 1965, 1973b] is as follows. In 1958/59 academic year the author was the second year student at the mathematical faculty of Moscow State University and attended a special seminar supervised by L.S.Pontryagin and his former students. In spring of 1959 M.I.Zelikin (who was then a postgraduate student) posed a problem to the author to analyze the integral curves near the degenerate stationary point of a two-dimensional system having non-simple zero eigenvalue, i.e. the Jordan cell. By the autumn of 1959 the author proposed to perform such an analysis with the help of a certain broken line. Moreover, the method was applicable to the analysis of any degenerate stationary point in the plane. M.I.Zelikin advised to the author to refer to Professor V.V.Nemytskii, who said that the broken line was similar to the Newton open polygon and gave the reference to the paper [Chebotarev 1943]. Through that paper the author got reference to the paper [Briot and Bouquet 1856], in which the method of the Newton open polygon was extended on differential equations from algebraic ones. Already by the autumn of 1959 the author contrived to study the multi-dimensional systems with the help of a polyhedron, that was called by Gindikin [1973] the "Newton polyhedron". Shestakov [1960, 1961] also was engaged in this geometry, but there was no polyhedron in his studies. In the beginning of 1960 the author proposed the power transformations and noted that they had a property of inducing the linear transformations of vector exponents and vector coefficients of a system of ODE. All this was presented at first in the undergraduate's theses of the third and the fourth year, and then it was published in a short paper [Bruno 1962]. When the author was the fourth year student (1960/61 academic year) Yu.N.Drozhzhinov (the fifth year student at the mathematical faculty) asked him to find asymptotics of solutions to the equation $\ddot{x} = t^\sigma x^\mu$. It was done then and here presented in Section 8. In spring of 1963, when the author was a postgraduate student at the Keldysh Institute of Applied Mathematics (KIAM), all these results were presented in a detailed article that was submitted to the magazine "Izvestiya of Akad. Nauk SSSR. Mathematical series". Referee requested to abridge the article significantly, and that was done at the cost of almost all figures and examples. In such a curtailed form the article was published [Bruno 1965], but it was difficult to understand. In 1973 all the discarded parts of the article were published as a preprint at KIAM [Bruno 1973b], after they had been rejected by magazines "Matematicheskii Sbornik" and "Izvestiya of AN SSSR. Mathematical series". In the book [Bruno 1979a] this method was exposed in detail for systems in the plane and briefly for multi-dimensional systems.

But still the works are being published, which propose methods of search of asymptotic solutions applicable only for isolated cases. Thus, in the book [Kozlov and Furta 1996] Chapter I is dedicated to solutions with power asymptotics, and Chapter III to solutions with the logarithmic one. In this book the group-theoretic approach is mostly used, although authors cannot do without the power geometry altogether (the Newton polyhedra are used), but its level remains approximately the same as it was in [Shestakov 1960]. In particular, the power transformations are absent. The linear transformations of coordinates, which may depend on time, are used instead. This restricts and complicates significantly the presentation of material and methods of computations. There are no such notions there as the dimension of a system and the normal cone of a truncation, since there is no apparatus of power transformations for their application. The book contains the references on the recent publications on the asymptotic solutions and a list of problems connected with such solutions.

Using the Bugaev [1891] method Lunev [1994a] computed asymptotic expansions for solutions to the problem of rotation of a rigid body with the fixed point. But the results of the paper [Lunev 1994a] are wrong.

Chapter 4

Hamiltonian truncations of a Hamiltonian system

1. The theory

Let a discrete set of points $\mathbf{S}$ be given in $\mathbb{R}^n$. We recall that in Chapter 1 the following objects were associated with it: the convex hull $\mathbf{\Gamma}$ of the set $\mathbf{S}$, its faces $\mathbf{\Gamma}_j^{(d)}$ of various dimensions $d\,(0 \le d < n)$, the boundary subsets $\mathbf{S}_j^{(d)} = \mathbf{\Gamma}_j^{(d)} \cap \mathbf{S}$. In the dual to $\mathbb{R}^n$ space $\mathbb{R}_*^n$ to each face $\mathbf{\Gamma}_j^{(d)}$ there corresponds the normal cone $\mathbf{U}_j^{(d)}$.

Let a linear projection

$$\Pi\colon \mathbb{R}^n \to \mathbb{R}^l, \quad 0 < l < n$$

be given, i.e. $\Pi R = R'$, where $R \in \mathbb{R}^n$ and $R' \in \mathbb{R}^l$. The projection maps the set $\mathbf{S}$ and associated with it sets of the space $\mathbb{R}^n$ into the following sets of the space $\mathbb{R}^l$:

$$\mathbf{S}' \stackrel{\text{def}}{=} \Pi\,\mathbf{S}, \quad \mathbf{\Gamma}' \stackrel{\text{def}}{=} \Pi\,\mathbf{\Gamma}, \quad \partial\mathbf{\Gamma}' = \partial\Pi\,\mathbf{\Gamma}.$$

To each face $\mathbf{\Gamma}_{j'}^{\prime(d')}$ of the polyhedron $\mathbf{\Gamma}'$ there corresponds in $\mathbb{R}^l$ some maximal face $\mathbf{\Gamma}_j^{(d)}$ of the polyhedron $\mathbf{\Gamma}$ in $\mathbb{R}^n$, such that $\Pi\,\mathbf{\Gamma}_j^{(d)} = \mathbf{\Gamma}_{j'}^{\prime(d')}$. Consequently, to each boundary subset $\mathbf{S}_{j'}^{\prime(d')}$ there corresponds some maximal boundary subset $\mathbf{S}_j^{(d)}$, such that $\Pi\mathbf{S}_j^{(d)} = \mathbf{S}_{j'}^{\prime(d')}$. Let the projection Π be made along some linear subspace $\mathbf{L} \in \mathbb{R}^n$. Let $\mathbf{L}_*$ be a subspace in $\mathbb{R}_*^n$, which is orthogonal to the subspace $\mathbf{L}$. Then the intersection of the normal cone $\mathbf{U}_j^{(d)}$ with the space $\mathbf{L}_*$ is uniquely reestablished by the normal cone $\mathbf{U}_{j'}^{\prime(d')}$ lying in the space $\mathbb{R}_*^l$ dual to $\mathbb{R}^l$. But, generally speaking, the normal cone $\mathbf{U}_j^{(d)}$ itself may be greater then this intersection.

To each point $R' \in \mathbb{R}^l$ there corresponds in $\mathbb{R}^n$ the linear manifold $\mathbf{L}(R') \stackrel{\text{def}}{=} \Pi^{-1}(R')$ formed by all the points $R \in \mathbb{R}^n$, which are projected into the point R'. In a similar way, to each set $\mathbf{\Delta}' \subset \mathbb{R}^l$ there corresponds in $\mathbb{R}^n$ the set $\mathbf{L}(\mathbf{\Delta}') \stackrel{\text{def}}{=} \Pi^{-1}(\mathbf{\Delta}')$ formed by all the points $R \in \mathbb{R}^n$, which are projected into the points of the set $\mathbf{\Delta}'$. Let now $\mathbf{S}$ be a set in $\mathbb{R}^n$ and the set $\mathbf{S}' \stackrel{\text{def}}{=} \Pi\,\mathbf{S}$ lie in $\mathbb{R}^l$. To each point $R' \in \mathbf{S}'$ there corresponds in $\mathbb{R}^n$ the subset $\mathbf{S}(R') \stackrel{\text{def}}{=} \mathbf{S} \cap \mathbf{L}(R')$ of the set $\mathbf{S}$. In a similar way, to each subset $\mathbf{T}'$ of the set $\mathbf{S}'$ there corresponds in $\mathbb{R}^n$ the subset $\mathbf{S}(\mathbf{T}') \stackrel{\text{def}}{=} \mathbf{S} \cap \mathbf{L}(\mathbf{T}')$ of the set $\mathbf{S}$ formed by all the points $R \in \mathbf{S}$, which are projected into points of the set $\mathbf{T}'$. In particular, to each boundary subset $\mathbf{S}'^{(d')}_{j'}$ there corresponds the subset $\mathbf{S}(\mathbf{S}'^{(d')}_{j'})$, that is the boundary one for the set $\mathbf{S}$: $\mathbf{S}(\mathbf{S}'^{(d')}_{j'}) = \mathbf{S}^{(d)}_j$. The inverse, generally speaking, is not true: not every boundary subset $\mathbf{S}^{(d)}_j$ of the set $\mathbf{S}$ is projected into the boundary subset of the set $\mathbf{S}'$. We note that the mentioned above maximal face $\mathbf{\Gamma}^{(d)}_j \subset \mathbf{\Gamma}$, which is projected into the face $\mathbf{\Gamma}'^{(d')}_{j'} \subset \mathbf{\Gamma}'$, is $\mathbf{\Gamma}^{(d)}_j = \mathbf{\Gamma}(\mathbf{\Gamma}'^{(d')}_{j'}) \stackrel{\text{def}}{=} \mathbf{\Gamma} \cap \mathbf{L}(\mathbf{\Gamma}'^{(d')}_{j'})$.

We consider now a polynomial or a power series

$$h(Z) \stackrel{\text{def}}{=} \sum h_R Z^R \quad \text{over} \quad R \in \mathbf{S}_1, \tag{1.1}$$

where $Z = (z_1, \ldots, z_n) \in \mathbb{C}^n$, $R = (r_1, \ldots, r_n) \in \mathbb{R}^n$, $h_R \in \mathbb{C}$, $Z^R = z_1^{r_1} \ldots z_n^{r_n}$. We assume that the similar terms in the sum (1.1) are collected, so that $\mathbf{S}_1 = \{R : h_R \neq 0\}$. The set $\mathbf{S}_1$ is called the *support of the sum* $h(Z)$. In Chapter 1 there were introduced and studied *truncations* $\hat{h}^{(d)}_j$ *of the sum* h:

$$\hat{h}^{(d)}_j(Z) \stackrel{\text{def}}{=} \sum h_R Z^R \quad \text{over} \quad R \in \mathbf{S}^{(d)}_{1j},$$

where the boundary subset $\mathbf{S}^{(d)}_{1j}$ of the set $\mathbf{S}_1$ is determined in the same way as it is described above for the set $\mathbf{S}$ (see Chapter 1).

We consider now a system of differential equations

$$dz_i/dt \stackrel{\text{def}}{=} \dot{z}_i = \varphi_i(Z), \quad i = 1, \ldots, n, \tag{1.2}$$

where φ_i are sums of the form (1.1). We put $\varphi_i = z_i f_i(Z)$ and according to Chapter 3 write System (1.2) in the form

$$(\log^{\cdot} Z) = F(Z) \stackrel{\text{def}}{=} \sum F_R Z^R \quad \text{over} \quad R \in \mathbf{S}_2. \tag{1.3}$$

Here $\log Z = (\log z_1, \ldots, \log z_n)$, $F = (f_1, \ldots, f_n)$, $F_R = (f_{1R}, \ldots, f_{nR}) \in \mathbb{C}^n$. As before, by the *support* $\mathbf{S}_2$ *of System* (1.3) the sets $\mathbf{\Gamma}_2$, $\mathbf{\Gamma}^{(d)}_{2j}$, $\mathbf{S}^{(d)}_{2j}$, $\mathbf{U}^{(d)}_{2j}$ are found, and the *truncated system*

$$(\log^{\cdot} Z) = \hat{F}^{(d)}_j(Z) \stackrel{\text{def}}{=} \sum F_R Z^R \quad \text{over} \quad R \in \mathbf{S}^{(d)}_{2j} \tag{1.4}$$

is determined.

Let now $n = 2m$, $Z = (X, Y)$ and $R = (P, Q)$, where $X, Y \in \mathbb{C}^m$ and $P, Q \in \mathbb{R}^m$. We consider the Hamiltonian system

$$\dot{x}_i = \partial h/\partial y_i, \quad \dot{y}_i = -\partial h/\partial x_i, \quad i = 1, \dots, m, \tag{1.5}$$

with the Hamiltonian function

$$h(X, Y) \stackrel{\text{def}}{=} \sum h_{PQ} X^P Y^Q \quad \text{over} \quad R = (P, Q) \in \mathbf{S}_1 \subset \mathbb{R}^{2m}. \tag{1.6}$$

If we write System (1.5) in the form (1.3), then for the system we obtain the support $\mathbf{S}_2 \subset \mathbb{R}^{2m}$. By the support $\mathbf{S}_2$ one can find for System (1.5), (1.3) all truncated Systems (1.4). Here a truncated system may not be a Hamiltonian system, and a Hamiltonian system with the truncated Hamiltonian function may not be a truncated system.

Example 1.1. Let $m = 2$. We put $\varrho_i = x_i y_i$, $i = 1, 2$ and consider the Hamiltonian function $h = \varrho_1 \varrho_2 + \varrho_1^2 \varrho_2 + \varrho_1 \rho_2^2$. Its support $\mathbf{S}_1$ consists of three points

$$R_1 = (1, 1, 1, 1), \quad R_2 = (2, 1, 2, 1), \quad R_3 = (1, 2, 1, 2).$$

The Newton polyhedron $\mathbf{\Gamma}_1$ is a triangle with three vertices and three edges. They along with the triangle $\mathbf{\Gamma}_1$ itself exhaust all its faces $\mathbf{\Gamma}_{1j}^{(d)}$. To each face there corresponds its own truncation $\hat{h}_j^{(d)}$ of the function h. Fig. 4.1, *a* shows the projection of the support $\mathbf{S}_1$ and of the triangle $\mathbf{\Gamma}_1$ in the plane $\tilde{q}_1 = p_1 + q_1$, $\tilde{q}_2 = p_2 + q_2$. The corresponding Hamiltonian system is

$$\dot{x}_i = \frac{\partial h}{\partial \varrho_i} x_i, \quad \dot{y}_i = -\frac{\partial h}{\partial \varrho_i} y_i, \quad i = 1, 2,$$

i.e. in the notation (1.3) it has

$$F = (\partial h/\partial \varrho_1, \partial h/\partial \varrho_2, -\partial h/\partial \varrho_1, -\partial h/\partial \varrho_2),$$

where

$$\partial h/\partial \varrho_1 = \varrho_2 + 2\varrho_1 \varrho_2 + \varrho_2^2, \quad \partial h/\partial \varrho_2 = \varrho_1 + \varrho_1^2 + 2\varrho_1 \varrho_2.$$

Consequently, the set $\mathbf{S}_2$ consists of five points

$$R_4 = (0, 1, 0, 1), \quad R_1, \; R_5 = (0, 2, 0, 2), \quad R_6 = (1, 0, 1, 0), \quad R_7 = (0, 2, 0, 2).$$

The polyhedron $\mathbf{\Gamma}_2$ is a trapezoid with four vertices R_4–R_7 and four edges. Along with the trapezoid $\mathbf{\Gamma}_2$ itself they exhaust all its faces $\mathbf{\Gamma}_{2j}^{(d)}$. To each of them there corresponds its own truncated System (1.4). Fig. 4.1, *b* shows the projections of the support $\mathbf{S}_2$ and of trapezoid $\mathbf{\Gamma}_2$ in the plane $\tilde{q}_1$, $\tilde{q}_2$. To the edge $\mathbf{\Gamma}_{21}^{(1)}$ connecting the vertices R_6 and R_7 there corresponds the truncated system

$$\dot{x}_1 = 0, \qquad \dot{y}_1 = 0,$$
$$\dot{x}_2 = (\varrho_1 + \varrho_1^2)x_2, \quad \dot{y}_2 = -(\varrho_1 + \varrho_1^2)y_2,$$

which is not a Hamiltonian system, since it cannot be written in the form (1.5) with a function h. Thus, not every truncation of a Hamiltonian system is a Hamiltonian system itself.

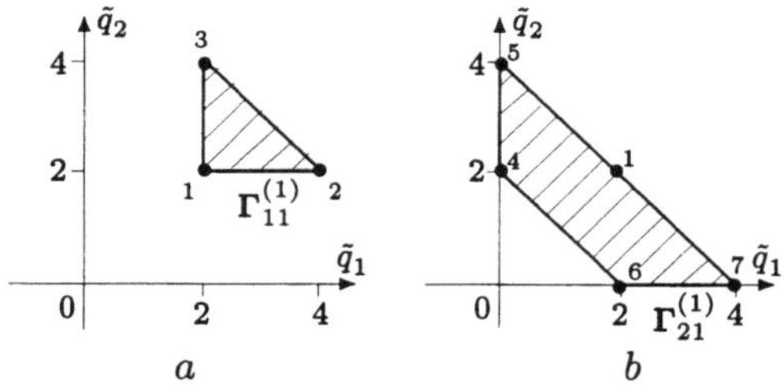

Figure 4.1: The Newton polyhedra $\mathbf{\Gamma}_1$ for the function h of Example 1.1 (a) and $\mathbf{\Gamma}_2$ for the corresponding Hamiltonian system (b) in coordinates $\tilde{q}_1 = p_1 + q_1$, $\tilde{q}_2 = p_2 + q_2$. The numbers j of points R_j are cited.

On the other hand, to the edge $\mathbf{\Gamma}_{11}^{(1)}$ connecting the points R_1 and R_2 of the triangle $\mathbf{\Gamma}_1$ there corresponds the truncation $\hat{h} = \varrho_1\varrho_2 + \varrho_1^2\varrho_2$ of the Hamiltonian function h. For the corresponding Hamiltonian system

$$\dot{x}_i = \partial\hat{h}/\partial y_i, \quad \dot{y}_i = -\partial\hat{h}/\partial x_i, \quad i = 1, 2$$

in the form (1.3), the support consists of four points R_4, R_1, R_6, R_7. Their convex hull is the parallelogram that cannot be a face of the trapezoid $\mathbf{\Gamma}_2$. □

Consequently, not all truncations $\hat{h}$ of the Hamiltonian function h have the corresponding Hamiltonian system

$$\dot{x}_i = \partial\hat{h}/\partial y_i, \quad \dot{y}_i = -\partial\hat{h}/\partial x_i, \quad i = 1, \ldots, m \tag{1.7}$$

as a truncation of System (1.5), (1.3).

So, the **problem** appears: to isolate all the truncated systems (1.4) for System (1.5), (1.3), which are the Hamiltonian systems, i.e. which can be written in the form (1.7) with a function $\hat{h}(X, Y)$. Below we give a solution of this problem. Here the function $\hat{h}$ is always a truncation of the Hamiltonian function h, i.e. the **inverse problem** is solved simultaneously: to find all the truncations $\hat{h}$ of the Hamiltonian function h, such that System (1.7) be a truncation of System (1.5) in the form (1.3).

In principle, we may look for solutions of the problems in the following way. Firstly, to compute polyhedrons $\mathbf{\Gamma}_1$ and $\mathbf{\Gamma}_2$, their faces $\mathbf{\Gamma}_{1j}^{(d)}$ and $\mathbf{\Gamma}_{2k}^{(e)}$, and normal cones $\mathbf{U}_{1j}^{(d)}$ and $\mathbf{U}_{2k}^{(e)}$. Then to consider for each nonempty intersection $\mathbf{U}_{1j}^{(d)} \cap \mathbf{U}_{2k}^{(e)}$, whether the truncation $\hat{h}_j^{(d)}$ and the truncated system $d \log Z/dt = \widehat{F}_k^{(e)}(Z)$ correspond to each other in the sense of the truncated System (1.7). But it is too bulky a way. Here we suggest a more simple algorithm of solution of the mentioned problems.

Let E_i be the ith unit vector. We define the projection Π from $\mathbb{R}^{2m}$ into $\mathbb{R}^{m+1}$ along vectors

$$E_i + E_{i+m} - E_m - E_{2m}, \quad i = 1, \ldots, m-1, \tag{1.8}$$

in the following way: $(P', q') = \Pi(P, Q)$, where $P' = (p'_1, \ldots, p'_m)$ and

$$\begin{aligned} p'_i &= p_i - q_i, & i &= 1, \ldots, m-1, \\ p'_m &= p_m + \sum_{i=1}^{m-1} q_i, & q' &= q_m + \sum_{i=1}^{m-1} q_i. \end{aligned} \tag{1.9}$$

We consider in $\mathbb{R}^{m+1}$ projections $\mathbf{S}'_1$ and $\mathbf{S}'_2$ of sets $\mathbf{S}_1$ and $\mathbf{S}_2$, which are supports of the Hamiltonian function (1.6) and of the Hamiltonian system (1.5) respectively. Here the point $R = 0$ is excluded from the set $\mathbf{S}_1$.

Theorem 1.1 *To each face* $\mathbf{\Gamma}_{1j'}^{'(d')}$ *of the polyhedron* $\mathbf{\Gamma}'_1$ *there correspond a truncated Hamiltonian function* $\hat{h}_j^{(d)}(X, Y)$ *and a truncated Hamiltonian system*

$$\dot{x}_i = \partial \hat{h}_j^{(d)} / \partial y_i, \quad \dot{y}_i = -\partial \hat{h}_j^{(d)} / \partial x_i, \quad i = 1, \ldots, m. \tag{1.10}$$

Conversely, to each truncated Hamiltonian system (1.7) *there corresponds a face* $\mathbf{\Gamma}_{1j'}^{'(d')}$ *of the polyhedron* $\mathbf{\Gamma}'_1$.

The proof of the theorem is based upon the following two lemmas.

Lemma 1.1 $\mathbf{S}'_1 = \mathbf{S}'_2 + (E_m, 1)$, *i.e. the set* $\mathbf{S}'_1$ *is obtained from the set* $\mathbf{S}'_2$ *by the parallel translation along the vector* $(E_m, 1)$.

Proof. Let the point $R = (P, Q) \in \mathbf{S}_1$, $R \neq 0$, i.e. the monomial $h_{PQ} X^P Y^Q \neq \text{const}$ belongs to the expansion (1.1) of the Hamiltonian function h. Then in System (1.5), (1.3) the series

$$f_i = x_i^{-1} \partial h / \partial y_i, \quad f_{i+m} = -y_i^{-1} \partial h / \partial x_i, \quad i = 1, \ldots, m$$

contain monomials with exponents

$$(P, Q) - (E_i, E_i), \quad i = 1, \ldots, m. \tag{1.11}$$

Let $(P', q') = \Pi(P, Q)$ under projection (1.9). Then

$$\Pi((P,Q) - (E_i, E_i)) = (P', q') - (E_m, 1), \quad i = 1, \ldots, m.$$

Consequently, the set $\mathbf{S}_1'$ for the Hamiltonian function h is obtained from the set $\mathbf{S}_2'$ for System (1.3), (1.5) by the parallel translation along the vector $(E_m, 1)$. □

Corollary 1.1 *The sets* $\mathbf{\Gamma}_{1j'}^{'(d')}$ *and* $\mathbf{S}_{1j'}^{'(d')}$ *are obtained from the sets* $\mathbf{\Gamma}_{2j'}^{'(d')}$ *and* $\mathbf{S}_{2j'}^{'(d')}$ *by the parallel translation along the vector* $(E_m, 1)$. *Their normal cones* $\mathbf{U}_{1j'}^{'(d')}$ *and* $\mathbf{U}_{2j'}^{'(d')}$ *coincide.*

Lemma 1.2 *To each face* $\mathbf{\Gamma}_{2j'}^{'(d')}$ *of the polyhedron* $\mathbf{\Gamma}_2'$ *there corresponds a truncated Hamiltonian system of the form* (1.10).

Proof. By virtue of linearity of the projection (1.9) for each face $\mathbf{\Gamma}'^{(d')}_{2j'}$ of the polyhedron $\mathbf{\Gamma}_2'$ there exists such a face $\mathbf{\Gamma}_{2j}^{(d)}$ of the polyhedron $\mathbf{\Gamma}_2$ that $\Pi\,\mathbf{\Gamma}_{2j}^{(d)} = \mathbf{\Gamma}'^{(d')}_{2j'}$. In general, such a face $\mathbf{\Gamma}_{2j}^{(d)}$ is not unique, but among all such faces there is the maximal one that include all the other faces. Let $\mathbf{\Gamma}_{2j}^{(d)}$ be the maximal face projected into the face $\mathbf{\Gamma}'^{(d')}_{2j'}$. To the face $\mathbf{\Gamma}_{2j}^{(d)}$ there correspond some truncated System (1.4). We show that it is a Hamiltonian system (1.10).

Let for an $i \leq m$ the point $(P,Q) - (E_i, E_i) \in \mathbf{S}_{2j}^{(d)}$, and $\Pi(P,Q) = (P', q')$. Then the point $(P', q') - (E_m, 1) \in \mathbf{S}'^{(d')}_{2j'}$, and the point $(P,Q) \in \mathbf{S}_{1j_1}^{(d_1)}$. According to Corollary to Lemma 1.1, to the face $\mathbf{\Gamma}'^{(d')}_{2j'}$ of the polyhedron $\mathbf{\Gamma}_2'$ there corresponds the face $\mathbf{\Gamma}'^{(d')}_{1j'}$ of the polyhedron $\mathbf{\Gamma}_1'$ obtained by the parallel translation along the vector $(E_m, 1)$. Let $\mathbf{\Gamma}'^{(d')}_{1j'}$ be such a maximal face of the polyhedron $\mathbf{\Gamma}_1'$ that $\Pi\,\mathbf{\Gamma}_{1j_1}^{(d_1)} = \mathbf{\Gamma}'^{(d')}_{1j'}$. To it there corresponds the truncation $\hat{h}_{j_1}^{(d_1)}$ of the Hamiltonian function h. In a similar way as in the proof of Lemma 1.1, one can prove that System (1.7) with $\hat{h} = \hat{h}_{j_1}^{(d_1)}$ is the truncated System (1.4). □

We note, that the space $\mathbf{L}$, along which the projection (1.9) is done, is spanned on vectors (1.8). Hence the space $\mathbf{L}_*$ is orthogonal to vectors (1.8).

Lemma 1.3 *To each truncated Hamiltonian System* (1.7) *there corresponds a face* $\mathbf{\Gamma}_{2j'}^{'(d')}$ *of the polyhedron* $\mathbf{\Gamma}_2'$.

Proof. Let the truncation $\hat{h}_j^{(d)}$ of the Hamiltonian function h be such that System (1.10) was the truncation (1.4) of System (1.5). Let $R = (P,Q) \in \mathbf{S}_{1j}^{(d)}$. Then the support $\tilde{\mathbf{S}}_2$ of System (1.10) includes m points (1.11), i.e. it is parallel to the space $\mathbf{L}$. Consequently, under the projection (1.9), the corresponding to $\tilde{\mathbf{S}}_2$ face $\tilde{\mathbf{\Gamma}}_2$ of the polyhedron $\mathbf{\Gamma}_2$ is projected into some face $\mathbf{\Gamma}'^{(d')}_{2j'}$ of the polyhedron $\mathbf{\Gamma}'_2$. □

Lemma 1.4 *Let $\mathbf{\Gamma}^{(d)}_{2j}$ be the maximal face of the polyhedron $\mathbf{\Gamma}_2$, which is projected into the face $\mathbf{\Gamma}'^{(d')}_{2j'}$ of the polyhedron $\mathbf{\Gamma}'_2$. Let the vectors*

$$(U_1, v_1), \ldots, (U_k, v_k) \tag{1.12}$$

form the skeleton of the normal cone $\mathbf{U}^{(d')}_{2j'}$ of the face $\mathbf{\Gamma}'^{(d')}_{2j'}$. Then vectors

$$(U_1, V_1), \ldots, (U_k, V_k), \tag{1.13}$$

where

$$V_l = (v_{1l}, \ldots, v_{ml}), \quad l = 1, \ldots, k, \tag{1.14}$$

$$v_{il} = v_l + u_{ml} - u_{il}, \quad i = 1, \ldots, m-1, \quad v_{ml} = v_l, \tag{1.15}$$

form the skeleton of the normal cone $\mathbf{U}^{(d)}_{2j}$ of the face $\mathbf{\Gamma}^{(d)}_{2j}$.

Proof. If the boundary subset $\mathbf{S}^{(d)}_{2j}$ includes one of the points (1.11), then it includes all points (1.11). The first $m-1$ points (1.11) are obtained from the last point $(P, Q) - (E_m, E_m)$ by the subtraction of vectors (1.8). Hence the face $\mathbf{\Gamma}^{(d)}_{2j}$ is parallel to the linear space $\mathbf{L}$ spanned on vectors (1.8). Consequently, its normal cone $\mathbf{U}^{(d)}_{2j}$ lies in the space $\mathbf{L}_*$ orthogonal to $\mathbf{L}$, and the skeleton of the cone $\mathbf{U}^{(d)}_{2j}$ is formed by vectors (1.13) orthogonal to vectors (1.8). The condition of orthogonality

$$\langle (U_l, V_l), E_i + E_{i+m} - E_m - E_{2m} \rangle = 0, \quad l = 1, \ldots, k; \quad i = 1, \ldots, m-1$$

for vectors (1.13), (1.14) takes the form

$$u_{il} + v_{il} - u_{ml} - v_{ml} = 0, \quad i = 1, \ldots, m-1, \quad l = 1, \ldots, k.$$

It is easy to see that the vectors (1.13), (1.14) determined by (1.15) satisfy to these relations. □

Lemma 1.5 *Let $\mathbf{\Gamma}^{(d)}_{1j}$ be the maximal face of the polyhedron $\mathbf{\Gamma}_1$, which is projected into the face $\mathbf{\Gamma}'^{(d')}_{1j'}$ of the polyhedron $\mathbf{\Gamma}'_1$. Let the vectors* (1.12) *form the skeleton of the normal cone $\mathbf{U}'^{(d')}_{1j'}$ of the face $\mathbf{\Gamma}'^{(d')}_{2j'}$. Then the vectors* (1.13)–(1.15) *form the skeleton of the intersection $\mathbf{U}^{(d)}_{1j} \cap \mathbf{L}_*$.*

The proof is analogous to that of Lemma 1.4.

Proof of Theorem 1.1. According to Corollary to Lemma 1.1, the faces $\mathbf{\Gamma}'^{(d')}_{1j'}$ and $\mathbf{\Gamma}'^{(d')}_{2j'}$ are obtained from each other by the parallel translation along the vector $(E_m, 1)$, and their normal cones $\mathbf{U}'^{(d')}_{1j'}$ and $\mathbf{U}'^{(d')}_{2j'}$ coincide. Let $\mathbf{\Gamma}^{(d)}_{1j}$

and $\mathbf{\Gamma}_{2k}^{(e)}$ be the maximal faces of the polyhedra $\mathbf{\Gamma}_1$ and $\mathbf{\Gamma}_2$ respectively, which are projected into $\mathbf{\Gamma}_{1j'}^{'(d')}$ and $\mathbf{\Gamma}_{2j'}^{'(d')}$. According to Lemma 1.4, the normal cone $\mathbf{U}_{2k}^{(e)}$ of the face $\mathbf{\Gamma}_{2k}^{(e)}$ is uniquely determined by the cone $\mathbf{U}_{1j'}^{'(d')}$, and according to Lemma 1.5, the intersection $\mathbf{U}_{1j}^{(d)} \cap \mathbf{L}_* = \mathbf{U}_{2k}^{(e)}$. Here the cone $\mathbf{U}_{1j}^{(d)}$ may be greater then the cone $\mathbf{U}_{2k}^{(e)}$. □

Thus, we have obtained three polyhedra:

$\mathbf{\Gamma}_1$, that determines truncations $\hat{h}$ of the Hamiltonian function h;

$\mathbf{\Gamma}_2$, that determines truncations of the Hamiltonian System (1.5);

$\mathbf{\Gamma}_1'$, that determines Hamiltonian truncations (1.7) of System (1.5), i.e. as it was the intersection of polyhedra $\mathbf{\Gamma}_1$ and $\mathbf{\Gamma}_2$.

The fourth polyhedron $\mathbf{\Gamma}_2'$ according to Lemma 1.1 practically coincides with the polyhedron $\mathbf{\Gamma}_1'$.

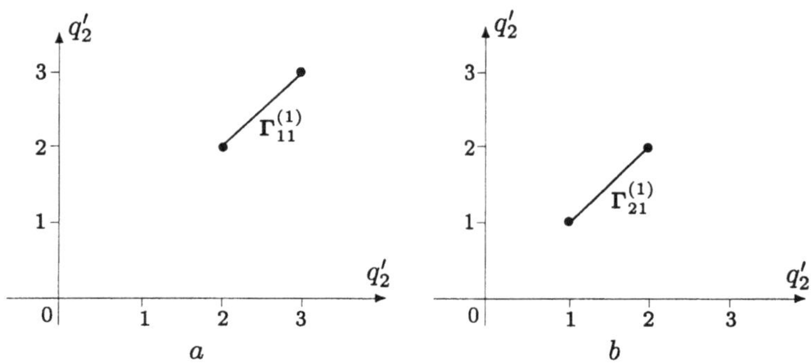

Figure 4.2: Polyhedra $\mathbf{\Gamma}_1'$ (a) and $\mathbf{\Gamma}_2'$ (b) in coordinates p_2', q' for the Hamiltonian function of Example 1.1.

Example 1.2 (continuation of Example 1.1). We compute the sets $\mathbf{S}_1'$ and $\mathbf{S}_2'$ for the Hamiltonian function of Example 1.1 according to formulae (1.9). For the points from the set $\mathbf{S}_1$ we have

$$\Pi(1,1,1,1) = (0,2,2), \quad \Pi(2,1,2,1) = (0,3,3), \quad \Pi(1,2,1,2) = (0,3,3).$$

Two last points coincide. Hence the set $\mathbf{S}_1'$ consists of two points, and the polyhedron $\mathbf{\Gamma}_1'$ is a segment. Fig. 4.2, a shows them in the plane with coordinates p_2', q'. For the points from the set $\mathbf{S}_2$ we have

$$\Pi(0,1,0,1) = (0,1,1), \quad \Pi(1,1,1,1) = (0,2,2), \quad \Pi(0,2,0,2) = (0,2,2),$$

$$\Pi(1,0,1,0) = (0,1,1), \quad \Pi(2,0,2,0) = (0,2,2).$$

So, the set $\mathbf{S}_2'$ consists of two points $(0,1,1)$ and $(0,2,2)$, and the polyhedron $\mathbf{\Gamma}_2'$ is a segment. Fig. 4.2, b shows them in the plane p_2', q'. It is clear that

$\mathbf{S}_1' = \mathbf{S}_2' + (1,1)$ in accordance with Lemma 1.1. So, according to Fig. 4.2, the Hamiltonian function h in Example 1.1 has only three truncations $\hat{h}_i$ giving the truncated Hamiltonian System (1.7). They are $\hat{h}_1 = \varrho_1\varrho_2$, $\hat{h}_2 = \varrho_1^2\varrho_2 + \varrho_1\varrho_2^2$ and $\hat{h}_3 = h$. □

We have obtained the **algorithm** for the finding of all Hamiltonian truncations (1.10) of the Hamiltonian System (1.5). It consists of four steps.

Step 1. Using the support $\mathbf{S}_1$ of the Hamiltonian function (1.6) we find the set $\mathbf{S}_1'$ with the help of Formula (1.9). Here to each point $R' = (P', q') \in \mathbf{S}_1'$ there corresponds a subset $\mathbf{S}_1(R') = \mathbf{S}_1 \cap \mathbf{L}(R')$. Let the set $\mathbf{S}_1$ consist of the finite number of points with some enumeration: $R_1, \ldots, R_s$. We enumerate the points of the set $\mathbf{S}_1'$: $R_1', \ldots, R_t'$. Now to each point $R_{j'}'$ there correspond several points R_j with indices $j = j_1, \ldots, j_k$, which are projected into $R_{j'}'$. By the set $\mathbf{S}_1$ we compose a list of points $R_1', \ldots, R_t'$ of the set $\mathbf{S}_1'$, and for each point we indicate the numbers $j = j_1, \ldots, j_k$ of the points R_j, which exhaust the subset $\mathbf{S}_1(R_{j'}')$. We obtain the *table of projections*.

Step 2. By the set $\mathbf{S}_1'$ we find faces $\mathbf{\Gamma}'^{(d')}_{1j'}$, their boundary subsets $\mathbf{S}'^{(d')}_{1j'}$ and normal cones $\mathbf{U}'^{(d')}_{1j'}$.

Step 3. We find the corresponding maximal boundary subsets $\mathbf{S}^{(d)}_{1j} = \mathbf{S}_1(\mathbf{S}'^{(d')}_{j'})$ and corresponding truncations $\hat{h}^{(d)}_j$ of the Hamiltonian function h. They give all Hamiltonian truncations (1.10) of System (1.5).

Step 4. By Lemma 1.4 and 1.5 we find the intersection of the normal cone $\mathbf{U}^{(d)}_{1j}$ with $\mathbf{L}_*$ equal to the normal cone $\mathbf{U}^{(e)}_{2k}$.

To compute the faces $\mathbf{\Gamma}'^{(d')}_{1j'}$ of the polyhedron $\mathbf{\Gamma}_1'$ and their normal cones one can use the methods described in Sections 3, 5 of Chapter 1 or the program of Section 7 of Chapter 1.

Remark 1.1. If the Hamiltonian system depends upon small parameters $\mathrm{M} = (\mu_1, \ldots, \mu_l)$, then in the expansion of the Hamiltonian function

$$h \stackrel{\text{def}}{=} \sum h_{PQT} X^P Y^Q \mathrm{M}^T$$

the support $\mathbf{S}_1 = \{(P, Q, T)\colon h_{PQT} \neq 0\}$. The points with $P = Q = 0$ are excluded from the support $\mathbf{S}_1$. All the other geometrical constructions should be considered in spaces $\mathbb{R}^{2m+l}$ and $\mathbb{R}^{m+1+l}$ in coordinates (P, Q, T) and (P', q', T) respectively (according to (1.9)). Then the Hamiltonian truncations are found with small parameters taken into account. Such an approach was used in [Bruno 1978, 1990b].

Example 1.3. In [Bruno 1978, 1990b] it was suggested that the Hamiltonian truncations of System (1.5) can be found in the following way. To each $(P, Q) \in \mathbb{R}^{2m}$ the point $\pi(P, Q) \stackrel{\text{def}}{=} (p, q) \in \mathbb{R}^2$ with coordinates $p = \sum_{i=1}^{m} p_i$,

$q = \sum_{i=1}^{m} q_i$ is put in correspondence. In the plane $\mathbb{R}^2$ the convex hull $\pi\mathbf{\Gamma}_1$ of such a projection $\pi\mathbf{S}_1$ of the support $\mathbf{S}_1$ is considered. To each face γ (i.e. to a vertex or to an edge) of its boundary $\partial\pi\mathbf{\Gamma}_1$ there corresponds the subset $\mathbf{S}_{1\gamma}$ of the set $\mathbf{S}_1$, which is projected into γ. The corresponding truncation $\hat{h}_\gamma$ of the Hamiltonian function h gives the truncated Hamiltonian system. These Hamiltonian truncations satisfy conditions of Theorem 1.1. Indeed, according to (1.9)

$$p = \sum_{i=1}^{m} p_i = \sum_{i=1}^{m} p'_i, \quad q = \sum_{i=1}^{m} q_i = q',$$

i.e. the projection π is the result of the two consecutive projections

$$\Pi(P,Q) = (P',q'), \quad \Pi_1(P',q') \stackrel{\text{def}}{=} (p',q') = (p,q).$$

Consequently, for each face γ the intersection $\mathbf{\Gamma}'_1 \cap \Pi_1^{-1}(\gamma)$ is the face of the polyhedron $\mathbf{\Gamma}'_1$, to which corresponds a Hamiltonian truncation by Theorem 1.1. The same truncation corresponds to the face γ. □

Theorem 1.2 *The truncation $\hat{h}_j^{(d)}(X,Y)$ of the Hamiltonian function $h(X,Y)$ corresponding to the face $\mathbf{\Gamma}_{1j}^{(d)}$ gives the truncated Hamiltonian System* (1.7) *if and only if the normal cone $\mathbf{U}_{1j}^{(d)}$ includes the vector (S,T), where $S = (s_1,\dots,s_m)$ and $T = (t_1,\dots,t_m)$, with the condition*

$$s_i + t_i = c = \text{const}\,, \quad i = 1,\dots,m. \tag{1.16}$$

Proof. Let such a vector exist, i.e. $\hat{h}_j^{(d)} = \hat{h}_{(S,T)}$. For every vector $R = (P,Q)$ according to (1.9) we have

$$\langle P',S\rangle + q't_m = \sum_{i=1}^{m-1}(p_i - q_i)s_i + \left(p_m + \sum_{i=1}^{m-1} q_i\right)s_m + \left(q_m + \sum_{i=1}^{m-1} q_i\right)t_m =$$

$$= \langle P,S\rangle - \sum_{i=1}^{m-1} q_i s_i + (s_m + t_m)\sum_{i=1}^{m-1} q_i + q_m t_m = \langle P,S\rangle + \sum_{i=1}^{m-1} q_i(s_m + t_m -$$

$$-s_i) + q_m t_m = \langle P,S\rangle + \sum_{i=1}^{m-1} q_i t_i + q_m t_m = \langle P,S\rangle + \langle Q,T\rangle.$$

Here the last but one equality uses the condition (1.16) in the form

$$t_i = s_m + t_m - s_i, \quad i = 1,\dots,m-1. \tag{1.17}$$

The obtained equality

$$\langle P',S\rangle + q't_m = \langle P,S\rangle + \langle Q,T\rangle \tag{1.18}$$

means that the intersection of the polyhedron $\mathbf{\Gamma}_1$ with its supporting hyperplane $\mathbf{L}_{(S,T)}$ in $\mathbb{R}^{2m}$ corresponds to the intersection of the polyhedron $\mathbf{\Gamma}'_1$ with its supporting hyperplane $\mathbf{L}'_{(S,t_m)}$ in $\mathbb{R}^{m+1}$.

Conversely, let $\mathbf{\Gamma}'^{(d')}_{1j}$ be some face of the polyhedron $\mathbf{\Gamma}'_1$ and vector (S, t_m) belong to the corresponding normal cone $\mathbf{U}'^{(d')}_{1j}$. We define the numbers $t_1, \ldots, t_{m-1}$ according to formula (1.17) and form the vector $T = (t_1, \ldots, t_m)$. Then the vector (S, T) has Property (1.16) and consequently Property (1.18). Under the projection (1.8), (1.9) the intersection of the supporting hyperplane $\mathbf{L}_{(S,T)}$ with the set $\mathbf{S}_1$ in $\mathbb{R}^{2m}$ entirely transforms into the intersection of the supporting hyperplane $\mathbf{L}'_{(S,t_m)}$ with the set $\mathbf{S}'_1$ in $\mathbb{R}^{m+1}$. □

The sufficiency of Property (1.16) for obtaining the Hamiltonian truncated systems follows from Lemma 3.3 of Chapter II [Kozlov and Furta 1996]. In truth, there was the additional restriction $\langle P, S\rangle + \langle Q, T\rangle = c+1$ over $(P, S) \in \mathbf{S}^{(d)}_{1j}$, but here it is not important.

Remark 1.2. If for System (1.5) there is the cone of the problem $\mathbf{K}$ (see Chapter 1) given by inequalities

$$\langle W, K_k\rangle < 0, \quad k = 1, \ldots, t, \tag{1.19}$$

then under projection (1.8), (1.9), it transforms into the cone $\mathbf{K}'$ given by inequalities

$$\langle W', K'_k\rangle < 0, \quad k = 1, \ldots, t, \tag{1.20}$$

where

$$K'_k = \Pi K_k, \quad k = 1, \ldots, t. \tag{1.21}$$

It is relevant for the first step of the algorithm. Since not only the cone $\mathbf{K}$ is projected into the cone $\mathbf{K}'$, then for the found by Lemma 1.4 normal cones $\mathbf{U}^{(d)}_{2j}$ one have to check whether they intersect the cone of the problem $\mathbf{K}$. It is relevant for the fourth step of the algorithm.

In the following Sections we consider some well known Hamiltonian systems, and with the help of the algorithm described in this Section we find all their Hamiltonian truncated systems.

2. The generalized Henon–Heiles system

Chang et al. [1982], Yoshida [1983], Lunev [1994b] considered the generalized Henon–Heiles system

$$\begin{aligned} \dot{x}_1 &= ay_1 + 2dy_1y_2, & \dot{y}_1 &= -x_1, \\ x_2 &= by_2 + dy_1^2y_2 - cy_2^2, & \dot{y}_2 &= -x_2, \end{aligned} \tag{2.1}$$

where $a, b, c, d \neq 0$. It is a Hamiltonian system with the Hamiltonian function

$$h = (1/2)(x_1^2 + x_2^2 + ay_1^2 + by_2^2) + dy_1^2 y_2 - cy_2^3/3. \tag{2.2}$$

The Hamiltonian function (2.2) has the support

$$\mathbf{S}_1 = \{R_1 = (2,0,0,0), \quad R_2 = (0,2,0,0), \quad R_3 = (0,0,2,0),$$
$$R_4 = (0,0,0,2), \quad R_5 = (0,0,2,1), \quad R_6 = (0,0,0,3)\}. \tag{2.3}$$

Following the algorithm of Section 1 step by step, we find all truncations $\hat{h}_j^{(d)}$ of the Hamiltonian function (2.2) giving all truncated Hamiltonian systems.

Step 1. After the projection (1.9) the set $\mathbf{S}_1$ transforms into the set in the space $\mathbb{R}^3$:

$$\mathbf{S}_1' = \{R_1' = (2,0,0), \; R_2' = (0,2,0), \; R_3' = (0,0,2),$$
$$R_4' = (-2,2,2), \; R_5' = (-2,2,3), \; R_6' = (0,0,3)\}. \tag{2.4}$$

Here

$$\Pi R_j = R_j', \quad j = 1, \ldots, 6. \tag{2.5}$$

Step 2. The corresponding polyhedron $\mathbf{\Gamma}_1'$ and the points of the set $\mathbf{S}_1'$ are shown in Fig. 4.3. In Tables 4.1 and 4.2 the results of computations of the polyhedron $\mathbf{\Gamma}_1'$ by the program of Section 7 of Chapter 1 are cited. The polyhedron $\mathbf{\Gamma}_1'$ contains 5 faces of dimension 2, 9 edges and 6 vertices. Altogether 20 different boundary subsets.

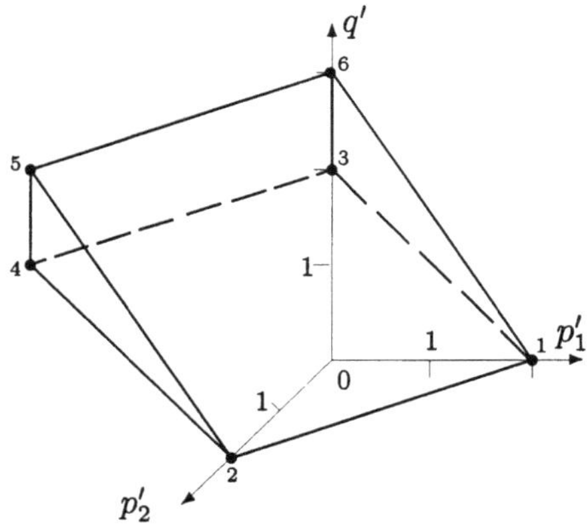

Figure 4.3: The polyhedron $\mathbf{\Gamma}_1'$ for the set (2.4). The numbers j of points R_j' are cited.

Table 4.1: Table of correspondence of the set (2.4).

					j	1	2	3	4	5	6
i	t_i	B_i,	N_i,	D_i	T_j	V	V	V	V	V	V
1	N	−1	−1	0		−	−	+	+	+	+
2	N	3	3	2		+	+	−	−	+	+
3	N	0	−1	0		+	−	+	−	−	+
4	N	0	1	0		−	+	−	+	+	−
5	N	−1	−1	−1		+	+	+	+	−	−

Table 4.2: Boundary subsets of the set (2.4).

d	i	j	k	m
2	1	3 4 5 6	1	1
2	2	1 2 5 6	2	1
2	3	1 3 6	3	1
2	4	2 4 5	4	1
2	5	1 2 3 4	5	1
1	1	5 6	1 2	1 2
1	2	3 6	1 3	1 3
1	3	4 5	1 4	1 4
1	4	3 4	1 5	1 5
1	5	1 6	2 3	2 3
1	6	2 5	2 4	2 4
1	7	1 2	2 5	2 5
1	8	1 3	3 5	3 5
1	9	2 4	4 5	4 5
0	1	6	1 2 3	1 2 5
0	2	5	1 2 4	1 3 6
0	3	3	1 3 5	2 4 8
0	4	4	1 4 5	3 4 9
0	5	1	2 3 5	5 7 8
0	6	2	2 4 5	6 7 9

Step 3. Taking into account the table of projections (2.5), we enumerate two-dimensional faces $\mathbf{\Gamma}_{1j}^{'(2)}$ of the polyhedron $\mathbf{\Gamma}_1^{'}$ and corresponding truncations $\hat{h}_j$ of the Hamiltonian function h giving truncated Hamiltonian systems. To the face $\mathbf{\Gamma}_{11}^{'(2)}$ (the first lines of Tables 4.1 and 4.2) there corresponds

$$\hat{h}_1 = (1/2)(x_1^2 + x_2^2 + ay_1^2 + by_2^2).$$

To the face $\mathbf{\Gamma}_{12}^{'(2)}$ (the second lines of Tables 4.1 and 4.2) there corresponds

$$\hat{h}_2 = (1/2)(x_2^2 + by_2^2) + dy_1^2 y_2.$$

To the face $\mathbf{\Gamma}_{13}^{'(2)}$ (the third lines of Tables 4.1 and 4.2) there corresponds

$$\hat{h}_3 = (1/2)(ay_1^2 + by_2^2) + dy_1^2 y_2 - cy_2^3/3.$$

To the face $\mathbf{\Gamma}_{14}^{'(2)}$ (the fourth lines of Tables 4.1 and 4.2) there corresponds

$$\hat{h}_4 = (1/2)(x_1^2 + x_2^2) + dy_1^2 y_2 - cy_2^3/3.$$

To the face $\mathbf{\Gamma}_{15}^{'(2)}$ (the fifth lines of Tables 4.1 and 4.2) there corresponds the truncation

$$\hat{h}_5 = (1/2)(x_1^2 + by_2^2) - cy_2^3/3.$$

The four-dimensional polyhedra $\mathbf{\Gamma}_1$ and $\mathbf{\Gamma}_2$ were also computed for the function (2.2) and System (2.1) by Bruno and Soleev [1995e] with the help of the program in Section 7 of Chapter 1. The results of computations are as follows. The polyhedron $\mathbf{\Gamma}_1$ has 6 faces of dimension 3, 13 faces of dimension 2, 13 edges and 6 vertices (i.e. all points (2.3) are vertices). The polyhedron $\mathbf{\Gamma}_2$ has 6 faces of dimension 3, 13 faces of dimension 2, 15 edges and 7 vertices. Here to the two-dimensional faces $\mathbf{\Gamma}_{11}^{'(2)}$ and $\mathbf{\Gamma}_{14}^{'(2)}$ of the polyhedron $\mathbf{\Gamma}_1'$ there correspond two three-dimensional faces of the polyhedron $\mathbf{\Gamma}_1$ (truncations $\hat{h}_1$ and $\hat{h}_4$), and to the other tree two-dimensional faces $\mathbf{\Gamma}_{12}^{'(2)}$, $\mathbf{\Gamma}_{13}^{'(2)}$ and $\mathbf{\Gamma}_{15}^{'(2)}$ of the polyhedron $\mathbf{\Gamma}_1'$ there correspond three two-dimensional faces of the polyhedron $\mathbf{\Gamma}_1$ (truncations $\hat{h}_2$, $\hat{h}_3$ and $\hat{h}_5$). To all edges and vertices of the polyhedron $\mathbf{\Gamma}_1'$ there correspond edges and vertices of the polyhedron $\mathbf{\Gamma}_1$. Thus, only truncations $\hat{h}_1$ and $\hat{h}_4$ correspond to the three-dimensional faces of the polyhedron $\mathbf{\Gamma}_1$. Other truncations $\hat{h}_2$, $\hat{h}_3$ and $\hat{h}_5$ correspond to its two-dimensional faces. The truncation $\hat{h}_1$ corresponds to the first linear approximation to System (2.1), and the truncation $\hat{h}_4$ is the Hamiltonian function of the Henon–Heiles system [Henon and Heiles 1964].

3. The Sokol'skii cases of zero frequencies

Sokol'skii [1981] considered the autonomous Hamiltonian system (1.5) with two degrees of freedom ($m = 2$) in the neighborhood of the stationary point $X = Y = 0$ in cases where the nonzero matrix of the linearized system has four-fold zero eigenvalue. In [Sokol'skii 1980, 1981] the several such cases were considered, and in all of them the truncation of the Hamiltonian function gave the truncated Hamiltonian system. We are going to study these cases separately (in our notation). Here we are interested only in such truncated systems, which have non-empty intersection of the normal cone with the negative orthant, i.e. the cone of the problem $\mathbf{K}$ is given by inequalities (1.19), where $K_k = E_k$, $k = 1, 2, 3, 4$. According to Remark 1.2, the cone $\mathbf{K}'$ is given by inequalities (1.20), (1.21), where according to (1.8), (1.9)

$$K_1' = (1,0,0), \quad K_2' = (0,1,0), \quad K_3' = (-1,1,1), \quad K_4' = (0,0,1). \tag{3.1}$$

The same vectors form the skeleton for the dual cone $\mathbf{K}'^*$. The Hamiltonian function is a power series over x_1, x_2, y_1, y_2 of the form

$$h = H_2 + H_3 + H_4 + \dots, \tag{3.2}$$

where $H_i = H_i(x_1, x_2, y_1, y_2)$ are homogeneous polynomials of power i.

1. The first case. Let in the Hamiltonian function (3.2)

$$H_2 = -x_1 x_2 + \delta y_1^2/2, \quad \delta = \pm 1, \tag{3.3}$$

and the form H_3 be generic. Obviously, the support $\mathbf{S}_1 \stackrel{\text{def}}{=} \mathbf{S}(h) \subset \mathbb{Z}_+^4$. We extract out of it the leading subset $\mathbf{S}_1/\mathbb{R}_+^4$ (see Section 8, Chapter 1). We take $K = (1,1,1,1)$. Since the form H_3 is generic, then according to Example 8.1 of Chapter 1 we may restrict ourselves with points $R \in \mathbf{S}_1$ with $\langle K, R\rangle \le 3$. According to (3.3), when $\langle K, R\rangle = 2$ the points of the set $\mathbf{S}_1$ are

$$R_1 = (0,0,1,1), \quad R_2 = (0,0,2,0). \tag{3.4}$$

If we exclude from points $R \in \mathbb{Z}_+^4$ with $\langle K, R\rangle = 3$ the points R for which

$$R - R_1 \ge 0 \quad \text{or} \quad R - R_2 \ge 0, \tag{3.5}$$

then the points $R_3 - R_{14}$ are left, which are cited in the upper part of Table 4.3. Below them in the middle part of Table 4.3 their projections $R_j' = \Pi R_j$ are cited according to (1.9). Now with the help of the program from Section 7 of Chapter 1, Problem 1$'$ of Section 6 of Chapter 1 is solved for the set of points $R_1' - R_{14}'$ and for the cone of the problem $\mathbf{K}'$. As the result we obtain the table of correspondence, which is cited in the lower part of Table 4.3. It shows that the surface $\partial_{\mathbf{K}'}\mathbf{\Gamma}_1'$ consists of three faces $\mathbf{\Gamma}_{11}'^{(2)} \supset \{R_1', R_2', R_8'\}$, $\mathbf{\Gamma}_{12}'^{(2)} \supset \{R_1', R_2', R_{14}'\}$, $\mathbf{\Gamma}_{13}'^{(2)} \supset \{R_1', R_8', R_9', R_{11}', R_{13}'\}$; of seven edges and five vertices R_1', R_2', R_8', R_{13}', R_{14}' (see Fig. 4.4). From Table 4.3 for every normal (U, v) according to Lemma 1.4, we find the normal (U, V) corresponding to the face of the polyhedron $\mathbf{\Gamma}_2$, where $V = (v_1, v_2)$ and

$$v_1 = v + u_2 - u_1, \quad v_2 = v. \tag{3.6}$$

Table 4.3: The first case of Section 3. Points R_j of the leading subset $\mathbf{S}/\mathbb{R}^4_+$, their projections R'_j, the table of correspondence for the cone of the problem $\mathbf{K}'$.

i	t_i	$N_i,\ D_i$															
			j	1	2	3	4	5	6	7	8	9	10	11	12	13	14
			R_j	1	0	0	1	0	2	0	0	1	0	2	0	3	0
				1	0	0	0	1	0	2	0	0	1	0	2	0	3
				0	2	1	1	1	1	1	0	0	0	0	0	0	0
				0	0	2	1	1	0	0	3	2	2	1	1	0	0
			T_j	V	V	V	d	d	d	V	V	d	d	d	d	V	V
			R^0_j	1	−2	−1	0	−1	1	−1	0	1	0	2	0	3	0
				1	2	1	1	2	1	3	0	0	1	0	2	0	3
				0	2	3	2	2	1	1	3	2	2	1	1	0	0
1	N	−1 −2 −1		+	−	−	−	−	−	−	+	+	−	+	−	+	−
2	N	−5 −7 −4		+	+	−	−	−	−	−	+	−	−	−	−	−	−
3	N	0 0 −1		+	−	−	−	−	−	−	−	−	−	−	−	+	+
4	N	−1 0 −1		−	+	−	−	−	−	+	−	−	−	−	−	−	+
5	N	−1 −1 0		−	+	+	−	−	−	−	+	−	−	−	−	−	−
6	N	−4 −2 −5		+	+	−	−	−	−	−	−	−	−	−	−	−	+
7	D	0 −1 0		−	−	−	−	−	−	−	+	+	−	+	−	+	−

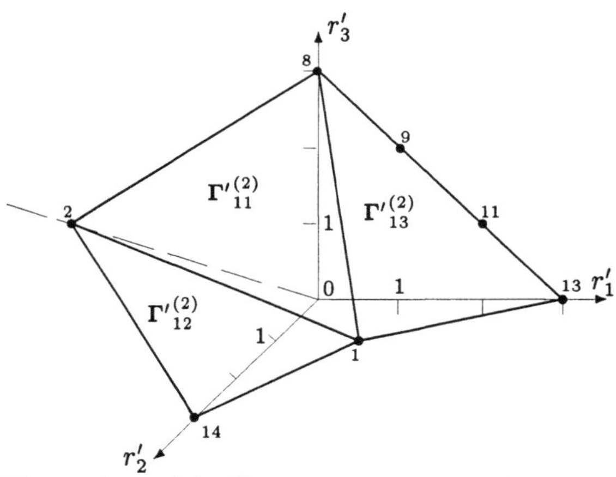

Figure 4.4: The surface $\mathbf{\Gamma}'_1/\mathbf{K}'^*$ for the first case of Section 3. Dashed line shows the ray spanned on K'_3.

Table 4.3 shows that the face $\mathbf{\Gamma}'^{(2)}_{11}$ has the normal vector $N'_1 \stackrel{\text{def}}{=} (U, v) = -(5,\ 7,\ 4)$. To it according to (3.6) there corresponds the normal vector $(U, V) = -(5, 7, 6, 4)$ lying in the cone of the problem $\mathbf{K}$. The corresponding truncated Hamiltonian function is

$$\hat{h}_1 = -x_1 x_2 + \delta y_1^2/2 + a y_2^3, \quad a \neq 0,$$

and the corresponding truncated Hamiltonian system

$$\dot{x}_1 = \delta y_1, \quad \dot{x}_2 = 3ay_2^2, \quad \dot{y}_1 = x_2, \quad \dot{y}_2 = x_1 \tag{3.7}$$

is a quasi-homogeneous one. If it has a solution of the form

$$x_1 = \alpha_1 t^{-5}, \quad x_2 = \alpha_2 t^{-7}, \quad y_1 = \beta_1 t^{-6}, \quad y_2 = \beta_2 t^{-4}, \tag{3.8}$$

then by virtue of System (3.7) coefficients of the solution are related by equalities

$$-5\alpha_1 = \delta\beta_1, \quad -7\alpha_2 = 3a\beta_2^2, \quad -6\beta_1 = \alpha_2, \quad -4\beta_2 = \alpha_1,$$

i.e.

$$\alpha_2 = -6\beta_1 = 30\delta\alpha_1 = -120\delta\beta_2 \neq 0, \quad \beta_2 = 280\delta a^{-1} \neq 0. \tag{3.9}$$

2. The second case. The problem of the motion of a dynamically symmetric satellite with respect to its mass center moving along a circular orbit is described by the Hamiltonian system with two degrees of freedom. We consider it near the stationary point corresponding to the conic precession for the values of parameters $\alpha = 4/3$, $\beta = 0$, $\theta_0 = \pi/3$. According to Formulae (2.7) and (2.8) from [Sokol'skii 1981], (5)–(7) from [Markeev et al. 1977], and to linear transformation with the matrix written in [Sokol'skii 1981] after Formula (2.8), the Hamiltonian function in this case has the form (3.2), (3.3) with

$$H_3 = \left(y_2 - \frac{1}{2}y_1\right)\left(\frac{1}{\sqrt{3}}y_1y_2 - \frac{\sqrt{3}}{2}y_1^2 - \frac{1}{\sqrt{3}}x_1^2 - \frac{1}{\sqrt{3}}x_1x_2 - \frac{1}{4\sqrt{3}}x_2^2\right), \tag{3.10}$$

$$H_4 = \frac{1}{6}y_2^4 + \frac{2}{9}x_1^4 + \frac{1}{72}x_2^4 + \tilde{H}_4, \tag{3.11}$$

where $\delta = 1$ and $\tilde{H}_4$ does not contain monomials y_2^4, x_1^4, x_2^4, $x_1y_2^3$, $x_2y_2^3$. We find at first the leading set $\mathbf{S}_1/\mathbb{R}_+^4$, where $\mathbf{S}_1 = \mathbf{S}(h)$. When $\langle K, R\rangle = 2$, there are points (3.4) in $\mathbf{S}_1$. According to (3.10) when $\langle K, R\rangle = 3$, there are nine points $(0,0,3,0)$, $(0,0,2,1)$, $(0,0,1,2)$, $(2,0,1,0)$, $(1,1,1,0)$, $(0,2,1,0)$, $(2,0,0,1)$, $(1,1,0,1)$, $(0,2,0,1)$ in $\mathbf{S}_1$. For the four of them (the first, second, fifth, eighth) one of inequalities (3.5) is satisfied. Hence we discard them. There remain five points R_3–R_7 cited in the upper part of Table 4.4. According to (3.11), all points R of the support of the form $\tilde{H}_4$ are discarded, since for all of them at least one of remainders $R - R_j \geq 0$, $j = 1, \ldots, 7$. There remain only points $R_8 = (0,0,0,4)$, $R_9 = (4,0,0,0)$, $R_{10} = (0,4,0,0)$. Points R_1–R_{10} are cited in the upper part of Table 4.4. In its middle part there cited their projections $R_j' = \Pi R_j$ according to (1.9). Using the program from Section 7 of Chapter 1, we compute the table of correspondence for points

R'_1–R'_{10} and for the cone of the problem $\mathbf{K}'$. This table of correspondence is cited in the lower part of Table 4.4 (see Fig. 4.5). This table shows that the surface $\partial_{\tilde{\mathbf{K}}'}\mathbf{\Gamma}'_1$ consists of three faces $\mathbf{\Gamma}^{'(2)}_{11} \supset \{R'_1, R'_2, R'_3, R'_6, R'_8\}$, $\mathbf{\Gamma}^{'(2)}_{12} \supset \{R'_1, R'_2, R'_7, R'_{10}\}$, $\mathbf{\Gamma}^{'(2)}_{13} \supset \{R'_1, R'_6, R'_9\}$, of eight edges and six vertices R'_1, R'_2, R'_6, R'_8, R'_9, R'_{10}.

Table 4.4: The points R_j, their projections R'_j and the table of correspondence for the second case of Section 3.

i	t_i	N_i, D_i		1	2	3	4	5	6	7	8	9	10
			j	1	2	3	4	5	6	7	8	9	10
			R_j	1	0	0	2	0	2	0	0	4	0
				1	0	0	0	2	0	2	0	0	4
				0	2	1	1	1	0	0	0	0	0
				0	0	2	0	0	1	1	4	0	0
			T_j	V	V	d	d	d	V	d	V	V	V
			R^0_j	1	−2	−1	1	−1	2	0	0	4	0
				1	2	1	1	3	0	2	0	0	4
				0	2	3	1	1	1	1	4	0	0
1	N	−1 −3 −2		+	−	−	−	−	+	−	−	+	−
2	N	−3 −5 −2		+	+	+	−	−	+	−	+	−	−
3	N	0 0 −1		+	−	−	−	−	−	−	−	+	+
4	N	0 −1 0		−	−	−	−	−	+	−	+	+	−
5	N	−3 −1 −4		+	+	−	−	+	−	−	−	−	+
6	D	−1 −1 0		−	+	+	−	−	−	−	+	−	−
7	D	−1 0 −1		−	+	−	−	+	−	−	−	−	+

Table 4.4 shows that the normal vector $N'_{11} = -(3,5,2)$ satisfies to all inequalities (1.20) with vectors (3.1), and in particular, all its components are negative. The corresponding to it normal vector $N_{11} = -(3,5,4,2) \in \mathbf{K}$ according to Lemma 1.4 and to Formula (3.6). The corresponding truncated Hamiltonian function is

$$\begin{aligned} \hat{h}_1 &= -x_1x_2 + \delta y_1^2/2 + by_1y_2^2 + cx_1^2y_2 + dy_2^4, \\ \delta &= 1, \quad b = 1/\sqrt{3}, \quad c = -1/\sqrt{3},\ d = 1/6, \end{aligned} \tag{3.11'}$$

and the truncated Hamiltonian system is

$$\begin{aligned} \dot{x}_1 &= \delta y_1 + by_2^2, & \dot{y}_1 &= x_2 - 2cx_1y_2, \\ \dot{x}_2 &= 2by_1y_2 + cx_1^2 + 4dy_2^3, & \dot{y}_2 &= x_1. \end{aligned} \tag{3.12}$$

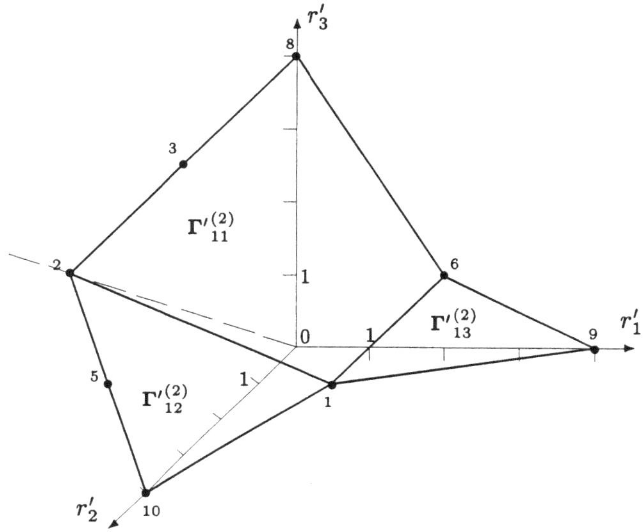

Figure 4.5: The surface $\mathbf{\Gamma}_1'/\mathbf{K}'^*$ for the second case of Section 3. Notation of Fig. 4.4.

It is a quasi-homogeneous one. If it has a solution of the form

$$x_1 = \alpha_1 t^{-3}, \quad x_2 = \alpha_2 t^{-5}, \quad y_1 = \beta_1 t^{-4}, \quad y_2 = \beta_2 t^{-2}, \tag{3.13}$$

then by virtue of System (3.12) coefficients of the solution are related by equations

$$-3\alpha_1 = \delta\beta_1 + b\beta_2^2, \quad -4\beta_1 = \alpha_2 - 2c\alpha_1\beta_2,$$
$$-5\alpha_2 = 2b\beta_1\beta_2 + c\alpha_1^2 + 4d\beta_2^3, \quad -2\beta_1 = \alpha_1.$$

Excluding from this system consecutively α_1, β_1 and α_2, we obtain

$$\alpha_1 = -3\beta_2, \quad \delta\beta_1 = 6\beta_2 - b\beta_2^2,$$
$$\alpha_2 = -4\beta_1 - 4c\beta_2^2 = -24\delta\beta_2 + 4(b\delta - c)\beta_2^2,$$
$$120\delta\beta_2 + 20(c - b\delta)\beta_2^2 = 2b\delta(6\beta_2 - b\beta_2^2) + 4c\beta_2^2 + 4d\beta_2^3.$$

Cancelling $2\beta_2$ from the last equation and collecting similar terms, we obtain the square equation

$$(2d - \delta b^2)\beta_2^2 + (16\delta b - 8c)\beta_2 - 60\delta = 0.$$

This equation has a real root $\beta_2 \neq 0, \infty$ only if its discriminant D satisfies relations

$$D \stackrel{\text{def}}{=} (8\delta b - 4c)^2 + 60\delta(2d - \delta b^2) > 0, \quad \text{or} \quad D = 0 \quad \text{and} \quad 2d - \delta b^2 \neq 0.$$

Cancelling 4, we obtain the condition

$$D/4 \overset{\text{def}}{=} b^2 - 16\delta bc + 4c^2 + 30\delta d > 0, \quad \text{or} \quad D/4 = 0 \quad \text{and} \quad 2d \neq \delta b^2. \quad (3.14)$$

System (3.12) has the real solution (3.13) with all $\alpha_i, \beta_i \neq 0$ only under Condition (3.14). When the values of coefficients are among (3.11′) the left hand part of Inequality (3.14) is equal to 12, i.e. Condition (3.14) is satisfied.

3. The third case. The Hamiltonian function is (3.2), where

$$H_2 = \delta_1 y_1^2 + \delta_2 y_2^2, \quad \delta_1 = \pm 1, \quad \delta_2 = \pm 1, \quad (3.15)$$

and the form H_3 is generic one. We denote $\mathbf{S}_1 = \mathbf{S}(h)$. According to (3.15), when $\langle K, R \rangle = 2$ there are two points in $\mathbf{S}_1$

$$R_1 = (0,0,2,0), \quad R_2 = (0,0,0,2).$$

From points $R \in \mathbb{Z}_+^4$ with $\langle K, R \rangle = 3$ we discard points with the property (3.5). The points that remain are R_3–R_{13}, they are represented in the upper part of Table 4.5. The middle part of the table cites projections of these points $\Pi R_j = R'_j$, $j = 1, \dots, 13$ according to (1.9). Using the program from Section 7 of Chapter 1, for the set of points R'_1–R'_{13} and the cone of the problem $\mathbf{K}'$ we compute the table of correspondence, which is cited in the lower part of Table 4.5. It shows that the surface $\mathbf{\Gamma}'_1/\mathbf{K}'$ consists of one face $\mathbf{\Gamma}'^{(2)}_{11} \supset \{R'_1, R'_2, R'_{10}\text{–}R'_{13}\}$, four edges and four vertices R'_1, R'_2, R'_{10}, R'_{13} (see Fig. 4.6).

Table 4.5: The points R_j, their projections R'_j and the table of correspondence for the third case of Section 3.

i	t_i	N_i, D_i														
			j	1	2	3	4	5	6	7	8	9	10	11	12	13
			R_j	0	0	1	0	2	1	0	2	0	3	2	1	0
				0	0	0	1	0	1	2	0	2	0	1	2	3
				2	0	1	1	1	1	1	0	0	0	0	0	0
				0	2	1	1	0	0	0	1	1	0	0	0	0
			T_j	V	V	V	V	d	d	V	V	d	V	d	d	V
			R_j^0	−2	0	0	−1	1	0	−1	2	0	3	2	1	0
				2	0	1	2	1	2	3	0	2	0	1	2	3
				2	2	2	2	1	1	1	1	1	0	0	0	0
1	N	−1 0 −1		+	−	−	−	−	−	+	−	−	−	−	−	+
2	N	0 −1 0		−	+	−	−	−	−	−	+	−	+	−	−	−
3	N	−2 −2 −3		+	+	−	−	−	−	−	−	−	+	+	+	+
4	D	−1 −1 0		+	+	−	−	−	−	−	−	−	−	−	−	−
5	D	0 0 −1		−	−	−	−	−	−	−	−	−	+	+	+	+

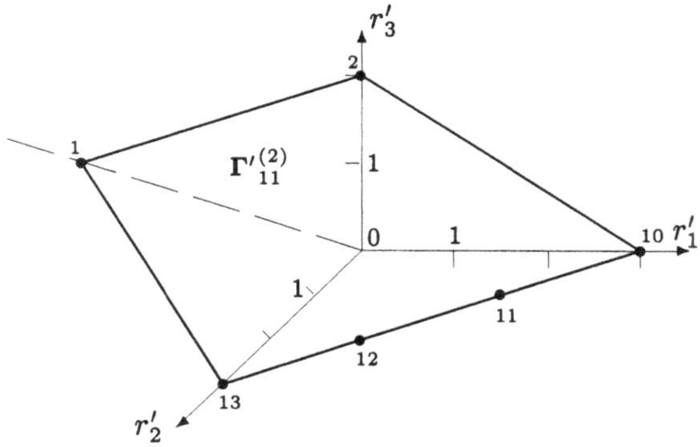

Figure 4.6: The surface $\Gamma'_1/\mathbf{K}'^*$ for the third case of Section 3. Notation of Fig. 4.4.

Table 4.5 shows that the normal vector $N'_{11} = -(2,2,3)$ lies in the cone $\mathbf{K}'$. According to Lemma 1.4 and Formula (3.6), to the cone there corresponds the normal vector $N_{11} = -(2,2,3,3) \in \mathbf{K}$. The corresponding truncated Hamiltonian function is

$$\begin{aligned} \hat{h}_1 &= \delta_1 y_1^2/2 + \delta_2 y_2^2/2 + \tilde{h}(x_1, x_2), \\ \tilde{h} &= a x_1^3 + b x_1^2 x_2 + c x_1 x_2^2 + d x_2^3, \end{aligned} \tag{3.16}$$

and the truncated Hamiltonian system is

$$\dot{x}_i = \delta_i y_i, \quad \dot{y}_i = -g_i(x_1, x_2), \quad i = 1, 2, \tag{3.17}$$

where $g_i \stackrel{\text{def}}{=} \partial \tilde{h}/\partial x_i$, $i = 1, 2$.

If this quasi-homogeneous system has solution of the form

$$x_i = \alpha_i t^{-2}, \quad y_i = \beta_i t^{-3}, \quad i = 1, 2, \tag{3.18}$$

then by virtue of System (3.17), coefficients of the solution are related by equations

$$-2\alpha_i = \delta_i \beta_i, \quad -3\beta_i = -g_i(\alpha_1, \alpha_2), \quad i = 1, 2. \tag{3.19}$$

Excluding β_1 and β_2, we obtain the system of two equations

$$-6\delta_i \alpha_i = g_i(\alpha_1, \alpha_2), \quad i = 1, 2. \tag{3.20}$$

From this system there follows one homogeneous equation of the third order $f(\alpha_1, \alpha_2) \stackrel{\text{def}}{=} \alpha_1 g_2 - \delta \alpha_2 g_1 = 0$, where $\delta \stackrel{\text{def}}{=} \delta_1 \delta_2$, i.e.

$$f(\alpha_1, \alpha_2) \stackrel{\text{def}}{=} b\alpha_1^3 + (2c - 3a\delta)\alpha_1^2 \alpha_2 + (3d - 2b\delta)\alpha_1 \alpha_2^2 - c\delta \alpha_2^3 = 0. \tag{3.21}$$

This cubic equation always has the real root $\varkappa = \alpha_1/\alpha_2$. If $\varkappa \neq 0, \infty$ and

$$g_i(\varkappa, 1) \neq 0, \quad i = 1, 2, \tag{3.22}$$

then System (3.19) has the non-zero real solution

$$\alpha_1 = -6\delta_2/g_2(\varkappa, 1), \;\; \alpha_2 = -6\delta_1/g_1(\varkappa, 1), \;\; \beta_i = -2\delta_i\alpha_i, \;\; i = 1, 2. \tag{3.23}$$

If $g_1(\varkappa, 1) = 0$ and $\varkappa \neq 0, \infty$, then $g_2(\varkappa, 1) = 0$, and to such a $\varkappa$ there is no corresponding non-zero solution to System (3.19)

Lemma 3.1 *If the sets of solutions to equations*

$$g_1(\varkappa, 1) = 0, \quad g_2(\varkappa, 1) = 0 \tag{3.24}$$

coincide in the complex sphere for $\varkappa$, then this set consists of one point, that is the two-fold root of each equation.

Proof. Under the terms of Lemma, the coefficients of both equations are proportional, i.e. $3a = \nu b$, $2b = \nu 2c$, $c = \nu 3d$. Hence $a = \nu^3 d$, $b = 3\nu^2 d$, $c = 3\nu d$, i.e. $\tilde{h}(\varkappa, 1) = d(\nu\varkappa + 1)^3$. If $\nu = 0$ or $\nu = \infty$, then $\tilde{h}(\varkappa, 1) = d$ or $\tilde{h}(\varkappa, 1) = a\varkappa^3$ respectively. Since $g_1(\varkappa, 1) = 3d\nu(\nu\varkappa + 1)^2$ and $g_2(\varkappa, 1) = 3d(\nu\varkappa + 1)^2$, then in all these cases $\varkappa = -\nu^{-1}$ is the unique two-fold root of each equation (3.24). □

We consider polynomials in coefficients a, b, c, d, δ:

$$D_1 = 9d^2 + 8c^2\delta - 12ac, \;\; D_2 = 9a^2 + 8b^2\delta - 12bd;$$

$$R = 27a^2d^2 + 4ac^3 + 4b^3d - b^2c^2 - 18abcd;$$

$$D_3 = 18ABCD + B^2C^2 - 4AC^3 - 4DB^3 - 27A^2D^2,$$

where $A = b$, $B = 2c - 3a\delta$, $C = 3d - 2b\delta$, $D = -c\delta$. Here $3R$ is the resultant of polynomials g_1 and g_2, $-R$ is the discriminant of the polynomial $\tilde{h}$, and D_3 is the discriminant of the polynomial f.

Theorem 3.1 *The system of equations* (3.20) *does not have the real solution $\alpha_1 \neq 0$, $\alpha_2 \neq 0$ only in ten cases*:

1) $b \neq 0, \; c \neq 0, \; R = 0, \; D_3 < 0;$
2) $\delta_1\delta_2 = -1, \; \pm 3a = b = \pm c = 3d \neq 0;$
3) $b = 0, \; 3a - 2c\delta_1\delta_2 \neq 0, \; D_1 < 0;$
4) $c = 0, \; 2b - 3d\delta_1\delta_2 \neq 0, \; D_2 < 0;$
5) $\delta_1\delta_2 = -1, \; b = 0, \; 3a = -2c, \; 2c^2 = 9d^2 \neq 0;$
6) $\delta_1\delta_2 = -1, \; c = 0, \; 3d = -2b, \; 9a^2 = 2b^2 \neq 0;$
7) $b = d = 0, \; 3a = 2c\delta_1\delta_2 \neq 0;$
8) $a = b = c = 0, \; d \neq 0;$
9) $b = c = d = 0, \; a \neq 0;$

10) $a = c = 0,\ 2b = 3d\delta_1\delta_2 \neq 0$.

Proof. We denote as k the number of different real roots $\varkappa$ of the equation

$$f(\varkappa, 1) = 0, \tag{3.25}$$

which are different from zero and infinity.

If $k > 1$, then according to Lemma 3.1, one of these roots satisfies the inequalities (3.22), and there is Solution (3.23). Hence in order to trace exceptional cases, it is sufficient to consider the cases when $k = 1$ and $k = 0$. The multiplicity of roots $\varkappa = 0$, $\varkappa \neq 0, \infty$ and $\varkappa = \infty$ of Equation (3.25) is denoted as l, m and n respectively. We consider different cases.

1^0. Let $l + n = 0$, i.e. $b \neq 0$, $c \neq 0$ according to (3.21). Then $k \neq 0$, i.e. $k = 1$. The multiplicity m of the root $\varkappa_0$ equals either one or three. If $m = 1$, then $D_3 < 0$ and in $\varkappa = \varkappa_0$ either both inequalities (3.22) or both equalities (3.24) are satisfied. In the first case System (3.20) has Solution (3.23), in the second case System (3.20) does not have real roots. But equations (3.24) have common root $\varkappa_0$ only if their resultant is $3R = 0$. It is Case 1.

If $m = 3$, then in the exceptional case $\varkappa_0$ is also the root of System (3.24), i.e. five quantities $a, b, c, d, \varkappa_0$ are related by five equations

$$f(\varkappa, 1) = f'(\varkappa, 1) = f''(\varkappa, 1) = g_1(\varkappa, 1) = g_2(\varkappa, 1) = 0, \tag{3.26}$$

where prime means derivative with respect to $\varkappa$. The equations of System (3.26) are linear homogeneous with respect to a, b, c, d. The determinant Δ of the last four equations (3.26) is $\Delta = -18\delta(\varkappa_0^2 + \delta)^2$, and the first equation linearly depends on the others. System (3.26) has a real solution only when $\Delta = 0$, i.e. when $\delta = -1$ and $\varkappa_0 = \pm 1$. With such δ and $\varkappa_0$ the fundamental solution to System (3.26) is $3a = -\varkappa_0 b = c = -3\varkappa_0 d$. It is Case 2.

2^0. We consider cases $l + n = 1$. Let $l = 0$, $n = 1$, i.e. $b = 0$, $B \neq 0$, $c \neq 0$. Then non-zero roots of Equation (3.25) satisfy to the square equation $B\varkappa^2 + 3d\varkappa - c\delta = 0$. In the exceptional case it does not have real roots, i.e. its discriminant $D_1 < 0$. It is Case 3.

Now let $l = 1$, $n = 0$, i.e. $c = 0$, $b \neq 0$, $C \neq 0$. Then in the exceptional case $D_2 < 0$. It is Case 4.

3^0. We consider cases $l + n = 2$. If $l = 0$, $n = 2$, i.e. $b = B = 0$, $d \neq 0$, $c \neq 0$, then the third root of Equation (3.25) is $\varkappa_0 = \delta_1\delta_2 c/(3d)$. In the exceptional case it is the root of System (3.24), which has the unique solution $\delta_1\delta_2 = -1$, $2c^2 = 9d^2$. It is Case 5.

If $l = 2$, $n = 0$, i.e. $c = D = 0$, $b \neq 0$, $a \neq 0$, then the third root of Equation (3.25) is $\varkappa_0 = 3a\delta_1\delta_2/b$. In the exceptional case it satisfies System (3.24), which has the unique solution $\delta_1\delta_2 = -1$, $9a^2 = 2b^2$. It is Case 6.

If $l = n = 1$, i.e. $b = c = 0$, $a \neq 0$, $d \neq 0$, then the third root of Equation (3.25) is $\varkappa_0 = \delta_1\delta_2 d/a$. Here $g_1(\varkappa_0, 1) = \delta_1\delta_2 d^3/a^2 \neq 0$. Consequently, here inequalities (3.22) are always satisfied, and there is solution (3.23).

4^0. Let $l+n=3$, then $k=0$, i.e. Equation (3.25) does not have real solution $\varkappa \neq 0, \infty$. Therefore all these cases are exceptional. If $l=0$, $n=3$, then $b=B=d=0$, $c\neq 0$; it is Case 7. If $l=1$, $n=2$, then $b=B=c=0$, $d\neq 0$; it is Case 8. If $l=2$, $n=1$, then $b=c=d=0$, $a\neq 0$; it is Case 9. If $l=3$, $n=0$, then $b\neq 0$, $a=c=C=0$; it is Case 10. □

Theorem 3.2 *System of equations* (3.20) *does not have real root* α_1, α_2 *with* $\alpha_1^2+\alpha_2^2\neq 0$ *only in six cases* 1–4, 7, 10 *of Theorem* 3.1.

Proof. In four cases of Theorem 3.1 there are the following solutions to System (3.20):

$\alpha_1=0$, $\alpha_2=-2\delta_2/d$ in Cases 5 and 8;

$\alpha_1=-2\delta_1/a$, $\alpha_2=0$ in Cases 6 and 9.

It is easy to see that in other cases of Theorem 3.1 System (3.20) does not have the non-zero real solution (α_1,α_2) with $\alpha_1=0$ or $\alpha_2=0$. □

4. Remarks. 1. In the neighborhood of the stationary point $Z=0$ of the Hamiltonian system (1.5) the Maclaurin series of the Hamiltonian function h has the form (3.2). To its quadratic part H_2 there corresponds the linear Hamiltonian system $\dot{Z}=AZ$. By means of the linear canonical change of coordinates $Z=BW$ it may be reduced to the Hamiltonian normal form [Bruno 1990a, Chapter I, § 1]. If all eigenvalues of the matrix A equal zero, then there are possible four types of systems with two degrees of freedom: (a) $r=3$; (b) $r=2$; (c) $r=1$; (d) $r=0$, i.e. $A=0$, where $r=\operatorname{rank} A$. For each of these types there is its own normal form. They are cited in [Sokol'skii 1981] along with the description of the method of computation of the normalizing change of variables. For the type (a) the Sokol'skii [1981] normal form is transformed into the normal form in [Bruno 1990a] by means of the change $q_2=-y_2$, $p_2=x_2$. For types (b) and (c) normal forms in [Bruno 1990a] and [Sokol'skii 1981] coincide. Here the normal forms from [Sokol'skii 1981] are used, but in coordinates (x_1,x_2,y_1,y_2) instead of (q_1,q_2,p_1,p_2). The first and the second cases analyzed in this Section are of the type (a), and the third is of the type (b).

2. If in System (1.5) the linear part is reduced to the normal form, then the non-linear part of the system (or the forms H_3, $H_4,\dots$ of the Hamiltonian function (3.2)) can be simplified by means of the non-linear canonical change of coordinates. Since here all eigenvalues equal zero, then the theory of the resonant Hamiltonian normal form [Bruno 1990a, Chapter I, § 2] does not prescribe here such a simplification. But it is possible to reduce the Hamiltonian function into the Belitskii normal form [Belitskii 1979b]: $h=h_2+\tilde{h}$ with the property $\langle A^*Z, \partial\tilde{h}/\partial Z\rangle=0$, where A^* is the transposed matrix of the linear part of System (1.5), or to reduce the Hamiltonian function into other simplified forms suggested in [Sokol'skii 1980, 1981]. But in all cases considered in [Sokol'skii 1980, 1981] and in the analogous cases of Subsections 1–3 of this Section the subsequent non-linear change of

coordinates is not necessary. The necessary analysis was done by means of the non-linear system, that was obtained after the linear normalization. It is note-worthy that the same situation appears in Section 20 of Chapter 5.

3. The normal form for H_4 suggested in formulae (20), (21) in [Sokol'skii 1980] is not complete. According to [Iooss 1995a], it must include at least one monomial from the polynomial

$$p_4 = 3\tilde{q}_1^2\tilde{p}_1^2 - 6\tilde{q}_1^3\tilde{q}_2 - 8\tilde{p}_1^3\tilde{p}_2 + 18\tilde{p}_1\tilde{p}_2\tilde{q}_1\tilde{q}_2 - 9\tilde{p}_2^2\tilde{q}_2^2$$

(in notation of [Sokol'skii 1980]).

4. The case of Subsection 1 corresponds to the case (2.2), (2.3) in [Sokol'skii 1981]. In the analogous to (3.8) solution (2.4) in [Sokol'skii 1981] the quantities $P_2(0)$ and $a\delta P_2(0)$ must be positive, since the root of the fourth degree is taken from them. But this means that the solution (2.4) in [Sokol'skii 1981] exists only when $a\delta > 0$. Indeed, Solution (3.8), (3.9) exists without this restriction on the coefficients of System (3.1), (3.2).

5. In [Sokol'skii 1981] at the end of Section 2 in pp. 445–446 there was studied the case of the conic precession of the satellite with $\alpha = 4/3$, $\beta = 0$, $\theta_0 = \pi/3$ considered here in Subsection 2. In this connection in [Sokol'skii 1981] it is erroneously stated that in this case the coefficient $a \neq 0$. Although from the formulae cited there it follows that $a = 0$. Therefore here in Subsection 2 this case was given a special treatment.

6. The case in Subsection 2 with $c = 0$ corresponds to the case (2.6) in [Sokol'skii 1981]. When $c = 0$ Condition (3.14) is $b^2 + 30\delta d > 0$. In [Sokol'skii 1981] instead of Condition (3.14) the false condition $b^2 + 6\delta d > 0$ is cited.

7. The case of the cylindrical precession of the symmetric satellite considered in [Sokol'skii 1980, 1981] with values of parameters $\alpha = 2/3$ and $\beta = 3/2$ is related to the type (a), and it has $H_3 \equiv 0$, i.e. $a = b = c = 0$. Nevertheless, the analysis of Subsection 2 is applicable for this case, although the leading polyhedron $\mathbf{\Gamma}_1'/\mathbb{R}_+^3$ differs from that in Fig. 4.5. We suggest to the reader to compute it as an exercise. For this case the result of Sokol'skii [1981, the end of Section 2] $d = 1/8$ is true.

8. The case of Subsection 3 here corresponds to the case of Section 3 in [Sokol'skii 1981]. In contradiction with Theorems 3.1 and 3.2, in [Sokol'skii 1981] it is stated that System (3.17) always has the real solution of the type (3.18), if

$$a^2 + b^2 + c^2 + d^2 \neq 0. \tag{3.27}$$

But it is not so. For example, when $a = b = d = 0$ and $c \neq 0$ Equation (3.21) takes the form $(2\alpha_1^2 - \delta\alpha_2^2)\alpha_2 = 0$. If $\delta \stackrel{\text{def}}{=} \delta_1\delta_2 < 0$, then this equation has only one real root $\alpha_2 = 0$. From the first equation (3.20) we obtain $\alpha_1 = 0$, and from System (3.14) we obtain $\beta_1 = \beta_2 = 0$. Thus, Condition (3.27) is satisfied, but the real solution to (3.18) is absent. It is easy to see that the case $a = b = d = 0$, $c \neq 0$, $\delta_1\delta_2 = -1$ is related to the case 3 of Theorem 3.1.

9. According to Theorem 5.1 of Chapter 3, the curves (3.8), (3.13), (3.18) when $t \to -\infty$ are asymptotics for some solutions to the complete System (1.5) with the Hamiltonian function (3.2) in cases of Subsections 1, 2, 3 respectively. That implies instability of the stationary point $X = Y = 0$. In [Sokol'skii 1980, 1981] in order to prove this instability the Chetaev functions were built for each case separately. In particular, from Subsection 2 there follows the instability of the conic precession of the symmetric satellite with $\alpha = 4/3$, $\beta = 0$, $\theta_0 = \pi/3$. Although it was stated in [Sokol'skii 1981] as well, but the statement was based upon the wrong proof (see Remark 5).

10. If in cases of Subsections 1–3 the coefficient of the monomial corresponding to the vertex of the face $\mathbf{\Gamma}_{11}^{'(2)}$ turns to zero, then the leading surface is changed. In the case of Subsection 1 when $a = 0$ the face $\mathbf{\Gamma}_{11}^{'(2)}$ disappears. In the case of Subsection 2 it is preserved when $b^2 + c^2 + d^2 \neq 0$, but its size is reduced. In the case of Subsection 3 the face $\mathbf{\Gamma}_{11}^{'(2)}$ with the same normal vector is present under Condition (3.27). But if $a = 0$ or $d = 0$, then the leading surface has other faces different from $\mathbf{\Gamma}_{11}^{'(2)}$.

4. The restricted three-body problem

Let the two bodies $\mathbf{P}_1$ and $\mathbf{P}_2$ with masses $1 - \mu$ and μ respectively turn in circular orbits around their common mass center with the period T. The plane circular restricted three-body problem consists in the study of the plane motion of the body $\mathbf{P}_3$ of infinitesimal mass under the influence of the Newton gravitation of bodies $\mathbf{P}_1$ and $\mathbf{P}_2$. In the rotating (synodical) standardized coordinate system the problem is described by the Hamiltonian system with two degrees of freedom and with one parameter μ. The Hamiltonian function has the form [Bruno 1990a]

$$h \stackrel{\text{def}}{=} \frac{1}{2}(y_1^2 + y_2^2) + x_2 y_1 - x_1 y_2 - \frac{1-\mu}{\sqrt{x_1^2 + x_2^2}} - \frac{\mu}{\sqrt{(x_1^2 - 1) + x_2^2}} + \mu x_1. \quad (4.1)$$

Here the body

$$\mathbf{P}_1 = \{X, Y : x_1 = x_2 = 0\}$$

and the body

$$\mathbf{P}_2 = \{X, Y : x_1 = 1, \quad x_2 = 0\}.$$

We consider the small values of the mass ratio $\mu \geq 0$.

When $\mu = 0$ the problem turns into the two-body problem for $\mathbf{P}_1$ and $\mathbf{P}_3$. But here the points corresponding to collisions of the bodies $\mathbf{P}_2$ and $\mathbf{P}_3$ must be excluded from the phase space. The points of collision split in parts solutions to the two-body problem for $\mathbf{P}_1$ and $\mathbf{P}_3$.

For small $\mu > 0$ there is a singular perturbation of the case $\mu = 0$ near the body $\mathbf{P}_2$. In order to find all the first approximations to the restricted three-body problem it is necessary to introduce the local coordinates near the body $\mathbf{P}_2$

$$\xi_1 = x_1 - 1, \quad \xi_2 = x_2, \quad \eta_1 = y_1, \quad \eta_2 = y_2 - 1$$

and to expand the Hamiltonian function in these coordinates. After the expansion of $1/\sqrt{(\xi_1+1)^2+\xi_2^2}$ in the Maclaurin series, the Hamiltonian function (4.1) takes the form

$$h + \frac{3}{2} - 2\mu \stackrel{\text{def}}{=} \frac{1}{2}(\eta_1^2 + \eta_2^2) + \xi_2\eta_1 - \xi_1\eta_2 - \xi_1^2 + \frac{1}{2}\xi_2^2 +$$

$$+ f(\xi_1, \xi_2^2) + \mu \left\{ \xi_1^2 - \frac{1}{2}\xi_2^2 - \frac{1}{\sqrt{\xi_1^2 + \xi_2^2}} - f(\xi_1, \xi_2^2) \right\}, \qquad (4.2)$$

where f is the convergent power series, where the terms of order less then three are absent. According to Remark 1.1 the support $\mathbf{S}_1$ of the series in the right hand side of (4.2) consists of points $R = (\operatorname{ord}\xi_1, \operatorname{ord}\xi_2, \operatorname{ord}\eta_1, \operatorname{ord}\eta_2, \operatorname{ord}\mu)$:

$$(0,0,2,0,0), \quad (0,0,0,2,0), \quad (0,1,1,0,0), \quad (1,0,1,0,0),$$

$$(2,0,0,0,0), \quad (0,2,0,0,0), \quad (k,2l,0,0,0), \quad (2,0,0,0,1),$$

$$(0,2,0,0,1), \quad (-1,0,0,0,1), \quad (0,-1,0,0,1), \quad (k,2l,0,0,1),$$

where $k, l \geq 0$, $k + 2l \geq 3$, and of the segment J connecting the points $(-1,0,0,0,1)$ and $(0,-1,0,0,1)$. That segment is the support of the root $1/\sqrt{\xi_1^2+\xi_2^2}$. Here the cone of the problem is

$$\mathbf{K} = \{W \in \mathbb{R}^5_*\colon w_1 < 0,\ w_2 < 0,\ w_5 < 0\}.$$

We make projection $\pi R = R'' \stackrel{\text{def}}{=} (p,q,s) \in \mathbb{R}^3$ of Example 1.3, where

$$p = r_1 + r_2 = \operatorname{ord}\xi_i, \quad q = r_3 + r_4 = \operatorname{ord}\eta_i, \quad s = r_5 = \operatorname{ord}\mu,$$

and use Remark 1.1 with $l = 1$. The set $\mathbf{S}''_1$ of these points R'' consists of

$$(0,2,0),\ (1,1,0),\ (2,0,0),\ (k,0,0),\ (2,0,1),\ (-1,0,1),\ (k,0,1),$$

where $k = 3,4,5,\ldots$ The outer convex hull $\operatorname{CNV}\mathbf{S}''_1$ of the set $\mathbf{S}''_1$ is the polyhedron $\mathbf{\Gamma} \subset \mathbb{R}^3$. The surface $\partial\mathbf{\Gamma}$ of the polyhedron $\mathbf{\Gamma}$ consists of faces $\mathbf{\Gamma}_j^{(2)}$, edges $\mathbf{\Gamma}_j^{(1)}$ and vertices $\mathbf{\Gamma}_j^{(0)}$. To each of the elements $\mathbf{\Gamma}_j^{(d)}$ there corresponds the truncated Hamiltonian $\hat{h}_j^{(d)}$, that is the sum of those terms of Series (4.2), the points R'' of which belong to $\mathbf{\Gamma}_j^{(d)}$. The truncated Hamiltonian functions $\hat{h}_j^{(d)}$ are various first approximations to the function (4.2) valid in different parts of the space $(\xi_1, \xi_2, \eta_1, \eta_2, \mu)$. Here the cone of the problem

$$\mathbf{K}'' = \{W'' \in \mathbb{R}^3_*\colon w_1 < 0,\ w_3 < 0,\}.$$

Fig 4.7 shows the polyhedron $\mathbf{\Gamma}$, which is the semi-infinite trihedral prism with an oblique base. It has four faces and six edges. Let us consider them.

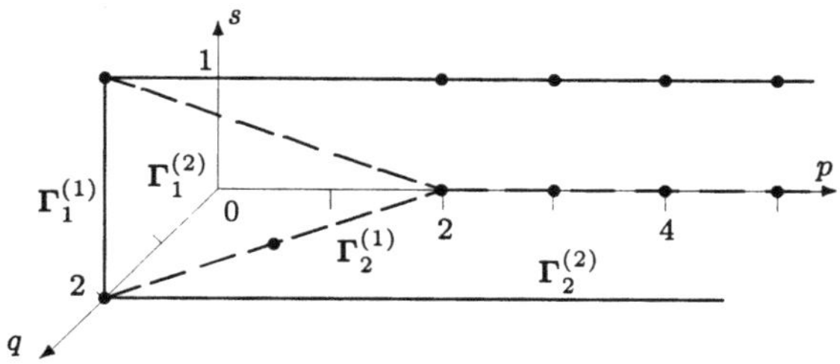

Figure 4.7: The polyhedron $\mathbf{\Gamma}$ for the Hamiltonian function (4.2) in coordinates p, q, s.

The face $\mathbf{\Gamma}_1^{(2)}$, which is the oblique base of the prism $\mathbf{\Gamma}$, contains vertices

$$(0,2,0),\ (2,0,0),\ (-1,0,1) \text{ and the point } (1,1,0).$$

Its normal vector is $N'' = -(1,1,1/3)$. To the face there corresponds the truncated Hamiltonian function

$$\hat{h}_1^{(2)} = \frac{1}{2}(\eta_1^2 + \eta_2^2) + \xi_2\eta_1 - \xi_1\eta_2 - \xi_1^2 + \frac{1}{2}\xi_2^2 - \frac{\mu}{\sqrt{\xi_1^2 + \xi_2^2}}. \tag{4.3}$$

It describes the Hill problem [Hill 1878], which is a non-integrable one. The power transformation

$$\tilde{\xi}_i = \xi_i\mu^{-1/3}, \quad \tilde{\eta}_i = \eta_i\mu^{-1/3}, \quad i = 1.2, \tag{4.4}$$

reduces the corresponding Hamiltonian system to the Hamiltonian system with the Hamiltonian function of the form (4.3), where ξ_i, η_i, μ must be substituted by $\tilde{\xi}_i, \tilde{\eta}_i, 1$ respectively.

The face $\mathbf{\Gamma}_2^{(2)}$ contains points

$$(0,2,0),\ (1,1,0),\ (2,0,0) \text{ and } (k,0,0).$$

Its normal vector is $U'' = (0,0,-1) \in \mathbf{K}''$. To the face there corresponds the truncated Hamiltonian function $\hat{h}_2^{(2)}$, which is obtained from the function h when $\mu = 0$. It describes the two-body problem for $\mathbf{P}_1$ and $\mathbf{P}_3$, which is an integrable one.

The remaining two faces have normal vectors $(0,-1,0)$ and $(0,1,2)$ lying outside the cone of the problem $\mathbf{K}''$. Hence the corresponding truncations are not suitable. We consider the edges. From the six edges one is unproper one. It goes through the point $(0,2,0)$ in parallel to the vector $(1,0,0)$. On three edges $q=0$, i.e. for them the truncated Hamiltonian function does not depend on η_1, η_2, and ξ_1, ξ_2 = const in the solutions to the corresponding Hamiltonian system. Such solutions are not interesting. There are remain two edges.

The edge $\mathbf{\Gamma}_1^{(1)}$. It includes points $(0,2,0)$ and $(-1,0,1)$ of the set $\mathbf{S}_1''$. The corresponding truncated Hamiltonian function is

$$\hat{h}_1^{(1)} = \frac{1}{2}(\eta_1^2 + \eta_2^2) - \frac{\mu}{\sqrt{\xi_1^2+\xi_2^2}}. \tag{4.5}$$

In describes the two-body problem for $\mathbf{P}_2$ and $\mathbf{P}_3$. The power transformation (4.4) transforms it into the Hamiltonian system with the Hamiltonian function of the form (4.5), where ξ_i, η_i, μ are substituted by $\tilde{\xi}_i, \tilde{\eta}_i, 1$ respectively.

The edge $\mathbf{\Gamma}_2^{(1)}$ includes points $(2,2,0)$, $(1,1,0)$, $(0,2,0)$ of the set $\mathbf{S}_1''$. To it there corresponds the truncated Hamiltonian function (4.3) with $\mu =$ 0. It describes the intermediate problem (between the Hill problem and the two-body problem for $\mathbf{P}_1$ and $\mathbf{P}_3$), which is an integrable one. This first approximation was introduced by Henon [1969].

Thus, the first approximation to the original restricted problem with the Hamiltonian function (4.2) depends on the distance from the body $\mathbf{P}_2$ in the following manner. Very close to the body $\mathbf{P}_2$, it is the two-body problem for bodies $\mathbf{P}_2$ and $\mathbf{P}_3$ with the Hamiltonian function (4.5); simply close, it is the Hill problem with the Hamiltonian function (4.3); farther from the body $\mathbf{P}_2$, it is the intermediate problem; and far from the body $\mathbf{P}_2$, it is the two-body problem for bodies $\mathbf{P}_1$ and $\mathbf{P}_3$. Near the body $\mathbf{P}_2$, the periodic solutions to the restricted problem are either perturbations of periodic solutions to all four mentioned first approximations or they are results of the matching of the hyperbolic orbits of the two-body problem for $\mathbf{P}_2$ and $\mathbf{P}_3$ with arc-solutions to the two-body problem for $\mathbf{P}_1$ and $\mathbf{P}_3$, or to the intermediate problem. In [Benest 1976; Kogan 1988; Lidov and Vashkov'yak 1990, 1993, 1994] the periodic solutions to the intermediate problem were used as the generating ones in order to find periodic quasi-satellite orbits of the restricted problem.

The results of Sections 1–3 of this Chapter were published in [Bruno and Soleev 1995e, 1996; Soleev and Bruno 1995], and of Section 4 in [Bruno 1978, 1999].

Chapter 5

Local analysis of singularities of a reversible system of ordinary differential equations

1. Introduction

In the book [Bruno 1979a] an apparatus for local analysis of a system of ordinary differential equations

$$dX/dt \stackrel{\text{def}}{=} \dot{X} = \Phi(X), \quad X \in \mathbb{R}^n \quad \text{or} \quad X \in \mathbb{C}^n$$

was developed.

It is based upon the two methods:

1. The normal form method. It is applicable to systems with non-zero linear part.

2. The method of truncated systems. It is applicable to systems with zero linear part (see Chapter 3).

It was shown that in arbitrary complicated cases combining the two methods it was possible to find both asymptotics of solutions to a system when $t \to \infty$ and its periodic and quasi-periodic solutions as well. However, the application of these methods to concrete multidimensional problems was hindered by the absence of an effective algorithm for computation of the Newton polyhedra. At present such an algorithm is developed and realized as the computer program (see Section 7 of Chapter 1).

In this Chapter these methods and algorithms are systematically applied for the local analysis of solutions to a reversible system of the fourth order depending on two small parameters $X'' = (x_5, x_6)$:

$$\begin{aligned} \dot{X}' &= \Phi(X', X''), \quad X' \in \mathbb{R}^4, \\ \dot{X}'' &= 0, \quad X'' \in \mathbb{R}^2. \end{aligned} \tag{1.1}$$

Here $X = (X', X'')$, $X' = (x_1, x_2, x_3, x_4)$ and the number of coordinates of the problem is $n = 4 + 2 = 6$. Namely, we consider System (1.1), where $\Phi(0, X'') = 0$, and when $X = 0$

$$\frac{\partial \Phi}{\partial X'} = \begin{pmatrix} 0 & 1 & 0 & 0 \\ 0 & 0 & 1 & 0 \\ 0 & 0 & 0 & 1 \\ 0 & 0 & 0 & 0 \end{pmatrix} \stackrel{\text{def}}{=} L, \tag{1.2}$$

i.e. the point $X' = 0$ is stationary, and the matrix of the unperturbed linear part L is a Jordan block with four-fold zero eigenvalue. In addition, System (1.1) is invariant under the substitution $X', X'', t \to SX', X'', -t$, where

$$S = \begin{pmatrix} 1 & 0 & 0 & 0 \\ 0 & -1 & 0 & 0 \\ 0 & 0 & 1 & 0 \\ 0 & 0 & 0 & -1 \end{pmatrix}, \tag{1.3}$$

i.e. the property of reversibility of System (1.1) means that

$$S\Phi(X', X'') = -\Phi(SX', X''). \tag{1.4}$$

Surveys on the general theory of reversible systems see in [Sevryuk 1986, Roberts and Quispel 1992].

Such a system arises in Hydrodynamics. For example, the problem of surface water waves for the Bond number b close to $1/3$ and the Froude number $\tilde{\lambda}$ close to 1 is transformed into System (1.1) as a result of reduction on the center manifold (see [Kirshgässner 1988; Ioose and Kirshgässner 1990, 1992; Amick and McLeod 1991; Iooss and Pérouème 1993; Iooss 1995a, b]).

The solutions to the analytic system (1.1) are studied in the neighborhood of the stationary point $X = 0$ in the following way. In Section 2, according to [Arnol'd 1971], the linear part of System (1.1), which depends on parameters, is reduced to the normal form. The lower triangular matrix of parameters is obtained, in such a way that each its diagonal consists of equal parameters. In the second diagonal from below and in the main diagonal there are only zeros because of the property (1.4). In Section 3 the support $\mathbf{S} \in \mathbb{R}^6$ of System (1.1) is isolated, i.e. the set of vector power exponents; and using it, the Newton polyhedron is computed in $\mathbb{R}^6$ by the program of Section 7 of Chapter 1. The table of correspondence is presented, which is used for extraction of the set of truncated systems. There are only five truncated systems of maximal "dimension" five. Among them there is the basic truncated system

$$\begin{aligned}
\dot{x}_1 &= x_2, \\
\dot{x}_2 &= \mu x_1 + x_3, \\
\dot{x}_3 &= \mu x_2 + x_4, \\
\dot{x}_4 &= \mu_2 x_1 + \mu x_3 + a x_1^2,
\end{aligned} \tag{1.5}$$

where μ and μ_1 are small parameters, $a \in \mathbb{R}$, which is non-integrable. The remaining truncated systems turn out to be integrable, and their explicit solutions are found. Iooss [1995a] wrote out System (1.5) as an approximate model of the Belitskii normal form of System (1.1), but he did not presented an algorithm for obtaining such systems. Section 4 is devoted to transformation and preliminary study of the basic truncated system (1.5). Firstly, the power transformation is applied to System (1.5) with $\mu \neq 0$. As a result the number of parameters is reduced to one and System (1.5) is reduced to the system

$$\begin{aligned}
\dot{y}_1 &= y_2, \\
\dot{y}_2 &= \sigma y_1 + y_3, \\
\dot{y}_3 &= \sigma y_2 + y_4, \\
\dot{y}_4 &= \nu y_1 + \sigma y_3 + a y_1^2,
\end{aligned} \tag{1.6}$$

where $\sigma = \pm 1$, parameters ν and $a \in \mathbb{R}$. It is shown that System (1.6) is a Hamiltonian one. In Section 5 the properties of the linear part of System (1.6) are studied. It is shown that it has only two independent quadratic first integrals. The set of parameter values $\sigma = \pm 1$ and $\nu \in \mathbb{R}$ is split into eight cases I–IV and γ_0–γ_3 corresponding to different disposition of eigenvalues $\lambda_i = \lambda_i(\sigma, \nu)$, $i = 1, 2, 3, 4$ of the matrix of the linear part of System (1.6). For these cases the linear changes of variables are written out, which reduce the linear part of System (1.6) to the normal form. In Section 6 we present the general facts about the nonlinear normal form, about convergence of the normalizing transformation, and about analytic sets of periodic and quasi-periodic solutions.

In Sections 7–15 we study solutions to System (1.6) near the point $Y = 0$ in the following order:

in Section 7: the case I, i.e. $\sigma = 1$ and $\nu \in (-5/4, 1)$, and the case γ_1, i.e. $\sigma = 1$ and $\nu = -5/4$, when all numbers λ_i are real and non-zero;

in Section 8: the case II, i.e. $\nu < -5/4$, and the case IV, i.e. $\nu > 1$, when the normalizing transformation converges;

in Sections 9–13: the case III, i.e. $\sigma = -1$ and $\nu \in (-5/4, 1)$, when the numbers λ_i are purely imaginary and simple; furthermore, in Section 10 a new theory of the structure of the resonant normal form is developed;

in Section 14: the case γ_2, i.e. $\sigma = -1$ and $\nu = -5/4$, when the eigenvalues are purely imaginary and multiple;

in Section 15: the case γ_0, i.e. $\sigma = 1$ and $\nu = 1$, and the case γ_3, i.e. $\sigma = -1$ and $\nu = 1$, when there is a two-fold zero eigenvalue.

Here the main attention is given to finding families of periodic solutions, families of quasi-periodic solutions, and homoclinic solutions. This is done with the help of the normal form of System (1.6), which is used to compute the sets $\mathcal{A}$ and $\mathcal{B}$ (see [Bruno 1974, 1979a, 1989a, b] and [Broer et al. 1996]).

In Section 16 we give a survey of solutions found for System (1.6). In Section 17 it is shown that System (1.1) may be considered as a regular perturbation of System (1.6), and we point out solutions to System (1.6), which are preserved. In Section 18 we compare the normal form of an arbitrary reversible system used in this Chapter with the Hamiltonian normal form that can be used for System (1.6) according to Theorem 4.1. In Section 19 we consider System (1.5) with $\mu = 0$.

In Section 20 we compute the Newton polyhedron for the Belitskii normal form [Belitskii 1975, 1979a, b] of System (1.1). The main result of this computation is the invariance of the basic truncated system after the Belitskii normalization of System (1.1). In Section 21 we demonstrate the relation of System (1.1) with the problem of surface water waves. We carry out a comparison of the methods and results of various papers. New families of periodic and quasi-periodic solutions are isolated. We point out inaccuracies in [Kirshgässner 1988, Iooss 1995b].

The main result of this Chapter is the existence of new families of periodic waves and of new families of quasi-periodic waves for $\mu < 0$. In addition it is shown that here the basic truncated system (1.5) is the same for the original system and for the Belitskii normal form, which simplifies computations. Finally, we create a new theory of the structure of the reversible resonant normal form, and for the Hamiltonian system we employ the non-Hamiltonian normal form, which simplifies computations as well. The results of this Chapter were announced in [Bruno and Soleev 1995f, g, 1997; Bruno 1999]. The first detailed version see in preprints [Soleev and Aranson 1995, Bruno and Soleev 1995b, c, d], the second version see in the paper [Bruno and Soleev 1998], the third version is the present Chapter.

In our analysis of System (1.1), we make up to three reductions of the dimension of the problem. The original system (1.1) has four variables X' and two parameters μ, μ_2, i.e. the number of its coordinates equals six. After the extraction of truncation and the power transformation, we obtain System (1.6) with four variables Y and one parameter ν (the number of its coordinates equals five). As a result of computation of the normal form of System (1.6) the problem either becomes integrable (cases II, IV and non-resonant sub-cases of the case III) or it is reduced to the system of the third order with one small parameter (the number of its coordinates equals four). The truncation of the last system is reduced by the power transformation to the problem with three coordinates, which is so simple that it is always integrable.

In this Chapter we denote vectors by capital letters and components of the vectors by corresponding small letters. For example, $X = (x_1, \dots, x_n)$, $Q = (q_1, \dots, q_n)$, $\Phi = (\varphi_1, \dots, \varphi_n)$.

The main purpose of this Chapter is to demonstrate the effectiveness of methods of Power Geometry for the investigation of rather complicated singularities on the example of System (1.1). Here we tried to use as simple algorithms of computations as possible, which allowed us to complete computations to the end. System (1.1) is suitable for this, because it has been an object of many investigations, which allows to compare our approach based upon the consecutive application of truncations and normal forms the necessary number of times with another approach based upon the single application of the Belitskii normalization, and which is effective only in the cases when the Belitskii normal form is integrable. But our approach works in all cases. Similar computations see in [Bruno 1988, 1998b; Edneral 1998; Sadov 1994a, b; Afendikov and Mielke 1999a, b].

2. Normal form of a linear system

Let the constant matrix L be a Jordan one. The linear system

$$\dot{X}' = (L + M)X', \quad X' \in \mathbb{C}^l \text{ or } X' \in \mathbb{R}^l, \tag{2.1}$$

where the matrix M depends only on parameters $X'' \in \mathbb{R}^m$, is called the *normal form* if

$$ML^* = L^*M, \tag{2.2}$$

where L^* is the transposed matrix L. Now, we formulate the main result of the paper [Arnol'd 1971] in the following way.

Theorem 2.1 [Arnol'd 1971] *For an analytic in zero $X'' = 0$ system*

$$\dot{Z} = (L + N(X''))Z, \quad N(0) = 0, \quad Z \in \mathbb{C}^l \text{ or } Z \in \mathbb{R}^l, \tag{2.3}$$

there exists an analytic in zero $X'' = 0$ substitution

$$Z = X' + B(X'')X', \quad B(0) = 0$$

reducing System (2.3) *to the normal form* (2.1), (2.2).

According to [Shih 1993] the reversibility of the original system is preserved in the normal form. In the case (1.1), (1.2) according to (2.2) the matrix $M = (m_{ij})$ has such a lower-triangular form that each diagonal consists of the same functions of the parameters X'': $m_{11} = m_{22} = m_{33} = m_{44}$, $m_{21} = m_{32} = m_{43}$, $m_{31} = m_{42}$. Here the reversibility of the normal form (2.1) implies $m_{11} = 0$ and $m_{31} = 0$. Denoting $\mu = m_{21}$ and $\mu_2 = m_{41}$ we obtain the explicit form of the matrix M:

$$M = \begin{pmatrix} 0 & 0 & 0 & 0 \\ \mu & 0 & 0 & 0 \\ 0 & \mu & 0 & 0 \\ \mu_2 & 0 & \mu & 0 \end{pmatrix}. \tag{2.4}$$

In what follows we suppose that linear in X' part of System (1.1) is already normalized.

3. The Newton polyhedron

According to Chapter 3 we write System (1.1) in the form

$$d(\log X')/dt = F(X) \stackrel{\text{def}}{=} \sum F_Q X^Q, \quad \dot{X}'' = 0, \tag{3.1}$$

where $X^Q = x_1^{q_1} x_2^{q_2} x_3^{q_3} x_4^{q_4} \mu^{q_5} \mu_2^{q_6}$, i.e. $Q = (q_1, q_2, q_3, q_4, q_5, q_6) \in \mathbb{R}^6$, $\log X' = (\log x_1, \ldots, \log x_4)$ and $F_Q = (f_{1Q}, \ldots, f_{4Q})$. In $\mathbb{R}^6$, we consider the set $\mathbf{S} \stackrel{\text{def}}{=} \mathbf{S}(F) = \{Q\colon F_Q \neq 0\}$, which is called the *support* of System (3.1). The closure of its convex hull $\mathbf{\Gamma} = \mathbf{\Gamma}(F)$ is called the *Newton polyhedron* of the system. The boundary $\partial\mathbf{\Gamma}$ of the polyhedron $\mathbf{\Gamma}$ consists of its *faces* $\mathbf{\Gamma}_j^{(d)}$ of various dimensions d (see Chapter 1). To each face $\mathbf{\Gamma}_j^{(d)}$ of the polyhedron $\mathbf{\Gamma}$ there corresponds the *boundary subset* $\mathbf{S}_j^{(d)} = \mathbf{\Gamma}_j^{(d)} \cap \mathbf{S}$, that determines the *truncated system*

$$d(\log X')/dt = \widehat{F}_j^{(d)}(X) \stackrel{\text{def}}{=} \sum F_Q X^Q \quad \text{over} \quad Q \in \mathbf{S}_j^{(d)}. \tag{3.2}$$

In the space $\mathbb{R}_*^6$ dual to $\mathbb{R}^6$, to each face $\mathbf{\Gamma}_j^{(d)}$ there corresponds its *normal cone* $\mathbf{U}_j^{(d)}$, i.e. the set of vectors $P \in \mathbb{R}_*^6$ for which the normal to P and supporting to $\mathbf{\Gamma}$ hyperplane $\langle P, Q\rangle = \text{const}$ intersect $\mathbf{\Gamma}$ exactly along the face $\mathbf{\Gamma}_j^{(d)}$. The brackets $\langle .\,, .\rangle$ denote scalar product. If $X \to 0$ along the curves of the form $x_i = b_i \tau^{p_i}(1 + o(1))$, $i = 1, \ldots, 6$, where constants $b_i \in \mathbb{C} \setminus \{0\}$, $\tau \to \infty$, and $P = (p_1, \ldots, p_6) \in \mathbf{U}_j^{(d)}$, then the truncated system (3.2) is the first approximation to System (3.1).

Let in $\mathbb{R}_*^6$ the polyhedral convex *cone of the problem* $\mathbf{K}$ (see Section 6 of Chapter 1) be given, i.e. we are interested only in faces $\mathbf{\Gamma}_j^{(d)}$ (and also in truncated systems (3.2)), the normal cones $\mathbf{U}_j^{(d)}$ of which intersect the cone of the problem $\mathbf{K}$. The intersection of two convex cones $\mathbf{U}_j^{(d)} \cap \mathbf{K}$ is also a polyhedral convex cone, and it may be given as the convex conic hull of the minimal finite number of vectors $V_1, \ldots, V_s$ forming the *skeleton* of that cone. We denote as N_j the normals to the hyperfaces $\mathbf{\Gamma}_j^{(5)}$. Obviously, the skeleton of the cone $\mathbf{U}_j^{(5)}$ consists of one vector N_j. On the other hand, the boundary $\partial\mathbf{K}$ of the cone of the problem $\mathbf{K}$ consist of a finite number of cones of smaller dimensions $\mathbf{K}_k^{(e)}$. The intersection $\mathbf{K}_k^{(e)} \cap \mathbf{U}_j^{(d)}$ of each of them with the normal cone is a polyhedral cone, and it may be given by some finite skeleton $D_1, \ldots, D_t$. It has been proved that the skeleton of the intersection $\mathbf{U}_i^{(d)} \cap \mathbf{K}$ consists of the vectors $N_j \in \mathbf{K}$ normal to the hyperfaces $\mathbf{\Gamma}_j^{(5)}$ and of the vectors D_j lying in the boundary $\partial\mathbf{K}$ of the cone of the problem $\mathbf{K}$ (see Section 6 of Chapter 1). For System (1.1) in the neighborhood of the stationary point $X = 0$, the cone of the problem is $\mathbf{K} = \{P \leq 0\}$.

We assume that the linear part of System (3.1) has the normal form in the sense of Section 2, and that the nonlinear part is not normalized. Then System (3.1) has the form

$$
\begin{aligned}
d(\log x_1)/dt &= x_1^{-1}x_2 + \sum f_{1Q}X^Q, \\
d(\log x_2)/dt &= x_2^{-1}x_3 + \mu x_2^{-1}x_1 + \sum f_{2Q}X^Q, \\
d(\log x_3)/dt &= x_3^{-1}x_4 + \mu x_3^{-1}x_2 + \sum f_{3Q}X^Q, \\
d(\log x_4)/dt &= \mu_2 x_4^{-1}x_1 + \mu x_4^{-1}x_3 + \sum f_{4Q}X^Q,
\end{aligned}
\tag{3.3}
$$

where in ith equation the vector of power exponents Q run through the set

$$
\mathbf{N}_i = \{Q\colon Q + E_i \in \mathbb{Z}_+^6,\ q_1 + \ldots + q_4 \geq 1,\ q_2 + q_4 \text{ is odd}\},\quad i = 1, 2, 3, 4.
$$

The oddness of $q_2 + q_4$ follows from the property (1.4). Let $\mathbf{N} = \mathbf{N}_1 \cup \ldots \cup \mathbf{N}_4$. For System (3.3), we will single out its support $\mathbf{S} \subset \mathbb{R}^6$, then we will compute its leading subset $\mathbf{S}' \stackrel{\text{def}}{=} \mathbf{S}/\mathbb{R}_+^6$ (see Section 8 of Chapter 1) and all objects accompanying the set $\mathbf{S}'$ until the collection of truncations is extracted.

Lemma 3.1 *The leading subset* $\mathbf{N}/\mathbb{R}_+^6$ *consists of* 22 *points*:

$$
\begin{aligned}
&2E_1 - E_2,\quad E_1 - E_2 + E_3,\quad -E_2 + 2E_3, \\
&\qquad 2E_1 - E_4,\quad E_1 + E_3 - E_4,\quad 2E_3 - E_4;
\end{aligned}
\tag{3.4}
$$

$$
\begin{aligned}
&E_1 + E_2 - E_3,\quad E_1 - E_3 + E_4,\quad 2E_2 - E_4,\quad E_2,\quad E_4,\quad 2E_4 - E_2, \\
&\qquad E_2 + E_3 - E_1,\quad E_3 + E_4 - E_1;
\end{aligned}
\tag{3.5}
$$

$$
3E_2 - E_1,\quad 2E_2 + E_4 - E_1,\quad E_2 + E_4 - E_1,\quad 3E_4 - E_1,\quad 3E_2 - E_3,
$$

$$2E_2 + E_4 - E_3, \quad E_2 + 2E_4 - E_3, \quad 3E_4 - E_3. \tag{3.6}$$

Proof. We denote $\mathbf{N}^0 = \mathbf{N} \cap \{q_5 = q_6 = 0\}$. Obviously, the set $\mathbf{N}^0$ is the leading subset of the set $\mathbf{N}$ with respect to the cone $\mathbb{R}^6_+$, since for every point $Q = (q_1, q_2, q_3, q_4, q_5, q_6) \in \mathbf{N}$ we have $Q - \tilde{Q} \geq 0$, $\tilde{Q} = (q_1, q_2, q_3, q_4, 0, 0)$.

We split the set $\mathbf{N}^0 \subset \mathbb{Z}^4$ of vectors $Q = (q_1, q_2, q_3, q_4)$ in three parts:

$$\mathbf{N}^1 = \{Q\colon q_1 + q_3 \geq 2,\ q_2 + q_4 = -1\},$$
$$\mathbf{N}^2 = \{Q\colon q_1 + q_3 = 0 \text{ or } 1,\ q_2 + q_4 \geq 1\},$$
$$\mathbf{N}^3 = \{Q\colon q_1 + q_3 = -1,\ q_2 + q_4 \geq 3\}.$$

Now to each point $Q \in \mathbf{N}^i$ we put in correspondence such a point $\tilde{Q} \in \mathbf{N}^i$ that $\tilde{q}_1 + \ldots + \tilde{q}_4$ equals 1 or 2 and $Q - \tilde{Q} \geq 0$.

For $Q \in \mathbf{N}^1$ we assume $(\tilde{q}_2, \tilde{q}_4) = (q_2, q_4)$. Since here q_2 or q_4 equals -1, then $q_1, q_3 \geq 0$. If $q_3 = 0$, then we assume $(\tilde{q}_1, \tilde{q}_3) = (2, 0)$; if $q_1, q_3 \geq 1$, then we assume $(\tilde{q}_1, \tilde{q}_3) = (1, 1)$; if $q_1 = 0$, then we assume $(\tilde{q}_1, \tilde{q}_3) = (0, 2)$. Obviously, $\tilde{Q} = (\tilde{q}_1, \tilde{q}_2, \tilde{q}_3, \tilde{q}_4) \in \mathbf{N}^1$, $\tilde{q}_1 + \ldots + \tilde{q}_4 = 1$ and $Q - \tilde{Q} \geq 0$. It is easy to check that in $\mathbf{N}^1$ there are altogether six points $\tilde{Q}$ cited in (3.4).

For $Q \in \mathbf{N}^2$, if $q_1 + q_3 = 0$, then we assume $(\tilde{q}_1, \tilde{q}_3) = (q_1, q_3)$; if $q_1 + q_3 = 1$, then we assume $(\tilde{q}_1, \tilde{q}_3) = (1, -1)$ for $(q_1, q_3) = (2, -1)$, $(\tilde{q}_1, \tilde{q}_3) = (0, 0)$ for $(q_1, q_3) = (1, 0)$ or (0,1) and $(\tilde{q}_1, \tilde{q}_3) = (-1, 1)$ for $(q_1, q_3) = (-1, 2)$. Besides, we put

$$(\tilde{q}_2, \tilde{q}_4) = \begin{cases} (-1, 2), & \text{if} \quad q_2 = -1, \\ (0, 1), & \text{if} \quad q_2 = 0, \\ (1, 0), & \text{if} \quad q_2 \text{ and } q_4 > 0, \\ (2, -1), & \text{if} \quad q_4 = -1. \end{cases}$$

Obviously, $\tilde{Q} = (\tilde{q}_1, \tilde{q}_2, \tilde{q}_3, \tilde{q}_4) \in \mathbf{N}^2$, $\tilde{q}_1 + \ldots + \tilde{q}_4 = 1$ and $Q - \tilde{Q} \geq 0$. It is easy to check that in $\mathbf{N}^2$ there are altogether eight points $\tilde{Q}$ cited in (3.5).

For $Q \in \mathbf{N}^3$ we assume $(\tilde{q}_1, \tilde{q}_3) = (q_1, q_3)$. Since here q_1 or q_3 equals -1, then $q_2, q_4 \geq 0$. If $q_4 = 0$, then we assume $(\tilde{q}_2, \tilde{q}_4) = (3, 0)$; if $q_2, q_4 \geq 1$ and $q_2 > q_4$, then we assume $(\tilde{q}_2, \tilde{q}_4) = (2, 1)$; if $q_2, q_4 \geq 1$ and $q_2 < q_4$, then we assume $(\tilde{q}_2, \tilde{q}_4) = (1, 2)$; if $q_2 = 0$, then we assume $(\tilde{q}_2, \tilde{q}_4) = (0, 3)$. Obviously, $\tilde{Q} = (\tilde{q}_1, \tilde{q}_2, \tilde{q}_3, \tilde{q}_4) \in \mathbf{N}^3$, $\tilde{q}_1 + \ldots + \tilde{q}_4 = 2$ and $Q - \tilde{Q} \geq 0$. It is easy to check that in $\mathbf{N}^3$ there are altogether eight points $\tilde{Q}$ cited in (3.6). □

In System (3.3) the linear terms without small parameters have vector powers

$$Q_1 = -E_1 + E_2, \quad Q_2 = -E_2 + E_3, \quad Q_3 = -E_3 + E_4. \tag{3.7}$$

Table 5.1: Table of correspondence for the set of points (3.7)–(3.9), and for the cone of the problem $\{P < 0\} \subset \mathbb{R}^6_*$.

i	t_i	B_i, N_i, D_i																		
			j	1	2	3	4	5	6	7	8	9	11	12	14	17	20	21	28	48
			T_j	V	V	V	V	V	V	V	V	V	V	V	d	V	V	V	V	V
			Q_j	−1	0	0	1	0	0	1	2	2	1	0	1	0	−1	0	0	−1
				1	−1	0	−1	1	0	0	−1	0	1	2	0	0	0	−1	3	0
				0	1	−1	0	−1	1	0	0	0	−1	0	1	2	1	0	−1	0
				0	0	1	0	0	−1	−1	0	−1	0	−1	−1	−1	1	2	0	3
				0	0	0	1	1	1	0	0	0	0	0	0	0	0	0	0	0
				0	0	0	0	0	0	1	0	0	0	0	0	0	0	0	0	0
1	D	0 0 0 -1 0 0		−	−	−	−	−	+	+	−	+	−	+	+	+	−	−	−	−
2	N	0 -1 0 -1 0 0		−	+	−	+	−	+	+	+	+	−	−	+	+	−	−	−	−
3	N	0 0 -1 -1 0 0		−	−	−	−	+	−	+	−	+	+	+	−	−	−	−	+	−
4	N	0 -1 -2 -1 0 0		−	−	+	+	+	−	+	+	+	+	−	−	−	−	−	−	−
5	N	0 0 0 0 -1 0		+	+	+	−	−	−	+	+	+	+	+	+	+	+	+	+	+
6	N	-2 -1 -4 -3 -2 0		+	−	+	−	+	−	+	−	−	+	+	−	−	−	−	+	−
7	N	0 0 0 0 0 -1		+	+	+	+	+	+	−	+	+	+	+	+	+	+	+	+	+
8	N	-1 -1 -1 -2 -1 -1		+	+	−	−	−	+	+	−	+	−	+	+	+	−	−	−	−
9	N	-1 -1 -2 -2 -1 -1		+	−	+	−	+	−	+	−	+	+	+	−	−	−	−	−	−
10	N	-2 -2 -3 -4 -1 -2		+	−	−	−	+	+	+	−	+	−	+	−	−	−	−	−	−
11	N	-1 -2 -2 -2 -1 -1		−	+	+	+	−	−	+	+	+	−	−	−	−	−	−	−	−
12	N	-4 -5 -6 -7 -2 -4		+	+	+	+	+	+	+	−	+	−	−	−	−	−	−	−	−
13	D	-2 -1 -2 -3 0 0		+	−	−	−	+	+	+	−	−	−	+	−	−	−	−	−	−
14	D	-1 -1 -1 -1 0 0		+	+	+	+	+	+	+	−	−	−	−	−	−	−	−	−	−
15	D	0 -1 0 0 0 0		−	+	−	+	−	−	−	+	−	−	−	−	−	−	+	−	−
16	D	0 0 -1 0 0 0		−	−	+	−	+	−	−	−	−	+	−	−	−	−	−	+	−
17	D	-1 0 -1 0 0 0		+	−	+	−	+	−	−	−	−	−	−	−	−	−	−	+	+
18	D	0 -1 -1 0 0 0		−	−	+	+	−	−	−	+	−	−	−	−	−	−	+	−	−
19	D	-1 0 0 -1 0 0		+	−	−	−	−	+	−	−	−	−	+	−	+	−	−	−	−
20	D	-2 -1 0 -1 0 0		+	+	−	−	−	+	−	−	−	−	−	−	+	+	−	−	−
21	D	-1 0 -1 -1 0 0		+	−	−	−	+	−	−	−	−	−	+	−	−	−	−	+	−
22	D	-1 0 0 0 0 0		+	−	−	−	−	−	−	−	−	−	−	−	−	+	−	−	+
23	D	-2 -1 -1 -3 -1 0		+	−	−	−	−	+	+	−	−	−	+	−	+	−	−	−	−
24	D	-4 -3 -2 -5 -2 0		+	+	−	−	−	+	+	−	−	−	−	−	+	−	−	−	−
25	D	-2 -5 -4 -3 -2 0		−	+	+	+	−	−	+	+	−	−	−	−	−	−	−	−	−
26	D	-1 -1 -2 -3 0 -1		−	−	−	−	+	+	+	−	+	−	+	−	−	−	−	−	−
27	D	-2 -3 -4 -5 0 -2		−	−	−	+	+	+	+	−	+	−	−	−	−	−	−	−	−
28	D	-1 -1 0 0 0 0		−	+	−	−	−	−	−	−	−	−	−	−	−	+	+	−	+
29	D	-1 -1 -1 0 0 0		−	−	+	−	−	−	−	−	−	−	−	−	−	−	+	−	+
30	D	-4 -3 -2 -1 0 0		+	+	+	−	−	−	−	−	−	−	−	−	−	+	+	−	+
31	D	-2 -3 -2 -1 0 0		−	+	+	+	−	−	−	−	−	−	−	−	−	−	+	−	−
32	D	-1 -3 -2 -1 -1 0		−	+	+	+	−	−	−	+	−	−	−	−	−	−	+	−	−

Lemma 3.2 *The leading subset* $(\mathbf{N}^0 \cup \{Q_1, Q_2, Q_3\})/\mathbb{R}^4_+$ *consists of points* (3.7) *and of* 10 *points*

$$2E_1 - E_2, \quad 2E_1 - E_4, \quad E_1 + E_3 - E_4, \quad 2E_3 - E_4, \quad E_1 + E_2 - E_3,$$
$$2E_2 - E_4, \quad 2E_4 - E_2, \quad E_3 + E_4 - E_1, \quad 3E_4 - E_1, \quad 3E_2 - E_3. \qquad (3.8)$$

Proof. Out of 22 points (3.4)–(3.6) we eliminate the points $\tilde{Q}$ for which $\tilde{Q} - Q_i \geq 0$ at least for one of $i = 1, 2, 3$. In (3.4) we have $E_1 - E_2 + E_3$ and $-E_2 + 2E_3 \geq Q_2$. In (3.5) we have $E_1 - E_3 + E_4$ and $E_4 \geq Q_3$, E_2 and $E_2 + E_3 - E_1 \geq Q_1$. In (3.6) we have $3E_2 - E_i$, $2E_2 + E_4 - E_i$ and $E_2 + 2E_4 - E_i \geq Q_i$, $i = 1, 3$. These 12 points are eliminated. There are remain only points (3.8). □

In System (3.3) there are four terms linear in $x_1, \ldots, x_4$ and depending on small parameters μ_1, μ_2. Their vector powers are

$$Q_4 = E_1 - E_2 + E_5, \quad Q_5 = E_2 - E_3 + E_5,$$
$$Q_6 = E_3 - E_4 + E_5, \quad Q_7 = E_1 - E_4 + E_6. \qquad (3.9)$$

Let $\mathbf{S}$ be the support of System (3.3) and $\mathbf{S}' = \mathbf{S}/\mathbb{R}^6_+$. We have proved the

Theorem 3.1 *If nonlinear terms of System* (3.3) *are generic, then the leading subset* $\mathbf{S}/\mathbb{R}^6_+$ *consists of* 17 *points* (3.7)–(3.9).

The points of the set $\mathbf{S}/\mathbb{R}^6_+$ are cited in the upper part of Table 5.1. Their numeration remains the same as it was in papers [Soleev and Aranson 1995, Bruno and Soleev 1998]. In these papers there was no extraction of the leading subset, and the points (3.7), (3.9) and all points of the set $\mathbf{N}^0$ with $q_1 + \ldots + q_4 = 1$ or 2 were used as the original set; altogether 49 points. Using the program from Section 7 of Chapter 1, the table of correspondence for the set $\mathbf{S}/\mathbb{R}^6_+$ and for the cone of the problem $\mathbf{K} = \{P < 0\} \subset \mathbb{R}^6_*$ was computed (Table 5.1). It contains the same set of vectors N_i and D_i as the table of correspondence computed in [Soleev and Aranson 1995, Bruno and Soleev 1998] for 49 points.

The results of computation had shown that for the Newton polyhedron corresponding to System (3.3), the cone of the problem $\{P \leq 0\}$ was intersected by the normal cones of 12 hyperfaces, 54 faces of dimension 4, 114 faces of dimension 3, 128 faces of dimension 2, 71 edges and 16 vertices.

We consider here only those truncations which correspond to hyperfaces (in Table 5.1 these lines are marked by sign N in the second column). There are 11 of such truncations. However six hyperfaces have normal vectors with some zero components, i.e. they lie in the boundary of the cone of the problem. By truncations corresponding to these hyperfaces it is impossible to analyze the behavior of solutions to System (3.3) in a neighborhood of the origin. Hence these truncations are not considered here either.

The remaining five hyperfaces have normal vectors N_i without zero components, and they lie inside the cone of the problem. The truncated system corresponding to the hyperface $\mathbf{\Gamma}_1^{(5)}$ with the normal vector $N_1 = (-4,-5,-6,-7,-2,-4)$ (12th line of Table 5.1) has the form

$$
\begin{aligned}
(\log \dot{x}_1) &= x_1^{-1}x_2,\\
(\log \dot{x}_2) &= x_2^{-1}x_3 + \mu x_2^{-1}x_1,\\
(\log \dot{x}_3) &= x_3^{-1}x_4 + \mu x_3^{-1}x_2,\\
(\log \dot{x}_4) &= \mu_2 x_4^{-1}x_1 + \mu x_4^{-1}x_3 + a x_4^{-1}x_1^2.
\end{aligned}
\tag{3.10}
$$

In the truncated systems corresponding to the remaining four hyperfaces one of equations of the truncated system has the right hand side as identical zero. We write out these truncated systems and find their solutions. The truncated system corresponding to the hyperface $\mathbf{\Gamma}_2^{(5)}$ with the normal vector $N_2 = (-1,-1,-2,-2,-1,-1)$ (9th line of Table 5.1) has the form

$$
\begin{aligned}
(\log \dot{x}_1) &= x_1^{-1}x_2,\\
(\log \dot{x}_2) &= 0,\\
(\log \dot{x}_3) &= x_3^{-1}x_4 + \mu x_3^{-1}x_2 + a_3 x_3^{-1}x_1x_2,\\
(\log \dot{x}_4) &= \mu_2 x_4^{-1}x_1 + a x_4^{-1}x_1^2 + b x_4^{-1}x_2^2,
\end{aligned}
$$

or in usual notation

$$
\begin{aligned}
\dot{x}_1 &= x_2,\\
\dot{x}_2 &= 0,\\
\dot{x}_3 &= \mu x_2 + x_4 + a_3 x_1 x_2,\\
\dot{x}_4 &= \mu_2 x_1 + a x_1^2 + b x_2^2.
\end{aligned}
$$

Integrating this system we obtain its explicit solutions

$$
\begin{aligned}
x_1 &= c_2 t + c_1,\\
x_2 &= c_2,\\
x_3 &= (1/12)ac_2t^4 + (1/6)(\mu_2 c_2 + 2ac_1c_2)t^3 +\\
&\quad + (1/2)(ac_1^2 + bc_2 + \mu_2 c_1 + a_3 c_3^2)t^2 + (c_1c_2 + \mu c_2 + c_4)t + c_3,\\
x_4 &= (1/3)ac_2^2t^3 + (1/2)(\mu^2 c_2 + 2ac_1c_2)t^2 + (\mu_2 c_1 + ac_1^2 + bc_2)t + c_4,
\end{aligned}
$$

where c_1, c_2, c_3, c_4 denote arbitrary constants. That notation is used further as well.

Now we consider the truncated system corresponding to the hyperface $\mathbf{\Gamma}_3^{(5)}$ with the normal vector $N_3 = (-1,-1,-1,-2,-1,-1)$ (8th line of Table 5.1), and we write out at once the system in the usual form

$$\begin{aligned}
\dot{x}_1 &= x_2,\\
\dot{x}_2 &= x_3,\\
\dot{x}_3 &= 0,\\
\dot{x}_4 &= \mu_2 x_1 + \mu x_3 + a x_1^2 + b x_2^2 + a_{13} x_1 x_3 + d x_3^2.
\end{aligned}$$

Its solutions are

$$\begin{aligned}
x_1 &= (1/2)c_3 t^2 + c_2 t + c_1,\\
x_2 &= c_3 t + c_2,\\
x_3 &= c_3,\\
x_4 &= (1/20)ac_3^3 t^5 + (1/4)ac_2c_3t^4 + (1/6)(\mu_2 c_3 + 2ac_2^2 + 2ac_1c_3 +\\
&\quad + a_{13}c_3^2 + bc_3^2)t^3 + (1/2)(\mu_2 c_2 + 2ac_1c_2 + a_{13}c_2c_3 + 2bc_3)t^2 +\\
&\quad + (\mu_2 c_1 + ac_1^2 + a_{13}c_1c_3 + bc_2^2 + dc_3^2)t + c_4.
\end{aligned}$$

To the normal vector $N_4 = (-2,-2,-3,-4,-1,-2)$ (10th line of Table 5.1) there correspond the hyperface $\mathbf{\Gamma}_4^{(5)}$ and the truncated system

$$\begin{aligned}
\dot{x}_1 &= x_2,\\
\dot{x}_2 &= 0,\\
\dot{x}_3 &= \mu x_2,\\
\dot{x}_4 &= \mu_2 x_1 + \mu x_3 + a x_1^2 + b x_2^2.
\end{aligned}$$

Its solutions are

$$\begin{aligned}
x_1 &= c_2 t + c_1,\\
x_2 &= c_2,\\
x_3 &= \mu c_2 t + c_3,\\
x_4 &= (1/3)ac_2^2 t^3 + (1/2)(\mu^2 c_2 + \mu_2 c_2 + 2ac_1c_2)t^2 +\\
&\quad + (\mu_2 c_1 + \mu c_3 + ac_1^2 + bc_2^2)t + c_4.
\end{aligned}$$

To the normal vector $N_5 = (-1,-2,-2,-2,-1,-1)$ (11th line of

Table 5.1) there correspond the hyperface $\mathbf{\Gamma}_5^{(5)}$ and the truncated system

$$\begin{aligned}
\dot{x}_1 &= 0,\\
\dot{x}_2 &= \mu x_1 + x_3 + a_2 x_1^2,\\
\dot{x}_3 &= x_4,\\
\dot{x}_4 &= \mu_2 x_1 + a x_1^2.
\end{aligned}$$

Its solutions are

$$\begin{aligned}
x_1 &= c_1,\\
x_2 &= (1/6)(ac_1^2 + \mu_2 c_1)t^3 + (1/2)c_4 t^2 + (c_3 + \mu c_1 + a_2 c_1^2)t + c_2,\\
x_3 &= (1/2)(ac_1^2 + \mu_2 c_1)t^2 + c_4 t + c_3,\\
x_4 &= (ac_1^2 + \mu_2 c_1)t + c_4.
\end{aligned}$$

We note that among 17 points $Q_j \in \mathbf{S}'$ only 13 points $Q_j \in \mathbf{S}'$ with numbers $j \le 17$ are lying in the leading hypersurface $\mathbf{\Gamma}/\mathbb{R}^6_+$ (see Table 5.1). Only terms corresponding to these 13 points belong to the enumerated truncated systems.

4. The reduction of System (3.10)

System (3.10) contains six coordinates $x_1, \ldots, x_4, \mu, \mu_2$. Since its *dimension* (i.e. the dimension of the convex hull $\mathbf{\Gamma}_1^{(5)}$ of the support $\mathbf{S}_1^{(5)}$ in $\mathbb{R}^6$) $d = 5$, then according to the general theory (see Section 2 of Chapter 3) by the power transformation of coordinates and the change of time the system may be reduced to the system with five coordinates. We use this possibility in order to reduce the number of parameters from two to one (see the beginning of Section 9 of Chapter 3). We perform this according to the second methods of solution of the Problem 3 in Section 11 of Chapter 1 in the following way. We take the vector N_1 normal to the face $\mathbf{\Gamma}_1^{(5)}$, and we normalize it by the fifth coordinate. We obtain the vector U having $u_5 = 1$. We form the matrix $\beta = (E_1\ E_2\ E_3\ E_4\ U\ E_6)$, where the column E_i is the ith unit vector. Now we make the power transformation $\log X = \beta \log \check{Y}$, where $\log \check{Y} = (\log y_1, \log y_2, \log y_3, \log y_4, \log|\mu|, \log\nu)$, and the corresponding change of time. In our case $N_1 = -(4, 5, 6, 7, 2, 4)$, i.e. $U = (2, 5/2, 3, 7/2, 1, 2)$, and the power transformation is

$$\begin{aligned}
&x_1 = y_1\mu^2, \quad x_2 = y_2|\mu|^{5/2}, \quad x_3 = y_3|\mu|^3, \quad x_4 = y_4|\mu|^{7/2},\\
&\mu = \mu, \quad \mu_2 = \nu\mu^2.
\end{aligned} \tag{4.1}$$

After the change of time $t_1 = \sqrt{|\mu|}\,t$, System (3.10) takes the form

$$\begin{aligned} dy_1/dt_1 &= y_2, \\ dy_2/dt_1 &= \sigma y_1 + y_3, \\ dy_3/dt_1 &= \sigma y_2 + y_4, \\ dy_4/dt_1 &= \nu y_1 + \sigma y_3 + a y_1^2, \end{aligned} \tag{4.2}$$

where $\sigma = \operatorname{sgn}\mu$, $\nu = \mu_2/\mu^2$. System (3.10) is reduced to System (4.2) for $\mu \neq 0$. System (3.10) with $\mu = 0$ will be considered in Section 19.

System (4.2) is equivalent to the fourth-order equation [Iooss 1995a]

$$y_1^{(4)} - 3\sigma y_1'' - (\nu - 1)y_1 - a y_1^2 = 0. \tag{4.3}$$

Lemma 4.1 [Iooss 1995a] *System* (4.2) *has the first integral*

$$J_1 \stackrel{\text{def}}{=} \nu y_1^2 + 2\sigma y_1 y_3 + y_3^2 - \sigma y_2^2 - 2y_2 y_4 + 2a y_1^3/3 = \text{const}\,. \tag{4.4}$$

The proof consists in checking that the derivative of J_1 by virtue of System (4.2) equals zero.

Theorem 4.1 *System* (4.2) *is a Hamiltonian one with the Hamiltonian function* $H = (1/2)J_1$.

Proof. System (4.2) can be written in the form

$$\begin{aligned} \dot y_4 &= \partial H/\partial y_1 = \nu y_1 + \sigma y_3 + a y_1^2, & \dot y_1 &= -\partial H/\partial y_4 = y_2, \\ \dot y_2 &= \partial H/\partial y_3 = \sigma y_1 + y_3, & \dot y_3 &= -\partial H/\partial y_2 = \sigma y_2 + y_4, \end{aligned}$$

where the dot means the differentiation with respect to t_1. Here coordinates (y_1, y_3) and (y_4, y_2) are canonically conjugate. The involution with the matrix S, i.e. $y_4, y_2, y_1, y_3, t \to -y_4, -y_2, y_1, y_3, -t$, is an anticanonical transformation. □

Further we are going to study System (4.2) by methods developed for generic systems (non-Hamiltonian), since for Hamiltonian systems the computations are essentially more complicated (see [Bruno 1990a, Chapter I]), but the results are the same (see Section 18).

System (4.2) has two stationary points $\mathcal{O}$ and $\mathcal{P}$: the first one is the origin $Y = (y_1, y_2, y_3, y_4) = 0$, and the second one is the point $Y = Y^0$, where

$$y_1^0 = (1 - \nu)/a, \quad y_2^0 = y_4^0 = 0, \quad y_3^0 = -\sigma(1 - \nu)/a. \tag{4.5}$$

Lemma 4.2 *With the fixed constant $\sigma = \pm 1$, the transformation*

$$y_1 = \tilde{y_1} + (\tilde{\nu} - 1)/a, \quad y_2 = \tilde{y_2}, \quad y_3 = \tilde{y_3} - \sigma(\tilde{\nu} - 1)/a, \quad y_4 = \tilde{y_4}, \quad \nu = 2 - \tilde{\nu} \tag{4.6}$$

transforms System (4.2) *into itself. Furthermore, the stationary points $\mathcal{O}$ and $\mathcal{P}$ are interchanged.*

Thus, all the results for the neighborhood of the zero stationary point $Y = 0$ remain valid for the second stationary point (4.5) after the mentioned change of parameter ν. Hence we are going to study System (4.2) mainly in a small neighborhood of the zero stationary point.

Remark 4.1. The linear substitution $aY = \tilde{Y}$ transforms System (4.2) in the same system with $a = 1$. But we keep a as the coefficient of the original system.

5. The classification of System (4.2)

If in the right hand side of the fourth equation of System (4.2) the unique nonlinear term ay_1^2 is dropped, then we obtain the linear system

$$\dot{Y} = AY, \tag{5.1}$$

which has the quadratic first integral

$$J_1^{(2)} \overset{\text{def}}{=} \nu y_1^2 + 2\sigma y_1 y_3 + y_3^2 - \sigma y_2^2 - 2y_2 y_4 = \text{const}\,, \tag{5.2}$$

which is the quadratic part of integral (4.4).

Theorem 5.1 *Linear System* (5.1) *has altogether two independent quadratic first integrals*: (5.2) *and*

$$J_2^{(2)} \overset{\text{def}}{=} \sigma y_1^2 + 2\nu y_1 y_3 - (1+\nu) y_2^2 + \sigma y_3^2 - y_4^2 = \text{const}\,. \tag{5.3}$$

Proof. Let System (5.1) have quadratic first integral

$$\langle Y, BY \rangle = \text{const}\,, \tag{5.3$'$}$$

where the matrix B is a symmetric one, i.e. the transposed matrix $B^* = B$. Further the brackets $\langle \cdot, \cdot \rangle$ mean the scalar product. From (5.3′) taking into account (5.1) we obtain

$$\langle \dot{Y}, BY \rangle + \langle Y, B\dot{Y} \rangle = \langle AY, BY \rangle + \langle Y, BAY \rangle =$$

$$= \langle Y, A^*BY \rangle + \langle Y, BAY \rangle = \langle Y, (A^*B + BA)Y \rangle = 0.$$

Hence $A^*B + BA = 0$, and since $A^*B = (BA)^*$, then $(BA)^* = -BA$. Consequently, the matrix BA is skew-symmetric. Let

$$B = \begin{pmatrix} b_{11} & b_{12} & b_{13} & b_{14} \\ b_{21} & b_{22} & b_{23} & b_{24} \\ b_{31} & b_{32} & b_{33} & b_{34} \\ b_{41} & b_{42} & b_{43} & b_{44} \end{pmatrix},$$

then

$$BA = \begin{pmatrix} \sigma b_{12} + \nu b_{14} & b_{11} + \sigma b_{13} & b_{12} + \sigma b_{14} & b_{13} \\ \sigma b_{22} + \nu b_{24} & b_{21} + \sigma b_{23} & b_{22} + \sigma b_{24} & b_{23} \\ \sigma b_{32} + \nu b_{34} & b_{31} + \sigma b_{33} & b_{32} + \sigma b_{34} & b_{33} \\ \sigma b_{42} + \nu b_{44} & b_{41} + \sigma b_{43} & b_{42} + \sigma b_{44} & b_{43} \end{pmatrix}.$$

From the condition of skew-symmetry of the matrix BA we obtain the system of linear equations

$$\sigma b_{12} + \nu b_{14} = 0, \quad b_{21} + \sigma b_{23} = 0, \quad b_{32} + \sigma b_{34} = 0, \quad b_{43} = 0,$$

$$\sigma b_{22} + \nu b_{24} = -b_{11} - \sigma b_{13}, \quad \sigma b_{32} + \nu b_{34} = -b_{12} - \sigma b_{14},$$

$$b_{31} + \sigma b_{33} = -b_{22} - \sigma b_{24}, \quad b_{23} = -b_{41} - \sigma b_{43},$$

$$b_{13} = -\sigma b_{42} - \nu b_{44}, \quad b_{33} = -b_{42} - \sigma b_{44}.$$

Since $b_{ij} = b_{ji}$, we obtain the system of 10 homogeneous equations in 10 variables b_{ij} with $i \le j$. From this system we obtain the system of two linear homogeneous equations:

$$b_{11} + \nu b_{24} + \sigma b_{44} = 0, \quad b_{13} + \sigma b_{24} + \nu b_{44} = 0, \tag{5.3''}$$

and the equations

$$b_{22} = -\sigma b_{11} - b_{13} - \sigma\nu b_{44}, \quad b_{33} = -b_{24} - \sigma b_{44}$$

with remaining $b_{ij} = 0$. The linear homogeneous system (5.3″) has two fundamental systems of solutions $b_{11} = \nu$, $b_{13} = \sigma$, $b_{24} = -1$, $b_{44} = 0$ and $b_{11} = \sigma$, $b_{13} = \nu$, $b_{24} = 0$, $b_{44} = -1$. To the first solution there corresponds the matrix

$$B = \begin{pmatrix} \nu & 0 & \sigma & 0 \\ 0 & -\sigma & 0 & -1 \\ \sigma & 0 & 1 & 0 \\ 0 & -1 & 0 & 0 \end{pmatrix},$$

which is the matrix of the integral (5.2). To the second solution there corresponds the matrix

$$B = \begin{pmatrix} \sigma & 0 & \nu & 0 \\ 0 & -1-\nu & 0 & 0 \\ \nu & 0 & \sigma & 0 \\ 0 & 0 & 0 & -1 \end{pmatrix}$$

of the integral (5.3). □

Now we find eigenvalues of System (5.1), i.e. the roots of the characteristic equation $\det(A - \lambda E) = 0$, which is

$$\lambda^4 - 3\sigma\lambda^2 - (\nu - 1) = 0. \tag{5.4}$$

Its roots are

$$\lambda_{1,3} = \pm\sqrt{\frac{3\sigma}{2} + \sqrt{\frac{5}{4} + \nu}}, \quad \lambda_{2,4} = \pm\sqrt{\frac{3\sigma}{2} - \sqrt{\frac{5}{4} + \nu}}. \tag{5.5}$$

The disposition of eigenvalues for various values of σ and ν is shown in Fig. 5.1. When $\nu = 1$, and when $\nu = -5/4$ there are multiple eigenvalues shown by crosses in Fig. 5.1. The number i in Fig. 5.1 coincides with the number of the eigenvalue λ_i. The enumeration of eigenvalues λ_i is chosen in such a way that they would change continuously with every value $\sigma = \pm 1$, and that when $-5/4 < \nu < 1$ the ratio λ_1/λ_2 had the same sign for both values $\sigma = \pm 1$. When $\nu < -5/4$, the eigenvalues λ_2 and λ_4 have a discontinuity for different values of σ. It is possible to choose the enumeration of λ_i, which is continuous in the whole cycle, but then $\operatorname{sgn}(\lambda_1/\lambda_2) = \sigma$ when $-5/4 < \nu < 1$.

We list all cases of disposition of eigenvalues depending on the values of parameters $\nu \in \mathbb{R}$ and $\sigma = \pm 1$ (see Fig. 5.1).

I. $\sigma = 1$, $-5/4 < \nu < 1$. In this case all eigenvalues (5.5) are real and different.

II. $\sigma = \pm 1$, $\nu < -5/4$. In this case all eigenvalues (5.5) are complex and different.

III. $\sigma = -1$, $-5/4 < \nu < 1$. In this case all eigenvalues (5.5) are purely imaginary and different.

IV. $\sigma = \pm 1$, $\nu > 1$. In this case two eigenvalues λ_1 and λ_3 are real and different, and two other eigenvalues λ_2 and λ_4 are purely imaginary and different.

γ_0. $\sigma = 1$, $\nu = 1$. In this case two eigenvalues $\lambda_1 = \sqrt{3}$ and $\lambda_3 = -\sqrt{3}$ are real and different, and there is the two-fold zero eigenvalue $\lambda_2 = \lambda_4 = 0$.

γ_1. $\sigma = 1$, $\nu = -5/4$. In this case there are two two-fold real eigenvalues: $\lambda_1 = \lambda_2 = \sqrt{3/2}$ and $\lambda_3 = \lambda_4 = -\sqrt{3/2}$, i.e. $\lambda_1 \neq \lambda_3$.

γ_2. $\sigma = -1$, $\nu = -5/4$. In this case there are two two-fold purely imaginary eigenvalues: $\lambda_1 = \lambda_2 = i\sqrt{3/2}$ and $\lambda_3 = \lambda_4 = -i\sqrt{3/2}$, i.e. $\lambda_1 \neq \lambda_3$.

γ_3. $\sigma = -1$, $\nu = 1$. In this case the two eigenvalues $\lambda_2 = i\sqrt{3}$ and $\lambda_4 = -i\sqrt{3}$ are purely imaginary and different, and there is the two-fold zero eigenvalue $\lambda_1 = \lambda_3 = 0$.

For $\nu \neq -5/4$ and $\nu \neq -1$, the eigenvectors corresponding to eigenvalues (5.5) are

$$\xi_i = (1,\ \lambda_i,\ \lambda_i^2 - \sigma,\ \lambda_i(\lambda_i^2 - 2\sigma)), \quad i = 1,2,3,4. \tag{5.6}$$

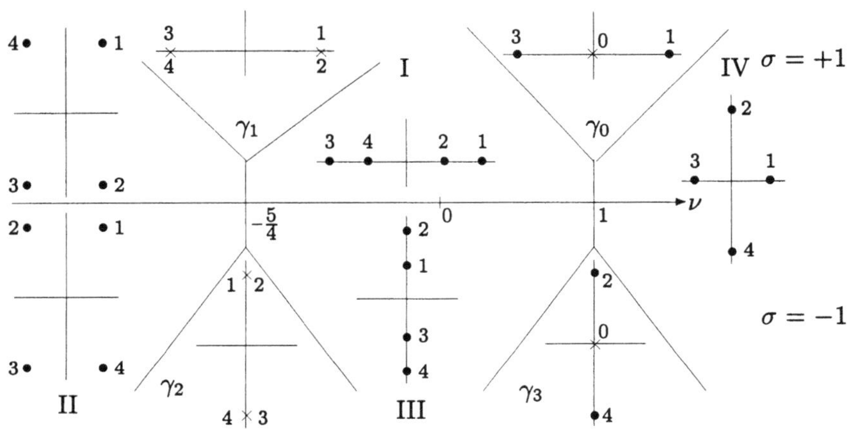

Figure 5.1: Cases I-IV and γ_0–γ_3 depending on the values $\nu \in \mathbb{R}$ and $\sigma = \pm 1$. For each case the disposition of eigenvalues λ_j from (5.5) is shown; they are represented by their numbers j.

Theorem 5.2 *For $\nu \neq -5/4$ and $\nu \neq -1$:*

a) all eigenvectors (5.6) *lie in the integral set*

$$J_1^{(2)} = 0 \ \text{and} \ J_2^{(2)} = 0, \tag{5.7}$$

determined by (5.2) *and* (5.3)*;*

b) in the set (5.7) *there lie two-dimensional invariant linear subspaces spanned on pairs of eigenvectors*

$$(\xi_1, \xi_2),\ (\xi_1, \xi_4),\ (\xi_2, \xi_3),\ (\xi_3, \xi_4); \tag{5.8}$$

c) the linear manifolds spanned on pairs of eigenvectors

$$(\xi_1, \xi_3),\ (\xi_2, \xi_4) \tag{5.9}$$

do not lie in the integral set (5.7).

Proof. By the direct substitution of coordinates of eigenvectors (5.6) in equations (5.7) one can verify that they satisfy these equations, i.e. the statement a) of Theorem 5.2 is easily proved.

From the first equation (5.7) (see (5.2)) we find

$$y_4 = (\nu y_1^2 + 2\sigma y_1 y_3 + y_3^2 - \sigma y_2^2)/(2y_2). \tag{5.10}$$

Substituting these values y_4 in the second equation (5.7) (see (5.8)), we obtain

$$4y_2^2(\sigma y_1^2 + 2\nu y_1 y_3 - (1+\nu)y_2^2 + \sigma y_3^2) - (\nu y_1^2 + 2\sigma y_1 y_3 + y_3^2 - \sigma y_2^2)^2 = 0. \tag{5.11}$$

In Equation (5.11) we put $y_1 = 1$ (taking into account that it is homogeneous, and that the first coordinate of eigenvectors (5.6) equals 1), then Equation (5.11) takes the form

$$4y_2^2(\sigma + 2\nu y_3 - (1+\nu)y_2^2 + \sigma y_3^2) = (\nu + 2\sigma y_3 + y_3^2 - \sigma y_2^2)^2, \tag{5.12}$$

i.e. it defines an algebraic curve of the fourth order in the plane y_2, y_3. In that plane to each eigenvector ξ_i there corresponds the point $(\lambda_i, \lambda_i^2 - \sigma)$. Let the straight line

$$y_3 = \alpha y_2 - \beta \tag{5.13}$$

pass through two such different points $(\lambda_i,\ \lambda_i^2 - \sigma)$ and $(\lambda_j,\ \lambda_j^2 - \sigma)$, then, as it is easy to observe,

$$\alpha = \lambda_i + \lambda_j, \quad \beta = \lambda_i \lambda_j + \sigma. \tag{5.14}$$

From (5.13) we have

$$y_3^2 = \alpha^2 y_2^2 - 2\alpha\beta y_2 + \beta^2. \tag{5.15}$$

Substituting (5.13) and (5.15) in (5.11) and collecting similar terms, we obtain

$$4y_2^2(\sigma - 2\nu\beta + \sigma\beta^2 + (2\nu\alpha - 2\sigma\alpha\beta)y_2 + (\sigma\alpha^2 - (1+\nu))y_2^2) =$$

$$= (\nu - 2\sigma\beta + \beta^2 + (2\sigma\alpha - 2\alpha\beta)y_2 + (\alpha^2 - \sigma)y_2^2)^2. \tag{5.16}$$

If the straight line (5.13) entirely lies in the curve (5.12), then the linear space spanned on ξ_i and ξ_j lies in the set (5.7). In this case the equality (5.16) is identically fulfilled for all values y_2, i.e. the coefficients at the same powers of y_2 in both parts of the equality (5.16) must be equal.

Comparing the coefficients at the same powers of y_2 in (5.16), we obtain the system of equations

$$\begin{aligned}
&\nu - 2\sigma\beta + \beta^2 = 0,\\
&\sigma - 2\nu\beta + \sigma\beta^2 = \alpha^2(\sigma - \beta)^2,\\
&2\alpha(\nu - \sigma\beta) = \alpha(\sigma - \beta)(\alpha^2 - \sigma),\\
&4(\sigma\alpha^2 - (1+\nu)) = (\alpha^2 - \sigma)^2.
\end{aligned} \tag{5.17}$$

From the first equation we obtain

$$\beta^2 = 2\sigma\beta - \nu. \tag{5.18}$$

The result of this is equality

$$(\sigma - \beta)^2 = 1 - \nu. \tag{5.19}$$

Changing β^2 in the second equation (5.17) by Formula (5.18) and taking into account (5.19), we obtain $(1-\nu)(\sigma+2\beta)^2 = (1-\nu)\alpha^2$. If $\nu \neq 1$, then it follows

$$\alpha^2 = \sigma + 2\beta. \tag{5.20}$$

It is easy to check that the third and the fourth equations (5.17) are derived from equalities (5.18) and (5.20). Thus, when $\nu \neq 1$ the principal equations of System (5.17) are (5.18) and (5.20). Substituting values of α and β from Formulae (5.14) into Formula (5.18), we obtain

$$\nu - 2\sigma(\lambda_i\lambda_j + \sigma) + (\lambda_i\lambda_j + \sigma)^2 = 0.$$

Hence $\lambda_i\lambda_j = \pm\sqrt{1-\nu}$. From Formulae (5.14) and (5.20) we have

$$(\lambda_i + \lambda_j)^2 = 3\sigma \pm 2\sqrt{1-\nu}. \tag{5.21}$$

On the other hand, if λ_i and λ_j are considered as eigenvalues (5.5), then from (5.21) it follows

$$\varkappa\sqrt{\frac{3\sigma}{2} + \sqrt{\frac{5}{4} + \nu}} \cdot \sqrt{\frac{3\sigma}{2} - \sqrt{\frac{5}{4} + \nu}} = \pm\sqrt{1-\nu}, \tag{5.22}$$

where $\varkappa$ is the product of signs at the outer radicals. Consequently, the pairwise combinations of eigenvectors (5.8) always satisfy the condition (5.22), and the combinations (5.9) do not satisfy it. □

For $\nu = -5/4$ and $\nu = 1$ there are multiple eigenvalues (see Fig. 5.1). When $\nu = -5/4$ (Cases γ_1 and γ_2) we have two two-fold eigenvalues $\lambda_{1,2} = \sqrt{3\sigma/2} = -\lambda_{3,4} = \lambda$. Eigenvectors ξ_1, ξ_3 are determined by (5.6), and the adjoint vectors are

$$\eta_2 = (0,\ 1,\ 2\lambda,\ 5\sigma/2), \quad \eta_4 = (0,\ -1,\ 2\lambda,\ -5\sigma/2).$$

Theorem 5.3 *When $\nu = -5/4$:*

a) both eigenvectors ξ_1, ξ_3 lie in the integral set (5.7)*;*

b) the set (5.7) *contains the two-dimensional invariant linear subspace spanned on these eigenvectors;*

c) adjoint vectors η_2, η_4 *do not lie in the integral set* (5.7), *and it does not contain other invariant linear subspaces of System* (5.1).

The proof of this Theorem is analogous to that of Theorem 5.2. The statement c) is proved by fact that the adjoint vectors do not satisfy Equations (5.7), and so there are no other invariant subspaces for System (5.1).

When $\nu = 1$ and $\sigma = 1$ (Case γ_0) eigenvalues are $\lambda_{1,3} = \pm\sqrt{3}$, $\lambda_{2,4} = 0$. Eigenvectors ξ_1, ξ_2, ξ_3 are determined by (5.6), and the adjoint vector is

$$\eta = (0, 1, 0, -2\sigma). \tag{5.23}$$

When $\nu = 1$ and $\sigma = -1$ (Case γ_3) eigenvalues are $\lambda_{1,3} = 0$, $\lambda_{2,4} = \pm i\sqrt{3}$. Eigenvectors ξ_1, ξ_2, ξ_4 are determined by (5.6), and the adjoint vector is (5.23).

Theorem 5.4 *When* $\nu = 1$*:*

a) all three eigenvectors ξ_1, ξ_2, ξ_3 *for* $\sigma = 1$*, or* ξ_1, ξ_2, ξ_4 *for* $\sigma = -1$ *lie in the integral set* (5.7)*;*

b) the set (5.7) *contains two two-dimensional invariant subspaces spanned on the following pairs of eigenvectors*

$$(\xi_1, \xi_2),\ (\xi_2, \xi_3)\ \textit{for}\ \sigma = 1\ \textit{and}\ \ (\xi_1, \xi_2),\ (\xi_1, \xi_4)\ \textit{for}\ \sigma = -1;$$

c) the adjoint vector η *in* (5.23) *does not lie in the integral set* (5.7), *and the set does not contain other invariant linear subspaces of System* (5.1).

Proof. It repeats the proof of Theorem 5.2 up to the obtaining of the system of equations (5.17). When $\nu = 1$ the second equation of System (5.17) follows from the first since $(\sigma - \beta)^2 = 1 - \nu = 0$ according to (5.19), i.e.

$$\beta = \sigma, \tag{5.24}$$

and the third equation takes the form $\alpha(1 - \sigma\beta) = 0$ and also follows from (5.24). Besides, when $\nu = 1$ the last equation of System (5.17) is $4(\sigma\alpha^2 - 2) = (\alpha^2 - \sigma)^2$, i.e. $(\alpha^2 - \sigma - 2\sigma)^2 = 0$. Consequently, $\alpha^2 = 3\sigma$. Finally, from (5.14) we have $\lambda_i + \lambda_j = \pm\sqrt{3\sigma}$ and $\lambda_i\lambda_j = 0$. These equalities prove the statement b).

The statement c) is proved by the fact that the adjoint vector (5.23) does not satisfy Equations (5.7). Consequently, there are no other invariant subspaces for System (5.1). □

Every matrix A is similar to some Jordan matrix [Gantmacher 1967], i.e. there exists such a non-singular matrix T that

$$T^{-1}AT = G, \tag{5.25}$$

where G is the Jordan normal form of the matrix A. By the linear change of coordinates

$$Y = TZ \tag{5.26}$$

we reduce the matrix A of System (5.1) to the Jordan form G. Here the matrices G and T and the new coordinates Z will be, generally speaking, complex. The linear transformation (5.26) transforms System (5.1) into the system

$$\dot{Z} = GZ, \tag{5.27}$$

here the involutory matrix (1.3) is transformed into

$$\tilde{S} = T^{-1}ST. \tag{5.28}$$

Since the matrix A is real, then

$$\overline{A} = A. \tag{5.29}$$

Let among its four eigenvalues be m real and l pairs of complex conjugate ones. Then the Jordan form G may be taken in the form of the block-diagonal matrix

$$G = \{G_1, G_2, G_3\}, \tag{5.30}$$

where $G_1 = \overline{G_2}$ is the Jordan matrix of the order l, and $G_3 = \overline{G_3}$ is the Jordan matrix of the order m. Obviously, $2l + m = 4$. The main diagonal of the matrix G contains the eigenvalues λ_1, λ_2, λ_3, λ_4. Here the values $\lambda_j = \overline{\lambda}_{l+j}$ are complex ($j = 1, \ldots, l$), the values λ_{2l+k} are real ($k = 1, \ldots, m$). Over the main diagonal there are either zeros or units. We denote as $E^{(k)}$ the square unit matrix of the order k. Since $\overline{Y} = Y$, then according to (5.26) $\overline{T}\,\overline{Z} = T\ Z$, i.e. $\overline{Z} = \overline{T}^{-1}TZ$.

Lemma 5.1 [Bruno 1979a, Chapter III, § 1] *If the matrix A is real, then there exists such a transformation* (5.26) *of System* (5.1) *to System* (5.27) *with the matrix* (5.30), *that*

$$I = \overline{T}^{-1}T, \tag{5.31}$$

where I is the block matrix

$$I = \begin{pmatrix} 0 & E^{(l)} & 0 \\ E^{(l)} & 0 & 0 \\ 0 & 0 & E^{(m)} \end{pmatrix}. \tag{5.32}$$

We note that $I^2 = E^{(4)}$, $I^{-1} = I$, $\overline{I} = I$, $\overline{G} = IGI$. We introduce the matrix

$$C = \begin{pmatrix} E^{(l)} & iE^{(l)} & 0 \\ iE^{(l)} & -iE^{(l)} & 0 \\ 0 & 0 & E^{(m)} \end{pmatrix}.$$

According to (5.32) $\overline{C} = IC$. Here the coordinates $W = C^{-1}Z$, connected with the complex coordinates Z by the *standard transformation*

$$Z = CW, \tag{5.33}$$

are real. That means that when the coordinates Y are real, coordinates W are also real, and coordinates Z satisfy the reality relation $\overline{Z} = IZ$.

Lemma 5.2 *In Cases* I, II, III, IV (*when all eigenvalues are different*) *the matrix G is diagonal:* $G = \{\lambda_1, \lambda_2, -\lambda_1, -\lambda_2\}$, *and the matrix T of the linear transformation* (5.26) *is*

$$\begin{pmatrix} 1 & 1 & 1 & 1 \\ \lambda_1 & \lambda_2 & -\lambda_1 & -\lambda_2 \\ \lambda_1^2 - \sigma & \lambda_2^2 - \sigma & \lambda_1^2 - \sigma & \lambda_2^2 - \sigma \\ \lambda_1(\lambda_1^2 - 2\sigma) & \lambda_2(\lambda_2^2 - 2\sigma) & -\lambda_1(\lambda_1^2 - 2\sigma) & -\lambda_2(\lambda_2^2 - 2\sigma) \end{pmatrix}. \quad (5.34)$$

The inverse matrix T^{-1} is

$$\frac{1}{2\lambda_1\lambda_2(\lambda_1^2 - \lambda_2^2)} \begin{pmatrix} -\lambda_1\lambda_2(\lambda_2^2 - \sigma) & -\lambda_2(\lambda_2^2 - 2\sigma) & \lambda_1\lambda_2 & \lambda_2 \\ \lambda_1\lambda_2(\lambda_1^2 - \sigma) & \lambda_1(\lambda_1^2 - 2\sigma) & -\lambda_1\lambda_2 & -\lambda_1 \\ -\lambda_1\lambda_2(\lambda_2^2 - \sigma) & \lambda_2(\lambda_2^2 - 2\sigma) & \lambda_1\lambda_2 & -\lambda_2 \\ \lambda_1\lambda_2(\lambda_1^2 - \sigma) & -\lambda_1(\lambda_1^2 - 2\sigma) & -\lambda_1\lambda_2 & \lambda_1 \end{pmatrix}. \quad (5.35)$$

The proof is obtained by direct checking of the validity of Formulae (5.25) and $TT^{-1} = E$ with the substitution of matrices (5.34) and (5.35). It is analogous for Lemmas 5.3–5.5 below.

Lemma 5.3 *In Cases* I, II, III, IV:

a) the involutory matrix S (see (1.3)) *is transformed into the matrix*

$$\tilde{S} \stackrel{\text{def}}{=} T^{-1}ST = \begin{pmatrix} 0 & 0 & 1 & 0 \\ 0 & 0 & 0 & 1 \\ 1 & 0 & 0 & 0 \\ 0 & 1 & 0 & 0 \end{pmatrix}; \quad (5.36)$$

b) the matrix $I \stackrel{\text{def}}{=} T^{-1}\overline{T}$ is E in Case I, *I_2 in Case* II, *$\tilde{S}$ in Case* III, *I_4 in Case* IV, *where*

$$I_2 \stackrel{\text{def}}{=} \begin{pmatrix} 0 & 1 & 0 & 0 \\ 1 & 0 & 0 & 0 \\ 0 & 0 & 0 & 1 \\ 0 & 0 & 1 & 0 \end{pmatrix}, \quad I_4 \stackrel{\text{def}}{=} \begin{pmatrix} 1 & 0 & 0 & 0 \\ 0 & 0 & 0 & 1 \\ 0 & 0 & 1 & 0 \\ 0 & 1 & 0 & 0 \end{pmatrix}. \quad (5.37)$$

Lemma 5.4 *When $\nu = -5/4$ (Cases γ_1 and γ_2) with* $\lambda \stackrel{\text{def}}{=} \lambda_1 = \lambda_2 = -\lambda_3 = -\lambda_4 = \sqrt{3\sigma/2}$

$$G = \begin{pmatrix} \lambda & \lambda & 0 & 0 \\ 0 & \lambda & 0 & 0 \\ 0 & 0 & -\lambda & -\lambda \\ 0 & 0 & 0 & -\lambda \end{pmatrix}, \quad T = \begin{pmatrix} 1 & 0 & 1 & 0 \\ \lambda & \lambda & -\lambda & -\lambda \\ \sigma/2 & 3\sigma & \sigma/2 & 3\sigma \\ -\sigma\lambda/2 & 5\sigma\lambda/2 & \sigma\lambda/2 & -5\sigma\lambda/2 \end{pmatrix},$$

$$T^{-1} = -\frac{\sigma}{54}\begin{pmatrix} -27\sigma & -15\lambda & 0 & 6\sigma\lambda \\ 9\sigma/2 & -3\lambda & -9 & -6\sigma\lambda \\ -27\sigma & 15\lambda & 0 & -6\sigma\lambda \\ 9\sigma/2 & 3\lambda & -9 & 6\sigma\lambda \end{pmatrix},$$

and the matrix I is equal to the identity matrix in Case γ_1, and to the matrix (5.36) *in Case γ_2; the matrix $\tilde{S}$ in these cases is* (5.36).

Lemma 5.5 *When $\nu = 1$ in Cases γ_0 ($\sigma = 1, \lambda \stackrel{\text{def}}{=} \lambda_1 = \sqrt{3}$) and γ_3 ($\sigma = -1, \lambda \stackrel{\text{def}}{=} \lambda_2 = i\sqrt{3}$)*

$$G = \begin{pmatrix} 0 & 0 & 1 & 0 \\ 0 & \lambda & 0 & 0 \\ 0 & 0 & 0 & 0 \\ 0 & 0 & 0 & -\lambda \end{pmatrix}, \quad T = \begin{pmatrix} 1 & 1 & 0 & 1 \\ 0 & \lambda & 1 & -\lambda \\ -\sigma & 2\sigma & 0 & 2\sigma \\ 0 & \sigma\lambda & -2\sigma & -\sigma\lambda \end{pmatrix},$$

$$T^{-1} = \frac{1}{6\lambda}\begin{pmatrix} 4\lambda & 0 & -2\sigma\lambda & 0 \\ \lambda & 2 & \sigma\lambda & \sigma \\ 0 & 2\lambda & 0 & -2\sigma\lambda \\ \lambda & -2 & \sigma\lambda & -\sigma \end{pmatrix}, \quad \tilde{S} = \begin{pmatrix} 1 & 0 & 0 & 0 \\ 0 & 0 & 0 & 1 \\ 0 & 0 & -1 & 0 \\ 0 & 1 & 0 & 0 \end{pmatrix};$$

$I = E$ in Case γ_0, and $I = I_4$ from (5.37) *in Case γ_3.*

In Lemmas 5.4 and 5.5 the matrix G is not the Jordan normal form, but it is more convenient in further computations and trivially transformed into the Jordan normal form.

6. The normal form of a nonlinear system

We recall the main aspects of the theory expounded in [Bruno 1971, 1974, 1979a, 1989a, b, 1992]. We consider the system of the order n

$$\dot{Y} = AY + \Phi(Y) \tag{6.1}$$

in the neighborhood of the stationary point $Y = 0$, supposing that the vector-function $\Phi(Y)$ is analytic in the point $Y = 0$, and that its Tailor series does not contain constant and linear terms.

Let the linear substitution $Y = TZ$ transform the matrix A to its Jordan normal form $G = T^{-1}AT$, and the whole system (6.1) into the form

$$\dot{Z} = GZ + \tilde{\Phi}(Z). \tag{6.2}$$

Let the formal change of coordinates

$$Z = U + B(U), \tag{6.3}$$

where $B = (b_1, \ldots, b_n)$ and $b_i(U)$ are formal power series without constant and linear terms, transform System (6.2) into the system

$$\dot{U} = GU + \tilde{\Psi}(U) \stackrel{\text{def}}{=} \Psi(U). \tag{6.4}$$

We write it out in the form

$$\dot{u}_j = u_j g_j(U) \stackrel{\text{def}}{=} u_j \sum g_{jQ} U^Q, \quad j = 1, \ldots, n. \tag{6.5}$$

Since G is a Jordan matrix, then its diagonal $\Lambda = (\lambda_1, \ldots, \lambda_n)$ consists of eigenvalues of the matrix A.

Definition [Bruno 1979a]. System (6.4), (6.5) is called the *resonant normal form*, if:

a) G is a Jordan matrix;

b) in expansions (6.5) there are only resonant terms for which the scalar product

$$\langle Q, \Lambda \rangle \stackrel{\text{def}}{=} q_1\lambda_1 + \ldots + q_n\lambda_n = 0. \tag{6.6}$$

Theorem on the normal form [Bruno 1964, 1971, 1979a]. *There exists a formal substitution* (6.3), *which transforms System* (6.2) *into the normal form* (6.4), (6.5), (6.6).

We note some **properties of the normal form and the normalizing transformation** [Bruno 1979a, Chapter III, § 1].

1. If equation (6.6) has exactly k linearly independent over $\mathbb{R}$ integer solutions $Q \in \mathbb{Z}^n$, then there exists a power transformation $\log U = \tilde{\alpha} \log X$ transforming the normal form (6.5) to the system

$$(\dot{\log x_i}) = f_i(x_1, \ldots, x_k), \quad i = 1, \ldots, n.$$

This system has an independent subsystem of the first k equations, which does not contain linear terms.

2. If in System (6.1) there are small parameters, then they do not change under the normalizing transformation, and the right hand sides in (6.5) may be written as power series in the variables, which have coefficients as power series in small parameters.

3. If System (6.2) has linear automorphism of the form $Z, t \to \tilde{S}Z, \delta t$, then the normal form has the same linear automorphism $U, t \to \tilde{S}U, \delta t$. This property was proven anew specially for reversible systems in [Gaeta 1994].

4. For a real system (6.1), its normal form (6.4) preserves the reality properties, i.e. the linear standard transformation $U = CV$ (see (5.33)) transforms it to the real in V system.

5. The normalizing transformation is far from being always analytic. In [Bruno 1971] the answer was given to the question: for which normal forms (6.4), (6.5) the analyticity of the original system (6.1) implies the analyticity of the normalizing transformation (6.3)? Below we will formulate this answer only for cases which we will encounter in further analysis of System (1.1).

Condition A. *In the normal form* (6.4)

$$\psi_i = \lambda_i u_i \alpha + \overline{\lambda}_i u_i \beta, \quad i = 1, \dots, n,$$

where $\alpha = \alpha(U)$ *and* $\beta = \beta(U)$ *are some power series.*

We put

$$\omega_k = \min |\langle Q, \Lambda \rangle| \text{ over } Q\colon \sum q_i < 2^k,\ Q + E_j \in \mathbb{Z}_+^n, \langle Q, \Lambda \rangle \neq 0,$$

where E_j is jth unit vector, and $\mathbb{Z}_+^n = \mathbb{Z}^n \cap \{Q \geq 0\}$.

Condition ω. $\sum\limits_{k=1}^{\infty} 2^{-k} \log \omega_k > -\infty$, *i.e. the series converges.*

It is a weak arithmetic restriction on the eigenvalues $\lambda_1, \dots, \lambda_n$, it is almost always satisfied. Conversely, Condition A is a rigid restriction on the normal form, since it demands that n series ψ_i were linearly expressed through the two series α and β.

Theorem 6.1 [Bruno 1971] *If for the analytic System* (6.1), *vector* Λ *satisfies Condition* ω *and the normal form* (6.4), (6.5) *satisfies Condition A, then there exists an analytic normalizing transformation of System* (6.1).

Furthermore, in [Bruno 1971] it is proved that Condition A may not be weakened, i.e. if the normal form (6.5) does not satisfy Condition A, then there exists an analytic original System (6.1), which is reduced to the normal form (6.5) by the divergent transformation. That is why the normalizing transformation is not analytic as a rule, and it is not possible to normalize analytically System (6.1) in the whole neighborhood of the stationary point $Y = 0$.

Even if the normalizing transformation (6.3) diverges, it may converge on some sets; in particular, on the sets of periodic and quasi-periodic solutions. We explain it in more detail for the case when all

$$\operatorname{Re} \lambda_j = 0. \tag{6.7}$$

Let functions $f_1(U), \dots, f_s(U)$ be analytic at the point $U = 0$ and vanish at that point. Then the system of equations

$$f_j(U) = 0, \quad j = 1, \dots, s, \tag{6.8}$$

defines the *analytic set* $\mathcal{M}$, which includes the point $U = 0$. To the set $\mathcal{M}$ in the ring of convergent power series there corresponds the ideal Υ with the basis $f_1, \ldots, f_s$. If $f_1, \ldots, f_s$ are formal power series, then we say that the system of equations (6.8) defines the *formal set* $\mathcal{M}$ with the basis $f_1, \ldots, f_s$. The set $\mathcal{M}$ is analytic, if in the ideal Υ there exists a basis of convergent power series. The set $\mathcal{M}$ is called the *manifold* (local one), if the system of equations (6.8) may be resolved with respect to s coordinates from $u_1, \ldots, u_n$. Under the formal invertible change of coordinates (6.3), the one to one correspondence is established between the formal sets. Hence we will consider the image and pre-image as the same set in different coordinate systems. The set $\mathcal{M}$ is an *integral* one (or invariant) for System (6.4), (6.5), if

$$\sum_{k=1}^{n} \frac{\partial f_j}{\partial u_k} \psi_k(U) \in \Upsilon, \quad j = 1, \ldots, s.$$

To the normal form (6.4) we put in correspondence the formal set

$$\mathcal{A} = \{U\colon \ \psi_i = \lambda_i u_i \alpha, \quad i = 1, \ldots, n\}, \tag{6.9}$$

where α is a free parameter. It may be excluded from the equations, and the definition of the set $\mathcal{A}$ may be obtained in the form

$$\mathcal{A} = \{U\colon \ f_1(U) = \ldots = f_s(U) = 0\}. \tag{6.10}$$

In other words, $\mathcal{A}$ is the set on which Condition A is satisfied. Each System (6.1) has its set $\mathcal{A}$, which is defined by the normal form of the system.

Theorem 6.2 [Bruno 1974, 1979a, 1989a] *If all eigenvalues λ_i of the analytic System* (6.2) *are pairwise commensurable, then its set $\mathcal{A}$ is analytic.*

We consider a situation when not all eigenvalues λ_i are pairwise commensurable, and there are small denominators $\langle Q, \Lambda\rangle$. Let $\tilde{\Lambda} = \{\lambda_1, \ldots, \lambda_n\}$ be a diagonal matrix. On the set $\mathcal{A}$ we consider the matrix $B = (\partial \Psi / \partial U) - \tilde{\Lambda}\alpha$, where α is the same parameter as in equations (6.9), which define the set $\mathcal{A}$. We define the formal *set* $\mathcal{B}$ as such a subset of the set $\mathcal{A}$ on which the matrix B is nilpotent, i.e.

$$\mathcal{B} = \{U : \ U \in \mathcal{A}, \ B^n = 0\}. \tag{6.11}$$

Theorem 6.3 [Bruno 1974, 1979a, 1989a] *If eigenvalues $\lambda_1, \ldots, \lambda_n$ of analytic System* (6.2) *satisfy Condition ω, then its set $\mathcal{B}$ is analytic.*

We list the **properties of the sets $\mathcal{A}$ and $\mathcal{B}$**. For the sake of simplicity we suppose that the normal form (6.4) is analytic, and we consider the properties of solutions to this system in the sets $\mathcal{A}$ and $\mathcal{B}$.

1) All solutions from the set Re$\mathcal{A}$ are quasi-periodic (including periodic and stationary solutions). Indeed, on each solution the value of the parameter α is constant, and we have

$$u_i = u_i^0 \exp \lambda_i \alpha t, \quad i = 1, \ldots, n. \tag{6.12}$$

2) Let $\mathcal{K}$ be the coordinate subspace $u_{i_1} = \ldots = u_{i_m} = 0$ $(0 < m \leq n)$. We denote as $r = r(\mathcal{K})$ such an integer that among $n - m$ numbers λ_j, $1 \leq j \leq n$, $j \neq i_1, \ldots, i_m$ there are r numbers which are linearly independent over the rational numbers, and every $r + 1$ numbers among these $n - m$ numbers are linearly dependent. If Solution (6.12) from the set Re$(\mathcal{A} \bigcap \mathcal{K})$ does not lie in a smaller coordinate subspace, then it is quasi-periodic with r basic frequencies $(0 \leq r \leq n - m)$.

3) For the quasi-periodic solution (6.12) from the set $\mathcal{A}$ the basis of frequencies is completely determined by the value of parameter α. Here eigenvalues may be arbitrary.

4) For the quasi-periodic solution (6.12) from the set $\mathcal{B}$ the value of parameter α defines not only basis of frequencies but the eigenvalues as well. This property isolates the set $\mathcal{B}$ in the set $\mathcal{A}$.

We denote $\tilde{\mathcal{A}} = \bigcup_{r(\mathcal{K}) \leq 1} (\mathcal{A} \bigcap \mathcal{K})$, where the union is taken over all coordinate subspaces $\mathcal{K}$ (over U), for which $r(\mathcal{K}) \leq 1$.

Theorem 6.4 [Bruno 1974, 1979a, 1989a] *The set $\tilde{\mathcal{A}}$ is analytic for the analytic system* (6.2).

Obviously, it is a generalization of Theorem 6.2. We note that all formal families of periodic $(r = 1)$ and stationary $(r = 0)$ solutions are in the set $\tilde{\mathcal{A}}$. By Theorem 6.4 they are always analytic. Thus, under Condition ω System (6.2) has analytic sets $\tilde{\mathcal{A}}$ and $\mathcal{B}$. Special cases of Theorems 6.2–6.4 for the reversible systems are in [Arnol'd 1984, Arnol'd and Sevryuk 1986, Sevryuk 1986].

Remark 6.1. If Condition (6.7) does not hold, but for some complex number τ, $|\tau| = 1$ the condition

$$\operatorname{Re}(\tau \lambda_i) = 0, \quad i = 1, \ldots, n \tag{6.13}$$

is satisfied, then the definitions of sets $\mathcal{A}$, $\tilde{\mathcal{A}}$, $\mathcal{B}$ and Theorems 6.2–6.4 are preserved. In particular, Condition (6.13) is satisfied if all λ_i are real. If Condition (6.13) does not hold, then System (6.2) has the formal central manifold corresponding to those λ_j for which (6.7) is satisfied. Then on this central manifold the sets $\mathcal{A}$, $\tilde{\mathcal{A}}$ and $\mathcal{B}$ are defined, and for them Theorems 6.2–6.4 are valid.

Theorem 6.5 [Grobman 1962; Samovol 1972, 1979] *If all* $\operatorname{Re} \lambda_j \neq 0$, $j = 1, \ldots, n$, *then the real solutions to System* (6.1) *are topologically equivalent* *to solutions to the corresponding linear system and smoothly equivalent to solutions to some generalized truncated normal form (which is obtained from the normal form* (6.4) *by omitting all terms the orders of which are greater then some s).*

7. The local analysis of System (4.2) in Cases I and γ_1

In Case I all eigenvalues (5.5) are real and $\lambda_1 > \lambda_2 > 0$, $\lambda_3 = -\lambda_1$, $\lambda_4 = -\lambda_2$. By Theorem 6.5 in Case I solutions to nonlinear System (4.2) are topologically equivalent to solutions to linear System (5.1). System (5.1) has one two-dimensional manifold $\mathcal{L}^u$ of unstable solutions, which is spanned on eigenvectors ξ_1, ξ_2 (since $\lambda_1, \lambda_2 > 0$), and the similar manifold $\mathcal{L}^s$ of stable solutions, which is spanned on vectors ξ_3, ξ_4 (since $\lambda_3, \lambda_4 < 0$). By Theorem 5.2 both manifolds $\mathcal{L}^u$ and $\mathcal{L}^s$ lie in the integral set (5.7). Therefore for the nonlinear System (4.2) some of solutions from the manifold $\mathcal{L}^u$ may turn into solutions from the manifold $\mathcal{L}^s$ forming loops. The question of existence of such homoclinic solutions will be considered in more detail later on. In the exterior of manifolds $\mathcal{L}^u$ and $\mathcal{L}^s$ all solutions to linear System (5.1) are transit, i.e. if they come into some neighborhood of the origin $Y = 0$, then they leave it in a finite time.

In Case I the normalizing transformation is apparently divergent, since the normal form does not satisfy Condition A. Let us show it. The equation $\langle Q, \Lambda \rangle = 0$ is $\lambda_1(q_1 - q_3) + \lambda_2(q_2 - q_4) = 0$, or

$$q_1 - q_3 + (\lambda_2/\lambda_1)(q_2 - q_4) = 0. \tag{7.1}$$

If the ratio of eigenvalues λ_1/λ_2 is irrational, then in integer q_i this equation has only trivial solutions

$$q_1 = q_3, \quad q_2 = q_4. \tag{7.2}$$

Consequently, when $\nu = \nu_0$ is fixed with the irrational λ_1/λ_2, the normal form (6.4) of System (4.2) has the form

$$\dot{u}_i = u_i(\lambda_i + g_i(\varrho_1, \varrho_2)), \quad i = 1, 2, 3, 4, \tag{7.3}$$

where $\varrho_j = u_j u_{j+2}$, $j = 1, 2$. Here all u_i, g_i, ϱ_j are real. By Property 3 of the normal form in Section 6, System (7.3) is reversible, i.e. according to Lemma 5.3 the substitution

$$u_1,\ u_2,\ u_3,\ u_4,\ t \to u_3,\ u_4,\ u_1,\ u_2,\ -t \tag{7.4}$$

transforms it into itself; consequently,

$$g_3 = -g_1, \quad g_4 = -g_2. \tag{7.5}$$

Theorem 7.1 *In Case* I *with irrational* λ_2/λ_1 *the normal form* (7.3) *of System* (4.2) *does not satisfy Condition* A.

Proof. Since in our case $\overline{\Lambda} = \Lambda$, then Condition A takes the form

$$\psi_i = \lambda_i u_i \alpha, \quad i = 1, 2, 3, 4, \tag{7.6}$$

where α is a power series. Since $\lambda_{j+2} = -\lambda_j$, then for the normal form (7.3), (7.5), Condition A takes the form $g_j = \lambda_j \alpha$, $j = 1, 2$, i.e.

$$g_1/\lambda_1 = g_2/\lambda_2 = \alpha. \tag{7.7}$$

In System (7.3), g_i are power series of ϱ_j without constant terms. Therefore, the identity $g_1/\lambda_1 = g_2/\lambda_2$ means the equality of all coefficients at the same powers of ϱ_1, ϱ_2. We write out their linear terms:

$$g_1 = c_1 \varrho_1 + c_2 \varrho_2 + \dots, \qquad g_2 = -d_1 \varrho_1 - d_2 \varrho_2 + \dots$$

In order to fulfill equalities (7.7) the following equalities are necessary

$$D_1 \stackrel{\text{def}}{=} \lambda_2 c_1 + \lambda_1 d_1 = 0, \quad D_2 \stackrel{\text{def}}{=} \lambda_2 c_2 + \lambda_1 d_2 = 0. \tag{7.8}$$

Coefficients c_i and d_i may be computed for System (6.2) with the help of the formula (35) in [Bruno 1979a and 1989a, § 1 of Chapter III]. Computations for System (4.2) through transformation $Y = TZ$ give the values

$$\begin{aligned}
c_1 &= \frac{a^2\left(24\lambda_1^2 - 5\lambda_2^2\right)}{3\lambda_1^3\lambda_2^2\left(\lambda_1^2 - \lambda_2^2\right)\left(4\lambda_1^2 - \lambda_2^2\right)}, \\
c_2 &= \frac{4a^2\left(2\lambda_1^4 - 11\lambda_1^2\lambda_2^2 + 2\lambda_2^4\right)}{\lambda_1^3\lambda_2^2\left(\lambda_1^2 - 4\lambda_2^2\right)\left(\lambda_1^2 - \lambda_2^2\right)\left(4\lambda_1^2 - \lambda_2^2\right)}, \\
d_1 &= \frac{-2a^2\left(4\lambda_1^4 - 7\lambda_1^2\lambda_2^2 + 4\lambda_2^4\right)}{\lambda_1^2\lambda_2^3\left(\lambda_1^2 - 4\lambda_2^2\right)\left(\lambda_1^2 - \lambda_2^2\right)\left(4\lambda_1^2 - \lambda_2^2\right)}, \\
d_2 &= \frac{-a^2\left(5\lambda_1^2 - 24\lambda_2^2\right)}{3\lambda_1^2\lambda_2^3\left(\lambda_1^2 - 4\lambda_2^2\right)\left(\lambda_1^2 - \lambda_2^2\right)}.
\end{aligned} \tag{7.9}$$

Lemma 7.1 *System of equations* (7.8) *is not satisfied in the interval* $-5/4 < \nu < 1$.

Proof. From Formulae (7.9) we obtain

$$D_1 = \frac{a^2\left(6\lambda_1^4 - 3\lambda_1^2\lambda_2^2 - 20\lambda_2^4\right)}{3\lambda_1^3\left(\lambda_1^2 - 4\lambda_2^2\right)\left(\lambda_1^2 - \lambda_2^2\right)\lambda_2^3},$$

(7.10)

$$D_2 = \frac{a^2\left(20\lambda_1^6 - 77\lambda_1^4\lambda_2^2 - 108\lambda_1^2\lambda_2^4 + 24\lambda_2^6\right)}{3\lambda_1^3\left(\lambda_1^2 - 4\lambda_2^2\right)\left(\lambda_1^2 - \lambda_2^2\right)\left(4\lambda_1^2 - \lambda_2^2\right)\lambda_2^3}.$$

We denote

$$s = \lambda_1^2/\lambda_2^2, \tag{7.11}$$

i.e. $s = (3+\sqrt{5+4\nu})/(3-\sqrt{5+4\nu})$ and

$$\nu = 1 - \frac{9s}{1+2s+s^2}. \tag{7.12}$$

Here the interval $-5/4 < \nu < 1$ is mapped one-to-one into the half-line $1 < s < \infty$. The equation $D_1 = 0$, i.e. $6s^2 - 3s - 20 = 0$, has the two solutions:

$$s_1 = \frac{3+\sqrt{489}}{12} \approx 2.09278, \quad s_2 = \frac{3-\sqrt{489}}{12} \approx -1.59278. \tag{7.13}$$

Among them only s_1 is in the half-line $s > 1$. The equation $D_2 = 0$, i.e. $20s^3 - 77s^2 - 108s + 24 = 0$, has three real roots:

$$s_3 \approx -1.24789, \quad s_4 \approx 0.196181, \quad s_5 \approx 4.90171. \tag{7.14}$$

Among them only s_5 is in the half-line $s > 1$. Since $s_1 \neq s_5$, then both equations $D_1 = 0$ and $D_2 = 0$ do not have a common root for $s > 1$. □

Since equalities (7.8) are impossible in the interval $-5/4 < \nu < 1$, then the statement of Theorem 7.1 follows from Lemma 7.1. □

Everything said above in Section 7 relates to System (4.2) with the fixed value ν. But since in the interval $\nu \in (-5/4, 1)$ the eigenvalues are smooth functions of ν, then manifolds $\mathcal{L}^u$ and $\mathcal{L}^s$ are also smoothly depend on ν.

In Case γ_1, when $\sigma = 1$ and $\nu = -5/4$, eigenvalues $\lambda_1 = \lambda_3 = -\lambda_2 = -\lambda_4 = \sqrt{3/2}$ are real. By Theorem 6.5, in a neighborhood of the point $Y = 0$, System (4.2) is smoothly equivalent to its linear part of (5.1). Although according to Lemma 5.3 the Jordan form of the matrix A has two Jordan blocks, topologically solutions to System (5.1) in Case γ_1 behave similarly to that of Case I (see the beginning of Section 7).

8. System (4.2) in Cases II and IV

In Case II all eigenvalues (5.5) are complex, and

$$\lambda_3 = -\lambda_1, \quad \lambda_4 = -\lambda_2 \quad \text{and} \quad \lambda_2 = \overline{\lambda_1}, \quad \lambda_4 = \overline{\lambda_2}. \tag{8.1}$$

Besides,

$$\mathrm{Re}\lambda_3 = \mathrm{Re}\lambda_4 < 0 < \mathrm{Re}\lambda_1 = \mathrm{Re}\lambda_2. \tag{8.2}$$

Therefore here Theorem 6.5 is applicable again, and System (5.1) has invariant two-dimensional manifolds $\mathcal{L}^u$, $\mathcal{L}^s$, which according to Theorem 5.2 also lie in the integral set (5.7). Here the homoclinic solutions are also possible.

Let $\nu = \nu_0 + \varepsilon$, where ν_0 is a fixed value of ν.

Theorem 8.1 *In Case* II *the normalizing transformation converges for sufficiently small* $|u_j|$ *and* $|\varepsilon|$.

Proof. In Case II vector $\Lambda = (\lambda_1, \overline{\lambda}_1, -\lambda_1, -\overline{\lambda}_1)$, and equation $\langle Q, \Lambda \rangle = 0$ takes the form $\lambda_1(q_1 - q_3) + \overline{\lambda}_1(q_2 - q_4) = 0$ or $q_1 - q_3 + \overline{\lambda}_1\lambda_1^{-1}(q_2 - q_4) = 0$. Since the number $\overline{\lambda_1}\lambda_1^{-1}$ is not real, then all real solutions $Q = (q_1, \dots, q_4)$ to the equation have

$$q_1 = q_3, \quad q_2 = q_4. \tag{8.3}$$

Consequently, the normal form (6.4) is

$$\dot{u}_i = u_i(\lambda_i + g_i(\varrho_1, \varrho_2, \varepsilon)), \quad i = 1, 2, 3, 4, \tag{8.4}$$

where $\lambda_i = \lambda_i(\nu_0)$, $i = 1, 2, 3, 4$, $\varrho_j = u_j u_{j+2}$, $j = 1, 2$. According to Lemma 5.3, in Case II the matrix I of the reality relation $\overline{U} = IU$ is I_2 from (5.37), i.e. $\overline{u}_1 = u_2$, $\overline{u}_3 = u_4$. Hence

$$\overline{\varrho}_1 = \varrho_2 \quad \text{and} \quad \overline{g}_1(\varrho_1, \varrho_2, \varepsilon) = g_2(\varrho_1, \varrho_2, \varepsilon), \tag{8.5}$$

where $\overline{g}_1$ means the series g_1 with the complex conjugate coefficients. According to Property 3 from Section 6, the normal form (8.4) is reversible with the involutory matrix $\tilde{S}$, which by Lemma 5.3 has the form (5.36), i.e. the substitution (7.4) transforms System (8.4) into itself. Here ϱ_j are preserved. Consequently, in (8.4)

$$g_3(\varrho_1, \varrho_2, \varepsilon) = -g_1(\varrho_1, \varrho_2, \varepsilon), \quad g_4(\varrho_1, \varrho_2, \varepsilon) = -g_2(\varrho_1, \varrho_2, \varepsilon). \tag{8.6}$$

If we put

$$\alpha = (g_1/\lambda_1 + g_2/\lambda_2)/2, \quad \beta = (g_1/\lambda_1 - g_2/\lambda_2)/2, \tag{8.7}$$

then according to (8.1)

$$g_i = \alpha\lambda_i + \beta\overline{\lambda}_i, \quad i = 1, 2, 3, 4. \tag{8.8}$$

Consequently, the normal form (8.4) satisfies Condition A. Condition ω here is trivially fulfilled, since for any integer Q with $\langle Q, \Lambda \rangle \neq 0$, we have

$$|\langle Q, \Lambda \rangle| > |\lambda_1||q_1 - q_3| + |\lambda_2||q_2 - q_4| \geq \min\{|\lambda_1|, |\lambda_2|\}.$$

By Theorem 6.1 there exists an analytic normalizing transformation. □

According to (8.6), System (8.4) has two first integrals: $\varrho_1 = \text{const}$ and $\varrho_2 = \text{const}$. They are not real, but according to (8.5) they are complex conjugate. Therefore, there are two real first integrals:

$$\varrho_1 + \varrho_2 = 2\,\mathrm{Re}\varrho_1 = \text{const}, \qquad -i(\varrho_1 - \varrho_2) = 2\,\mathrm{Im}\varrho_1 = \text{const}.$$

According to Theorem 4.1, the analyticity of the normalizing transformation in Case II with the fixed ν follows from

Corollary to the Moser Theorem [Moser 1958] *If in a Hamiltonian system with two degrees of freedom the ratio of eigenvalues λ_1/λ_2 is not real, then its normalizing transformation converges.*

In Case IV the eigenvalues $\lambda_1 = -\lambda_3$ are real, and $\lambda_2 = -\lambda_4$ are purely imaginary. Therefore here Theorem 6.5 is not applicable.

Theorem 8.2 *In Case* IV *the normalizing transformation converges for sufficiently small $|u_j|$ and $|\varepsilon|$.*

Proof. It is analogous to that of Theorem 8.1. Namely, the normal form has the form (8.4), but by Lemma 5.3 the reality relation is $\overline{u}_1 = u_1$, $\overline{u}_2 = u_4$, $\overline{u}_3 = u_3$. Consequently, $\overline{\varrho}_1 = \varrho_1$, $\overline{\varrho}_2 = \varrho_2$ and $\overline{g}_2(\varrho_1, \varrho_2, \varepsilon) = g_4(\varrho_1, \varrho_2, \varepsilon)$, and g_1 and g_3 are real, i.e. the integrals ϱ_i = const are real. As in Case II, the reversibility implies the property (8.6). This implies the property (8.8) with the series α and β defined in (8.7), i.e. Condition A is satisfied. □

When ν is fixed, here the Moser Theorem [1958] is applicable as well.

We consider now the behavior of solutions to the normal form (8.4) for Case IV. The axes u_1 and u_3 are respectively unstable and stable one-dimensional invariant manifolds. They lie in the set (5.7), i.e. $J_1^{(2)} = J_2^{(2)} = 0$, and they may form homoclinic solutions. The plane $u_1 = u_3 = 0$ is filled by periodic solutions ϱ_2 = const with the frequency $\operatorname{Im}(\lambda_2 + g_2(0, \varrho_2))$. We denote this family of periodic solutions as $\mathcal{F}$. It is analytic in Y and ν. With the fixed ν, every periodic solution $\mathcal{M}$ of the family $\mathcal{F}$ is an intersection of two two-dimensional manifolds $\mathcal{M}^u$ and $\mathcal{M}^s$, the solutions in which tend to $\mathcal{M}$ when $t \to -\infty$ and $t \to +\infty$ respectively.

9. The non-resonant case III

In Case III all eigenvalues (5.5) are purely imaginary and different, furthermore

$$\lambda_{1,3} = \pm i \left| \sqrt{\frac{3\sigma}{2} + \sqrt{\frac{5}{4} + \nu}} \right|, \quad \lambda_{2,4} = \pm i \left| \sqrt{\frac{3\sigma}{2} - \sqrt{\frac{5}{4} + \nu}} \right|,$$

$$\lambda_1 = -\lambda_3, \quad \lambda_2 = -\lambda_4, \quad 0 < |\lambda_1| < |\lambda_2|. \tag{9.1}$$

In Case III Theorems 6.1 and 6.5 are not applicable, since the normal form does not satisfy Condition A, as we will see below, and all $\operatorname{Re}\lambda_i = 0$. Therefore we will compute sets $\mathcal{A}$, $\tilde{\mathcal{A}}$ and $\mathcal{B}$ for various values of λ_1/λ_2.

Let ν_0 be the fixed value of ν. We put $\nu = \nu_0 + \varepsilon$, where ε is a small parameter. Then the matrix A of System (4.2) has the form $A_0 + \tilde{A}\varepsilon$, where

A_0 is the matrix A with $\nu = \nu_0$, and the matrix $\tilde{A} = (\tilde{a}_{ij})$ has all elements as zeros except $\tilde{a}_{41} = 1$. According to Lemma 5.2, in Case III the linear transformation $Y = T(\nu_0)Z$ transforms System (4.2) into the system

$$\dot{z}_1 = \lambda_1 z_1 + c_0 \varepsilon x_1 + c_0 a x_1^2, \quad \dot{z}_2 = \lambda_2 z_2 - d_0 \varepsilon x_1 - d_0 a x_1^2 \tag{9.2}$$

and the conjugate equations. Here

$$x_1 = z_1 + z_2 + z_3 + z_4, \quad \lambda_i = \lambda_i(\nu_0), \quad i = 1, 2,$$
$$c_0 = \frac{1}{2\lambda_1(\lambda_1^2 - \lambda_2^2)}, \quad d_0 = \frac{1}{2\lambda_2(\lambda_1^2 - \lambda_2^2)}. \tag{9.3}$$

By Lemma 5.3, in System (9.2) the coordinates Z are connected by the reality relation

$$\bar{z}_1 = z_3, \quad \bar{z}_2 = z_4, \tag{9.3'}$$

and System (9.2) is invariant with respect to the involution

$$z_1, z_2, z_3, z_4, t \to z_3, z_4, z_1, z_2, -t. \tag{9.3''}$$

By Property 2 from Section 6, the normal form of System (9.2) has the form

$$\begin{gathered} \dot{u}_i = u_i(\lambda_i + g_i(U, \varepsilon)), \quad i = 1, 2, 3, 4, \\ g_i \stackrel{\text{def}}{=} \sum g_{iQ}(\varepsilon) U^Q \text{ over } Q + E_i \geq 0, \quad Q \in \mathbb{Z}^n, \quad \langle Q, \Lambda \rangle = 0. \end{gathered} \tag{9.4}$$

By Properties 3 and 4, it preserves the linear automorphism and the reality relation, i.e. coordinates U are connected by the reality relation

$$\bar{u}_1 = u_3, \quad \bar{u}_2 = u_4, \tag{9.5}$$

and System (9.4) is invariant with respect to the involution

$$u_1, u_2, u_3, u_4, t \to u_3, u_4, u_1, u_2, -t. \tag{9.6}$$

We consider such ν_0 that $\beta = \beta(\nu_0) \stackrel{\text{def}}{=} \lambda_1(\nu_0)/\lambda_2(\nu_0)$ is an irrational number (*the non-resonant case*). Here all integer solutions to the equation $\langle Q, \Lambda \rangle = 0$ have the form $q_j = q_{j+2}$, $j = 1, 2$. Therefore in the normal form (9.4) $g_i = g_i(\varrho_1, \varrho_2, \varepsilon)$, where $\varrho_j = u_j u_{j+2}$. We introduce the vector $\rho = (\varrho_1, \varrho_2)$. For ϱ_j, the reality relation (9.5) gives $\bar{\rho} = \rho$, and the involution (9.6) gives $\rho \to \rho$. Therefore in the normal form (9.4)

$$g_j(\rho, \varepsilon) = \bar{g}_{j+2}(\rho, \varepsilon), \quad g_j(\rho. \varepsilon) = -g_{j+2}(\rho, \varepsilon), \quad j = 1, 2, \tag{9.7}$$

i.e. all coefficients of the series g_i are purely imaginary. By virtue of (9.7), $\varrho_1 = \text{const}$ and $\varrho_2 = \text{const}$ are formal first integrals of System (9.4).

By the definition (6.8), the set $\mathcal{A} = \mathcal{A}(\nu_0)$ is described by the system of equations

$$u_i(\lambda_i + g_i) = u_i \lambda_i \alpha, \quad i = 1, 2, 3, 4. \tag{9.8}$$

In three-dimensional space with coordinates ϱ_1, ϱ_2, ε it consists of three components:

$$\mathcal{A}_1 = \mathcal{A}_1(\nu_0) \stackrel{\text{def}}{=} \{\rho, \varepsilon \colon \varrho_2 = 0\}, \quad \mathcal{A}_2 = \mathcal{A}_2(\nu_0) \stackrel{\text{def}}{=} \{\rho, \varepsilon \colon \varrho_1 = 0\}, \tag{9.9}$$
$$\mathcal{A}_3 = \mathcal{A}_3(\nu_0) \stackrel{\text{def}}{=} \{\rho, \varepsilon \colon g_1/\lambda_1 = g_2/\lambda_2 = \alpha\} = \{\rho, \varepsilon \colon \lambda_2 g_1 - \lambda_1 g_2 = 0\}.$$

Lemma 9.1 *Components $\mathcal{A}_1$ and $\mathcal{A}_2$ belong to the set $\tilde{\mathcal{A}}$. The component $\mathcal{A}_i$ is filled by periodic solutions $u_i = \sqrt{\varrho_i}\exp(\lambda_i \alpha t)$, $u_{i+2} = \sqrt{\varrho_i}\exp(-\lambda_i \alpha t)$, $u_j = u_{j+2} = 0$, with frequencies $|\lambda_i \alpha| = |\lambda_i + g_i|$ and eigenvalues $\pm(\lambda_j + g_j)$, where g_i and g_j are computed with $\varrho_j = 0$ as functions of ϱ_i, ε; here $i, j = 1, 2$, $i \neq j$.*

Proof. With the fixed $i = 1, 2$ eigenvalues $\pm\lambda_i$ are commensurable. Hence the subspace $\varrho_j = 0$ lies in the set $\tilde{\mathcal{A}}$, and Theorem 6.4 is applicable to it, i.e. components $\mathcal{A}_1$ and $\mathcal{A}_2$ are analytic in the original system. The properties of periodic solutions that fill them are checked directly. □

We denote as $\mathcal{F}_i$ the family of periodic solutions lying in the set $\mathcal{A}_i$, $i = 1, 2$. It is the Lyapunov family [Lyapunov 1892].

Lemma 9.2 *The component $\mathcal{A}_3$ is the set $\mathcal{B} = \mathcal{B}(\nu_0)$.*

Proof. We introduce the (2×2)-matrices

$$\Pi = \frac{\partial g}{\partial \rho}, \quad U = \begin{pmatrix} u_1 & 0 \\ 0 & u_2 \end{pmatrix},$$

where $g = (g_1, g_2)$, then according to (9.5)

$$U\bar{U} = \bar{U}U = \begin{pmatrix} \varrho_1 & 0 \\ 0 & \varrho_2 \end{pmatrix}. \tag{9.10}$$

The matrix B from Section 6 may be written in the block form

$$B = \begin{pmatrix} U\Pi\bar{U} & U\Pi U \\ -\bar{U}\Pi\bar{U} & -\bar{U}\Pi U \end{pmatrix}.$$

Its square is the block matrix

$$B^2 = \begin{pmatrix} U\Pi\bar{U}U\Pi\bar{U} - U\Pi U\bar{U}\Pi\bar{U} & U\Pi\bar{U}U\Pi U - U\Pi U\bar{U}\Pi U \\ -\bar{U}\Pi\bar{U}U\Pi\bar{U} + \bar{U}\Pi U\bar{U}\Pi\bar{U} & -\bar{U}\Pi\bar{U}U\Pi U + \bar{U}\Pi U\bar{U}\Pi U \end{pmatrix},$$

which has all blocks equal to zero according to (9.10). Consequently, the matrix B is a nilpotent one. □

In coordinates ε, ϱ_1, ϱ_2, to the real values of the original coordinates ε, Y there corresponds the part of their real values with $\varrho_1 \geq 0$, $\varrho_2 \geq 0$ and an arbitrary ε. According to (9.9), in coordinates ε, ϱ_1, ϱ_2, the set $\mathcal{B} = \mathcal{B}(\nu_0)$ is determined by one equation $\lambda_2 g_1 - \lambda_1 g_2 = 0$. The expansion of series g_i begins with the linear terms over ε, ϱ_1, ϱ_2:

$$g_1 = c_0\varepsilon + c_1\varrho_1 + c_2\varrho_2 + \dots, \qquad g_2 = -d_0\varepsilon - d_1\varrho_1 - d_2\varrho_2 + \dots \tag{9.11}$$

Hence (compare with (7.8))

$$\lambda_2 g_1 - \lambda_1 g_2 = D_0\varepsilon + D_1\varrho_1 + D_2\varrho_2 + \dots, \tag{9.12}$$

where

$$D_i = \lambda_2 c_i + \lambda_1 d_i, \quad i = 0, 1, 2. \tag{9.13}$$

Here the vector $D = (D_0, D_1, D_2)$ is normal to the tangent plane of the set $\mathcal{B}$ at the origin $\varepsilon = \varrho_1 = \varrho_2 = 0$, i.e. it completely determines the disposition of the set $\mathcal{B}$ near the origin. Coefficients c_0 and d_0 are obtained from the linear part of System (9.2) in the form (9.3). According to (9.13), (9.3) and (9.1), we obtain

$$D_0 = \frac{\lambda_1^2 + \lambda_2^2}{2\lambda_1\lambda_2(\lambda_1^2 - \lambda_2^2)} = \frac{3\sigma}{-2|\sqrt{1-\nu}|\sqrt{5+4\nu}}, \tag{9.14}$$

and the values of D_1 and D_2 are cited in (7.10). As in (7.11), we denote $s = \lambda_1^2/\lambda_2^2 \overset{\text{def}}{=} (3 - \sqrt{5+4\nu})/(3+\sqrt{5+4\nu})$, i.e.

$$\nu = 1 - \frac{9s}{(1+s)^2}. \tag{9.15}$$

Here the interval $-5/4 < \nu < 1$ is mapped one-to-one into the interval $0 < s < 1$ of the parameter s. According to (9.14), when $\sigma = -1$ and $-5/4 < \nu < 1$, it is always $D_0 > 0$. According to Fig. 5.1 and Formulae (7.10), (7.11), (7.13) and (7.14), in the interval $-5/4 < \nu < 1$ the coefficient D_1 does not change sign, and $D_1 > 0$; but $D_2 = 0$ when $s = s_4$, where $\nu_4 = -0.233973$. Besides, the denominator of the fraction D_2 changes sign when $\lambda_2 = 2\lambda_1$, where $\nu = -0.44$. Consequently,

$$D_2 = \begin{cases} > 0 & \text{when } -5/4 < \nu < -0.44, \\ < 0 & \text{when } -0.44 < \nu < -0.233973 = \nu_4, \\ > 0 & \text{when } \nu_4 < \nu < 1. \end{cases}$$

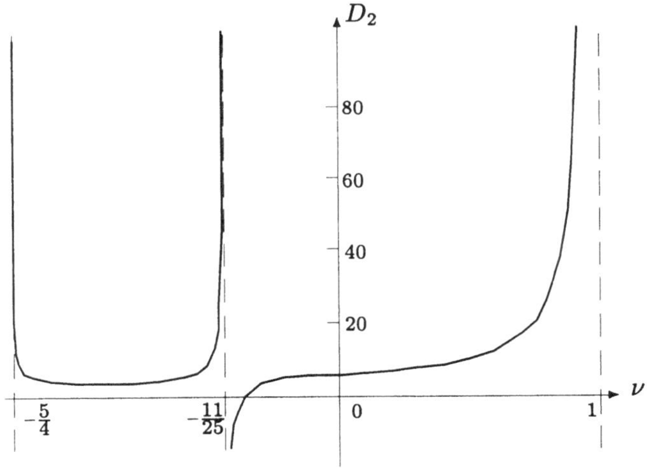

Figure 5.2: The non-resonant Case III. The graph of the function D_2 for $\nu \in (-5/4, 1)$.

Fig. 5.2 shows the graph of the function $D_2(\nu)$ for $\nu \in (-5/4, 1)$. It is clear from the graph that here we encounter two subcases.

(1) $D_0, D_1, D_2 > 0$. For this subcase, the disposition of projection of sections $\varepsilon = \text{const}$ of the set $\mathcal{B} = \mathcal{B}(\nu_0)$ into the plane ϱ_1, ϱ_2 is shown in Fig. 5.3, *a*. The real part of the set $\mathcal{B}$ exists only for $\varepsilon < 0$, and it forms the bridge between periodic solutions of the families $\mathcal{F}_1$ and $\mathcal{F}_2$. We recall that the family $\mathcal{B}$ consists of the invariant tori with the ratio of frequencies $\beta(\nu_0)$ and with zero eigenvalues.

(2) $D_0, D_1 > 0$, $D_2 < 0$. For this subcase, the disposition of projection of sections $\varepsilon = \text{const}$ of the set $\mathcal{B} = \mathcal{B}(\nu_0)$ into the plane ϱ_1, ϱ_2 is shown in Fig. 5.3, *b*. Here the real part of the set $\mathcal{B}$ exists for any small ε, and it ends on a periodic solution of the family $\mathcal{F}_1$ if $\varepsilon < 0$, or of the family $\mathcal{F}_2$ if $\varepsilon > 0$.

Thus, we have proved

Theorem 9.2 *In Case* III *with the fixed* ν_0 *with an irrational* $\beta = \lambda_1(\nu_0)/\lambda_2(\nu_0)$ *there are two analytic families* $\mathcal{F}_1$ *and* $\mathcal{F}_2$ *of periodic solutions. If the number* β *satisfies Condition* ω, *then there is also the analytic family* $\mathcal{B} = \mathcal{B}(\nu_0)$ *of quasi-periodic solutions with the frequency ratio* β. *For* $\nu_0 \in (-1.25, -0.44)$ *and* $\nu_0 \in (-0.233973, 1)$ *it exists for* $\nu < \nu_0$ *(Subcase* (1)), *and for* $\nu_0 \in (-0.44, -0.233973)$ *for arbitrary* ν *(Subcase* (2)).

Remark 9.1. Since in Case III equations $D_1 = 0$ and $D_2 = 0$ do not have common roots for $-5/4 < \nu < 1$, then the normal form does not satisfy Condition A (compare with Theorem 7.1). Since for $\nu_0 = \nu_4$ the value $D_2 = 0$ and the ratio $\lambda_1(\nu_4)/\lambda_2(\nu_4) = \sqrt{s_4}$ is an irrational number, this case does not fit in Subcases (1) and (2).

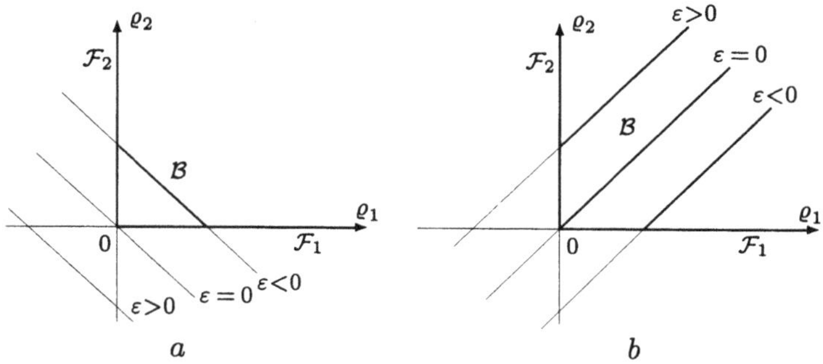

Figure 5.3: The non-resonant Case III. Projections of sections $\varepsilon =$ const of the set $\mathcal{B}(\nu_0)$ in Subcase (1) (a) and in Subcase (2) (b). Bold lines correspond to real families, thin lines correspond to complex ones.

10. The normal form in the resonant Case III

We consider the system $\dot{U} = \Psi(U)$, where $U \in \mathbb{C}^4$, and the vector Ψ expands in the power series of $U = (u_1, u_2, u_3, u_4)$. If the system is written in the form

$$\dot{u}_i = u_i g_i(U) \stackrel{\text{def}}{=} u_i \sum g_{iQ} U^Q \quad \text{over } Q \in \mathbf{N}_i, \quad i = 1,2,3,4, \tag{10.1}$$

then

$$\mathbf{N}_i \stackrel{\text{def}}{=} -E_i + \mathbb{Z}_+^4, \quad i = 1,2,3,4,$$

where E_i is the ith unit vector, $\mathbb{Z}_+^4 = \{Q \in \mathbb{Z}^4, Q \geq 0\}$, and the sum of sets means the set of vectors, which are the sum of vectors belonging to the sets being summed. We denote $\mathbf{N} = \mathbf{N}_1 \cup \ldots \cup \mathbf{N}_4$.

If System (10.1) is the resonant normal form, then according to Section 6 the power exponents Q satisfy the equation $\langle Q, \Lambda \rangle = 0$. Let $\mathbf{L}$ be the set of points $Q \in \mathbb{R}^4$ satisfying this equation. Then the set $\mathbf{S}_i \stackrel{\text{def}}{=} \mathbf{L} \cap \mathbf{N}_i$ is the support of the series g_i in the normal form (10.1), $i = 1,2,3,4$.

Let the number $\lambda_1(\nu_0)/\lambda_2(\nu_0)$ be rational, i.e.

$$\lambda_1(\nu_0)/\lambda_2(\nu_0) = r/s, \quad 1 \leq r \leq s, \tag{10.1'}$$

where r and s are integer mutually simple numbers. Then the equation $\langle Q, \Lambda \rangle = 0$ for integer vectors $Q = (q_1, q_2, q_3, q_4)$ is the equation

$$r(q_1 - q_3) + s(q_2 - q_4) = 0. \tag{10.2}$$

All its integer solutions have the form $q_1 - q_3 = ks$, $q_2 - q_4 = -kr$, where k is an integer, i.e.

$$q_1 = q_3 + ks, \quad q_4 = q_2 + kr. \tag{10.3}$$

Consequently,

$$q_1 + q_3 = 2q_3 + ks, \quad q_2 + q_4 = 2q_2 + kr. \tag{10.4}$$

The vector $Q \in \mathbb{R}^4$ with $\langle Q, \Lambda \rangle = 0$ may be uniquely written in the form

$$Q = q_2(E_2+E_4)+q_3(E_1+E_3)+k(sE_1+rE_4) =$$
$$= q_1(E_1 + E_3) + q_4(E_2 + E_4) - k(rE_2 + sE_3). \tag{10.5}$$

We will use the first expression if $k \geq 0$, and the second one if $k \leq 0$. We introduce vectors

$$R = sE_1 + rE_4 = (s, 0, 0, r), \quad S = rE_2 + sE_3 = (0, r, s, 0). \tag{10.6}$$

We denote as $\mathbf{L}^0$ the set of vectors $Q = (q_1, q_2, q_1, q_2) \in \mathbb{R}^4$, i.e. with $q_3 = q_1$ and $q_4 = q_2$, and as $\mathbf{M}^0$ its subset of integer vectors $Q \geq 0$. We introduce the sets

$$\begin{gathered} \mathbf{L}^k \stackrel{\text{def}}{=} kR + \mathbf{L}^0 \text{ for } k > 0, \quad \mathbf{L}^k \stackrel{\text{def}}{=} -kS + \mathbf{L}^0 \text{ for } k < 0, \\ \mathbf{S}_i^k \stackrel{\text{def}}{=} \mathbf{L}^k \cap \mathbf{N}_i \stackrel{\text{def}}{=} \mathbf{L}^k \cap \mathbf{S}_i, \quad i = 1, 2, 3, 4, \quad k \in \mathbb{Z}. \end{gathered} \tag{10.7}$$

Obviously, $\mathbf{S}_i = \bigcup \mathbf{S}_i^k$ over $k \in \mathbb{Z}$. Let us study the structure of supports $\mathbf{S}_i$ of the series g_i of the normal form (10.1).

Lemma 10.1 *With the resonance* (10.1′):

$$\mathbf{S}_i^0 = \mathbf{M}^0, \quad i = 1, 2, 3, 4;$$

$$\begin{gathered} \mathbf{S}_1^k = \mathbf{S}_4^k = kR+\mathbf{M}^0, \quad \mathbf{S}_2^k = kR+\mathbf{M}^0-(E_2+E_4), \\ \mathbf{S}_3^k = kR + \mathbf{M}^0 - (E_1 + E_3) \ \textit{for} \ k > 0; \\ \mathbf{S}_1^k = -kS + \mathbf{M}^0 - (E_1 + E_3), \quad \mathbf{S}_2^k = \mathbf{S}_3^k = -kS + \mathbf{M}^0, \\ \mathbf{S}_4^k = -kS + \mathbf{M}^0 - (E_2 + E_4) \ \textit{for} \ k < 0. \end{gathered} \tag{10.8}$$

Proof. It is based on equalities (10.5). For $k = 0$ the points $Q \in \mathbf{L}^0 \cap \mathbf{N}_i \stackrel{\text{def}}{=} \mathbf{S}_i^0$ are integer ones, they have $q_1 = q_3 \geq 0$ and $q_2 = q_4 \geq 0$, i.e. they belong to the set $\mathbf{M}^0$.

For $k > 0$ the points $Q \in \mathbf{S}_i^k$ are represented by the first equality (10.5). Here $q_2, q_3 \geq 0$ for $Q \in \mathbf{S}_1^k$ and $Q \in \mathbf{S}_4^k$; $q_2 \geq -1$, $q_3 \geq 0$ for $Q \in \mathbf{S}_2^k$; $q_2 \geq 0$, $q_3 \geq -1$ for $Q \in \mathbf{S}_3^k$. This implies the statement of Lemma for $k > 0$.

For $k < 0$ the points $Q \in \mathbf{S}_i^k$ are represented by the second equality (10.5). Here $q_1 \geq -1$, $q_4 \geq 0$ for $Q \in \mathbf{S}_1^k$; $q_1, q_4 \geq 0$ for $Q \in \mathbf{S}_2^k$ and $Q \in \mathbf{S}_3^k$; $q_1 \geq 0$, $q_4 \geq -1$ for $Q \in \mathbf{S}_4^k$. This implies the statement of Lemma for $k < 0$. □

Now we introduce new variables

$$\varrho_i = u_i u_{i+2}, \quad i = 1, 2, \quad w = u_1^s u_4^r, \qquad \bar{w} = u_2^r u_3^s. \tag{10.9}$$

Here $w\bar{w} = \varrho_1^s \varrho_2^r$. According to (10.5), every monomial U^Q with $\langle Q, \Lambda \rangle = 0$ and $Q \in \mathbf{N}$ may be uniquely written either in the form $\varrho_1^{q_3} \varrho_2^{q_2} w^k$ if $k \geq 0$ or in the form $\varrho_1^{q_1} \varrho_2^{q_4} \bar{w}^{-k}$ if $k \leq 0$.

Lemma 10.2 *With the resonance* (10.1′) *in the normal form* (9.4)

$$g_i = g_i(\rho, w, \bar{w}) = g_{i0}(\rho) + \sum_{k=1}^{\infty} f_{ik}(\rho) w^k + \sum_{k=1}^{\infty} h_{ik}(\rho) \bar{w}^k, \quad i = 1, 2, 3, 4, \tag{10.10}$$

where g_{i0}, f_{1k}, $\varrho_2 f_{2k}$, $\varrho_1 f_{3k}$, f_{4k}, $\varrho_1 h_{1k}$, h_{2k}, h_{3k}, $\varrho_2 h_{4k}$ are power series in ρ. Here the series $\varrho_1 h_{1k}$ and $\varrho_2 f_{2k}$ begin with the constant terms $\tilde{\gamma}_{1k}$ and γ_{2k} respectively, and the series g_{i0} begins with linear terms (9.11).

Proof. It follows from Lemma 10.1, since the set $\mathbf{S}_i^0$ is the support of the series g_{i0}, the set $\mathbf{S}_i^k$ is the support of the series $f_{ik} w^k$ for $k > 0$ and the series $h_{i|k|} \bar{w}^{|k|}$ for $k < 0$. According to Lemma 10.1, the series f_{1k}, f_{4k}, h_{2k}, h_{3k} have the set $\mathbf{M}^0$ as the support; it is also the support for the series $\varrho_2 f_{2k}$, $\varrho_1 f_{3k}$, $\varrho_1 h_{1k}$, $\varrho_2 h_{4k}$. According to (10.9), if the support of a series is the set $\mathbf{M}^0$, then the series is over non-negative powers of ϱ_1, ϱ_2. □

From the reality relation (9.5) we have

$$\bar{\varrho}_i = \varrho_i, \quad \bar{w} = \overline{(w)}, \quad i = 1, 2. \tag{10.11}$$

Consequently, in (10.10)

$$\overline{g_i(\rho, w, \bar{w})} = g_{i+2}(\rho, w, \bar{w}), \quad i = 1, 2. \tag{10.12}$$

With the involution (9.6)

$$\varrho_i \to \varrho_i, \quad w \to \bar{w}, \quad t \to -t, \quad i = 1, 2. \tag{10.13}$$

Consequently, in (10.10)

$$-g_i(\rho, \bar{w}, w) = g_{i+2}(\rho, w, \bar{w}), \quad i = 1, 2. \tag{10.14}$$

From equalities (10.12) and (10.14) we obtain

$$\overline{g_i(\rho, w, \bar{w})} = -g_i(\rho, \bar{w}, w), \quad i = 1, 2, 3, 4. \tag{10.15}$$

Lemma 10.3 *In expansions* (10.10) *all coefficients of the series g_{i0}, f_{ik} and g_{ik} are purely imaginary. Moreover*

$$g_{j+2,0} = -g_{j,0}, \quad f_{j+2,k} = -h_{j,k}, \quad h_{j+2,k} = -f_{j,k}, \quad j = 1, 2, \quad k \in \mathbb{Z}_+.$$

Proof. Using the complex-conjugate values in expansions (10.10), we have

$$\overline{g_i(\rho, w, \bar{w})} = \bar{g}_{i0}(\rho) + \sum_{k=1}^{\infty} \bar{f}_{ik}(\rho)\bar{w}^k + \sum_{k=1}^{\infty} \bar{h}_{ik}(\rho)w^k, \quad i = 1,2,3,4. \quad (10.16)$$

By virtue of the involution (10.13), according to (10.14) we obtain

$$-g_i(\rho, \bar{w}, w) = -g_{i0}(\rho) - \sum_{k=1}^{\infty} f_{ik}(\rho)\bar{w}^k - \sum_{k=1}^{\infty} h_{ik}(\rho)w^k, \quad i = 1,2,3,4. \quad (10.17)$$

Since by the equality (10.15) the left hand parts of the equalities (10.16) and (10.17) are equal, then equating the coefficients of the right hand sides at w^k and $\bar{w}^k$, we obtain

$$\bar{g}_{i0} = -g_{i0}, \quad \bar{f}_{ik} = -f_{ik}, \quad \bar{h}_{ik} = -h_{ik}, \quad i = 1,2,3,4, \quad k \in \mathbb{Z}_+. \quad (10.18)$$

The last statement of Lemma follows from the equalities (10.14). $\square$

We consider now the projection Π from $\mathbb{R}^4$ into $\mathbb{R}^2$:

$$\Pi Q = \tilde{Q} \stackrel{\text{def}}{=} (\tilde{q}_1, \tilde{q}_2), \quad \text{where} \quad q_1 = q_1 + q_3, \; \tilde{q}_2 = q_2 + q_4.$$

We mark by tilde the result of this projection:

$$\Pi R \stackrel{\text{def}}{=} \tilde{R} = (s, r) = \tilde{S} \stackrel{\text{def}}{=} \Pi S,$$

$$\Pi(E_1 + E_2) = 2\tilde{E}_1 = (2, 0),$$

$$\Pi(E_2 + E_4) = 2\tilde{E}_2 = (0, 2).$$

We denote $\Pi \mathbf{S}_i^k = \tilde{\mathbf{S}}_i^k$ and so on.

Lemma 10.4 *Under the projection* Π

$$\Pi \mathbf{M}^0 \stackrel{\text{def}}{=} \tilde{\mathbf{M}}^0 = \{\tilde{Q} = (2l, 2m) \geq 0; \;\; l, m \in \mathbb{Z}_+\},$$

$$\Pi \mathbf{S}_1 \stackrel{\text{def}}{=} \tilde{\mathbf{S}}_1 = \tilde{\mathbf{M}}^0 \cup \{\tilde{R} + \tilde{\mathbf{M}}^0 - 2\tilde{E}_1\} = \tilde{\mathbf{S}}_3 \stackrel{\text{def}}{=} \Pi \mathbf{S}_3,$$

$$\Pi \mathbf{S}_2 \stackrel{\text{def}}{=} \tilde{\mathbf{S}}_2 = \tilde{\mathbf{M}}^0 \cup \{\tilde{R} + \tilde{\mathbf{M}}^0 - 2\tilde{E}_2\} = \tilde{\mathbf{S}}_4 \stackrel{\text{def}}{=} \Pi \mathbf{S}_4,$$

$$\Pi \mathbf{S} \stackrel{\text{def}}{=} \tilde{\mathbf{S}} = \tilde{\mathbf{M}}^0 \cup \{\tilde{R} + \tilde{\mathbf{M}}^2\}, \;\; \text{where} \;\; \tilde{\mathbf{M}}^2 = \{\tilde{\mathbf{M}}^0 - 2\tilde{E}_1\} \cup \{\tilde{\mathbf{M}}^0 - 2\tilde{E}_2\}.$$

Proof. It is based on Lemma 10.1. The statement for $\mathbf{M}^0$ follows from the definition of $\mathbf{M}^0 = \{Q = (q_1, q_2, q_1, q_2)\}$.

Now we prove the statement for $\mathbf{S}_1 = \bigcup \mathbf{S}_1^k$ over $k \in \mathbb{Z}$. By Lemma 10.1, we have $\Pi \mathbf{S}_1^0 = \tilde{\mathbf{M}}^0$, $\Pi \mathbf{S}_1^k = k\tilde{R} + \tilde{\mathbf{M}}^0$ for $k > 0$, and $\Pi \mathbf{S}_1^k = -k\tilde{R} + \tilde{\mathbf{M}}^0 - 2\tilde{E}_1$ for $k < 0$. Let an integer $l \geq 1$, then $2l\tilde{R} - 2\tilde{E}_1 \geq 0$, i.e. $2l\tilde{R} - 2\tilde{E}_1 \in \tilde{\mathbf{M}}^0$.

Consequently, $2l\tilde{R}-2\tilde{E}_1+\tilde{\mathbf{M}}^0 \subset \tilde{\mathbf{M}}^0$ and $(2l+1)\tilde{R}-2\tilde{E}_1+\tilde{\mathbf{M}}^0 \subset \tilde{R}+\tilde{\mathbf{M}}^0 \subset \tilde{R}+\tilde{\mathbf{M}}^0-2\tilde{E}_1 \stackrel{\text{def}}{=} \tilde{\mathbf{S}}_1^{-1}$. Hence $\Pi\mathbf{S}_1 \stackrel{\text{def}}{=} \tilde{\mathbf{S}}_1 = \tilde{\mathbf{S}}_1^0 \cup \tilde{\mathbf{S}}_1^{-1} = \tilde{\mathbf{M}}^0 \cup \{\tilde{R}+\tilde{\mathbf{M}}^0-2\tilde{E}_1\}$.

The statements of Lemma 10.4 for the rest of sets $\tilde{\mathbf{S}}_i$ and for their union $\tilde{\mathbf{S}}$ is derived from Lemma 10.1 in a similar way. □

Lemma 10.5 *If in* (10.1′) *the integer* $s > 1$, *then the normal form* (9.4) *has the formal invariant manifold*

$$\mathcal{A}_2 = \{U, \varepsilon\colon\ u_1 = u_3 = 0\};$$

and if $r > 1$, *then it has also the formal invariant manifold*

$$\mathcal{A}_1 = \{U, \varepsilon\colon\ u_2 = u_4 = 0\}.$$

In System (4.2), *to manifolds* $\mathcal{A}_1$ *and* $\mathcal{A}_2$ *there correspond analytic families* $\mathcal{F}_1$ *and* $\mathcal{F}_2$ *of stable periodic solutions.*

Proof. It is based on Lemma 10.4. If $s > 1$, i.e. $s \geq 2$, then for $\tilde{Q} \in \tilde{\mathbf{S}}_1 = \tilde{\mathbf{S}}_3$ we have $\tilde{q}_1 \geq 0$, since in this case $\tilde{R} - 2\tilde{E}_1 = (s-2, r)$ and $\tilde{q}_1 = s - 2 + 2l$, where $l \geq 0$. Hence $\tilde{q}_1 \geq 0$. Consequently, in the manifold $\mathcal{A}_2$ the identities $u_1(\lambda_1 + g_1) \equiv 0$ and $u_3(\lambda_3 + g_3) \equiv 0$ are fulfilled, i.e. $\dot{u}_1 \equiv 0$ and $\dot{u}_3 \equiv 0$. That means that the manifold $\mathcal{A}_2$ is invariant in System (9.4). By similar reasoning we obtain that when $r > 1$ the manifold $\mathcal{A}_1$ is invariant in System (9.4). When the ratio λ_1/λ_2 is rational, the set $\tilde{\mathcal{A}} = \mathcal{A}$. It is described by the system of equations (9.8), and it includes the components $\mathcal{A}_1$ and $\mathcal{A}_2$. The further reasoning is the same as in the proof of Lemma 9.1. □

From the normal form (9.4), (10.10), we obtain the system for ϱ_i, w, $\bar{w}$:

$$\begin{aligned}
\dot{\varrho}_i &= \varrho_i(g_i + g_{i+2}) \stackrel{\text{def}}{=} b_i(\rho, w, \bar{w}), \quad i = 1, 2,\\
\dot{w} &= w(sg_1 + rg_4) \stackrel{\text{def}}{=} b_3(\rho, w, \bar{w}),\\
\dot{\bar{w}} &= \bar{w}(rg_2 + sg_3) = -b_3(\rho, \bar{w}, w),
\end{aligned} \tag{10.19}$$

where according to Lemmas 10.2 and 10.3

$$b_i = \varrho_i \sum_{k=1}^{\infty} (f_{ik} - h_{ik})(w^k - \bar{w}^k), \quad i = 1, 2, \tag{10.20}$$

$$b_3 = w\left[sg_{10} - rg_{20} + \sum_{k=1}^{\infty}(sf_{1k} - rh_{2k})w^k + \sum_{k=1}^{\infty}(sh_{1k} - rf_{2k})\bar{w}^k\right] =$$

$$= (sg_{10} - rg_{20})w + \sum_{k=1}^{\infty}(sf_{1k} - rh_{2k})w^{k+1} + \varrho_1^s\varrho_2^r \sum_{k=1}^{\infty}(sh_{1k} - rf_{2k})\bar{w}^{k-1}.$$

Let us find the stationary points $\rho = (\rho_1, \rho_2)$ of System (10.19), with $\rho_1 \neq 0$ and $\rho_2 \neq 0$. Equating the series b_1 and b_2 from (10.20) to zero, we obtain the system of equations

$$\varrho_i \sum_{k=1}^{\infty} (f_{ik} - h_{ik})(w^k - \bar{w}^k) = 0, \quad i = 1, 2. \tag{10.21}$$

In these series, isolating terms of the lowest order in ρ, according to Lemma 10.2, we obtain the system of equations

$$-\tilde{\gamma}_{11}(w - \bar{w}) = 0, \quad \gamma_{21}(w - \bar{w}) = 0. \tag{10.22}$$

If

$$\gamma_{21} \neq 0 \text{ or } \tilde{\gamma}_{11} \neq 0, \tag{10.23}$$

then all solutions to the truncated system (10.22) with $\varrho_1 \neq 0$ and $\varrho_2 \neq 0$ are

$$w = \bar{w} = \delta\varrho_3, \quad \delta = \pm 1, \tag{10.24}$$

where $\varrho_3 = |w| = \varrho_1^{s/2}\varrho_2^{r/2}$. They are solutions to the complete system of equations (10.21). Consequently, if (10.23) is satisfied, then for small ϱ_1 and ϱ_2 different from zero, all solutions to equations (10.21) are (10.24). Then on the values of w with the property (10.24), equating the series b_3 from (10.19) to zero, we obtain the equation

$$\chi_\delta(\varepsilon, \rho) \stackrel{\text{def}}{=} s g_1(\rho, \delta\varrho_3, \delta\varrho_3) + r g_4(\rho, \delta\varrho_3, \delta\varrho_3) = 0. \tag{10.25}$$

The set of solutions to the equation (10.25) for the values δ equal to -1, 0 and 1, we denote as $\mathcal{F}_-$, $\mathcal{F}_0$ and $\mathcal{F}_+$ respectively. Now we study the stability (in linear approximation) of stationary points of the families $\mathcal{F}_\pm$ in System (10.19), (10.20). By virtue of (10.24), on the families $\mathcal{F}_\pm$

$$\frac{\partial b_i}{\partial \varrho_j} = 0, \quad \frac{\partial b_i}{\partial w} = -\frac{\partial b_i}{\partial \bar{w}}, \quad i, j = 1, 2.$$

Consequently, the Jacobi matrix of System (10.19) is

$$\begin{pmatrix} 0 & 0 & \partial b_1/\partial w & -\partial b_1/\partial w \\ 0 & 0 & \partial b_2/\partial w & -\partial b_2/\partial w \\ \partial b_3/\partial \varrho_1 & \partial b_3/\partial \varrho_2 & \partial b_3/\partial w & \partial b_3/\partial \bar{w} \\ -\partial b_3/\partial \varrho_1 & -\partial b_3/\partial \varrho_2 & -\partial b_3/\partial \bar{w} & -\partial b_3/\partial w \end{pmatrix}.$$

Its characteristic polynomial is $\tilde{\lambda}^4 + \omega\tilde{\lambda}^2$, where

$$\omega = -2\frac{\partial b_1}{\partial w}\frac{\partial b_3}{\partial \varrho_1} - 2\frac{\partial b_2}{\partial w}\frac{\partial b_3}{\partial \varrho_2} - \left(\frac{\partial b_3}{\partial w}\right)^2 + \left(\frac{\partial b_3}{\partial \bar{w}}\right)^2. \tag{10.26}$$

Its roots are

$$\tilde{\lambda}_{1,2} = \pm\sqrt{-\omega}, \quad \tilde{\lambda}_{3,4} = 0. \tag{10.27}$$

The point ρ, w from $\mathcal{F}_\pm$ is stable if $\omega > 0$ in it, and unstable if $\omega < 0$.

It remains to note that according to [Bruno 1974; 1989a, Part II] the families $\mathcal{F}_1$, $\mathcal{F}_2$, $\mathcal{F}_\pm$ form the set $\mathcal{A}$. To each stationary point of these families there corresponds a periodic solution lying in the set $\mathcal{A}$. Indeed, according to Section 6 the set $\mathcal{A}$ is determined by the system of equations (9.8). According to Lemma 10.5, when $s > 1$ the manifold $\mathcal{A}_2 = \{u_1 = u_3 = 0\}$ always satisfies it, and when $r > 1$ the manifold $\mathcal{A}_1 = \{u_2 = u_4 = 0\}$ as well. For remaining part $\mathcal{A}_3$ of the set $\mathcal{A}$ with $\varrho_1, \varrho_2 \neq 0$, we have

$$\lambda_i + g_i = \lambda_i \alpha, \quad i = 1, 2, 3, 4. \tag{10.28}$$

Since $\lambda_i + \lambda_{i+2} = 0$ and $s\lambda_1 - r\lambda_2 = 0$, then on the set $\mathcal{A}_3$ we have

$$g_i + g_{i+2} = 0, \quad i = 1, 2, \quad sg_1 + rg_4 = 0, \quad rg_2 + sg_3 = 0.$$

Consequently, on the set $\mathcal{A}_3$ all right hand sides of System (10.19) vanish. In the stationary point ϱ_1^0, ϱ_2^0, w^0 of System (10.19), according to (10.28) the parameter α takes the definite value α^0. Hence there

$$u_j = u_j^0 \exp(i\alpha^0 t), \quad j = 1, 2, 3, 4,$$

i.e. coordinates u_j are periodic with respect to t with frequencies $|\lambda_j \alpha|$ and periods $2\pi/|\lambda_j \alpha|$. The common period of the whole solution is

$$2\pi r/|\lambda_1 \alpha^0| = 2\pi s/|\lambda_2 \alpha^0|.$$

It is closed after r rotations of $\theta_1 = \arg u_1$ and s rotations of $\theta_2 = \arg u_2$. The eigenvalues of the periodic solution are given in (10.27). Thus, we have proved

Theorem 10.1 *Under the condition* (10.23) *for* $\varrho_1, \varrho_2 \neq 0$ *the set* $\mathcal{A}_3$ *consists of periodic solutions forming the families* $\mathcal{F}_\pm$*, which are described by the system of equations* (10.24), (10.25). *A periodic solution in these families is stable if* $\omega > 0$*, and unstable if* $\omega < 0$*, where* ω *is defined in* (10.26).

To describe the families $\mathcal{F}_\pm$ with small $\varrho_1, \varrho_2 \neq 0$, ε, we need to find the first approximations to solutions to Equation (10.25). According to Chapter 2, they are solutions to the truncated equations corresponding to those faces of the Newton polyhedron of the series $\chi_\delta(\varepsilon, \rho)$, which have the negative normal vector $P \in \mathbb{R}^3_*$: $P < 0$. Let $f(\varepsilon, \rho)$ be the series in powers of ε, ϱ_1, ϱ_2. We denote as $f^0(\rho)$ and $\varepsilon f^1(\rho)$ its parts containing ε in the zero and in the first powers respectively. If the series χ_δ^1 is generic, then the mentioned truncations $\hat{\chi}_\delta$ of the series χ_δ are truncations of the series $\chi_\delta^0 + \varepsilon\chi_\delta^1$, i.e. $\hat{\chi}_\delta = \hat{\chi}_\delta^0 + \varepsilon\hat{\chi}_\delta^1$.

Consequently, it is sufficient to find the first approximations $\hat{\chi}_{\delta j}^{0(d)}$, $\hat{\chi}_{\delta k}^{1(e)}$ and to choose their combinations that give truncations of the series $\chi_\delta^0 + \varepsilon\chi_\delta^1$. For this, in the plane $\tilde{q}_1 = \operatorname{ord}\varrho_1$, $\tilde{q}_2 = \operatorname{ord}\varrho_2$ one needs to consider the supports of the series χ_δ^0 and χ_δ^1, and their *Newton's open polygons* (see Chapter 2 and [Bruno 1979a, Chapter I, § 2]), i.e. those parts of boundaries $\partial\mathbf{\Gamma}(\chi_\delta^0)$ and $\partial\mathbf{\Gamma}(\chi_\delta^1)$ of two-dimensional polyhedrons (polygons) $\mathbf{\Gamma}(\chi_\delta^0)$ and $\mathbf{\Gamma}(\chi_\delta^1)$, for which the normal cones intersect the negative quadrant $\tilde{P} = (\tilde{p}_1, \tilde{p}_2) < 0$. According to (10.19) and (9.11), in the expansion for χ_δ, among terms of the first order in ε there is the constant term $C_0\varepsilon$ the power in ρ of which is the smallest when $r + s > 2$. Here $C_0 = sc_0 + rd_0$. Therefore in order to find the truncation $\hat{\chi}_\delta$, it is sufficient now to find the truncation $\hat{\chi}_\delta^0$. To each edge of the open polygon $\partial\mathbf{\Gamma}(\chi_\delta^0)$ with the normal vector $\tilde{P} < 0$ there corresponds the truncation $\hat{\chi}_\delta^0$ when $\varepsilon = 0$, and truncation $\hat{\chi}_\delta = C_0\varepsilon + \hat{\chi}_\delta^0$ when $\varepsilon \neq 0$. Consequently, the families $\mathcal{F}_\pm$ in the first approximation are described by the equation

$$C_0\varepsilon + \hat{\chi}_\delta^0(\rho) = 0. \tag{10.29}$$

The Newton's open polygon for χ_δ^0 may have more then one edge. Then there will be several different truncated equations (10.29).

Let us find the first approximation to the value ω in (10.26), which defines the stability. According to Lemma 10.2,

$$\partial b_1/\partial w = -\tilde{\gamma}_{11} + o(1), \quad \partial b_2/\partial w = \gamma_{21} + o(1).$$

Hence by replacing all values in (10.20) by their first approximations, we obtain the first approximation to the ω:

$$\hat{\omega} = 2\tilde{\gamma}_{11}\frac{\partial\hat{b}_3}{\partial\varrho_1} - 2\gamma_{21}\frac{\partial\hat{b}_3}{\partial\varrho_2} - \left(\frac{\partial\hat{b}_3}{\partial w}\right)^2 + \left(\frac{\partial\hat{b}_3}{\partial\bar{w}}\right)^2. \tag{10.30}$$

It remains to note that according to (10.19), (10.14), (10.24), (10.25), $\hat{\chi}_\delta = s\hat{g}_1 - r\hat{g}_2$, i.e. to compute the truncation $\hat{\chi}_\delta^0$ one needs to find truncations $\hat{g}_i^0$ in coordinates ϱ_1, ϱ_2; for that, the coordinates w and $\bar{w}$ in g_i^0 must be replaced by $\delta\varrho_3 = \delta\varrho_1^{s/2}\varrho_2^{r/2}$. That means that to each monomial U^Q in series g_i, we put in correspondence the point $\tilde{Q} = (\tilde{q}_1, \tilde{q}_2)$ with coordinates

$$\tilde{q}_1 = q_1 + q_3, \quad \tilde{q}_2 = q_2 + q_4. \tag{10.31}$$

The truncations $\hat{g}_i$ for g_i are found by the points $\tilde{Q}$ of supports $\tilde{\mathbf{S}}_i$ of the series g_i. Lemma 10.4 may be used for this.

11. The resonances of higher order

We consider resonances (10.1′) with $r + s > 4$. According to (10.20) and (10.25)

$$\chi_\delta(\varepsilon, \rho) \stackrel{\text{def}}{=} \delta F + G, \tag{11.1}$$

where

$$\begin{aligned} F &\stackrel{\text{def}}{=} \sum_{l=1}^{\infty} (s(f_{1,2l-1} + h_{1,2l-1}) - r(f_{2,2l-1} + h_{2,2l-1}))\varrho_3^{2l-1}, \\ G &\stackrel{\text{def}}{=} sg_{10} - rg_{20} + \sum_{l=1}^{\infty} (s(f_{1,2l} + h_{1,2l}) - r(f_{2,2l} + h_{2,2l}))\varrho_3^{2l}. \end{aligned} \tag{11.2}$$

In the series G the expansion begins with the first power of ϱ_1 and ϱ_2 if $r + s \geq 2$, and in the sum F it begins with the power $-1 + (r + s)/2$, which is present for $l = 1$, since series f_{2l} and $h_{1,2l-1}$ begin with ϱ_2^{-1} and ϱ_1^{-1} respectively. Therefore if $r + s > 4$, then $-1 + (r + s)/2 > 1$ and the character of solutions to Equation (10.25) is determined by the "secular" part $sg_{10} - rg_{20}$. Otherwise the resonant terms play an important part (see Sections 12–14 below). The sets of solutions to Equation (10.25) for values δ equal to -1, 0 and 1 were denoted as $\mathcal{F}_-$, $\mathcal{F}_0$ and $\mathcal{F}_+$ respectively. Let us study the disposition of sets $\mathcal{F}_-$, $\mathcal{F}_0$, $\mathcal{F}_+$ near the point $\varepsilon = \varrho_1 = \varrho_2 = 0$. We consider various generic subcases with $r + s > 4$, i.e. assuming that all necessary coefficients differ from zero.

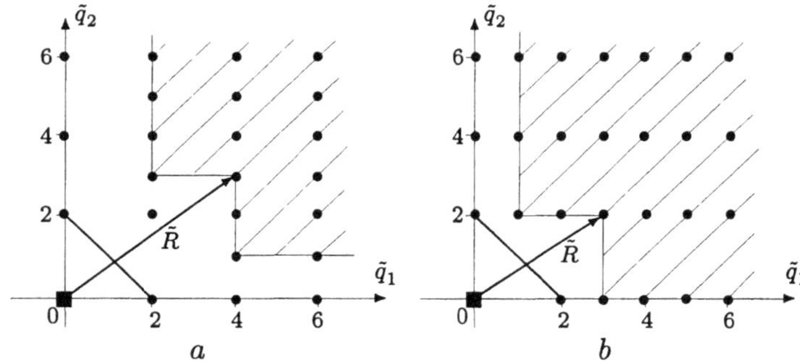

Figure 5.4: The resonant Case III with $r = 3$, $s = 4$ (a), and with $r = 2$, $s = 3$ (b). Supports of the series χ_δ^0 and χ_δ^1. Circles mark common points of supports $\mathbf{S}(\chi_\delta^0)$ and $\mathbf{S}(\chi_\delta^1)$, the square marks the point from $\mathbf{S}(\chi_\delta^1)$. The set $\mathbf{R}^2 + \tilde{R}$ is hatched. The vector $\tilde{R}$ is shown, and the edge $\mathbf{\Gamma}_1^{(1)}$ of the Newton's open polygon for χ_δ^0 is distinguished by the bold line.

1. $r \geq 3$. Then $s \geq 4$, i.e. $r/2 - 1 > 0$ and $s/r - 1 > 0$. Consequently, $F = 0$ when $\varrho_1 = 0$ and when $\varrho_2 = 0$. Hence on the axes $\varrho_1 = 0$ and $\varrho_2 = 0$ all three sets $\mathcal{F}_\delta$ coincide. We shall use notation of Lemma 10.4, and furthermore we put

$$\mathbf{R}^2 \stackrel{\text{def}}{=} \{\mathbb{R}^2_+ - 2\tilde{E}_1\} \cup \{\mathbb{R}^2_+ - 2\tilde{E}_2\}.$$

According to Lemma 10.4 the support with respect to ρ of the series χ_δ is $\tilde{\mathbf{M}}^0 \cup \{\tilde{\mathbf{M}}^2 + \tilde{R}\}$. Here $\tilde{\mathbf{M}}^0 \subset \mathbb{R}^2_+ \stackrel{\text{def}}{=} \{\tilde{Q} \geq 0\}$ and $\tilde{\mathbf{M}}^2 \subset \mathbf{R}^2$, i.e. $\tilde{\mathbf{M}}^2 + \tilde{R} \subset \mathbf{R}^2 + \tilde{R}$.

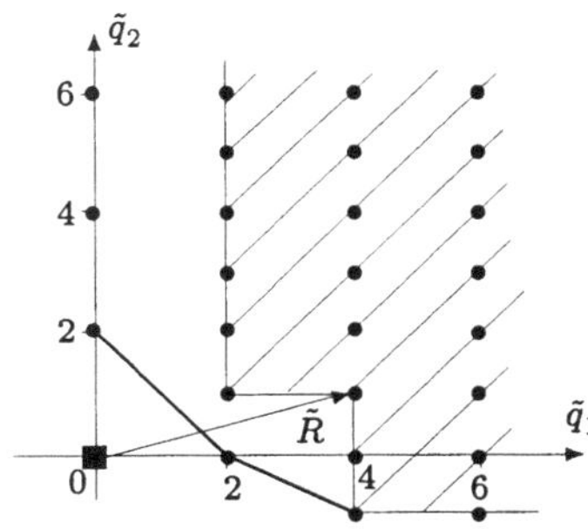

Figure 5.5: The resonant Case III with $r = 1$, $s = 4$. The supports of the series χ^0_δ and χ^1_δ. Notation of Fig. 5.4. The two edges of the Newton's open polygon for χ^0_δ are visible.

Fig. 5.4, *a* shows supports of the series χ^0_δ, χ^1_δ for $r = 3$, $s = 4$, here the set $\mathbf{R}^2 + \tilde{R}$ is hatched; circles mark points of the support $\mathbf{S}(\chi^0_\delta)$ (they belong also to the support $\mathbf{S}(\chi^1_\delta)$); the point (0,0) is marked by the square, since it belongs only to $\mathbf{S}(\chi^1_\delta)$; the unique edge $\mathbf{\Gamma}^{(1)}_1$ of the Newton's open polygon of the series χ^0_δ is distinguished. Thus, when $r \geq 3$, for the normal form (9.4) the Newton's open polygon of the series χ^0_δ in coordinates (10.31) consists of one edge with vertices $(2, 0)$ and $(0, 2)$. Hence there is the unique truncation

$$\hat{\chi}_\delta \stackrel{\text{def}}{=} C_0\varepsilon + C_1\varrho_1 + C_2\varrho_2, \tag{11.3}$$

where

$$C_i \stackrel{\text{def}}{=} sc_i + rd_i, \quad i = 0, 1, 2. \tag{11.4}$$

Here in the first approximation three families $\mathcal{F}_0$, $\mathcal{F}_\pm$ coincide, and they are determined by the equation $C_0\varepsilon + C_1\varrho_1 + C_2\varrho_2 = 0$. By (10.30), from (11.3) we obtain

$$\hat{\omega} = 2\tilde{\gamma}_{11}C_1\delta\varrho_3 - 2\gamma_{21}C_2\delta\varrho_3 = 2\delta(\tilde{\gamma}_{11}C_1 - \gamma_{21}C_2)\varrho_3.$$

Since $\varrho_3 > 0$ and the constant $\tilde{\gamma}_{11}C_1 - \gamma_{21}C_2 \neq 0$, then the sign of the quantity $\hat{\omega}$ changes along with the sign of δ. Consequently, for small ρ one of the families $\mathcal{F}_+$ and $\mathcal{F}_-$ is stable, and the other is unstable.

2. $r = 2$. Then $s \geq 3$, i.e. $F = 0$ when $\varrho_1 = 0$, and $F \neq 0$ when $\varrho_2 = 0$. Hence all three sets $\mathcal{F}_\delta$ coincide on the axis $\varrho_1 = 0$, and they all have different intersections with the axis $\varrho_2 = 0$. In Fig. 5.4, *b* for $r = 2$, $s = 3$ the sets $\mathbf{R}^2 + \tilde{R}$, the supports $\mathbf{S}(\chi_\delta^0)$ and $\mathbf{S}(\chi_\delta^1)$, and the unique edge of the Newton's open polygon of the series χ_δ^0 are shown. Thus, when $r \geq 2$ there is also the unique truncation (11.3). Its analysis is the same as in Subcase 1.

3. $r = 1$. Then $s > 3$, i.e. $F = 0$ when $\varrho_1 = 0$. When $\varepsilon = 0$ in the plane $\tilde{q}_1, \tilde{q}_2$, the support of the series F has the point $(s, -1)$, and the support of the series G has points $(2, 0)$ and $(0, 2)$. Hence for $\delta = \pm 1$ the Newton's open polygon of the series χ_δ^0 has two edges with these vertices (see Fig. 5.5). To them there correspond two truncations of the series χ_δ: one of them is (11.3), and the other is

$$\hat{\chi}_\delta \stackrel{\text{def}}{=} C_0\varepsilon + C_1\varrho_1 - \gamma_{21}\delta\varrho_1^{s/2}\varrho_2^{-1/2}. \tag{11.5}$$

Therefore each of the sets $\mathcal{F}_\pm$ consists of two components: the component $\mathcal{F}'_\pm$ corresponding to the truncation (11.3), and the component $\mathcal{F}''_\pm$ corresponding to the truncation (11.5). Here components $\mathcal{F}'_\pm$ have the same structure as the sets $\mathcal{F}_\pm$ in Subcase 2. Let us study the structure of components $\mathcal{F}''_\pm$. From (11.5) and (10.25) we obtain their first approximations in the form

$$\varrho_2^{1/2} = \frac{\delta\gamma_{21}\varrho_1^{s/2}}{C_0\varepsilon + C_1\varrho_1}. \tag{11.6}$$

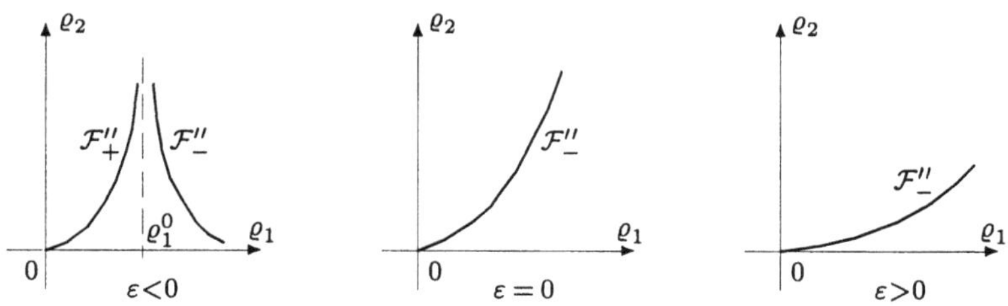

Figure 5.6: The resonant Case III with $r = 1$, $s = 4$. The disposition of families $\mathcal{F}''_\pm$ for $\varrho \geq 0$ and different ε.

The zeros of the denominator in (11.6) are $\varrho_1^0 = -\varepsilon C_0/C_1$. For $C_0/\gamma_{21} > 0$ and $C_1/\gamma_{21} > 0$, and the fixed values of ε, the disposition of curves (11.6) is shown in Fig. 5.6 for $\rho \geq 0$. Consequently, the real family $\mathcal{F}''_-$ exists only when $\varepsilon \geq 0$, and the family $\mathcal{F}''_+$ exists only when $\varepsilon < 0$. Indeed, families $\mathcal{F}''_\pm$

are the substitutes for a family lying in the invariant subspace $u_2 = u_4 = 0$ (or $\varrho_2 = 0$), which exists when $r > 1$.

For the truncation (11.5), according to (10.30) we have

$$\hat{\omega} = 2\tilde{\gamma}_{11}(C_1\delta\varrho_3 - s\gamma_{21}\varrho_1^{s-1}) - (C_0\varepsilon + C_1\varrho_1)^2.$$

Here $\varrho_2 \sim \varrho_1^{s-2}$. Replacing $C_0\varepsilon + C_1\varrho_1$ by $\gamma_{21}\delta\varrho_3^{-1}\varrho_1^s$ and computing powers in ϱ_1, we obtain $\varrho_3 \sim \varrho_1^{s-1}$. Hence the addend $-(C_0\varepsilon + C_1\varrho_1)^2 = -\gamma_{21}^2\varrho_1^s\varrho_2^{-1}$ has the smallest power in ϱ_1. Since γ_{21} is purely imaginary, then $\hat{\omega} > 0$, and the families $\mathcal{F}''_\pm$ are stable. Thus, we have proved

Theorem 11.1 *In the generic case, for the resonance $\lambda_1(\nu_0)/\lambda_2(\nu_0) = r/s < 1$ with $r + s > 4$ the set $\mathcal{A}$ is filled by periodic solutions, and it consists of four components:*

$$\mathcal{F}_1 = \{\varrho_2 = 0\} \quad \textit{for} \quad r > 1 \quad \textit{and} \quad \mathcal{F}_1 = \mathcal{F}''_\pm \quad \textit{for} \quad r = 1;$$
$$\mathcal{F}_2 = \{\varrho_1 = 0\};$$
$$\mathcal{F}_3 = \mathcal{F}_- \quad \textit{for} \quad r > 1 \quad \textit{and} \quad \mathcal{F}_3 = \mathcal{F}'_- \quad \textit{for} \quad r = 1;$$
$$\mathcal{F}_4 = \mathcal{F}_+ \quad \textit{for} \quad r > 1 \quad \textit{and} \quad \mathcal{F}_4 = \mathcal{F}'_+ \quad \textit{for} \quad r = 1.$$

In the first approximation they are described by Formulae (11.3)–(11.6).

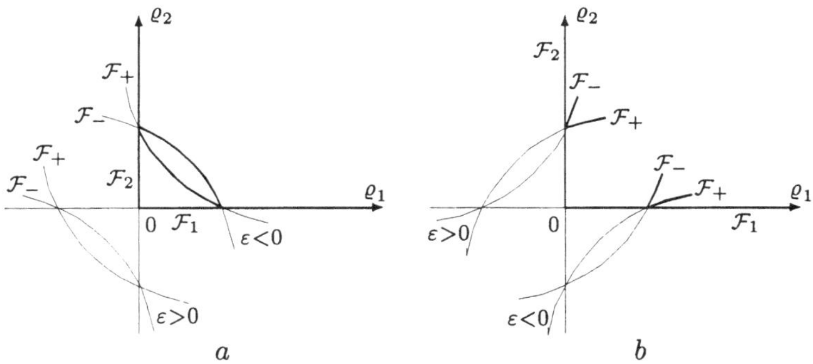

Figure 5.7: The resonant Case III with $r + s > 4$. The disposition of families $\mathcal{F}_1$–$\mathcal{F}_4$ in Subcase (1) from Section 9 (a) and in Subcase (2) from Section 9 (b). Notation of Fig. 5.3.

We note that families $\mathcal{F}_3$ and $\mathcal{F}_4$ are tangent to the family $\mathcal{F}_0$, which is tangent to the component $\mathcal{A}_3$, the disposition of which were studied in Section 9. Furthermore, they coincide with $\mathcal{F}_0$ (and with each other) in the axis ϱ_2 and in the axis ϱ_1 if $r > 2$. Periodic solutions of families $\mathcal{F}_3$ and $\mathcal{F}_4$ near their intersections with families $\mathcal{F}_1$ and $\mathcal{F}_2$ are r- and s-multiple periodic solutions of families $\mathcal{F}_1$ and $\mathcal{F}_2$ respectively. In Subcase (1) from Section 9, families $\mathcal{F}_3$ and $\mathcal{F}_4$ form a sort of bridges between periodic solutions of families $\mathcal{F}_1$ and $\mathcal{F}_2$ (compare with Figs. 5.7, *a* and 5.3, *a*). In Subcase (2) from Section 9, when ε is fixed, families $\mathcal{F}_3$ and $\mathcal{F}_4$ intersect only one of the families $\mathcal{F}_1$ and $\mathcal{F}_2$ (compare with Figs. 5.7, *b* and 5.3, *b*).

The structure of the set $\mathcal{A}$ is analogous for Hamiltonian systems (see [Bruno 1974, § 3; 1989a, Part II, § 3]). Since the set $\mathcal{A}$ does not fill the neighborhood of the stationary point, then Condition A is not satisfied, and the normalizing transformation apparently diverges (compare with Section 7).

12. The resonance 1:3 in Case III

We isolate those values of ν_0, for which there are resonances of lower order, i.e. $\lambda_1(\nu_0)/\lambda_2(\nu_0) = r/s$ with $r + s \leq 4$. According to (7.11), (7.12)

$$\nu_0 = 1 - \frac{9\tilde{s}}{1 + 2\tilde{s} + \tilde{s}^2},$$

where $\tilde{s} = \lambda_1^2/\lambda_2^2$. Consequently,

$$\nu_0 = 0.19 \quad \text{for} \quad r/s = 1/3,$$

$$\nu_0 = -0.44 \quad \text{for} \quad r/s = 1/2,$$

$$\nu_0 = -5/4 \quad \text{for} \quad r/s = 1/1.$$

These cases will be analyzed one-by-one in Sections 12–14.

Here we consider the case $r/s = 1/3$, i.e. $\nu_0 = 0.19$. According to (9.1) here $\lambda_1 = i\sqrt{30}/10$. Let us find all integer solutions Q to equation $\langle Q, \Lambda\rangle = 0$, which has the form

$$q_1 - q_3 + 3(q_2 - q_4) = 0. \tag{12.1}$$

In this case the theory of construction of the normal form of Section 10 can be applied. According to Fig. 5.8, the Newton's open polygon of the series χ_δ^0 consists of one edge with the normal vector $-(1,1)$ and with four points of the support: $(3,-1)$, $(2,0)$, $(1,1)$, $(0,2)$, which have $\tilde{q}_1 + \tilde{q}_2 = 2$. According to equalities (10.5), to them there correspond points Q of the sets $\mathbf{S}_i^k$ with $q_1 + q_2 + q_3 + q_4 = \tilde{q}_1 + \tilde{q}_2 = 2$ only for $k = -1, 0, 1$. Namely, when $k = -1$ we obtain $(-1,1,2,0)$, $(0,0,3,-1)$; when $k = 0$ we obtain $(1,0,1,0)$, $(0,1,0,1)$; and when $k = 1$ we obtain $(3,-1,0,0)$, $(2,0,-1,1)$. Let us write the corresponding truncations $\hat{g}_i$. Introducing new coordinates by (10.9)

$$\varrho_i = u_i u_{i+2}, \quad i = 1, 2, \quad w = u_1^3 u_4, \quad \overline{w} = u_2 u_3^3, \tag{12.2}$$

from Formula (10.10) for truncated polynomials $\hat{g}_i^0$ we obtain

$$\begin{aligned} \hat{g}_1^0(\rho, w, \overline{w}) &= c_1 \varrho_1 + c_2 \varrho_2 + c_3 \varrho_1^{-1} \overline{w}, \\ \hat{g}_2^0(\rho, w, \overline{w}) &= -d_1 \varrho_1 - d_2 \varrho_2 - d_3 \varrho_2^{-1} w, \end{aligned} \tag{12.3}$$

and the complex-conjugate polynomials. In notation of Lemma 10.2 $\tilde{\gamma}_{11} = c_3$, $\gamma_{21} = -d_3$.

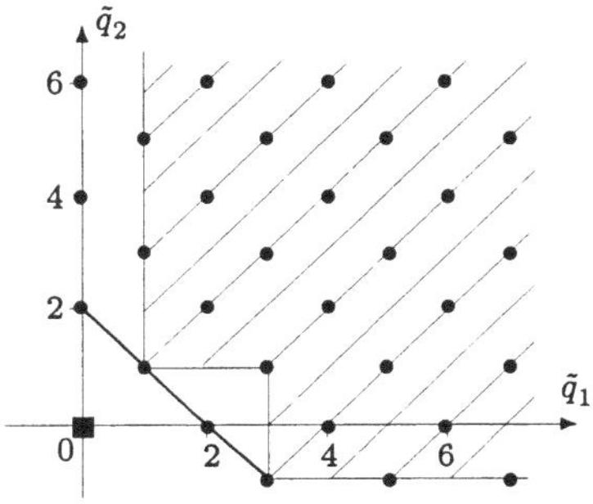

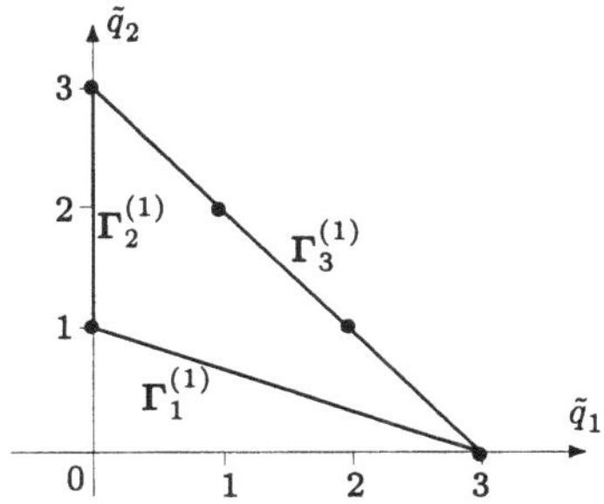

Fig. 5.8 (left): The resonant Case III with $r = 1$, $s = 3$. The supports of the series χ_δ^0 and χ_δ^1. Notation of Fig. 5.4.

Fig. 5.9 (right) : The Newton polygon for Equation (12.7).

Thus, the truncated with respect to ρ, ε series g_i for the normal form (9.4) of System (4.2) in Case III with $\lambda_1/\lambda_2 = 1/3$, with (9.2), (9.3) and (12.3) taken into account, are

$$\begin{aligned} \hat{g}_1 &= c_0 \varepsilon + c_1 \varrho_1 + c_2 \varrho_2 + c_3 \varrho_1^{-1} \overline{w}, \\ \hat{g}_2 &= -d_0 \varepsilon - d_1 \varrho_1 - d_2 \varrho_2 - d_3 \varrho_2^{-1} w, \end{aligned} \tag{12.4}$$

and the complex-conjugate equalities, where $\varepsilon = \nu - \nu_0$. Coefficients c_0, d_0 are given in (9.3); and c_i, d_i, $i = 1, 2, 3$ are computed by the general formula (35) for coefficients of the normal form in [Bruno 1979a, 1989a, Chapter III, Subsection 1.7]. As a result, c_i, d_i, $i = 1, 2$ are obtained in the form (7.9) and

$$c_3 = \frac{-a^2 \left(12\lambda_1^4 - 18\lambda_1^3\lambda_2 + 17\lambda_1^2\lambda_2^2 + 5\lambda_1\lambda_2^3 - 6\lambda_2^4\right)}{2\lambda_1^3 \left(\lambda_1 - 2\lambda_2\right) \left(\lambda_1^2 - \lambda_2^2\right)^2 \left(4\lambda_1^2 - \lambda_2^2\right) \lambda_2},$$

$$d_3 = \frac{-a^2 \left(7\lambda_1^2 + 7\lambda_1\lambda_2 + 2\lambda_2^2\right)}{2\lambda_1^2\lambda_2^2 \left(2\lambda_1 + \lambda_2\right) \left(\lambda_1^2 - \lambda_2^2\right)^2}. \tag{12.5}$$

Here Lemma 10.3 is valid. From the truncations (12.4), we obtain the truncation of System (10.19) for $\varrho_i, w, \overline{w}$:

$$\dot{\varrho}_i = \varrho_i(\hat{g}_i + \hat{g}_{i+2}), \quad i = 1, 2,$$
$$\dot{w} = w(3\hat{g}_1 + \hat{g}_4),$$
$$\dot{\overline{w}} = \overline{w}(\hat{g}_2 + 3\hat{g}_3),$$

or, taking into account (11.4),

$$\begin{aligned} \dot{\varrho}_1 &= c_3(\overline{w} - w), \\ \dot{\varrho}_2 &= d_3(\overline{w} - w), \\ \dot{w} &= w(C_0\varepsilon + C_1\varrho_1 + C_2\varrho_2) + 3c_3\varrho_1^2\varrho_2 + d_3\varrho_1^3 \stackrel{\text{def}}{=} \hat{b}_3 \end{aligned} \tag{12.6}$$

and the complex-conjugate equation for $\dot{\overline{w}}$.

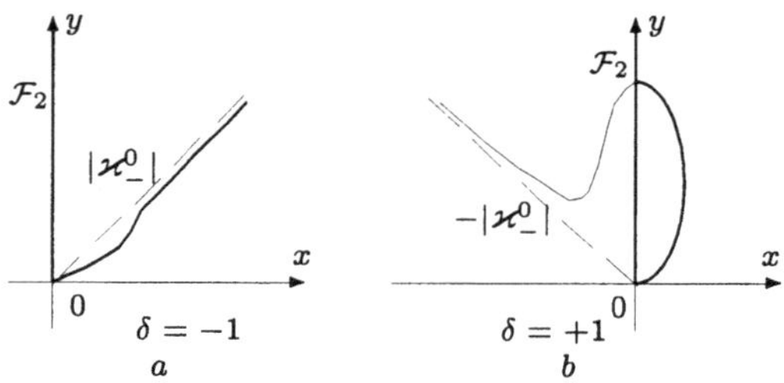

Figure 5.10: The disposition of solutions to Equation (12.7) with $y > 0$ when $\delta = -1$, $\varepsilon > 0$ (a), and when $\delta = 1$, $\varepsilon < 0$ (b). Dashed lines show asymptotes $x = \pm|\varkappa_-^0|y$.

The stationary points $\rho = (\rho_1, \rho_2)$ of this system with $\rho_1 \neq 0$ and $\rho_2 \neq 0$ satisfy to the system of equations (10.24) and (10.25), i.e.

$$C_0\varepsilon y + C_1x^2y + C_2y^3 + 3\delta c_3xy^2 + \delta d_3x^3 = 0, \tag{12.7}$$

where

$$x = \sqrt{\varrho_1}, \quad y = \sqrt{\varrho_2} \geq 0, \tag{12.8}$$

and C_i are cited in (11.4). If in Formulae (7.9), (12.5), (11.4) we put $\lambda_2 = 3\lambda_1$, $\lambda_1 = i\sqrt{30}/10$, $\nu_0 = 0.19$, then we obtain the numerical values of coefficients:

$$\begin{aligned} C_0 = -1.267888i, \ C_1 = -4.89269ia^2, \ C_2 = -4.8151ia^2, \\ c_3 = 1.6905ia^2, \ d_3 = -0.540021ia^2. \end{aligned} \tag{12.9}$$

When $\varepsilon = 0$, Equation (12.7) is a cubic one with respect to $\varkappa = x/y = \sqrt{\varrho_1/\varrho_2}$. It has two complex roots and the real one $\varkappa = \varkappa_-^0 \approx -\delta \cdot 10.0795$.

To find roots of Equation (12.7) when $\varepsilon \neq 0$, we consider its Newton polygon in the plane $q_1 = \operatorname{ord} x$, $q_2 = \operatorname{ord} y$ (Fig. 5.9), i.e. to each monomial $x^{q_1}y^{q_2}$ from (12.7), we put in correspondence the point (q_1, q_2). The convex hull of these points is a triangle. To the edge $\mathbf{\Gamma}_3$ there corresponds the case $\varepsilon = 0$, its truncated equation gives the asymptotics of solutions when $x, y \to \infty$. To the edge $\mathbf{\Gamma}_2$ there corresponds the value $x = 0$ and the truncated equation $y(C_0\varepsilon + C_2y^2) = 0$. It has three roots:

$$
\begin{aligned}
&y_1 = 0, \\
&y_{2,3} = \pm\sqrt{-C_0\varepsilon/C_2} \approx \pm 0.51314\sqrt{-\varepsilon}/a,
\end{aligned}
\tag{12.10}
$$

i.e. $y_{2,3}$ are real only when $\varepsilon \leq 0$. To the edge $\mathbf{\Gamma}_1$ there corresponds the truncated equation $C_0\varepsilon y + \delta d_3 x^3 = 0$. For any values of $\delta = \pm 1$, $\varepsilon \neq 0$ it has the single real root

$$
y = -\delta(d_3/C_0)x^3\varepsilon^{-1} = -0.42592\,\delta x^3/\varepsilon.
$$

Hence for $y > 0$ the solutions to Equation (12.7) are placed in such a way as it is shown in Fig. 5.10 when $\delta = -1$ and $\delta = +1$. To the real values of the original coordinates there corresponds only the first quadrant $x, y \geq 0$. In it by Lemma 10.5 there is always the family $\mathcal{F}_2 = \{\varrho_1 = 0\}$. Furthermore, families $\mathcal{F}_\pm$, which are placed as shown in Fig. 5.11, are there. When $\varepsilon \neq 0$ they are tangent to the axis x. It is clear from Fig. 5.11 that two families $\mathcal{F}_2$ and $\mathcal{F}_\pm$ are going from the stationary point $\varrho_1 = \varrho_2 = 0$. When $\varepsilon < 0$ the second one is the family $\mathcal{F}_+$; it crosses the family $\mathcal{F}_2$ again when $x = 0$, $y = y_2$ (see (12.10)) and after the crossing it turns into the family $\mathcal{F}_-$.

Let us determine now the stability (in linear approximation) of periodic solutions of the families $\mathcal{F}_\pm$, since solutions of the families $\mathcal{F}_2$ are stable. For $\hat{b}_3$, we have from (12.6) and (10.30) that

$$
\hat{\omega} = 2c_3(C_1w + 6c_3\varrho_1\varrho_2 + 3d_3\varrho_1^2) + 2d_3(wC_2 + 3c_3\varrho_1^2) - (C_0\varepsilon + C_1\varrho_1 + C_2\varrho_2)^2.
$$

Taking into account (12.7), we obtain

$$
\hat{\omega} = 2(c_3C_1 + d_3C_2)\delta\varrho_1^{3/2}\varrho_2^{1/2} + 12c_3^2\varrho_1\varrho_2 + 12c_3d_3\varrho_1^2 - \varrho_1(3c_3\varrho_2 + d_3\varrho_1)^2/\varrho_2,
$$

i.e. in notation (12.8)

$$
\varrho_2\hat{\omega}/\varrho_1 = 3c_3^2y^4 + 2(c_3C_1 + d_3C_2)\delta y^3x + 6c_3d_3y^2x^2 - d_3^2x^4.
$$

Coefficients are computed by (12.9).

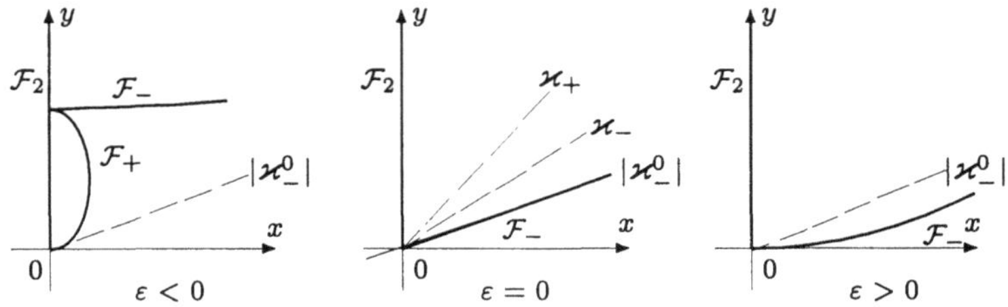

Figure 5.11: The resonance 1:3 in Case III. The disposition of real families $\mathcal{F}_2$ and $\mathcal{F}_\pm$ in coordinates $x = \sqrt{\varrho_1}$, $y = \sqrt{\varrho_2}$ for various ε. Dashed lines show rays $x/y = |\varkappa_-^0|, \varkappa_-, \varkappa_+$.

The equation $\hat{\omega} = 0$ is of the fourth order with respect to $\varkappa = x/y$. When $\delta = +1$ it has the unique positive root $\varkappa_+ = 0.586666$. When $\delta = -1$ it also has the unique positive root $\varkappa_- = 2.20721$. That means that the family $\mathcal{F}_+$ is stable when $x > \varkappa_+ y$ and unstable when $x < \varkappa_+ y$. The family $\mathcal{F}_-$ is stable when $x > \varkappa_- y$ and unstable when $x < \varkappa_- y$. Since $\varkappa_+ < \varkappa_- < \varkappa_-^0$ (see Fig. 5.11), then when $\varepsilon < 0$ the family $\mathcal{F}_+$ goes from zero as stable one, then crosses the ray $x = \varkappa_+ y$ and becomes unstable. After crossing the family $\mathcal{F}_2$ the unstable family $\mathcal{F}_+$ turns into the unstable family $\mathcal{F}_-$, that becomes stable after crossing the ray $x = \varkappa_- y$. In the place of intersection of families $\mathcal{F}_\pm$ with $\mathcal{F}_2$, the periodic solutions of families $\mathcal{F}_\pm$ have periods three times greater than the periods of solutions of families $\mathcal{F}_2$. Thus, we have proved

Theorem 12.1 *Near the value $\nu = \nu_0 = 0.19$ the set $\mathcal{A}$ consists of three families of periodic solutions: $\mathcal{F}_2 = \{\varrho_1 = 0\}$ and $\mathcal{F}_\pm$. The disposition of their projections in the plane $\sqrt{\varrho_1}$, $\sqrt{\varrho_2}$ is shown in Fig . 5.11. The family $\mathcal{F}_2$ is stable, and $\mathcal{F}_\pm$ have both stable and unstable parts.*

13. The resonance 1:2 in Case III

Now we consider the resonance $\lambda_1/\lambda_2 = 1/2$ in Case III. Here $\nu_0 = -0.44$, and according to (9.1) $\lambda_1 = i\sqrt{60}/10$. Let us find integer solutions Q to the equation $\langle \Lambda, Q \rangle = 0$, which has the form

$$q_1 - q_3 + 2(q_2 - q_4) = 0. \tag{13.1}$$

For such Q the remainder $q_1 - q_3$ is an even number; consequently, the number $q_1 + q_3$ is also even.

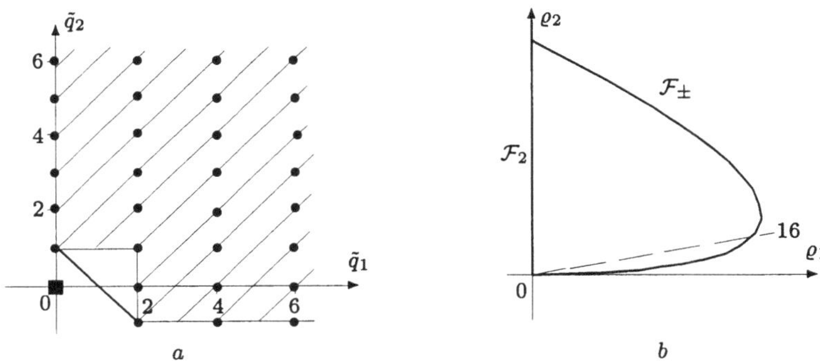

Figure 5.12: The resonance $r = 1$, $s = 2$ in Case III. The supports of the series χ_δ^0 and χ_δ^1 (a); notation of Fig. 5.4. The families $\mathcal{F}_2$ and $\mathcal{F}_\pm$ in coordinates $\varrho \geq 0$ (b). Both families $\mathcal{F}_+$ and $\mathcal{F}_-$ are projected in the same curve $\mathcal{F}_\pm$. Dashed line shows the ray that determines the boundary of stability in families $\mathcal{F}_\pm$.

In this case, let us apply the theory of construction of the normal form developed in Section 10. According to Fig. 5.12, a, the Newton's open polygon of the series χ_δ^0 consists of one edge with the normal vector $-(1,1)$ and with two points $(2,-1)$ and $(0,1)$, which have $\tilde{q}_1 + \tilde{q}_2 = 1$. According to equalities (10.5), to them there correspond the points Q of sets $\mathbf{S}_i^k$ with $q_1+q_2+q_3+q_4 = \tilde{q}_1 + \tilde{q}_2 = 1$ only for $k = -1, 1$. Namely, when $k = -1$ we obtain $(-1,1,1,0)$, $(0,0,2,-1)$; and when $k = 1$ we obtain $(1,0,-1,1)$, $(2,-1,0,0)$. Hence the truncated polynomials $\hat{g}_i^0$ (see (10.10)) are

$$\hat{g}_1^0(\rho, w, \overline{w}) = 2ac_0\varrho_1^{-1}\overline{w}, \quad \hat{g}_2^0(\rho, w, \overline{w}) = -ad_0\varrho_2^{-1}w, \tag{13.2}$$

where $\varrho_i = u_i u_{i+2}$, $i = 1, 2$, $w = u_1^2 u_4$, $\overline{w} = u_2 u_3^2$. Consequently,

$$\tilde{\gamma}_{11} = 2ac_0, \quad \gamma_{21} = -ad_0. \tag{13.3}$$

The truncations $\hat{g}_1 = c_0\varepsilon + \hat{g}_1^0$, $\hat{g}_2 = -d_0\varepsilon + \hat{g}_2^0$ with $\hat{g}_i^0$ from (13.2). The coefficients c_0, d_0 are given in (9.3). Now we compose the truncated system for $\varrho_i, w, \overline{w}$ with (11.4) taken into account:

$$\begin{aligned}
\dot{\varrho}_1 &= 2ac_0(\overline{w} - w), \\
\dot{\varrho}_2 &= ad_0(\overline{w} - w), \\
\dot{w} &= w((2c_0 + d_0)\varepsilon + (4ac_0\varrho_1^{-1} + ad_0\varrho_2^{-1})\overline{w}) = \\
&\stackrel{\text{def}}{=} wC_0\varepsilon + 4ac_0\varrho_1\varrho_2 + ad_0\varrho_1^2 \stackrel{\text{def}}{=} \hat{b}_3.
\end{aligned} \tag{13.4}$$

We find the stationary points ρ of System (13.4) with $\rho_1 \neq 0$ and $\rho_2 \neq 0$. For them $w = \overline{w} = \delta \varrho_1 \varrho_2^{1/2}$. Equating to zero the right hand side of the third equation (13.4), we obtain the equation

$$\delta C_0 \varepsilon \varrho_1 \varrho_2^{1/2} + 4ac_0 \varrho_1 \varrho_2 + ad_0 \varrho_1^2 = 0. \tag{13.5}$$

Hence $\varrho_1 = 0$ is a solution to Equation (13.5). According to (9.13), $c_0\lambda_1 = d_0\lambda_2$, i.e. in our case $c_0 = 2d_0$. Hence Equation (13.5) takes the form

$$5\varepsilon\delta\sqrt{\varrho_2} + 8a\varrho_2 + a\varrho_1 = 0, \tag{13.6}$$

whence

$$\varrho_1 = -8\varrho_2 - \frac{5\varepsilon\delta}{a}\sqrt{\varrho_2}. \tag{13.7}$$

The disposition of these solutions in the first quadrant of the plane ϱ_1, ϱ_2 is shown in Fig. 5.12, b for

$$\varepsilon\delta/a < 0. \tag{13.8}$$

For $\delta\varepsilon/a \geq 0$ Equation (13.6) has no solutions in the first quadrant. Thus, we have obtained two families of stationary points $\mathcal{F}_\pm$ described by Formula (13.7). Each of the families $\mathcal{F}_+$ and $\mathcal{F}_-$ exists only for the appropriate sign of ε according to (13.8).

We consider now the stability of these stationary points of System (13.4). Taking into account (13.3), (13.5) and the equality $c_0 = 2d_0$, from (13.4) by (10.30) we compute

$$\hat{\omega} = d_0^2[32a^2(2\varrho_2 + \varrho_1) - 25\varepsilon^2].$$

From (13.6) we obtain $5\varepsilon\delta\sqrt{\varrho_2} = -a(8\varrho_2 + \varrho_1)$. Taking the square of both parts, we obtain

$$25\varepsilon^2/a^2 = (8\varrho_2 + \varrho_1)^2/\varrho_2.$$

Substituting that into the expression for $\hat{\omega}$, we obtain

$$\hat{\omega} = d_0^2 a^2[32(2\varrho_2 + \varrho_1) - (8\varrho_2 + \varrho_1)^2/\varrho_2] = d_0^2 a^2 \varrho_1(16\varrho_2 - \varrho_1)/\varrho_2.$$

Since d_0 is a purely imaginary number, then $d_0^2 < 0$. Consequently,

$$\hat{\omega} \begin{cases} > 0, & \text{if} \quad \varrho_1 > 16\varrho_2, \\ < 0, & \text{if} \quad \varrho_1 < 16\varrho_2, \end{cases}$$

and families $\mathcal{F}_\pm$ are stable when $\varrho_1 > 16\varrho_2$ and unstable when $\varrho_1 < 16\varrho_2$.

In Fig. 5.12, b the boundary of stability $\varrho_1 = 16\varrho_2$ is shown by dashed line. Under that line the families $\mathcal{F}_\pm$ are stable, and over it they are unstable. The families $\mathcal{F}_\pm$ go out of the origin being tangent to the axis ϱ_1; at first they are stable; then they become unstable and turn towards the axis ϱ_2, where they terminate. To the families of stationary points $\mathcal{F}_2$ and $\mathcal{F}_\pm$ of System (13.4) there correspond families of periodic solutions of the normal form (9.4) forming the set $\mathcal{A}$. Solutions of families $\mathcal{F}_\pm$ are in a certain way two-fold with respect to periodic solutions of families $\mathcal{F}_2$. Thus, we have proved

Theorem 13.1 *Near the value $\nu = \nu_0 = -0.44$ the set $\mathcal{A}$ consists of three families of periodic solutions: two families $\mathcal{F}_\pm$ and the Liapunov family $\mathcal{F}_2 = \{\varrho_1 = 0\}$. The disposition of their projections on the plane ϱ_1, ϱ_2 is shown in Fig. 5.12, b. The families $\mathcal{F}_\pm$ are stable near the axis ϱ_1 and unstable at the intersection with the family $\mathcal{F}_2$.*

Remark 13.1. In the generic case in a Hamiltonian system the intersection of families of periodic solutions for the resonance 1/2 is such that the main family intersects one two-fold family and changes its stability on a solution where the families intersect (see [Bruno 1990a]). In our case with the resonance 1/2 the main family $\mathcal{F}_2$ intersects both two-fold families $\mathcal{F}_\pm$ simultaneously, and the stability of the main family $\mathcal{F}_2$ is not changed.

14. The normal form in Case γ_2

In cases γ_1, γ_2 the parameter $\nu_0 = -5/4$, and eigenvalues

$$\lambda_1 = \lambda_2 = -\lambda_3 = -\lambda_4 = \sqrt{3\sigma/2} \stackrel{\text{def}}{=} \lambda \tag{14.1}$$

are two-fold, and, as it was mentioned in Section 12, $\lambda_1/\lambda_2 = 1/1$, i.e. $r = s = 1$. The case γ_1 was considered in Section 7. Here we will consider Case γ_2, when $\sigma = -1$ and λ is purely imaginary. After the linear transformation

$$Y = TZ \tag{14.2}$$

with the matrix T from Lemma 5.4, System (4.2) turns into a system consisting of equations

$$\begin{aligned} \dot{z}_1 &= \lambda z_1 + \lambda z_2 - (\lambda\varepsilon/9)(z_1 + z_3) - (\lambda a/9)(z_1 + z_3)^2, \\ \dot{z}_2 &= \lambda z_2 + (\lambda\varepsilon/9)(z_1 + z_3) + (\lambda a/9)(z_1 + z_3)^2, \end{aligned} \tag{14.3}$$

and the complex-conjugate ones. According to Lemma 5.4, in System (14.3) coordinates Z are connected by the reality relation (9.3′), and System (14.3) is invariant under the involution (9.3″).

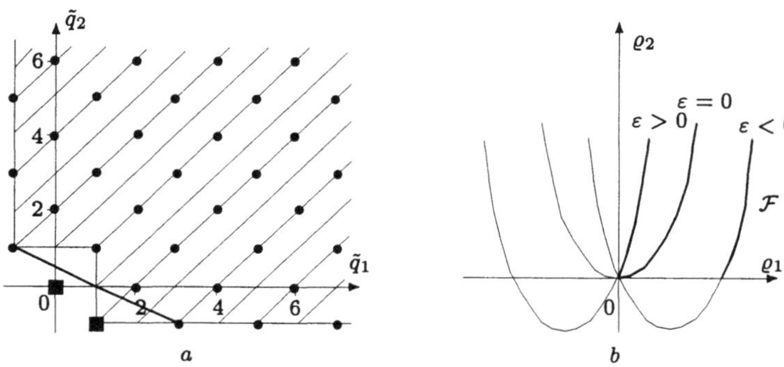

Figure 5.13: Case γ_2. Picture (a) shows supports of series χ^0_δ and χ^1_δ. Notation of Fig. 5.4. Picture (b) shows the families $\mathcal{F}_\pm$ in coordinates ρ for various ε. Into the one family $\mathcal{F}$ there are projected two families $\mathcal{F}_+$ and $\mathcal{F}_-$, which are united into one family at $\varrho_2 = 0$.

We consider now the normal form (9.4) of System (14.3). By Properties 3 and 4 of the normal form from Section 6, a normal form preserves the linear automorphism and the reality relation; so System (9.4) is invariant under the involution (9.6), and coordinates U are connected by the reality relation (9.5). Therefore, in Case γ_2, the theory of construction of the normal form developed in Section 10 is applicable. The linear in U part of the normal form (9.4) consists of equations

$$\dot u_1 = \lambda u_1 + \lambda u_2 - (\lambda\varepsilon/9)u_1, \quad \dot u_2 = \lambda u_2 + (\lambda\varepsilon/9)u_1 \tag{14.4}$$

and the complex-conjugate ones. Hence the support of the series χ^0_δ does not contain the point $(1,-1)$ belonging to $\tilde{\mathbf{M}}^2 + \tilde R$ and corresponding to points $Q = (1,-1,0,0)$ and $Q = (0,0,1,-1)$. According to Fig. 5.13, a the Newton's open polygon of the series χ^0_δ consists of one edge with the normal vector $-(1,2)$ and with two points of its support $(3,-1)$ and $(-1,1)$.

Let $\tilde q_3 = \operatorname{ord}\varepsilon$. Then the support of the series χ_δ includes points $\tilde Q_1 = (3,-1,0)$, $\tilde Q_2 = (-1,1,0)$, $\tilde Q_3 = (1,-1,1)$, $\tilde Q_4 = (0,0,1)$, $\tilde Q_5 = (-1,1,1)$ lying in the boundaries of convex hulls of supports of the series χ^0_δ and $\varepsilon\chi^1_\delta$. Here the supporting planes to these points are the following two:

the plane $-\tilde q_1 - 2\tilde q_2 - 2\tilde q_3 = -1$ passing through points $\tilde Q_1$, $\tilde Q_2$, $\tilde Q_3$,

and the plane $-\tilde q_1 - \tilde q_2 = 0$ passing through points $\tilde Q_2$, $\tilde Q_3$, $\tilde Q_4$, $\tilde Q_5$.

But the second plane has the normal vector $-(1,1,0)$ with the zero component. The corresponding truncation here does not suit us. There remains the truncation with the support $\{\tilde Q_1, \tilde Q_2, \tilde Q_3\}$.

For vectors $Q \in \mathbf{N}$ the equation $\langle Q, \Lambda\rangle = 0$ has the form

$$q_1 - q_3 + q_2 - q_4 = 0, \tag{14.5}$$

i.e. $q_1 + q_2 = q_3 + q_4$. Consequently, $q_1 + q_2 + q_3 + q_4$ is an even number. From Equalities (10.3) with $r = s = 1$ when $k = -1$, we obtain integer solutions Q from $\mathbf{N}$ corresponding to points $\tilde{Q}_1$, $\tilde{Q}_2$: $Q = (2,-1,1,0)$, $(1,0,2,-1)$, $(-1,1,0,0)$, $(0,0,-1,1)$ for the zero power of ε, and solutions $Q = (0,0,1,-1)$, $(1,-1,0,0)$ corresponding to the point $\tilde{Q}_3$ for the first power of ε. We introduce new coordinates by (10.9), i.e.

$$\varrho_i = u_i u_{i+2}, \quad i = 1, 2, \quad w = u_1 u_4, \quad \bar{w} = u_2 u_3. \tag{14.6}$$

By Formula (10.10), for the truncated polynomials $\hat{g}_i$ we obtain

$$\hat{g}_1 = \lambda \varrho_1^{-1} \bar{w}, \quad \hat{g}_2 = (1/9)\lambda\varepsilon\varrho_2^{-1} w + \varkappa\lambda\varrho_1 w \varrho_2^{-1}. \tag{14.7}$$

Here the coefficient

$$\varkappa = 20a^2/243 \tag{14.8}$$

is found by Formula (35) for coefficients of the normal form in [Bruno 1979a, 1989a, Chapter III, Subsection 1.7].

The truncation of System (10.19) has equations

$$\begin{aligned} \dot{\varrho}_1 &= \lambda(\bar{w} - w), \\ \dot{\varrho}_2 &= \lambda(\varepsilon/9 + \varkappa\varrho_1)(w - \bar{w}), \\ \dot{w} &= -\lambda w((\varepsilon/9)\varrho_2^{-1}\bar{w} - \varrho_1^{-1}\bar{w} + \varkappa\varrho_1\bar{w}\varrho_2^{-1}) = -\lambda(\varepsilon/9)\varrho_1 + \lambda\varrho_2 - \lambda\varkappa\varrho_1^2 \stackrel{\text{def}}{=} \hat{b}_3 \end{aligned} \tag{14.9}$$

and the complex-conjugate equation for $\dot{\bar{w}}$. The stationary points ρ of System (14.9) with $\varrho_1 \neq 0$) and $\varrho_2 \neq 0$ satisfy the equation $(\varepsilon/9)\varrho_1 - \varrho_2 + \varkappa\varrho_1^2 = 0$, whence

$$\varrho_2 = \varepsilon\varrho_1/9 + \varkappa\varrho_1^2. \tag{14.10}$$

For various values of ε the disposition of curves of the type (14.10) in the plane ϱ_1, ϱ_2 is shown in Fig. 5.13, *b*. To each curve there correspond two families $\mathcal{F}_\pm$ of stationary points of System (14.9). When $\varepsilon < 0$, families $\mathcal{F}_+$ and $\mathcal{F}_-$ turn into each other in the axis $\varrho_2 = 0$ with $\varrho_1 > 0$.

We consider now the stability of these stationary points of System (14.9). By (14.7), here $\tilde{\gamma}_{11} = \lambda$, $\gamma_{21} = \varkappa\lambda$. We compute $\hat{\omega}$ by (10.30) using (14.9). In the curves (14.10)

$$\hat{\omega} = -\lambda^2(\varepsilon/9 + 2\varkappa\varrho_1 + \varkappa) = 3\varkappa/2 + \ldots > 0,$$

i.e. solutions of families $\mathcal{F}_\pm$ are stable, since $\varkappa > 0$ by Formula (14.8). Thus, we have proved

Theorem 14.1 *Near the value $\nu = \nu_0 = -5/4$ the set $\mathcal{A}$ consists of two families of periodic solutions $\mathcal{F}_\pm$. When $\varepsilon \geq 0$ they enter into the stationary*

point, and when $\varepsilon < 0$ they compose one family including the solution lying in the half-axis $\varrho_1 > 0$, $\varrho_2 = 0$. The disposition of their projections on the plane ϱ_1, ϱ_2 is shown in Fig. 5.13, b. The families $\mathcal{F}_\pm$ are stable.

Corollary 14.1 *When $\nu \geq -5/4$ families $\mathcal{F}_\pm$ enter into the stationary point. When $\nu < -5/4$ there exist families of stable periodic solutions $\mathcal{F}_\pm$, which do not pass through the stationary point. They unite in one family at some distance from the stationary point.*

It is possible to track the existence of these families for finite $\nu < -5/4$, but only numerically, extending them by the parameter ν.

System (14.9) has two first integrals

$$w + \bar{w} = \text{const} = c_1, \quad \varrho_2 + 9^{-1}\varepsilon\varrho_1 + 2^{-1}\varkappa\varrho_1^2 = \text{const} = c_2. \tag{14.11}$$

Integral curves of System (14.9) are intersection of integral surfaces (14.11). In order to clarify their structure, we introduce the polar coordinates

$$u_j = \sqrt{\varrho_j}\exp i\theta_j, \quad u_{j+2} = \sqrt{\varrho_j}\exp(-i\theta_j), \quad j = 1, 2.$$

Then

$$w = u_1 u_4 = \sqrt{\varrho_1\varrho_2}\exp i(\theta_1 - \theta_2).$$

We introduce the resonant phase $\theta_* = \theta_1 - \theta_2$ and put $\sqrt{\varrho_2}\exp i\theta_* = x + iy$. Then the first integral (14.11) takes the form

$$2x\sqrt{\varrho_1} = c_1. \tag{14.12}$$

We consider at first the case $c_1 = 0$, i.e. $x = 0$. In Fig. 5.14 the disposition of integral curves in the plane ϱ_1, y is shown for various ε. When $\varepsilon < 0$ there is a stationary point with

$$\varrho_1 = -\varepsilon/(9\varkappa), \quad y = 0, \quad (c_2 = -\varepsilon^2/(162\varkappa))$$

and the doubly asymptotic (homoclinic) solution $\mathcal{H}$ with $c_2 = 0$, i.e. $\varrho_2 = -9^{-1}\varepsilon\varrho_1 - 2^{-1}\varkappa\varrho_1^2$. The remaining curves are either closed and lie inside the curve $\mathcal{H}$ or opened and lie outside the curve $\mathcal{H}$; to them there correspond periodic solutions to System (14.9).

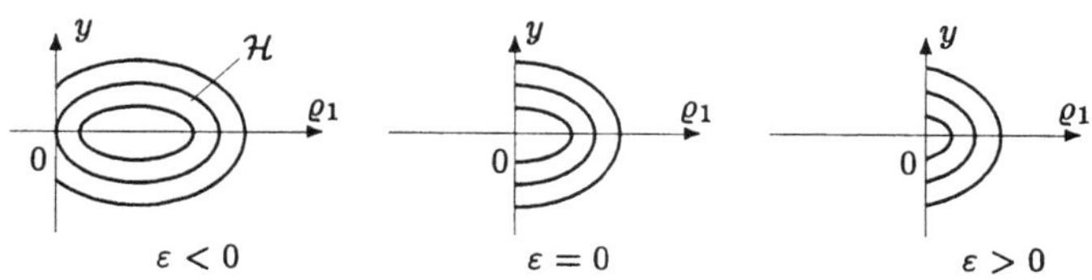

Figure 5.14: Case γ_2. The disposition of integral curves (14.11) in the half-plane $\varrho_1 \geq 0$, y when $c_1 = x = 0$.

If in the first integral from (14.11) $c_1 \neq 0$, then we consider intersections of the level surfaces of the second integral (14.11) with the surface (14.12). Their projections on the plane ϱ_1, y again form the pictures shown in Fig. 5.14. The only difference from the case $c_1 = 0$ is that now to the curve, projection of which passes through the point $\varrho_1 = 0$, $y = 0$, there corresponds a non-homoclinic solution, since the integral curve itself does not pass through the point $\varrho_1 = \varrho_2 = 0$.

Now we note that to the stationary points of System (14.9) there correspond periodic solutions to System (9.4) mentioned in Theorem 14.1. To the homoclinic solution $\mathcal{H}$ to System (14.9) there corresponds the family of homoclinic solutions to System (9.4), which has as the parameter the value θ_1^0 of the angle θ_1, taken in the point of the solution, where $\varrho_1 = -\varepsilon(9\varkappa)^{-1}$, $\varrho_2 = 0$. Finally, to periodic solutions to System (14.9) there correspond invariant tori of System (9.4); they are filled either by periodic solutions or by quasi-periodic solutions.

Now among solutions to System (9.4), we isolate the symmetric ones, i.e. the integral curves of which turn into themselves under the involution (9.6). These are:

1) all periodic solutions of the set $\mathcal{A}$;
2) two homoclinic solutions with $\theta_1^0 = 0$ and $\theta_1^0 = \pi$;
3) a whole invariant torus filled by quasi-periodic solutions;
4) from one up to four periodic solutions lying in an invariant torus filled by periodic solutions.

It may be proved in the same way as in [Bruno 1990a, Chapters VII and IX], that symmetric solutions are preserved under the symmetric perturbations, i.e. perturbations which are invariant with respect to the involution (9.6). Since System (4.2) may be considered as a symmetric perturbation of System (9.4), then the following theorems are valid.

Theorem 14.2 [Iooss and Kirshgässner 1992] *When $\nu + 5/4 < 0$ is small, System* (4.2) *has a pair of homoclinic solutions.*

Theorem 14.3 *When $|\nu + 5/4|$ is small, System* (4.2) *has invariant tori with an irrational frequency ratio satisfying Condition ω.*

Besides, there are families of periodic solutions, which are different from the families $\mathcal{F}_\pm$ cited in Theorem 14.1. All these additional solutions lie in the extensions with respect to ν of families $\tilde{\mathcal{A}}(\nu_0)$ and $\mathcal{B}(\nu_0)$, which were found in Sections 11–13 for $\nu_0 > -5/4$.

15. The normal form in Cases γ_0 and γ_3

In Cases γ_0 and γ_3 we have $\sigma = 1$ and -1 respectively and $\nu_0 = 1$. When $\nu_0 = 1$, according to Formulae (5.5), $\lambda_1 = \lambda_3 = 0$, $\lambda_2 = -\lambda_4 = \sqrt{3\sigma} \stackrel{\text{def}}{=} \lambda$.

After the substitution $Y = TZ$ with the matrix T from Lemma 5.5 and $\nu = 1 + \varepsilon$, System (4.2) takes the form

$$\begin{aligned}
\dot{z}_1 &= z_3, \\
\dot{z}_2 &= \lambda z_2 + \sigma\varepsilon y_1 (6\lambda)^{-1} + \sigma a y_1^2 (6\lambda)^{-1}, \\
\dot{z}_3 &= \sigma(\varepsilon y_1 - a y_1^2)/3, \\
\dot{z}_4 &= -\lambda z_4 - \sigma\varepsilon y_1^2 (6\lambda)^{-1} - \sigma a y_1^2 (6\lambda)^{-1},
\end{aligned} \tag{15.1}$$

where $y_1 = z_1 + z_2 + z_4$. In Cases γ_0 and γ_3 the equation $\langle Q, \Lambda \rangle = 0$ is $\lambda(q_2 - q_4) = 0$. For its solutions, q_1 and q_3 are arbitrary, and $q_2 = q_4$. Hence the normal form (9.4) has the form

$$\dot{u}_i = u_i(\lambda + g_i(u_1, \varrho_2, u_3, \varepsilon)), \quad i = 1, 2, 3, 4, \tag{15.2}$$

where $\varrho_2 = u_2 u_4$. Here u_1, ϱ_2, u_3 are resonant coordinates. For them the normal form (15.2) gives the system

$$\begin{aligned}
\dot{u}_1 &= u_1 g_1 \stackrel{\text{def}}{=} b_1(u_1, \varrho_2, u_3, \varepsilon), \\
\dot{\varrho}_2 &= \varrho_2(g_2 + g_4) \stackrel{\text{def}}{=} \varrho_2 g_0(u_1, \varrho_2, u_3, \varepsilon) \stackrel{\text{def}}{=} b_2, \\
\dot{u}_3 &= u_3 g_3 \stackrel{\text{def}}{=} b_3.
\end{aligned} \tag{15.3}$$

Lemma 15.1 *The reduced normal form* (15.3) *is a real system. In it b_1, g_0, b_3 are series in non-negative powers of u_1, ϱ_2, u_3, ε; here g_1, g_0, g_3 contain only odd powers of u_3. To real values of Y there correspond arbitrary real u_1, u_3 and arbitrary real ϱ_2 in Case γ_0, but $\varrho_2 \geq 0$ in Case γ_3.*

Proof. Firstly, we consider Case γ_3. According to Lemma 5.5, coordinates Z are connected by the reality relation

$$\bar{z}_1 = z_1, \quad \bar{z}_2 = z_4, \quad \bar{z}_3 = z_3, \tag{15.4}$$

and System (15.1) is invariant with respect to the involution

$$z_1, z_2, z_3, z_4, t \to z_1, z_4, -z_3, z_2, -t. \tag{15.5}$$

By Properties 3 and 4 from Section 6, the normal form (15.2) preserves the reality relation

$$\bar{u}_1 = u_1, \quad \bar{u}_2 = u_4, \quad \bar{u}_3 = u_3 \tag{15.6}$$

and the linear automorphism

$$u_1, u_2, u_3, u_4, t \to u_1, u_4, -u_3, u_2, -t. \tag{15.7}$$

In coordinates u_1, ϱ_2, u_3, it gives the reality relation

$$\bar{u}_1 = u_1, \quad \bar{\varrho}_2 = \varrho_2, \quad \bar{u}_3 = u_3 \tag{15.8}$$

and the automorphism

$$u_1, \varrho_2, u_3, t \to u_1, \varrho_2, -u_3, -t. \tag{15.9}$$

Consequently, System (15.3) is real, and in it

$$g_i(u_1, \varrho_2, u_3) = -g_i(u_1, \varrho_2, -u_3), \quad i = 1, 0, 3,$$

i.e. all series g_i contain u_3 only in odd powers.

In Case γ_0, by Lemma 5.5, instead of (15.4) we have the reality relation

$$\bar{z}_i = z_i, \quad i = 1, 2, 3, 4.$$

The automorphism (15.5) is preserved here. In the normal form (15.2) we have the reality relation

$$\bar{u}_i = u_i, \quad i = 1, 2, 3, 4,$$

and the automorphism (15.7). In coordinates u_1, ϱ_2, u_3 for System (15.3), it gives the reality relation (15.8) and automorphism (15.9). The further discussion is the same as in Case γ_3.

It remains to note that in the normal form (15.2) the series $u_1 g_1$, $u_3 g_3$, g_2 and g_4 contain only non-negative powers of u_1, u_3, ϱ_2, ε. It is valid for $g_0 = g_2 + g_4$ as well. □

Lemma 15.2 *For small u_1, u_3, ϱ_2, ε the stationary points of System* (15.3) *satisfy equations*

$$u_3 = 0, \quad b_3(u_1, \varrho_2, 0, \varepsilon) = 0. \tag{15.10}$$

Such a point is stable (in linear approximation) if in it

$$\omega \stackrel{\text{def}}{=} -\frac{\partial b_1}{\partial u_3}\frac{\partial b_3}{\partial u_1} - \frac{\partial b_2}{\partial u_3}\frac{\partial b_3}{\partial \varrho_2} > 0, \tag{15.11}$$

and it is unstable if $\omega < 0$.

Proof. Stationary points of System (15.3) are determined by the system of equations

$$b_1 = 0, \quad b_2 = 0, \quad b_3 = 0.$$

According to Lemma 15.1, the series b_1 is divisible by u_3. According to Lemma 5.5, the series $b_1 = u_3 + \dots$ Consequently, for small u_1, u_3, ϱ_2, ε all solutions to the equation $b_1 = 0$ are $u_3 = 0$. According to Lemma 15.1, $b_2 = 0$ when $u_3 = 0$. Hence stationary points are determined by System (15.10).

In order to determine their stability (in linear approximation), we compute the Jacobi matrix of the right hand sides of System (15.3) when $u_3 = 0$:

$$\begin{pmatrix} 0 & 0 & \partial b_1/\partial u_3 \\ 0 & 0 & \partial b_2/\partial u_3 \\ \partial b_3/\partial u_1 & \partial b_3/\partial \varrho_2 & 0 \end{pmatrix}.$$

Its characteristic polynomial is $-\tilde{\lambda}^3 - \omega\tilde{\lambda}$, where ω is determined in (15.11). Its roots are $\tilde{\lambda}_{1,2} = \pm\sqrt{-\omega}$, $\tilde{\lambda}_3 = 0$. □

We denote the set of solutions to System (15.10) as $\mathcal{F}$, and its subset with $\varrho_2 = 0$ as $\mathcal{F}_0$. It is easy to see that, according to Remark 6.1, to the set $\mathcal{F}$ there corresponds the set $\mathcal{A}$ in the normal form (15.2), since in it

$$u_i(\lambda_i + g_i) = \lambda_i \alpha u_i, \quad i = 1, 2, 3, 4;$$

and to the subset $\mathcal{F}_0$ there corresponds the set $\mathcal{A}_0$ of stationary points of System (15.2), since in it

$$u_i g_i = 0, \quad i = 1, 2, 3, 4.$$

Here solutions from the set $\mathcal{A}$ have the form

$$u_i = u_i^0 \exp(\lambda_i \alpha t), \quad i = 1, 2, 3, 4,$$

i.e.

$$u_1 = u_1^0, \quad u_3 = u_3^0 = 0, \quad u_2 = u_2^0 \exp(\lambda \alpha t), \quad u_4 = u_4^0 \exp(-\lambda \alpha t).$$

In Case γ_0, we have real $\lambda = \sqrt{3}$. Hence the set $\mathcal{A}_1 = \mathcal{A} \setminus \mathcal{A}_0$ consists of hyperbolic integral curves.

In Case γ_3, we have purely imaginary $\lambda = i\sqrt{3}$. Hence the set $\mathcal{A}_1$ consists of periodic solutions with the frequency $|i\sqrt{3} + g_2(u_1^0, u_2^0 u_4^0, 0, \varepsilon)|$. Thus, the set $\mathcal{A}$ consists of periodic solutions (the set $\mathcal{A}_1$) and of stationary points (the set $\mathcal{A}_0$). The eigenvalues (5.5) are pair-wise commensurable. By Theorem 6.2, the set $\mathcal{A}$ is analytic in the original system (15.1), i.e. the system has the family $\mathcal{F}_1$ of periodic solutions.

We compute now the first approximations for System (15.3), for the set $\mathcal{F}$ and for the quantity ω in (15.11). According to Lemma 15.1, the initial parts of series b_1, g_0 and b_3, when $\varepsilon = 0$, are as follows:

$$\begin{aligned} u_1 g_1 &= c u_3 + \ldots = u_1 c u_1^{-1} u_3 + \ldots, \\ g_0 &= c u_3 + \ldots, \\ u_3 g_3 &= c u_1^2 + c\varrho_2 + c u_3^2 + \ldots = u_3 (c u_1^2 u_3^{-1} + c\varrho_2 u_3^{-1} + c u_3) + \ldots, \end{aligned}$$

where various constants are denoted as c. To these terms there correspond exponents $\tilde{Q}_1 = (-1,0,1)$, $\tilde{Q}_2 = (0,0,1)$, $\tilde{Q}_3 = (2,0,-1)$, $\tilde{Q}_4 = (0,1,-1)$ of monomials $u_1^{\tilde{q}_1} \varrho_2^{\tilde{q}_2} u_3^{\tilde{q}_3}$. The polyhedron spanned on these points and on the other points of supports of the series g_1, g_0, g_3 has three two-dimensional faces passing through the point $\tilde{Q}_1$ (Fig. 5.15):

$\mathbf{\Gamma}_1 \supset \{\tilde{Q}_1, \tilde{Q}_3, \tilde{Q}_4\}$ in the supporting plane $-2\tilde{q}_1 - 4\tilde{q}_2 - 3\tilde{q}_3 = -1$;
$\mathbf{\Gamma}_2 \supset \{\tilde{Q}_1, \tilde{Q}_2, \tilde{Q}_4\}$ in the supporting plane $-2\tilde{q}_1 - \tilde{q}_3 = 1$;
$\mathbf{\Gamma}_3 \supset \{\tilde{Q}_1, \tilde{Q}_2, \tilde{Q}_3\}$ in the supporting plane $\tilde{q}_2 = 0$.

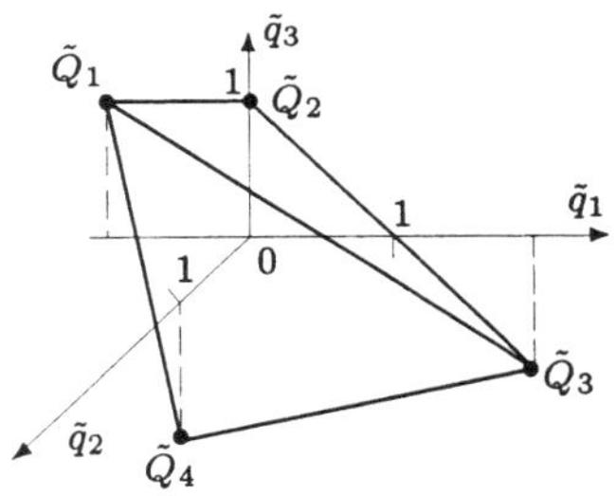

Figure 5.15: Cases γ_2 and γ_3. The leading part of the boundary of the Newton polyhedron for the reduced normal form (15.3).

The normal vectors to faces $\mathbf{\Gamma}_2$ and $\mathbf{\Gamma}_3$ include zero components, i.e. they are not in correspondence with truncations near the point $u_1 = \varrho_2 = u_3 = 0$. Thus, to a small neighborhood of the point there corresponds only the truncation given by the face $\mathbf{\Gamma}_1$. In $\mathbf{\Gamma}_1$ there are only the points $\tilde{Q}_1, \tilde{Q}_3, \tilde{Q}_4$, which give the basic truncation (when $\varepsilon = 0$)

$$\hat{b}_1 = cu_3, \quad g_2 = 0, \quad \hat{b}_3 = cu_1^2 + c_2\varrho_2.$$

Hence it is sufficient to compute the normal form (15.2) up to the square terms in right hand sides. It is clear from System (15.1), that within that precision

$$\begin{aligned}
\dot{u}_1 &= u_3 + \ldots, \\
\dot{u}_2 &= \lambda u_2 + \sigma\varepsilon u_2(6\lambda)^{-1} + \sigma a u_2 u_1 (3\lambda)^{-1} + \ldots, \\
\dot{u}_3 &= -\sigma(\varepsilon u_1 + a u_1^2 + 2a\varrho_2 + \ldots)/3, \\
\dot{u}_4 &= -\lambda u_4 - \sigma\varepsilon u_4(6\lambda)^{-1} - \sigma a u_1 u_4 (3\lambda)^{-1} + \ldots,
\end{aligned} \tag{15.12}$$

i.e. the truncation of System (15.3) corresponding to the face $\mathbf{\Gamma}_1$ is

$$\dot{u}_1 = u_3 \stackrel{\text{def}}{=} \hat{b}_1, \quad \dot{\varrho}_2 = 0, \quad \dot{u}_3 = -\sigma(\varepsilon u_1 + a u_1^2 + 2a\varrho_2)/3 \stackrel{\text{def}}{=} \hat{b}_3. \tag{15.13}$$

This system has two first integrals

$$\varrho_2 = \text{const} = c_1, \tag{15.14}$$

$$J \stackrel{\text{def}}{=} 3\sigma u_3^2 + \varepsilon u_1^2 + (2/3)au_1^3 + 4au_1\varrho_2. \tag{15.15}$$

The integral (15.15) is the first approximation to the integral (4.14).

Now we find stationary points of System (15.13), i.e. solutions to the system of equations $u_3 = 0, \quad \hat{b}_3 = 0$. We obtain

$$\varrho_2 = -\frac{\varepsilon}{2a}u_1 - \frac{1}{2}u_1^2. \tag{15.16}$$

Fig. 5.16 shows these curves in the plane u_1, ϱ_2 for various values of ε.

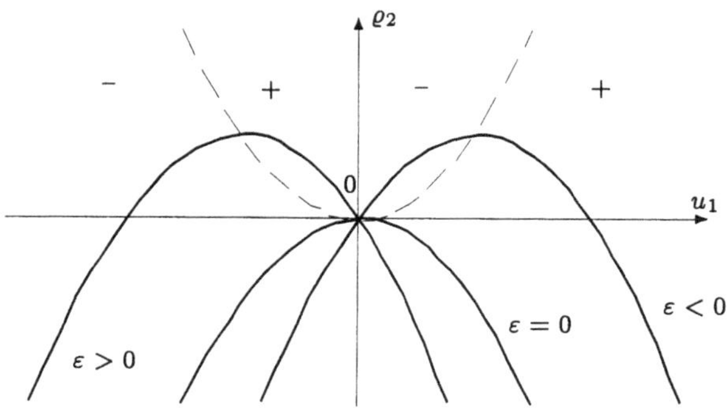

Figure 5.16: Cases γ_2 and γ_3. The section $\varepsilon =$ const of the set $\mathcal{A}$ in coordinates u_1, ϱ_2. The sign of $\hat{\omega}$ for $\sigma a > 0$ is cited.

According to (15.11) and (15.13)

$$\hat{\omega} = -\frac{\partial \hat{b}_1}{\partial u_3} \cdot \frac{\partial \hat{b}_3}{\partial u_1} = \frac{\sigma}{3}(\varepsilon + 2au_1). \tag{15.16$'$}$$

On the curves (15.16) $\varepsilon = -a(u_1 + \varrho_2/u_1)$, hence on them

$$\hat{\omega} = \frac{\sigma a}{3u_1}(u_1^2 - 2\varrho_2). \tag{15.17}$$

Fig. 5.16 shows the line of the change of sign of $\hat{\omega}$ by dashed line, and for $\sigma a > 0$ it shows signs of $\hat{\omega}$. It is remarkable that in the curve $\varrho_2 = u_1^2/2$, the change of stability and the attainment of maximum of values of ϱ_2 on curves (15.16) with fixed ε takes place simultaneously. These results are of the most interest in Case γ_3, when the set $\mathcal{A} \setminus \mathcal{A}_1$ consists of periodic solutions. Thus, we have proved

Theorem 15.1 *In Case γ_3 System* (4.2) *has the family $\mathcal{F}_1$ of periodic solutions, and the family $\mathcal{F}_0$ of stationary solutions. In the first approximation the family $\mathcal{F}_1$ is described by curves* (15.16) *with $\varrho_2 > 0$, and the family $\mathcal{F}_0$ is described by their points with $\varrho_2 = 0$. They are stable if $\hat{\omega} > 0$ in* (15.17)*, and unstable if $\hat{\omega} < 0$.*

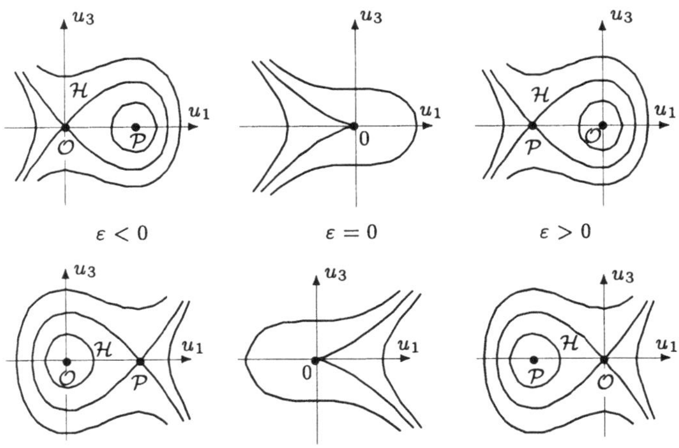

Figure 5.17: Cases γ_2 ($\sigma = 1$) (above) and γ_3 ($\sigma = -1$) (below). The integral curves of System (5.19) for $a > 0$ and various ε.

In the normal form (15.2), the series g_2 and g_4 contain only non-negative powers of ϱ_2. Hence System (15.2) has the formal invariant manifold

$$u_2 = u_4 = 0. \tag{15.18}$$

In it System (15.13) induces the formal system

$$\dot{u}_1 = u_3, \quad \dot{u}_3 = -\sigma(\varepsilon u_1 + a u_1^2)/3. \tag{15.19}$$

This system has the first integral $J \stackrel{\text{def}}{=} \varepsilon u_1^2 + 3\sigma u_3^2 + (2/3) a u_1^3 = \text{const}$, which is the restriction of the integral (15.15) on the manifold (15.18). For any $\varepsilon \neq 0$, System (15.19) has two stationary points:

$$\mathcal{O}: \quad u_1 = u_3 = 0; \tag{15.20}$$

$$\mathcal{P}: \quad u_1 = -\varepsilon/a, \quad u_3 = 0. \tag{15.21}$$

For $\varepsilon = 0$ they merge into one point. According to (15.16′), in these points

$$\hat{\omega} = \sigma(\varepsilon + 2au_1)/3,$$

i.e. in the point (15.20) $\hat{\omega} = \sigma\varepsilon/3$, and in the point (15.21) $\hat{\omega} = -\sigma\varepsilon/3$. Consequently, for a fixed $\varepsilon \neq 0$, one of the points (15.20) and (15.21) is stable and the other unstable. For $\sigma = +1$, $a > 0$ the behavior of the integral curves of System (15.19) in the plane u_1, u_3 is shown in the upper Fig. 5.17 for various values ε. It shows that for $\varepsilon < 0$, the point $\mathcal{P}$ is a center lying inside the homoclinic loop $\mathcal{H}$ going from the point $\mathcal{O}$. For $\varepsilon > 0$ it is the reverse: the point $\mathcal{O}$ is a center lying inside the homoclinic loop $\mathcal{H}$ going from the point $\mathcal{P}$. For $\sigma = -1$ and $a > 0$, the phase pictures are shown in the lower Fig. 5.17. We note that the stationary point $\mathcal{O}$ of System (15.19) is the stationary point $Y = 0$ of System (4.2), and the stationary point $\mathcal{P}$ is the point $Y = Y^0$ cited in (4.5).

Generally speaking, the manifold (15.18) is non-analytic in System (4.2), but only a formal one. Nevertheless, there exist analytic families of periodic solutions contracting to the points $\mathcal{O}$ and $\mathcal{P}$. Indeed, near the point $\mathcal{O}$, when $\sigma = 1$ and $\varepsilon > 0$, there is the family of periodic solutions found in Case IV; and when $\sigma = -1$ and $\varepsilon < 0$, there is the family $\mathcal{F}_1$ found in Case III. Near the point $\mathcal{P}$ we also obtain periodic solutions, since it turns into the point $\mathcal{O}$ under the substitution (4.6).

The homoclinic loops $\mathcal{H}$ are symmetric, i.e. they turn into themselves under the involution (15.9) (and (15.7)). Nevertheless, the question of their existence is resolved in different ways for $\sigma = 1$ and $\sigma = -1$. When $\sigma = 1$ the manifold (15.18) is a center one for System (4.2). Therefore by usual methods of [Bruno 1990a, Chapters VII and IX], one can prove the existence of the loops in the complete system (4.2), that may be considered as a symmetric perturbation of System (15.12). In [Iooss and Kirshgässner 1992] there is such a proof suitable both for System (4.2) and for the original reversible system (1.1).

When $\sigma = -1$ the manifold (15.18) is only the formal submanifold of the center manifold. Hence the splitting of the separatrix under the perturbations is possible here as it was indicated in [Melnikov 1963]. In [Hammersley and Mazarino 1989], for System (4.2) it was shown the absence of the simple homoclinic loop. The small effects of the splitting of separatrices may be computed with the help of the resurgent functions [Ecalle 1993]. In every level $\varrho_2 = \text{const} > 0$ System (15.13) has the same phase picture as in Fig. 5.17. Only the origin $u_1 = u_3 = 0$ is not a stationary point. Here there are three types of bounded solutions:

1) stationary points;
2) periodic solutions;
3) homoclinic solutions.

In System (15.12), to them there correspond:

1) periodic solutions of the set $\mathcal{A}$;

2) invariant tori filled either by periodic or by quasi-periodic solutions;

3) solutions which are homoclinic to periodic solutions of the type 1).

In System (4.2) considered as a symmetric perturbation of System (15.12), there are preserved: solutions of the types 1) and 3); symmetric periodic solutions lying in tori of the type 2); and tori of the type 2) filled by quasi-periodic solutions with the frequency ratio satisfying Condition ω (see [Bruno 1974, 1989a, 1990a; Iooss and Kirshgässner 1992]).

16. The review of the results for System (4.2)

In System (4.2) there are two stationary points: $\mathcal{O} = \{Y = 0\}$ and $\mathcal{P} = \{Y = Y^0\}$ (see (4.5)).

Theorem 16.1 *In System* (4.2) *there are no heteroclinic solutions connecting the points $\mathcal{O}$ and $\mathcal{P}$.*

Proof. The first integral (4.4) is analytic in points $\mathcal{O}$ and $\mathcal{P}$. In the point $\mathcal{O}$ the integral (4.4) is equal to zero. In the point $\mathcal{P}$ the integral (4.4) is equal to $-2(1-\nu)^3/3a^2$, i.e. it is equal to zero only when $\nu = 1$, when points $\mathcal{O}$ and $\mathcal{P}$ coincide. But on each solution the first integral is constant. Consequently, there are no solution that is an asymptotic one to both points $\mathcal{O}$ and $\mathcal{P}$ simultaneously. □

Now we collect together our knowledge about solutions to System (4.2) when $\sigma = 1$ and $\sigma = -1$. For this it is sufficient to consider the values $\nu \leq 1$, since when $\nu > 1$ System (4.2) is transformed into itself by the substitution (4.6). Here ν is changed into $2 - \nu$, and points $\mathcal{O}$ and $\mathcal{P}$ are interchanged.

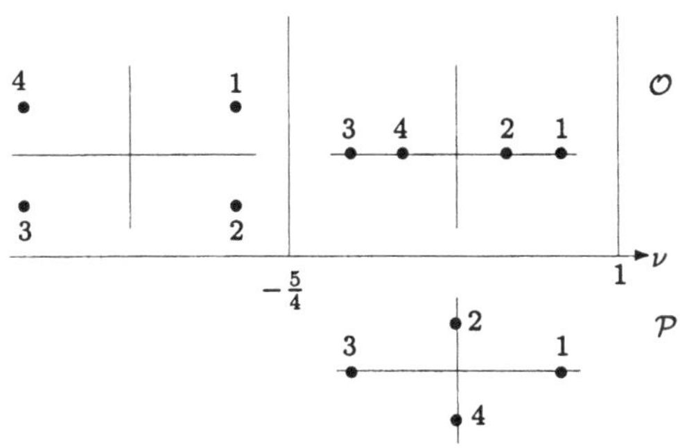

Figure 5.18: $\sigma = 1$. The disposition of eigenvalues λ_i of the stationary points $\mathcal{O}$ and $\mathcal{P}$ for $\nu \leq 1$ (compare with Fig. 5.1).

$\sigma = 1$. In Fig. 5.18, for $\nu \leq 1$ the disposition of eigenvalues is compared for the stationary points $\mathcal{O}$ and $\mathcal{P}$. By Theorem 6.5, for all $\nu < 1$ there are the unique two-dimensional manifold $\mathcal{M}_2^u$ of solutions going from the point $\mathcal{O}$ and the other two-dimensional manifold $\mathcal{M}_2^s$ of solutions coming into the point. There are the unique one-dimensional manifold $\mathcal{M}_1^u$ of solutions going from the point $\mathcal{P}$ and the other one-dimensional manifold $\mathcal{M}_1^s$ of solutions coming into the point. Besides there is a two-dimensional analytic manifold passing through the point $\mathcal{P}$, which is filled by periodic solutions forming the Liapunov family $\mathcal{F}(\nu)$. When ν is close to 1, this family ends at the homoclinic solution $\mathcal{H}(\nu)$ going from the point $\mathcal{O}$ and coming into it. It is possible that the situation continues from $\nu = 1$ up to some $\nu = \nu_0 < 1$. But it is difficult to determine this by the local analysis. It is possible that there are other homoclinic solutions to each of the points $\mathcal{O}$ and $\mathcal{P}$, but they cannot be found by the local analysis.

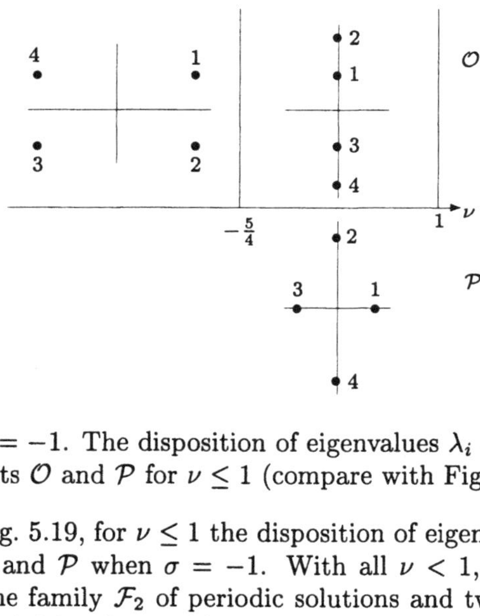

Figure 5.19: $\sigma = -1$. The disposition of eigenvalues λ_i of the stationary points $\mathcal{O}$ and $\mathcal{P}$ for $\nu \leq 1$ (compare with Fig. 5.1).

$\sigma = -1$. In Fig. 5.19, for $\nu \leq 1$ the disposition of eigenvalues is compared for the points $\mathcal{O}$ and $\mathcal{P}$ when $\sigma = -1$. With all $\nu < 1$, through the point $\mathcal{P}$ there passes one family $\mathcal{F}_2$ of periodic solutions and two one-dimensional manifolds $\mathcal{M}_1^u$ and M_1^s of unstable and stable to $\mathcal{P}$ trajectories. For $\nu \in (-5/4, 1)$ and $\nu \neq -0.44$, 0.19, two families $\mathcal{F}_1$ and $\mathcal{F}_2$ of stable periodic solutions contract to the point $\mathcal{O}$. There are also families of periodic solutions $\mathcal{F}_3(\nu_0)$ and $\mathcal{F}_4(\nu_0)$, which contract to the point $\mathcal{O}$ when $\nu = \nu_0$ and exist only for such ν_0, for which $\lambda_1(\nu_0)/\lambda_2(\nu_0)$ is a rational number. Besides there are families $\mathcal{B}(\nu_0)$ of invariant tori contracting to the point $\mathcal{O}$ when $\nu = \nu_0$, and which exist only for such ν_0, for which the number $\lambda_1(\nu_0)/\lambda_2(\nu_0)$ is irrational and satisfies Condition ω (see Section 6). For ν close to 1, the family of stable periodic solutions $\mathcal{F}_1$ going out from the point $\mathcal{O}$ is bounded by the splitted solutions homoclinic to the point $\mathcal{P}$. Besides there are solutions, which are homoclinic to periodic ones (solutions of the type 3) by classification of Section 15). It is possible that there are other homoclinic to $\mathcal{P}$ solutions for different $\nu < 1$. When $\nu < -5/4$ there are homoclinic to $\mathcal{O}$ solutions, and there exists the family of stable periodic solutions, which does not contract to the stationary point. It is possible that there exist continuations of some families $\mathcal{F}_i(\nu_0)$ of periodic solutions and of some families $\mathcal{B}_i(\nu_0)$ of invariant tori. But the local analysis is not sufficient to trace their existence and their boundaries with respect to ν, and the numeric continuation of these families with respect to the parameter ν should be used.

17. The transference of results to the original system

Theorem 17.1 *Under the power transformation* (4.1) *and the change of time* $t_1 = |\mu|^{1/2}t$, *System* (1.1) *turns into the system*

$$dY/dt_1 = F(Y,\nu) + \mu\tilde{F}(Y,\nu,\mu), \tag{17.1}$$

where $dY/dt_1 = F(Y,\nu)$ *is the system* (4.2), *and* $\tilde{F}$ *is its perturbation with the small parameter* μ.

Proof. On account of the reversibility (1.3), (1.4) of System (1.1), series φ_2 and φ_4 contain x_2 and x_4 in even powers summed together, and φ_1 and φ_3 in odd powers summed together. After the substitution of (4.1) in (1.1), φ_2 and φ_3 include μ in integer powers, but φ_1 and φ_3 include μ only in half-integer powers. After the conversion to Y in derivatives and the mentioned change of time, we obtain system, the right hand sides of which contain only integer powers of μ. □

We are going to study solutions to System (17.1) in the neighborhood of the point $Y = 0$, $\varepsilon = \mu = 0$, when $\sigma = \sigma_0$ and $\nu = \nu_0 + \varepsilon$, where ε and μ are small parameters. The eigenvalues of the linear part of System (17.1) when $\varepsilon = \mu = 0$ (or of System (4.2) when $\varepsilon = 0$) are (5.5).

According to Section 6, the formal non-linear change of coordinates

$$Y = T(\nu_0)U + \Xi(U,\varepsilon) + \mu\tilde{\Xi}(U,\varepsilon,\mu)$$

transforms System (17.1) into the normal form

$$\dot{u}_i = \lambda_i u_i + \psi_i(U,\varepsilon) + \mu\tilde{\psi}_i(U,\varepsilon,\mu)) \stackrel{\text{def}}{=} \lambda_i u_i + u_i g_i \stackrel{\text{def}}{=}$$

$$\stackrel{\text{def}}{=} \lambda_i u_i + u_i \sum g_{iQ} U^Q, \quad i = 1,2,3,4, \tag{17.2}$$

where $Q = (q_1, q_2, q_3, q_4)$, $U^Q = u_1^{q_1} u_2^{q_2} u_3^{q_3} u_4^{q_4}$, and there are only resonant terms U^Q, for which $(q_1 - q_3)\lambda_1 + (q_2 - q_4)\lambda_2 = 0$. Here g_{iQ} are power series in ε, μ. In this connection the system

$$\dot{u}_i = \lambda_i u_i + \psi_i(U, \varepsilon), \quad i = 1, 2, 3, 4, \tag{17.3}$$

is the normal form of System (4.2).

Both normal forms (17.2) and (17.3) are reversible, and System (17.2) is a perturbation of System (17.3) in small parameter μ. All theorems of Sections 7–15 proved for System (4.2) by its normal form (17.3) are valid also for System (17.1) by the following reasons.

In Cases I and γ_1, an analog of Theorem 6.5 for a system with small parameters is applicable to System (17.1) [Samovol 1982]. In Cases II and IV the normal form (17.2) satisfies Condition A. In Cases III, γ_2, γ_0, γ_3, the sets $\mathcal{A}$, $\tilde{\mathcal{A}}$ and $\mathcal{B}$ computed for the normal form (17.2) are only perturbations of the same sets computed for the normal form (17.3), i.e. all local branches of these sets are already isolated and described in Sections 9–15.

Returning to System (1.1) from System (17.1) by means of the transformation (4.1), we obtain for System (1.1) the theorems analogous to Theorems of Sections 7–15. The whole procedure is applicable for an arbitrary smooth system (1.1) with the change of analyticity by smoothness [Bruno 1974, 1989a].

18. The comparison with the Hamiltonian normal form

By Theorem 4.1, System (4.2) is Hamiltonian and reversible. Computing its normal form, we have neglected its Hamiltonian structure, and used only its reversibility. It was sufficient for computation of additional analytic integrals in Cases II and IV, and of the sets $\mathcal{A}$, $\tilde{\mathcal{A}}$ and $\mathcal{B}$ in Case III.

System (4.2) may be reduced to the Hamiltonian normal form by means of the canonical change of coordinates [Bruno 1990a, Chapter I, II]. But it is a narrower class of substitutions, which leads to additional computational difficulties. First of all, the linear change $Y = TZ$ must be canonical.

Theorem 18.1 [Bruno 1990a, Chapter I] *A Hamiltonian system*

$$\dot{x}_i = -\partial H/\partial y_i, \quad \dot{y}_i = \partial H/\partial x_i, \quad i = 1, 2, \tag{18.1}$$

by means of the formal canonical change of coordinates

$$x_i = b_i(U, V), \quad y_i = d_i(U, V), \quad i = 1, 2,$$

may be reduced to the Hamiltonian normal form

$$\dot{u}_i = -\partial h/\partial v_i, \quad \dot{v}_i = \partial h/\partial u_i, \quad i = 1, 2, \tag{18.2}$$

where

$$h = \sum h_{PQ} U^P V^Q \quad over \quad \langle P - Q, \Lambda' \rangle = 0, \tag{18.3}$$

the vector $\Lambda' = (\lambda_1, \lambda_2)$ *is composed from eigenvalues of the matrix of the linear part of System* (18.1).

If λ_1/λ_2 is irrational, then $h = h(\varrho_1, \varrho_2)$, where $\varrho_i = u_i v_i$. Here the normal form (18.2) has two formal integrals: $\varrho_i =$ const, $i = 1, 2$. But the normal form of Section 9 also has two formal integrals.

If $\lambda_1/\lambda_2 = r/s$, then in (18.3)

$$h = h_0(\rho) + \sum_{k=1}^{\infty} f_k(\rho) w^k + \sum_{k=1}^{\infty} g_k(\rho) \bar{w}^k,$$

where $w = u_1^s v_2^r$, $\bar{w} = u_2^r v_1^s$ and $\rho = (\varrho_1, \varrho_2)$. But the transition to coordinates $\rho, w, \bar{w}$ is not canonical. Therefore it is necessary to introduce the polar coordinates $\varrho_j = u_j v_j$, $\theta_j = (2i)^{-1} \log(u_j/v_j)$, so that

$$u_j = \sqrt{\varrho_j} \exp(i\theta_j), \quad v_j = \sqrt{\varrho_j} \exp(-i\theta_j), \quad j = 1, 2,$$

and to make the linear canonical transformation

$$\begin{aligned} \varrho_1^* &= \varrho_1 + s\varrho_2/r, & \theta_1^* &= \theta_1, \\ \varrho_2^* &= \varrho_2, & \theta_2^* &= -s\theta_1/r + \theta_2. \end{aligned}$$

Then $h = h(\varrho_1^*, \varrho_2^*, \theta_2^*)$ and

$$\dot{\varrho}_1^* = -\partial h/\partial \theta_1^* = 0, \quad \dot{\theta}_1^* = \partial h/\partial \varrho_1^*, \tag{18.4}$$

$$\dot{\varrho}_2^* = -\partial h/\partial \theta_2^*, \quad \dot{\theta}_2^* = \partial h/\partial \varrho_2^*. \tag{18.5}$$

To the set $\mathcal{A}$ there correspond stationary points of Subsystem (18.5), i.e. the set $\mathcal{A}$ is determined by the system of two equations:

$$\partial h/\partial \theta_2^* = 0, \quad \partial h/\partial \varrho_2^* = 0.$$

For a reversible system the Fourier series of the Hamiltonian function h over the angle θ_2^* contains only $\cos kr\theta_2^*$. Hence $\partial h/\partial \theta_2^*$ is divisible by $\sin r\theta_2^*$, and in a generic case outside $\varrho_1 = 0$ and $\varrho_2 = 0$ all its roots are $r\theta_2^* = k\pi$. The relations between ϱ_1 and ϱ_2 are found from the equation $\partial h/\partial \varrho_2^* = 0$, i.e. the description of the set $\mathcal{A}$ is the same as in Section 10.

The difference is outside the set $\mathcal{A}$: for a Hamiltonian normal form there are always two formal first integrals: h and $\lambda_1\varrho_1 + \lambda_2\varrho_2$ (i.e. ϱ_1^*), and a smaller number of coefficients to be computed in any segment of the normal form. But this difference is negligible in first approximations, since the truncation of the transformed normal form (10.19) apparently does not depend on the normalizing transformation.

19. The case $\mu = 0$

If in System (3.10) μ is equal to zero, then the power transformation (4.1) transforming System (3.10) into System (4.2) cannot be made. Therefore here we consider separately System (3.10) when $\mu = 0$, i.e. the system

$$\dot{x}_1 = x_2, \quad \dot{x}_2 = x_3, \quad \dot{x}_3 = x_4, \quad \dot{x}_4 = \mu_2 x_1 + a x_1^2. \tag{19.1}$$

Let $\mu_2 \neq 0$. The power transformation

$$x_1 = \mu_2 y_1, \quad x_2 = |\mu_2|^{5/4} y_2, \quad x_3 = |\mu_2|^{6/4} y_3, \quad x_4 = |\mu_2|^{7/4}$$

and the change of time $t = |\mu_2|^{-1/4} t_1$ reduce System (19.1) to the form

$$dy_1/dt_1 = y_2, \; dy_2/dt_1 = y_3, \; dy_3/dt_1 = y_4, \; dy_4/dt_1 = \sigma y_1 + a y_1^2, \tag{19.2}$$

where $\sigma = \mathrm{sgn}\mu_2 = \pm 1$. System (19.2) is System (4.2) with $\sigma = 0$ and $\nu = \sigma$. It is equivalent to the equation $y_1^{(4)} - \sigma y_1 - a y_1^2 = 0$; it has the first integral $J_1 \stackrel{\text{def}}{=} \sigma y_1^2 + y_3^2 - 2y_2 y_4 + 2a y_1^3/3 = \text{const}$; it is a Hamiltonian system with the Hamiltonian function $H = J_1/2$:

$$\begin{aligned} \dot{y}_4 &= \partial H/\partial y_1, \quad \dot{y}_1 = -\partial H/\partial y_4, \\ \dot{y}_2 &= \partial H/\partial y_3, \quad \dot{y}_3 = -\partial H/\partial y_2; \end{aligned}$$

and it has two stationary points: $\mathcal{O} = (0,0,0,0)$ and $\mathcal{P} = (-\sigma/a, 0, 0, 0)$.

Lemma 19.1 *Under the transformation*

$$y_1 = \tilde{y}_1 + \tilde{\sigma}/a, \quad y_i = \tilde{y}_i, \quad i = 2,3,4, \quad \sigma = -\tilde{\sigma}$$

System (19.2) *preserves its form, and the stationary points* $\mathcal{O}$ *and* $\mathcal{P}$ *are interchanged.*

The proof is an immediate checking.

The characteristic equation of the linear part of System (19.2) is $\lambda^4 - \sigma = 0$, i.e. $\lambda = \sqrt[4]{\sigma}$. For $\sigma = 1$ we have $\lambda_{1,3} = \pm 1$, $\lambda_{2,4} = \pm i$. For $\sigma = -1$ we have $\lambda_{1,3} = \pm(1+i)/\sqrt{2}$, $\lambda_{2,4} = \pm(1-i)/\sqrt{2}$. In both cases the ratio λ_1/λ_2 is not a real number. Hence, in the same way as in Section 8, the normal form satisfies Condition A, and the normalizing transformation is convergent. Consequently, System (19.2) has two analytic first integrals: ϱ_1 and ϱ_2. When $\sigma = 1$, near the point $\mathcal{O}$ it has an analytic family of periodic solutions.

20. The Belitskii normal form

We consider the formal system

$$\dot{X} = GX + \Phi(X), \quad X \in \mathbb{R}^n \text{ or } X \in \mathbb{C}^n, \tag{20.1}$$

where the matrix G is the Jordan normal form, and the series $\Phi(X)$ does not contain constant and linear terms. Let the formal change

$$X = Y + \Xi(Y) \tag{20.2}$$

reduce System (20.1) to the form

$$\dot{Y} = GY + \Psi(Y). \tag{20.3}$$

System (20.3) is called the *Belitskii normal form* if

$$\frac{\partial \Psi(Y)}{\partial Y} G^* Y = G^* \Psi(Y), \tag{20.4}$$

where G^* is the transposed matrix G.

Theorem 20.1 [Belitskii 1975, 1979a, b] *For every System* (20.1) *there exists the substitution* (20.2) *reducing System* (20.1) *to the Belitskii normal form* (20.3).

The normalizing transformation (20.2) has Properties 2–4 from Section 6. The question of its convergence for an analytical System (20.1) has not been studied.

We write System (20.3) in the form

$$\dot{y}_j = y_j g_j(Y) \stackrel{\text{def}}{=} y_j \sum g_{jQ} Y^Q, \quad j = 1, \ldots, n. \tag{20.5}$$

If we introduce the diagonal matrix $\tilde{\Lambda} = \{\lambda_1, \ldots, \lambda_n\}$, then the definition (6.6) of the resonant normal form may be written for System (20.3) in the form

$$\frac{\partial \Psi}{\partial Y} \tilde{\Lambda} Y - \tilde{\Lambda} \Psi(Y) = 0. \tag{20.6}$$

If System (20.3) has the property (20.4), then the property (20.6) is satisfied for it, and besides

$$\frac{\partial \Psi}{\partial Y} (G^* - \tilde{\Lambda}) Y - (G^* - \tilde{\Lambda}) \Psi(Y) = 0.$$

Thus, the right hand sides of the Belitskii normal form (20.3), (20.5) contain only resonant terms $y_j g_{jQ} Y^Q$ with the property (6.6); but some terms are missing, and the non-zero coefficients g_{jQ} with the same values $\|Q\| \stackrel{\text{def}}{=} q_1 + \ldots + q_n$ are connected by some relations.

If the Jordan form G is diagonal, then $G^* = G$, and the Belitskii normal form (20.2) coincides with the resonant normal form from Section 6. But if the matrix G has non-trivial Jordan blocks, then the Belitskii normal form (20.3) has fewer independent coefficients than the resonant normal form. For example, the Arnol'd's normal form (2.1) for the linear system with small parameters (2.3) is a special case of the Belitskii normal form (20.3). We note that the original system (20.1) may be considered as a perturbation of the normal form (20.3).

The computation of the Belitskii normal form is more complicated then that of the resonant one, but in some cases it may be simplified or even omitted. We write System (20.1) in the form

$$\dot{x}_j = x_j f_j(X) \stackrel{\text{def}}{=} x_j \sum f_{jQ} X^Q, \quad j = 1, \ldots, n. \tag{20.7}$$

Theorem 20.2 *Let for System* (20.1), (20.7) *exist such a vector* $K \in \mathbb{R}^n$, $K \geq 0$, $K \neq 0$ *that*

1) $\langle Q, K \rangle = r \geq 0$ *for all linear terms* $x_j f_{jQ} X^Q$ *in* (20.7) *with* $Q \neq 0$;
2) $\langle Q, K \rangle \geq r$ *for all non-linear terms* $x_j f_{jQ} X^Q$ *in* (20.7);
3) *the generalized truncation of System* (20.7)

$$\dot{x}_j = x_j \sum f_{jQ} X^Q \quad \textit{over} \ \langle Q, K \rangle \leq s, \quad j = 1, \ldots, n,$$

with a fixed $s \geq r$ *is the Belitskii normal form.*

Then in the Belitskii normal form (20.3), (20.5) *of System* (20.1), (20.7) *the following equalities are satisfied*

$$g_{jQ} = f_{jQ} \quad for \quad \langle Q, K \rangle \leq s.$$

Proof. At first we reduce System (20.1) to the resonant normal form. According to Theorem 5 from [Bruno 1979a, Chapter III, § 1], up to terms of the order s (i.e. for $\langle Q, K \rangle \leq s$) the resonant normal form coincides with the original System (20.1). Hence we may assume that System (20.1) itself is the resonant normal form, and the transformation (20.2) reduces it to the Belitskii normal form (20.3). Then, according to Theorem 2 from [Bruno 1979a, Chapter III, § 1], the transformation (20.2) includes only resonant terms $y_j h_{jQ} Y^Q$ with $\langle Q, \Lambda \rangle = 0$, if we write it in the form

$$x_j = y_j h_j(Y) \stackrel{\text{def}}{=} y_j \sum h_{jQ} Y^Q, \quad j = 1, \ldots, n. \tag{20.8}$$

Let us take a point Q' corresponding to the power exponent of the linear term $x_j f_{jQ'} X^{Q'}$ of the right hand side of System (20.1), (20.7), and denote as $\mathbf{K}_{Q'}$ the convex conic hull of the vectors $Q - Q'$, where $Q \in \mathbf{S}(F)$, $Q \neq 0$. Let the cone $\mathbf{K}$ be the convex linear hull of all cones $\mathbf{K}_{Q'}$ corresponding to vectors $Q' \neq 0$ of the linear terms in System (20.1), (20.7). From the proof of the Belitskii Theorem about his normal form [Belitskii 1979a, b], it is clear that coefficients g_{jQ} and h_{jQ} depend only on such g_{jP} and h_{jP} that $Q - P \in \mathbf{K}$. Now we note that by properties of the vector K cited in the statement of

Theorem, for every vector $Q \in \mathbf{K}$ we have $\langle Q, K\rangle \geq 0$. Consequently, for every pair of admissible vectors Q and P, such that $Q - P \in \mathbf{K}$ and $\langle Q, K\rangle \leq s$, we have $\langle P, K\rangle \leq s$. By the condition 3) of Theorem, all coefficients f_{jP} correspond to the Belitskii normal form, and in (20.8) all $h_{jP} = 0$ for $\langle P, K\rangle \leq s$, $P \neq 0$. □

Corollary 20.1 *If* $\Lambda = 0$, *and the usual truncation*

$$\dot{X} = GX + \hat{\Phi}(X)$$

of System (20.1) *contains all its linear terms, and is the Belitskii normal form, then the Belitskii normal form* (20.3) *has the usual truncation*

$$\dot{Y} = GY + \hat{\Phi}(Y).$$

Now we consider System (3.3) obtained from System (1.1)–(1.3) after the linear normalization of Section 2. According to Section 3, System (3.3) has the truncation (3.10) satisfying conditions of Corollary 20.1. Consequently, the Belitskii normal form of System (3.3) has the same truncation (3.10). Let us consider its other truncations.

For System (1.1)–(1.3), with conditions (1.4) and (20.4) taken into account, the Belitskii normal form has the form (see Formula (19) in [Iooss 1995b])

$$\begin{aligned}\dot{Y} &= LY + P_4(y_1, p_2, p_4)E_4 + P_2(y_1, p_2, p_4)(y_1E_2 + y_2E_3 + \\ &+ y_3E_4) + Q(p_2, p_4)(p_2E_2 + q_2E_3 + r_2E_4) + P_1(y_1, p_2, p_4)\times \\ &\times (p_3E_1 + q_3E_2 + r_3E_3 + s_3E_4) + P_3(y_1, p_2, p_4)(p_3E_3 + q_3E_4),\end{aligned} \tag{20.9}$$

where

$$\begin{aligned}&p_2 = y_2^2 - 2y_1y_3, \quad p_3 = y_2^3 - 3y_1y_2y_3 + 3y_1^2y_4, \\ &p_4 = 3y_2^2y_3^2 - 6y_2^3y_4 - 8y_1y_3^3 + 18y_1y_2y_3y_4 - 9y_1^2y_4^2, \\ &q_2 = -3y_1y_4 + y_2y_3, \quad q_3 = 3y_1y_2y_4 - 2y_1y_3^2 + y_2^2y_3, \\ &r_2 = -3y_2y_4 + 2y_3^2, \quad r_3 = -3y_1y_3y_4 + 3y_2^2y_4 - y_2y_3^2, \\ &s_3 = 3y_2y_3y_4 - (4/3)y_3^3 - 3y_1y_4^2,\end{aligned} \tag{20.10}$$

and $P_1 - P_4$ and Q are polynomials in their arguments. Their Maclaurin series begin with the constant terms, except

$$\begin{aligned}&P_4(y_1, p_2, p_4) = \mu_2y_1 + ay_1^2 + bp_2 + O(|Y|^3), \\ &P_2(y_1, p_2, p_4) = \mu + cy_1 + O(|Y|^2).\end{aligned}$$

Table 5.2: Table of correspondence for the set of points (3.7), (3.9), (20.16), and for the cone of the problem $\{P < 0\} \subset \mathbb{R}^6_*$.

			j	1	2	3	4	5	6	7	8	9	11	12	14	17	28
			T_j	V	V	V	V	V	V	V	V	V	V	V	d	V	V
				−1	0	0	1	0	0	1	2	2	1	0	1	0	0
				1	−1	0	−1	1	0	0	−1	0	1	2	0	0	3
			Q_j	0	1	−1	0	−1	1	0	0	0	−1	0	1	2	−1
				0	0	1	0	0	−1	−1	0	−1	0	−1	−1	−1	0
				0	0	0	1	1	1	0	0	0	0	0	0	0	0
i	t_i	B_i, N_i, D_i		0	0	0	0	0	0	1	0	0	0	0	0	0	0
1	D	0 0 0 −1 0 0		−	−	−	−	−	+	+	−	+	−	+	+	+	−
2	N	0 −1 0 −1 0 0		−	+	−	+	−	+	+	+	+	−	−	+	+	−
3	N	0 0 −1 −1 0 0		−	−	−	−	+	−	+	−	+	+	+	−	−	+
4	N	0 −1 −2 −1 0 0		−	−	+	+	+	−	+	+	+	+	−	−	−	−
5	N	0 0 0 0 −1 0		+	+	+	−	−	−	+	+	+	+	+	+	+	+
6	N	−2 −1 −4 −3 −2 0		+	−	+	−	+	−	+	−	−	+	+	−	−	+
7	N	0 0 0 0 0 −1		+	+	+	+	+	+	−	+	+	+	+	+	+	+
8	N	−1 −1 −1 −2 −1 −1		+	+	−	−	−	+	+	−	+	−	+	+	+	−
9	N	−1 −1 −2 −2 −1 −1		+	−	+	−	+	−	+	−	+	+	+	−	−	−
10	N	−2 −2 −3 −4 −1 −2		+	−	−	−	+	+	+	−	+	−	+	−	−	−
11	N	−1 −2 −2 −2 −1 −1		−	+	+	+	−	−	+	+	+	−	−	−	−	−
12	N	−4 −5 −6 −7 −2 −4		+	+	+	+	+	+	+	−	+	−	−	−	−	−
13	D	−2 −1 −2 −3 0 0		+	−	−	−	+	+	+	−	−	−	+	−	−	−
14	D	−1 −1 −1 −1 0 0		+	+	+	+	+	+	+	−	−	−	−	−	−	−
15	D	0 −1 0 0 0 0		−	+	−	+	−	−	−	+	−	−	−	−	−	−
16	D	0 0 −1 0 0 0		−	−	+	−	+	−	−	−	−	+	−	−	−	+
17	D	−1 0 −1 0 0 0		+	−	+	−	+	−	−	−	−	−	−	−	−	+
18	D	0 −1 −1 0 0 0		−	−	+	+	−	−	−	+	−	−	−	−	−	−
19	D	−1 0 0 −1 0 0		+	−	−	−	−	+	−	−	−	−	+	−	+	−
20	D	−2 −1 0 −1 0 0		+	+	−	−	−	+	−	−	−	−	−	−	+	−
21	D	−1 0 −1 −1 0 0		+	−	−	−	+	−	−	−	−	−	+	−	−	+
22	D	−1 0 0 0 0 0		+	−	−	−	−	−	−	−	−	−	−	−	−	−
23	D	−2 −1 −1 −3 −1 0		+	−	−	−	−	+	+	−	−	−	+	−	+	−
24	D	−4 −3 −2 −5 −2 0		+	+	−	−	−	+	+	−	−	−	−	−	+	−
25	D	−2 −5 −4 −3 −2 0		−	+	+	+	−	−	+	+	−	−	−	−	−	−
26	D	−1 −1 −2 −3 0 −1		−	−	−	−	+	+	+	−	+	−	+	−	−	−
27	D	−2 −3 −4 −5 0 −2		−	−	−	+	+	+	+	−	+	−	−	−	−	−
28	D	−3 −2 −1 0 0 0		+	+	+	−	−	−	−	−	−	−	−	−	−	−
29	D	−2 −1 0 0 0 0		+	+	−	−	−	−	−	−	−	−	−	−	−	−
30	D	−1 −2 −1 0 0 0		−	+	+	+	−	−	−	−	−	−	−	−	−	−
31	D	−1 −4 −2 0 −1 0		−	+	+	+	−	−	−	+	−	−	−	−	−	−

Let $\tilde{\mathbf{S}}$ be the support of System (20.9) in notation (20.8). In order to compute the leading subset $\tilde{\mathbf{S}}' \overset{\text{def}}{=} \tilde{\mathbf{S}}/\mathbb{R}^6_+$, we compose the table of supports of polynomials (20.10). We obtain

$$
\begin{aligned}
&\mathbf{S}(p_2)=\{2E_2, E_1+E_3\},\ \mathbf{S}(p_3)=\{3E_2, E_1+E_2+E_3, 2E_1+E_4\},\\
&\mathbf{S}(p_4)=\{2E_2+2E_3, 3E_2+E_4, E_1+3E_3, E_1+E_2+E_3+E_4, 2E_1+2E_2\},\\
&\mathbf{S}(q_2)=\{E_1+E_4, E_2+E_3\},\ \mathbf{S}(q_3)=\{E_1+E_2+E_4, E_1+2E_3, 2E_2+E_3\},\\
&\mathbf{S}(r_2)=\{E_2+E_4, 2E_3\},\ \mathbf{S}(r_3)=\{E_1+E_3+E_4, 2E_2+E_4, E_2+2E_3\},\\
&\mathbf{S}(s_3)=\{E_2+E_3+E_4, 3E_3, E_1+2E_4\}.
\end{aligned}
\tag{20.11}
$$

We compose the table of leading power exponents of polynomials $P_1 - P_4$, Q when $\mu = \mu_2 = 0$, assuming that they are generic:

$$
\begin{aligned}
&\mathbf{S}(P_4)/\mathbb{R}^4_+ = \{2E_1, 2E_2, E_1+E_3\}, \quad \mathbf{S}(P_2)/\mathbb{R}^4_+ = \{E_1, 2E_2\},\\
&\mathbf{S}(P_1)/\mathbb{R}^4_+ = \mathbf{S}(P_3)/\mathbb{R}^4_+ = \mathbf{S}(Q)/\mathbb{R}^4_+ = \{0\}.
\end{aligned}
$$

Now, for the right hand side of each equation of System (20.9), we write the leading exponents for terms that do not depend on μ and μ_2. Here we use the fact that the leading set of the product lies in the sum of the leading sets of factors. For the first equation we obtain that its leading exponents lie in the set $\{E_2,\ 3E_2,\ E_1+E_2+E_3,\ 2E_1+E_4\}$. But the second and the third points are greater then the first, hence they may be omitted. We obtain that the leading exponents of the first equation are E_2 and $2E_1+E_4$. We subtract from them E_1 in order to obtain the exponents for recording them in the form (20.5):

$$
-E_1+E_2, \quad E_1+E_4. \tag{20.12}
$$

For the second equation (20.9), from (20.11) we obtain the set containing the leading exponents of the right hand side: $\{E_3,\ 2E_1,\ E_1+2E_2,\ 2E_2,\ E_1+E_3,\ E_1+E_2+E_3,\ E_1+2E_3,\ 2E_2+E_3\}$. Here the third point is greater then the fourth, and the fifth, seventh and eighth are greater then the first. Hence we omit the third, fifth, seventh and eighth points. Subtracting E_2 from the remaining points, we obtain

$$
-E_2+E_3, \quad 2E_1-E_2, \quad E_2, \quad E_1+E_4. \tag{20.13}
$$

For the third equation (20.9), from (20.11) we obtain the set containing the leading exponents of the right hand side: $\{E_4,\ E_1+E_2,\ 3E_2,\ E_1+E_4,\ E_2+E_3,\ E_1+E_3+E_4,\ 2E_2+E_4,\ E_2+2E_3,\ 3E_2,\ E_1+E_2+E_3,\ 2E_1+E_4\}$. Here the fourth, sixth, seventh and eleventh points are greater then the first, and the eighth and tenth are greater then the fifth. Besides, the ninth point coincides with the third. Subtracting E_3 from the remaining points, we obtain

$$-E_3 + E_4, \quad -E_3 + E_1 + E_2, \quad -E_3 + 3E_2, \quad E_2. \tag{20.14}$$

After an analogous procedure for the fourth equation (20.9), we obtain points

$$\begin{aligned} &-E_4 + 2E_1, \quad -E_4 + 2E_2, \quad -E_4 + E_1 + E_3, \quad E_2, \\ &-E_4 + 2E_3, \quad E_1, \quad E_1 + E_4. \end{aligned} \tag{20.15}$$

From the set of points (20.12)–(20.15) we exclude the points E_2 and $E_1 + E_4$, since they are greater then $-E_1+E_2$ and $-E_3+E_4$ respectively. There remain three points (3.7), seven points

$$\begin{aligned} &2E_1 - E_2, \quad -E_3 + E_1 + E_2, \quad -E_3 + 3E_2, \quad -E_4 + 2E_1, \\ &-E_4 + 2E_2, \quad -E_4 + E_1 + E_3, \quad -E_4 + 2E_3 \end{aligned} \tag{20.16}$$

and four points (3.9) corresponding to the linear in Y terms depending on μ, μ_2. For this set $\tilde{\mathbf{S}}' = \tilde{\mathbf{S}}/\mathbb{R}^6_+$ of 14 points, and for the cone of the problem $\mathbf{K} = \{P \leq 0\}$, the table of correspondence was computed with the help of the program of Section 7 in Chapter 1.

The results of computation are shown in Table 5.2 organized in the same way as Table 5.1. The set $\tilde{\mathbf{S}}'$ contains only 14 points Q_j from the 17 points isolated in Section 3. In Table 5.2 the points $Q_j \in \tilde{\mathbf{S}}'$ keep the same numbers j as in Table 5.1; here the points Q_j with numbers $j = 20, 21, 48$ are absent. Besides, Table 5.2 has only 31 lines instead of 32 that are in Table 5.1. Here in both Tables, the vectors N_i and D_i placed in lines with numbers i from 1 to 27 coincide. In the remaining lines there are only vectors D_i, which are different in different Tables. Therefore in both cases the essential truncations are determined by five lines 8–12. To these five lines there correspond five truncations of the Belitskii normal form, which differ from truncations of Section 3 only in more specific coefficients, i.e. between the coefficients of the Belitskii normal form there are some relations. Here the basic truncated system (3.10) is the same both for System (3.3) and for its Belitskii normal form (20.9). Moreover, in the leading hypersurface $\tilde{\mathbf{\Gamma}}/\mathbb{R}^6_+$ there lie the same 13 points $Q_j \in \tilde{\mathbf{S}}' \subset \mathbf{S}'$ with $j \leq 17$ as in the case of Section 3. According to

(20.9), the Belitskii normal form is (see also Formula (21) in [Iooss 1995b])

$$\begin{aligned}
\dot{y}_1 &= y_2 + \zeta_1(Y), \\
\dot{y}_2 &= \mu y_1 + y_3 + b y_1^2 + c(y_2^2 - 2y_1y_3) + \zeta_2(Y), \qquad (20.17)\\
\dot{y}_3 &= \mu y_2 + y_4 + b y_1 y_2 + c(y_2y_3 - 3y_1y_4) + \zeta_3(Y), \\
\dot{y}_4 &= \mu_2 y_1 + \mu y_3 + a y_1^2 + b y_1 y_3 + c(2y_3^2 - 3y_2y_4) + d(y_2^2 - 2y_1y_3) + \zeta_4(Y),
\end{aligned}$$

where series $\zeta_i(Y)$, $i = 1, 2, 3, 4$ contain terms of the order higher then two. From System (20.17), with the help of Table 5.2, one can obtain all its truncations.

In [Soleev and Aranson 1995] the coincidence of basic truncated systems for System (3.3) and its Belitskii normal form was noted. Theorem 20.2 and its Corollary 20.1 explain this phenomenon.

We note that for System (4.2), near the stationary point $Y = 0$ the Belitskii normal form differs from the resonant normal form of Section 6 only in Cases $\gamma_0 - \gamma_3$.

21. The problem of surface waves

The two-dimensional problem of the surface waves on the ideal fluid with a finite depth, and with the gravitation and the surface tension taken into account was studied in papers [Kirshgässner 1988; Iooss and Kirshgässner 1990, 1992; Amick and McLeod 1991; Iooss and Pérouème 1993; Iooss 1995a, b]. In these papers, the Euler system of equations for the ideal fluid was written in the form of an evolution equation, i.e. as an infinitely dimensional system of ordinary differential equations. The role of the time variable t was played by the spatial coordinate directed along the unperturbed constant surface of the fluid. This problem is reversible, and it has two parameters: the Bond number b and the Froude number $\tilde{\lambda}$. The spectrum of the linear part of the evolution equation is discrete. It is symmetric with respect to the both axis (the real and the imaginary ones), and it depends on both parameters.

The points of the spectrum are on the imaginary axes, and Jordan blocks exist simultaneously only in four cases:

1) $\tilde{\lambda} = 1$, $b > 1/3$; two-fold zero;

2) $\tilde{\lambda} = 1$, $b < 1/3$; two-fold zero and one pair of imaginary points;

3) the point $(b, \tilde{\lambda})$ lies in a curve Γ going from the point $b = 1/3$, $\tilde{\lambda} = 1$; two two-fold imaginary points;

4) $\tilde{\lambda} = 1$, $b = 1/3$; four-fold zero eigenvalue.

In each case the reduction on the center manifold was made, and the system on the center manifold was reduced to the Belitskii normal form. In Cases 1)–3), this form is integrable [Elphick et al. 1987; Iooss and

Kirshgässner 1992; Iooss and Adelmeyer 1992; Iooss 1995a, b]. Therefore they succeeded in finding its periodic, quasi-periodic, and homoclinic solutions. For some of them it was possible to prove their preservation under the reversible perturbations, i.e. on the center manifold. In Case 4) the system on the center manifold has the form (1.1), and its Belitskii normal form is not integrable. Comparing the Belitskii normal forms of Cases 1)–3) computed near Case 4) with the Belitskii normal form in Case 4) Iooss [1995a] had noted, that in all these cases the decisive role was played by the term ax_1^2 in the fourth equation of Case 4). Therefore he proposed to consider System (1.5) as a model system for Case 4). Besides, for Cases 1)–3) near Case 4), he proposed the transformation (4.1).

In the present Chapter, System (1.1) corresponding to the system on the center manifold of Case 4) have been studied with the help of methods expounded in books [Bruno 1979a, 1989a]: the Newton polyhedron, the resonant normal form, and the sets $\tilde{\mathcal{A}}$, $\mathcal{B}$. Here to Cases 1), 2), 3) there correspond Cases γ_0, γ_3, γ_2; and instead of the Belitskii normalization the truncation is used. That made it possible to study solutions in the complete neighborhood of the point $b = 1/3$, $\tilde{\lambda} = 1$, and not only in the parts of it placed along curves of Cases 1)–3). In particular, in the resonant subcases of Case III, the Belitskii normal form coincides with the resonant normal form, and it is not integrable. But the power transformation and the subsequent truncation give the integrable system. Thus, all the results obtained here for Cases I–IV are new; and among them the most interesting are the existence of periodic solutions and quasi-periodic solutions in Case III. It is worthy to note that in the evolution equation in variable t all solutions of the center manifold are roughly unstable. Hence, the statements about stability of periodic solutions in Theorems of Sections 9–15 have no sense for the problem of surface waves. Besides, since the role of t is played by the spatial coordinate, the stability with respect to t is not at all a stability in time.

In [Kirshgässner 1988, p. 159] the initial terms of expansions of the right hand side of System (1.1) corresponding to Case 4) of the problem of the surface waves are written. In Formula (21) from [Iooss 1995b] the initial terms of the Belitskii normal form of the system are given. Simple computations based upon formulae from [Kirshgässner 1988] give values of small parameters $\mu_1 = 15(b - 1/3) + 10/7(\tilde{\lambda} - 1) + \ldots$ and $\mu_2 = -45(\tilde{\lambda} - 1) + \ldots$ At the same time in [Iooss 1995b] in the line under Formula (21) it is written $\mu_1 = -45(\tilde{\lambda} - 1) + \ldots$ and $\mu_2 = 15(b - 1/3) + 5/2(\tilde{\lambda} - 1) + \ldots$ The correct values of μ_i are obtained as the local roots of the equation (9) from [Iooss 1995b]:

$$\mu_1 = 15(b - 1/3) + 5/2(\tilde{\lambda} - 1) + \ldots, \quad \mu_2 = -45(\tilde{\lambda} - 1) + \ldots$$

Thus, in [Iooss 1995b] the indices were incorrect, and in [Kirshgässner 1988] the initial coefficients of the original system were computed incorrectly.

22. On the supernormal form

Near the stationary point $X = 0$, we consider the formal system

$$\dot{X} = \Phi(X), \quad X \in \mathbb{C}^n, \tag{22.1}$$

where $\Phi(X)$ is a vector power series, $\Phi(0) = 0$. Let the formal invertible change of coordinates

$$X = \Xi(Y), \quad \Xi(0) = 0, \tag{22.2}$$

$$\det(\partial\Xi/\partial Y) \neq 0 \quad \text{at} \quad Y = 0 \tag{22.3}$$

transform System (22.1) into the formal system

$$\dot{Y} = \Psi(Y). \tag{22.4}$$

System (22.4) is called the *supernormal form* of System (22.1) if all coefficients of the vector series $\Psi(Y)$ are invariants of System (22.1) with respect to the invertible formal changes of the form (22.2).

Let $A = (\partial\Phi/\partial X)_0$ be the matrix of the linear part of System (22.1) and $\Lambda = (\lambda_1, \ldots, \lambda_n)$ be the vector of its eigenvalues. Let $\mathbf{N}_i = \{\mathbb{Z}_+^n - E_i\}$ and $\mathbf{N} = \mathbf{N}_1 \cup \ldots \cup \mathbf{N}_n$. If the equation

$$\langle Q, \Lambda \rangle = 0 \tag{22.5}$$

has exactly k linearly independent solutions $Q \in \mathbf{N}$, then System (22.1) is in the *k-multiple resonance.*

If $k = 0$, then the resonant normal form of System (22.1) is

$$\dot{y}_i = \lambda_i y_i, \quad i = 1, \ldots, n.$$

It is the supernormal form. On the other hand, if the multiplicity k of the resonance is large, then according to Example 1 in [Bruno 1971, § 1], the finding of the complete system of formal invariants of System (22.1) still remains an unresolved problem. However, in some cases with $k > 0$, the supernormal form can be found. Below we cite a brief review of the results in this direction.

At first, we consider the case with $k = 1$. We denote as R the solution $Q \in \mathbf{N}$, $Q \neq 0$ to Equation (22.5) with the smallest $\|Q\|$. Let $R \geq 0$. Then the resonant normal form has the form

$$\dot{y}_i = y_i \sum_{l=0}^{\infty} g_{il} Y^{lR}, \quad i = 1, \ldots, n. \tag{22.6}$$

We denote $G_l = (g_{1l}, \ldots, g_{nl})$. Evidently, $G_0 = \Lambda$, hence $\langle R, G_0 \rangle = 0$. Let $\langle R, G \rangle \not\equiv 0$. Since $\langle R, G \rangle \equiv \sum \langle R, G_l \rangle Y^{lR}$, then for a natural m we have

$$\langle R, G_j \rangle = 0 \text{ for } j < m, \quad \langle R, G_m \rangle \neq 0. \tag{22.7}$$

Theorem 22.1 [Bruno 1973c; 1979a, Chapter III, Subsection 2.3] *If the property* (22.7) *of the formal system* (22.6) *is satisfied, then there exists the formal reversible transformation* $Y \to Z$ *reducing System* (22.6) *to the form*

$$\dot{z}_i = z_i \left(\sum_{l=0}^{m} \tilde{g}_{il} Z^{lR} + a r_i Z^{2mR} \right), \quad i = 1, \dots, n. \tag{22.8}$$

Here $R = (r_1, \dots, r_n)$, all coefficients of System (22.8) are invariants to within the similarity transformation $z_i = c_i \tilde{z}_i$, i.e. System (22.8) is the supernormal form.

Example 22.1. Let $n = 2$, the system

$$\dot{x}_1 = x_1^3, \quad \dot{x}_2 = x_2(1 + x_1 + x_1^2) \tag{22.9}$$

have the form (22.8). Here $\Lambda = (0,1)$, $R = (1,0)$, $G_0 = (0,1)$, $G_1 = (0,1)$, $G_2 = (1,1)$. Consequently, $m = 2$, $a = 0$ in Formula (22.8), and System (22.9) is the supernormal form. □

Another approach to the problem of the supernormal form was suggested by Belitskii [1979a, b]. Let us write System (22.4) in the form

$$\dot{Y} = \sum_{k=0}^{\infty} \Psi_k(Y), \tag{22.10}$$

where Ψ_k are homogeneous polynomials of the degree $k+1$. Let $\tilde{\Psi}_l = \sum_{k=0}^{l} \Psi_k$. We compose the operator

$$\tilde{\tilde{\Psi}}_l(H) \stackrel{\text{def}}{=} \frac{\partial \tilde{\Psi}_l}{\partial Y} H - \frac{\partial H}{\partial Y} \tilde{\Psi}_l, \tag{22.11}$$

and we denote as $\tilde{\tilde{\Psi}}_l^*$ the operator adjoint to the operator (22.11). Belitskii [1979a, b] stated that with the help of the invertible formal change (22.2) any System (22.1) could be reduced to such a System (22.4), which in the writing (22.10) has the property

$$\tilde{\tilde{\Psi}}_{k-1}^*(\Psi_k) = \tilde{\tilde{\Psi}}_{k-1}^*(\tilde{V}_{k-1}), \quad k = 1, 2, \dots \tag{22.12}$$

for some $\tilde{V}_j = \sum_{l=0}^{j} V_{jl}$, where V_{jl} are homogeneous polynomials of the degree $l+1$. The uniqueness of such a normal form was also proved there, i.e. the mentioned System (22.4) is the supernormal form. But the statements and proofs are very difficult to understand. Besides, even in simple cases the Belitskii supernormal form may be unjustifiably complicated.

Example 22.2 (the continuation of Example 22.1). For System (22.9), which is the supernormal form itself in the Bruno's sense, the Belitskii supernormal form is

$$\dot{y}_1 = y_1^3 - y_1^4 + y_1^5, \quad \dot{y}_2 = y_2(1 + y_1). \ \square$$

In the recent cycle of studies [Gaeta 1997, 1999; Cicogna and Gaeta 1999] Gaeta has stated that System (22.1) could always be reduced to such a System (22.4), which in the writing (2.10) has the property

$$\tilde{\tilde{\Psi}}^{*}_{k-1}(\Psi_k) = 0, \quad k = 1, 2, \ldots$$

This property is more restrictive than the property (22.12). The statement of Gaeta contradicts the uniqueness of the Belitskii supernormal form. It is easy to show that the statement of Gaeta is wrong.

Example 22.3 (the continuation of Examples 22.1 and 22.2). For System (22.9) the renormal form of Gaeta is

$$\dot{y}_1 = a_2 y_1^3 + \alpha y_1^5, \quad \dot{y}_2 = y_2(1 + \beta y_1), \quad a_2 \neq 0. \tag{22.13}$$

Let us find an invertible change

$$\begin{aligned} x_1 &= h_1(y_1) \stackrel{\text{def}}{=} b_1 y_1 + b_2 y_1^2 + \ldots, \\ x_2 &= y_2 h_2(y_1) \stackrel{\text{def}}{=} y_2(c_0 + c_1 y_1 + \ldots), \quad b_1 \neq 0, \ c_0 \neq 0, \end{aligned} \tag{22.14}$$

transforming System (22.9) into System (22.13). Differentiating equality (22.14) with respect to t, expressing the derivatives with respect to t according to (22.9) and (22.13), and expressing X through Y by (22.14), we obtain the system of differential equations for h_1 and h_2

$$\begin{aligned} h_1^3 &= h_1'(a_2 y_1^3 + \alpha y_1^5), \\ h_2(h_1 + h_1^2 - \beta y_1) &= h_2'(a_2 y_1^3 + \alpha y_1^5), \end{aligned} \tag{22.15}$$

where the prime means the derivative with respect to y_1. Let us write in (22.15) the terms up to the fourth power in the first equation and up to the second power in the second one. Comparing coefficients, we obtain equalities

$$\begin{aligned} &b_1^3 = b_1 a_2, \quad 3b_1^2 b_2 = 2b_2 a_2, \\ &c_0(b_1 - \beta) = 0, \quad c_1(b_1 - \beta) + c_0(b_2 + b_1^2) = 0. \end{aligned} \tag{22.16}$$

Since $c_0 \neq 0$, from the third equality (22.16), we obtain $b_1 = \beta$; and from the fourth one, we obtain $b_2 = -b_1^2$. Since $b_1 \neq 0$, then $b_2 \neq 0$. From the first equality (22.16), we obtain $a_2 = b_1^2$; and from the second one, we obtain $3b_1^2 b_2 = 2b_2 b_1^2$. Since $b_1 \neq 0$ and $b_2 \neq 0$, then the last equality is impossible. Consequently, the change (22.14) is impossible. $\square$

Chapter 6

Singularities of systems of arbitrary equations

1. Truncated systems

Let an n-vector X be split into three parts: parameters X_1, independent variables X_2 and dependent variables X_3 with dimensions n_1, n_2 and n_3 respectively, i.e. $X = (X_1, X_2, X_3)$, $n = n_1 + n_2 + n_3$, $n_i \geq 0$. We denote as a *differential monomial* $a(X)$ the product of a constant, of powers of coordinates x_i of the vector X, and of derivatives of the form

$$\partial^{||K_2||} x_j / \partial X_2^{K_2}, \tag{1.1}$$

where $K_2 = (k_{n_1+1}, \ldots, k_{n_2}) \geq 0$, $||K_2|| = k_{n_1+1} + \ldots + k_{n_2}$, and $j > n_1 + n_2$, i.e. the coordinate x_j belongs to X_3. To each differential monomial $a(X)$ we put in correspondence its *vector power* $Q = Q(a) \in \mathbb{R}^n$ by the following rule. To a constant there corresponds $Q = 0$. To a monomial X^Q there corresponds its vector power Q. To the derivative (1.1) there corresponds the vector power exponent $Q = (Q_1, Q_2, Q_3)$, where $Q_1 = 0$, $Q_2 = -K_2$, $Q_3 = E_j$. Here E_k denotes the kth unit vector. When the monomials are multiplied, their vector powers Q are summed, i.e. $Q(a_1 a_2) = Q(a_1) + Q(a_2)$. We denote as a *differential polynomial* $f(X)$ the finite sum of differential monomials:

$$f(X) = \sum_{r=1}^{s} a_r(X). \tag{1.2}$$

The set of vector powers of its monomials $\{Q(a_r), r = 1, \ldots, s\}$ is called the *support* of the polynomial (1.2), and it is denoted as $\mathbf{S}(f)$ or $\operatorname{supp} f$. Here to different monomials a_r there may correspond one point Q, but to different points Q there correspond different monomials. In the same way as in Chapter 1, according to the set $\mathbf{S} = \mathbf{S}(f)$, we determine its polyhedron $\mathbf{\Gamma} = \mathbf{\Gamma}(f)$, its faces $\mathbf{\Gamma}_k^{(d)}$, and the boundary subsets $\mathbf{S}_k^{(d)}$ in $\mathbb{R}^n$, and also their normal cones $\mathbf{U}_k^{(d)}$ in the dual space $\mathbb{R}_*^n$. To each boundary subset $\mathbf{S}_k^{(d)}$ we put in correspondence its *truncated polynomial* (or truncation)

$$\hat{f}_k^{(d)}(X) = \sum a_r(X) \quad \text{over} \quad r\text{: } Q(a_r) \in \mathbf{S}_k^{(d)}. \tag{1.3}$$

This truncation is the first approximation to the polynomial (1.2) if $\log|X| \to \infty$ near the normal cone $\mathbf{U}_k^{(d)}$.

Now we consider the aggregate of differential polynomials

$$f_1, \ldots, f_m. \tag{1.4}$$

To each f_j there corresponds its support and all the accompanying objects $\mathbf{\Gamma}_j$, $\mathbf{\Gamma}_{jk_j}^{(d_j)}$, $\mathbf{U}_{jk_j}^{(d_j)}$, $\mathbf{S}_{jk_j}^{(d_j)}$, $\hat{f}_{jk_j}^{(d_j)}$. Besides, to each non-empty intersection

$$\mathbf{U}_{1k_1}^{(d_1)} \cap \ldots \cap \mathbf{U}_{mk_m}^{(d_m)} \subset \mathbb{R}_*^n \tag{1.5}$$

there corresponds the aggregate of truncations of the form (1.3)

$$\hat{f}_{1k_1}^{(d_1)}, \ldots, \hat{f}_{mk_m}^{(d_m)}, \tag{1.6}$$

which is the first approximation to the aggregate (1.4), when $\log|X| \to \infty$ near the intersection (1.5); and it is named the *truncation of the aggregate* (1.4).

We consider now the system of equations

$$f_j = 0, \quad j = 1 \ldots, m, \tag{1.7}$$

corresponding to the aggregate (1.4). To System (1.7) there correspond all objects indicated for the aggregate (1.4), and also the *truncated systems of equations*

$$\hat{f}_{jk_j}^{(d_j)} = 0, \quad j = 1 \ldots, m, \tag{1.8}$$

each of which corresponds to one aggregate of truncations (1.6). We say that the truncated system (1.8) is the *truncation of System* (1.7) *with respect to the order* $P \neq 0$ if the vector P lies in the cone (1.5). Every truncated system (1.8) is the first approximation to the complete system (1.7). Let us specify this statement. Let $Z_1 = (z_1, \ldots, z_{l_1})$, where $l_1 \leq n_1$, $Z_2 = X_2$ and $Z = (Z_1, Z_2)$ are l-vectors where $l_2 = n_2$ and $l = l_1 + l_2$. Let the system of equations (1.7) have a solution of the form

$$X_1 = G_1(Z_1), \quad X_2 = Z_2, \quad X_3 = G_3(Z). \tag{1.9}$$

We assume that the components of vectors G_1 and G_3 expand in power series in their arguments:

$$g_i = g_i(Z) = \sum g_{iR} Z^R \quad \text{over} \quad R \in \mathbf{G}_i \subset \mathbb{R}^l,$$
$$i = 1, \ldots, n_1, n_1 + n_2 + 1, \ldots, n, \tag{1.10}$$

where $\mathbf{G}_i$ are supports of series g_i lying in $\mathbb{R}^l$. To them there correspond their polyhedrons, their faces, boundary subsets etc. Let truncations of series g_i with respect to the vector order $T \stackrel{\text{def}}{=} (t_1, \ldots, t_l) \stackrel{\text{def}}{=} (T_1, T_2) \in \mathbb{R}^l_*$ according to Chapter 2 be

$$\hat{g}_i = \sum g_{iR} Z^R \quad \text{over} \quad R \in \mathbf{G}_{ik_i}^{(\gamma_i)}. \tag{1.11}$$

That means that substituting

$$z_i = b_i \tau^{t_i}(1 + o(1)), \quad b_i \stackrel{\text{def}}{=} \text{const} \neq 0, \quad i = 1, \ldots, l, \quad \tau \to \infty \tag{1.12}$$

in the series (1.7), we obtain

$$x_i = \tau^{p_i}(\hat{g}_i(B) + o(1)), \quad i = 1, \ldots, n,$$

where $B = (b_1, \ldots, b_l)$ and

$$\begin{aligned} &p_i = \langle T, R \rangle, \ R \in \mathbf{G}_{ik_i}^{(\gamma_i)}, \quad i = 1, \ldots, n_1, n_1 + n_2 + 1, \ldots, n, \\ &p_i = t_i, \ i = n_1 + 1, \ldots, n_1 + n_2, \end{aligned} \tag{1.13}$$

We form the vector

$$P = (p_1, \ldots, p_n) \in \mathbb{R}^n. \tag{1.14}$$

Thus, for the system of equations (1.7) and for its solution (1.9), (1.10), to each vector order T of coordinates Z there uniquely correspond:

(a) the first approximation of the solution

$$X_1 = \hat{G}_1(Z_1), \quad X_2 = Z_2, \quad X_3 = \hat{G}_3(Z) \tag{1.15}$$

determined by (1.11);

(b) its order (1.14) determined by (1.13);

(c) the truncated with respect to the order P system of equations (1.8).

Theorem 1.1 [Bruno 1994] *Let the system of equations* (1.7) *have solution* (1.9), (1.10). *Let the truncation of the solution with respect to the order* $T \in \mathbb{R}^l_*$ *be* (1.15), *and let it have the order* (1.14), (1.13). *Let System* (1.8) *be the truncation of System* (1.7) *with respect to the order* (1.14). *Then the truncation* (1.15) *of solution* (1.9) *is a solution to the truncated system* (1.8).

Proof. We make it under the additional assumption that coordinates X_1 and X_3 are only in integer powers in the polynomials (1.4). Then we may assume that their powers are integer and non-negative, since we can attain this multiplying polynomials (1.4) by corresponding powers of X_1 and of X_3.

According to (1.10)–(1.13), the support of the series $g_i(Z)$ lies in the half-space $\langle T, R\rangle \le p_i$ of the space $\mathbb{R}^l = \{R\}$, i.e.

$$\mathbf{S}(g_i) \subset \{R\colon \langle T, R\rangle \le p_i\},$$

besides in the boundary hyperplane $\langle T, R\rangle = p_i$ there lies the support $\mathbf{S}(\hat{g}_i)$. For an integer non-negative q the power g_i^q is the power series in Z, and its support is also placed in the half-space $\mathbb{R}^l$:

$$\mathbf{S}(g_i^q) \subset \{R\colon \langle T, R\rangle \le p_i q\},$$

besides in the boundary hyperplane $\langle T, R\rangle = p_i q$ there lies the support $\mathbf{S}(\hat{g}_i^q)$. Consequently, for Q: $Q_1 \in \mathbb{Z}_+^{n_1}$, $Q_3 \in \mathbb{Z}_+^{n_3}$, we have

$$\mathbf{S}(G^Q) \subset \{R\colon \langle T, R\rangle \le \langle P, Q\rangle\},$$

besides in the boundary hyperplane $\langle T, R\rangle = \langle P, Q\rangle$ there lies the support $\mathbf{S}(\hat{G}^Q)$. Since $Z_2 = X_2$, then for the derivative of the form (1.1) we have

$$\mathbf{S}\left(\frac{\partial^{\|K_2\|} g_j}{\partial X_2^{K_2}}\right) \subset \{R\colon \langle T, R\rangle \le p_j - \langle T_2, K_2\rangle = p_j - \langle P_2, K_2\rangle\},$$

since $T_2 = P_2$ according to (1.12). Besides in the boundary hyperplane $\langle T, R\rangle = p_j - \langle P_2, K_2\rangle$ there lies the support $\mathbf{S}\left(\partial^{\|K_2\|}\hat{g}_j/\partial X_2^{K_2}\right)$. Let a differential monomial $a(X)$ have the vector power $Q = Q(a)$. Under the substitution (1.9) it becomes a function of Z. Since it is the product of the series considered above, it is also a power series, and its support is

$$\mathbf{S}(a(G)) \subset \{R\colon \langle T, R\rangle \le \langle P, Q\rangle\},$$

besides in the boundary hyperplane there lies the support $\mathbf{S}(a(\hat{G}))$. Hence for the differential polynomial (1.2), on curves of the form (1.12)

$$f(G(Z)) = \hat{f}(\hat{G}(Z))\tau^{\langle P,Q\rangle} + \tau^{\langle P,Q\rangle} o(\tau^{-\varepsilon}),$$

where $\varepsilon > 0$. By the statement of Theorem $f(G(Z)) = 0$, hence $\hat{f}(\hat{G}(Z)) = 0$. And it is valid for every differential polynomial from the aggregate (1.4). □

Theorem 1.1 means that for some class of solutions, the first approximation to a solution to the system of equations is the solution to the corresponding first approximation to the system of equations. Apparently, this property should be considered as characteristic with the definition of the first approximation. The reverse statement, generally speaking, is not correct, i.e. not every solution to the first approximation (the truncation) of a system of equations is the first approximation to some solution to the system.

Remark 1.1. In the extraction of truncations only the local and the asymptotic coordinates x_i are essential, which tend either to zero or to infinity. If differential polynomials also depend on the neutral coordinates Y in such a way that the coefficients in monomials are functions of Y, which remain bounded away from 0 and ∞, then the described above procedure of extraction of truncations is applicable in this case as well. Here the coefficients g_{iR} and h_{iS} in series (1.10) are also bounded functions of Y. The neutral coordinates Y are also split into parameters, independent variables and dependent variables. In the derivatives only the contribution of asymptotic variables is taken into account, i.e. to the derivatives

$$\frac{\partial^{||K_2||+||L||} x_j}{\partial X_2^{K_2}\, \partial Y^L} \quad \text{and} \quad \frac{\partial^{||K_2||+||L||} y_j}{\partial X_2^{K_2}\, \partial Y^L}$$

there correspond points $Q = (0, -K_2, R)$ with $R = E_j$ and $R = 0$ respectively.

Remark 1.2. If the coordinate x_i occurs in System (1.7) only as the derivative dx_i (or ∂x_i), then the asymptotic property (tending to zero) is assumed for Δx_i, i.e. for the displacement along the coordinate x_i, and not for the coordinate itself.

Example 1.1 [Bruno 1995b, c]. Let us find asymptotics of solutions to the equation

$$y'' \stackrel{\text{def}}{=} d^2y/dx^2 = 3y^2 + x$$

when $x \to 0$ or $x \to \infty$. Here $n_1 = 0$, $n_2 = n_3 = 1$, $n = 2$, $X_2 = x$, $X_3 = y$. The set $\mathbf{S} = \{Q_1 = (-2, 1), Q_2 = (0, 2), Q_3 = (1, 0)\}$. The polyhedron $\mathbf{\Gamma}$ is the triangle with the vertices $\mathbf{\Gamma}_k^{(0)} = Q_k$, $k = 1, 2, 3$ (see Fig. 6.1). To the edge $\mathbf{\Gamma}_1^{(1)}$ there corresponds the truncated equation $y'' = 3y^2$. Its solutions with $P \in \mathbf{U}_1^{(1)}$ have the form $y = bx^{-2}$, i.e. $6bx^{-4} = 3b^2x^{-4}$, whence $b = 2$. To the edge $\mathbf{\Gamma}_2^{(1)}$ there corresponds the truncated equation $3y^2 + x = 0$, whence $y = \pm\sqrt{-x/3}$. To the edge $\mathbf{\Gamma}_3^{(1)}$ there corresponds the truncated equation $y'' = x$. Its solutions with $P \in \mathbf{U}_3^{(1)}$ have the form $y = bx^3$, i.e. $6bx = x$. Consequently, $b = 1/6$. To the vertex $\mathbf{\Gamma}_1^{(0)}$ there corresponds the truncated equation $y'' = 0$. Its solutions $y = c_0 + c_1 x$ when $\tau^{-1} = x \to 0$ have $p_1 = -1$ and

$$p_2 = \begin{cases} 0, & \text{if} \quad c_0 \neq 0, \\ -1, & \text{if} \quad c_0 = 0. \end{cases}$$

Both vectors $P = (p_1, p_2)$ lie in the cone $\mathbf{U}_1^{(0)}$. To the vertex $\mathbf{\Gamma}_2^{(0)}$ there corresponds the truncated equation $y^2 = 0$, the solution $y = 0$ of which cannot be represented in the form (1.11). The same is true for the vertex $\mathbf{\Gamma}_3^{(0)}$. □

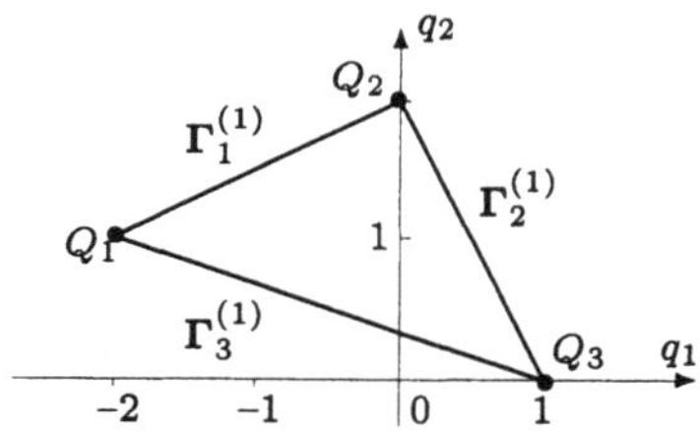

Figure 6.1: The support and the triangle of Example 1.1.

Example 1.2. Let us find asymptotics of solutions to the Emden–Fowler equation (compare with Examples 2.1, 5.1, 6.4 and 7.1 of Chapter 3)

$$d^2y/dx^2 - ax^\sigma y^\mu = 0, \quad a = \pm 1, \tag{1.15'}$$

when $x \to \infty$. Here $n_1 = 0$, $n_2 = n_3 = 1$, $X_2 = x$, $X_3 = y$. The support of the equation consists of two points $Q = (q_1, q_2)$: $Q_1 = (-2, 1)$ and $Q_2 = (\sigma, \mu)$ if $Q_1 \neq Q_2$ (i.e. $\mu \neq 1$ and $\sigma \neq -2$), which we assume to be fulfilled. Then $\mathbf{\Gamma}$ is the segment $[Q_1, Q_2]$. Since the normal vector to the segment $\mathbf{\Gamma}$ is $N = (1 - \mu, \sigma + 2)$, then for $\mu \neq 1$ the equation has a solution of the form $y = bx^\alpha$, where $\alpha = (\sigma + 2)/(1 - \mu)$, and the coefficient b is found from the equation $\alpha(\alpha - 1)b = ab^\mu$. Here we have to exclude the values $\alpha = 0, 1$, when the equation has only the trivial solution $b = 0$.

We consider now truncated equations corresponding to ends of the segment $\mathbf{\Gamma}$. To the point Q_2 there corresponds the truncated equation $x^\sigma y^\mu = 0$, the solutions to which are trivial. To the point Q_1 there corresponds the truncated equation $d^2y/dx^2 = 0$. Its solutions have the form

$$y_1 = \text{const} \cdot x^\nu, \quad \nu = 0, 1; \tag{1.16}$$

their order is $P = (1, \nu)$. The equation $d^2y/dx^2 = 0$ is the truncation of the original equation with respect to the order P if $\langle P, Q_1 \rangle > \langle P, Q_2 \rangle$, i.e. $\nu(\mu - 1) + \sigma + 2 < 0$. With such values of parameters, we obtain the first approximations y_1 of the form (1.16) for solutions y. The second approximation y_2 is obtained by the integration from the equation $d^2y_2/dx^2 = ax^\sigma y_1^\mu$. These results were obtained in Chapter 3 by transition from the original equation to the system of three equations resolved with respect to derivatives, and by the consideration of the support of that system in the three-dimensional space (compare Formulae (5.14′)–(5.16) of Chapter 3). □

Example 1.3. In the singularly perturbed system

$$\varepsilon dx/dt - \varphi_1(x,y) = 0, \quad dy/dt - \varphi_2(x,y) = 0 \tag{1.17}$$

with the small parameter ε the differential dt may also be a small value. According to Remarks 1.1 and 1.2, here $n_1 = n_2 = 1$, $n_3 = 0$, $n = 2$, $X_1 = \varepsilon$, $X_2 = t$; coordinates x, y are neutral. The support $\mathbf{S}_1$ of the first equation (1.17) consists of two points $Q_1 = (1,-1)$ and $Q_2 = (0,0)$. The support $\mathbf{S}_2$ of the second equation (1.17) consists of two points $Q_3 = (0,-1)$ and Q_2. For $P = (-1,-1)$, we have $\langle P, Q_1\rangle = 0$, $\langle P, Q_2\rangle = 0$, $\langle P, Q_3\rangle = 1$. Hence the corresponding truncation of the first equation coincides with the equation itself, and the truncation of the second equation is $dy/dt = 0$, i.e. we obtain the truncated system

$$\varepsilon dx/dt = \varphi_1(x,y), \quad dy/dt = 0. \tag{1.18}$$

For $P = (-1,0)$, we have $\langle P, Q_1\rangle = -1$, $\langle P, Q_2\rangle = 0 = \langle P, Q_3\rangle$. The corresponding truncation of the first equation is $\varphi_1 = 0$, and the truncation of the second equation is the equation itself, i.e. we have the truncated system

$$\varphi_1(x,y) = 0, \quad dy/dt = \varphi_2(x,y). \tag{1.19}$$

The solutions to System (1.18) are horizontal segments $y = \text{const}$, along which x moves with great velocity $\varphi_1(x,y)/\varepsilon$, and stationary points (x^0, y^0) satisfying the equation $\varphi_1(x^0,y^0) = 0$. Along these points, solutions to System (1.19) move with moderate velocity. When $\varepsilon \to 0$, solutions to the complete system tend to the limit positions, which are composed from parts of solutions to truncated systems (1.18) and (1.19). In particular, such is the limit cycle, which describes the relaxational oscillations [Mishchenko and Rozov 1975, Mishchenko et al. 1994]. □

Example 1.4. Let the mass center of a satellite move along an elliptic orbit of eccentricity e. In the plane of the orbit the motion of the satellite around its mass center is described by the Beletskii [1959] equation

$$(1 + e\cos\nu)\frac{d^2\delta}{d\nu^2} - 2e\sin\nu\frac{d\delta}{d\nu} + \mu\sin\delta = 4e\sin\nu, \tag{1.20}$$

where μ is the inertia parameter, ν is the independent variable (the true anomaly on the orbit). Equation (1.20) is singular when $e = 1$, $\nu = \pi$, since in the point the coefficient at the highest derivative turns into zero. We introduce new (local) coordinates $x = \nu - \pi$, $\varepsilon = 1 - e$. Then Equation (1.20) takes the form

$$\left[\varepsilon + \frac{1}{2}x^2 + o(\varepsilon, x^2)\right]\frac{d^2\delta}{dx^2} + 2[x + o(\varepsilon,x)]\frac{d\delta}{dx} + \mu\sin\delta = -4[x + o(\varepsilon,x)]. \tag{1.21}$$

Here $n_1 = n_2 = 1$, $n_3 = 0$, $n = 2$, $X_1 = \varepsilon$, $X_2 = x$. Coordinates μ and δ are neutral. The corresponding set $\mathbf{S}$ is shown in Fig. 6.2, *a* in coordinates $q_1 = \operatorname{ord}\varepsilon$, $q_2 = \operatorname{ord} x$. The boundary $\partial\mathbf{\Gamma}$ of the set $\mathbf{\Gamma}$ consists of three edges: $\mathbf{\Gamma}_1^{(1)}$, $\mathbf{\Gamma}_2^{(1)}$, $\mathbf{\Gamma}_3^{(1)}$, and two vertices. Each element of the boundary gives its own first approximation to Equation (1.21). On the edge $\mathbf{\Gamma}_1^{(1)}$ the unit vector is $(0, 1)$, which corresponds to the variable x. On the edge $\mathbf{\Gamma}_2^{(1)}$ the unit vector is $(-1/2, 1)$, which corresponds to the variable $x/\sqrt{\varepsilon}$. Using the variable with such type of behavior (the eccentric anomaly for example), we can regularize Equation (1.20) and compute its solutions as the relaxational oscillations [Bruno and Petrovich 1994a, b, 1997; Varin 1996, 1997, 1999a, b]. In [Bruno and Varin 1995, 1996, 1997a, b, 1999], the limit equations were studied corresponding to the vertex $\mathbf{\Gamma}_1^{(0)}$ and edges $\mathbf{\Gamma}_1^{(1)}$, $\mathbf{\Gamma}_2^{(1)}$. From solutions to these equations the limits of solutions to Equation (1.20) can be matched when $e \to 1$. The theory of singular perturbations for solutions to Equation (1.20) with small μ is constructed in [Sadov 1995, 1996a, b, 1997a, b, 1998, 1999]. It also uses Power Geometry. □

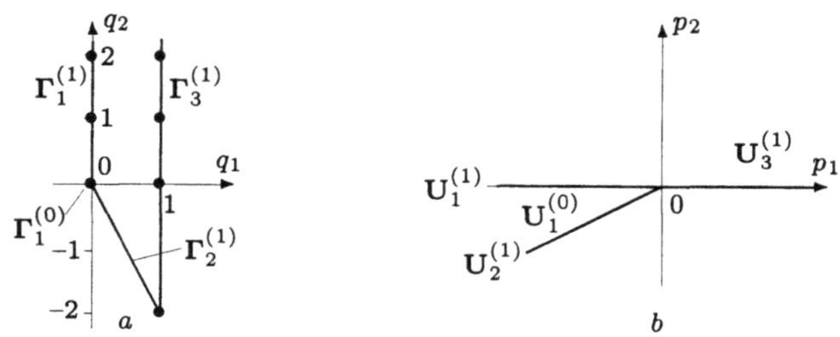

Figure 6.2: The polyhedron $\mathbf{\Gamma}$ (*a*) and the normal cones (*b*) of Example 1.4.

Example 1.5. In Chapter 3 the system of ordinary differential equations of the form

$$\frac{dy_i}{y_i\, dt} - f_i(Y) = 0, \quad i = 1, \ldots, m, \quad Y \in \mathbb{R}^m \text{ or } Y \in \mathbb{C}^m \tag{1.22}$$

was considered. Here $n_1 = 0$, $n_2 = 1$, $n_3 = m$, $n = m+1$, $X_2 = t$, $X_3 = Y$. Let $F = (f_1, \ldots, f_m) = \sum F_{Q_3} Y^{Q_3}$ be the vector polynomial. Let $\hat{F} = (\check{f}_1, \ldots, \check{f}_m)$ be its truncation with respect to the order $P_3 \in \mathbb{R}^m_*$. Then $\hat{F} \not\equiv 0$ as vector, but some of its components $\check{f}_i$ may be identical zeros. If the unique support $\mathbf{S}(F) \in \mathbb{R}^m$ is put in correspondence to System (1.22), as it was done in Chapter 3, then according to the definition of this Section, the support $\mathbf{S}_1$ of System (1.22) in $\mathbb{R}^{m+1}$ consists of the point $Q_0 = (-1, 0, \ldots, 0)$ and of the support $\mathbf{S}(F)$ placed in the coordinate hyperplane $q_1 = 0$. Hence System

(1.22) has three truncated systems corresponding to the one vector order $P_3 \in \mathbb{R}_*^m$:

$$
\begin{aligned}
&-\check{f}_i(Y) = 0, \quad i = 1, \dots, m; \quad \text{if} \quad -p_0 < \langle P_3, Q_3 \rangle; \\
&\frac{dy_i}{y_i\, dt} - \check{f}_i(Y) = 0, \quad i = 1, \dots, m; \quad \text{if} \quad -p_0 = \langle P_3, Q_3 \rangle; \\
&\frac{dy_i}{y_i\, dt} = 0, \quad i = 1, \dots, m; \quad \text{if} \quad -p_0 > \langle P_3, Q_3 \rangle.
\end{aligned}
\tag{1.23}
$$

Here the vector $P = (p_0, P_3) \in \mathbb{R}_*^m$ is the normal vector to the hyperplane $\mathbf{H}_P$, which is the supporting one to the mentioned support $\mathbf{S}_1 \in \mathbb{R}^{m+1}$; and Q_3 is an arbitrary point in $\mathbf{S}(\hat{F})$. Among the truncated systems (1.22), the first one corresponds to the case when the point Q_0 does not lie in the hyperplane $\mathbf{H}_P$; for the second one the point $Q_0 \in \mathbf{H}_P$ and $\mathbf{S}(\hat{F}) \in \mathbf{H}_P$; for the third one the point $Q_0 \in \mathbf{H}_P$, but $\mathbf{S}(\hat{F})$ does not lie in $\mathbf{H}_P$. If $\check{f}_i \not\equiv 0$ for all $f_i \not\equiv 0$, then by definition of the present Section, only these three systems are the meaningful truncated systems with respect to the vector order P_3. But if $\check{f}_i \equiv 0$ for some $f_i \not\equiv 0$, then by definition of the present Section, the truncated systems for $-p_1 < \langle P_3, Q_3 \rangle$ are also the systems of the form

$$
\begin{aligned}
&-\hat{f}_i(Y) = 0, \quad \text{over} \quad i: \ c_i > -p_0; \\
&\frac{dy_j}{y_j\, dt} - \hat{f}_j(Y) = 0, \quad \text{over} \quad j: \ c_j = -p_0; \\
&\frac{dy_k}{y_k\, dt} = 0, \quad \text{over} \quad k: \ c_k < -p_0,
\end{aligned}
$$

where

$$c_l = \langle P_3, Q'_{3l} \rangle \quad \text{for} \quad Q'_{3l} \in \mathbf{S}(\hat{f}_l), \quad l = 1, \dots, m$$

and $\hat{f}_i$ is the truncation of the scalar polynomial f_i with respect to the order P_3, which is always different from the identical zero if $f_i \not\equiv 0$. Hence, these truncated systems are more meaningful than the first System (1.23). □

Example 1.6. Let us consider the differential polynomial of the form $f(D)u$, where $f(D)$ is the polynomial in the differential operator $D = (\partial/\partial x_1, \dots, \partial/\partial x_m)$. Here $n_1 = 0$, $n_2 = m$, $n_3 = 1$, $n = m + 1$, $X_2 = X$, $X_3 = u$. The support $\mathbf{S}_1 \subset \mathbb{R}^{1+m}$ of the polynomial $f(D)u$ consists of points of the form $Q = (Q_2, 1)$, where the point $-Q_2 \in \mathbb{R}^m$ belongs to the support $\mathbf{S}(f) \subset \mathbb{R}^m$ of the polynomial $f(\Xi)$. Since all points of the support $\mathbf{S}_1$ have the same last coordinate, so it does not participate in the extraction of truncations. Hence truncations of the polynomial $f(D)u$ with respect to the order $P = (-T, p_{m+1})$ have the form $\hat{f}_j^{(d)}(D)u$, where $\hat{f}_j^{(d)}(\Xi)$ is the truncation of the polynomial $f(\Xi)$ with respect to the order T. In [Mikhailov 1963, 1965, 1967a, b; Volevich and Gindikin 1968, 1985; Gindikin 1973, 1974; Volevich 1974; Gindikin and Volevich 1992], the first approximations to the differential polynomial $f(D)u$ were determined exactly in the same manner. □

Example 1.7. Near the point $t = x = 0$, we consider the equation

$$\varepsilon^2 \frac{\partial u}{\partial t} - \varepsilon^2 \frac{\partial^2 u}{\partial x^2} - \frac{\partial^2 u}{\partial y^2} = \varepsilon f(t, x, u) \tag{1.24}$$

with the small parameter ε and regular function f. Here $n_1 = 1$, $n_2 = 2$, $n_3 = 0$, $n = 3$, $X_1 = \varepsilon$, $X_2 = (t, x)$; coordinates y and u are neutral. Correspondingly the support $\mathbf{S}$ of the equation consists of the points $Q_1 = (2, -1, 0)$, $Q_2 = (2, 0, -2)$, $Q_3 = 0$, $Q_4 = (1, 0, 0)$, and points of the form (q_1, q_2, q_3), where $q_1 \geq 1$, $q_2, q_3 \geq 0$ representing the support of the expansion for εf. The polyhedron $\mathbf{\Gamma}$ includes the face $\mathbf{\Gamma}_1^{(2)}$ spanned on vertices Q_1, Q_2, Q_3. For $P < 0$ all truncations of Equation (1.23) correspond to faces $\mathbf{\Gamma}_j^{(d)}$ lying in that face $\mathbf{\Gamma}_1^{(2)}$. The directing vectors of its edges $[Q_1, Q_3]$ and $[Q_2, Q_3]$ are $(-2, 1, 0)$ and $(-1, 0, 1)$ respectively. They form the basis in the face $\mathbf{\Gamma}_1^{(2)}$. To them there correspond "variables of the boundary layer" $\varepsilon^{-2}t$ and $\varepsilon^{-1}x$. In Section 15 in [Vasil'eva and Butusov 1990] it was proved that near the point $t = x = 0$, the solution u to Equation (1.24) is the regular function in these variables and in variables ε, t, x, y.

2. Power transformations

Let an n-vector X be split into three parts: parameters X_1, independent variables X_2 and dependent variables X_3, with dimensions n_1, n_2 and n_3 respectively: $X = (X_1, X_2, X_3)$, $n = n_1 + n_2 + n_3$, $n_i \geq 0$. We consider the *power transformation*

$$\begin{aligned} \log X_1 &= W_{11} \log Y_1, \\ \log X_2 &= W_{21} \log Y_1 + W_{22} \log Y_2, \\ \log X_3 &= W_{31} \log Y_1 + W_{32} \log Y_2 + W_{33} \log Y_3. \end{aligned} \tag{2.1}$$

Here $\log X_1 = (\log x_1, \ldots, \log x_{n_1})^*$, W_{ii} are non-degenerate square matrices, W_{ij} are rectangular real matrices, $W = (W_{ij})$ is the square block matrix, and the asterisk means the transposition.

Let the change of coordinates (2.1) transform the differential polynomial $f(X)$ in $g(Y) = f(X)$. We want to study the relation of their supports $\mathbf{S}(f) = \{Q_j\}$ and $\mathbf{S}(g) = \{S_j\}$.

Theorem 2.1 [Bruno 1996b] *Under the power transformation* (2.1), *the differential polynomial* $f(X)$ *transforms into the differential polynomial* $g(Y)$, *their supports* $\mathbf{S}(g) = \{S_j\}$ *and* $\mathbf{S}(f) = \{Q_j\}$ *are related by the linear transformation*

$$S = W^* Q, \tag{2.2}$$

and vectors of the dual space $\mathbb{R}^n_*$ *are transformed as*

$$R = W^{-1} P. \tag{2.3}$$

Proof. We write the power transformation (2.1) in the form $\log X = W \log Y$. The reverse transformation is

$$\log Y = A \log X, \tag{2.4}$$

where $A = (a_{ij}) = W^{-1}$. For the monomial $X^Q = Y^S$ the equality (2.2) was proved in Chapter 3. Let us prove it for the derivative (1.1).

From (2.4) it follows that $\partial \log y_i / \partial \log x_j = a_{ij}$, i.e.

$$\partial y_i / \partial x_j = a_{ij} y_i / x_j. \tag{2.5}$$

1^0. Let $b(Y)$ be a differential monomial, $\mathbf{S}(b) = S$, $n_1 < i \le n_1 + n_2$. Then $\partial b / \partial y_i$ is a differential polynomial with the support consisting of one point $S - E_i$. Indeed, it is true for $b = Y^S$, for $b = \partial^{||K_2||} y_j / \partial Y_2^{K_2}$, and for the product of such monomials, i.e. for every monomial.

2^0. Now we prove that the derivative (1.1) is the differential polynomial in Y, i.e.

$$h(Y) = \sum b_r(Y), \tag{2.6}$$

where b_r are differential monomials, and its support consists of one point

$$\mathbf{S}(h) = S \stackrel{\text{def}}{=} W^*(E_j - K_2). \tag{2.7}$$

We make the proof by the induction on the increasing $||K_2||$. When $||K_2|| = 0$ the derivative (1.1) is $x_j = Y^{W^* E_j}$, i.e. Formula (2.6) is valid. Let Formulae (2.6) and (2.7) be valid for all K_2 with $||K_2|| \le k$. We prove it for $||K_2|| = k + 1$. Obviously, it is sufficient to prove this for one monomial b_r. So, let $\mathbf{S}(b_r) = S$, $||S_2|| = k$ and $n_1 < j \le n_2$. According to (2.5)

$$\frac{\partial b_r}{\partial x_j} = \sum_{i=n_1+1}^{n_2} \frac{\partial b_r}{\partial y_i} \frac{\partial y_i}{\partial x_j} = \sum_{i=n_1+1}^{n_2} \frac{\partial b_r}{\partial y_i} a_{ij} \frac{y_i}{x_j} =$$

$$= \frac{1}{x_j} \sum_{i=n_1+1}^{n_2} \frac{\partial b_r}{\partial y_i} a_{ij} y_i \stackrel{\text{def}}{=} \frac{1}{x_j} \beta(Y). \tag{2.8}$$

According to Statement 1^0, $\beta(Y)$ is a differential polynomial, the support of which consists of one point S. Consequently, the derivative (2.8) is also the differential polynomial in Y, and its support is $S - W^* E_j$. That proves Statement 2^0 and Formulae (2.6), (2.7).

Thus, each differential monomial $a(X)$ of the differential polynomial $f(X)$ is the product of some elementary monomials of the form X^Q and (1.1), for which the statement (2.2) of Theorem is proved. Since when the monomials are multiplied, their supports are summed, hence it follows the statement (2.2) of Theorem for every monomial and for every differential polynomial. Since (2.2) is the transformation in the main space $\mathbb{R}^n$, then the transformation in the dual space $\mathbb{R}^n_*$ is (2.3), because the scalar product $\langle P, Q\rangle = \langle R, S\rangle$ must be preserved. □

According to Theorem 2.1, all sets in $\mathbb{R}^n$ and $\mathbb{R}^n_*$ are also changed by the corresponding linear transformations (2.2) and (2.3). Consequently, the extraction of a truncation commutes with any power transformation. Besides, the power transformation can be used for the simplification of the truncated equation making its support parallel to the coordinate subspace.

Theorem 2.2 [Bruno 1996b] *Let $d = \dim \mathbf{\Gamma}(f) < n$. Then there exists such a matrix W that after the transformation* (2.1) *the values of $n - d$ coordinates s_j are constant for all $S = (s_1, \dots, s_n) \in \mathbf{S}(g)$, where $g(Y) = f(X)$. Let $s_j = \text{const}$. In $y_j^{-s_j} g(Y)$, the coordinate y_j is absent if $j \le n_1$, and it is present only in the form $\partial \log y_j$ if $j > n_1$.*

Proof. For the first statement of Theorem it follows from Theorem 2.1 and the results of Section 11 of Chapter 1. The second statement for $j \le n_1$ follows from the results of Section 3 of Chapter 2. Let us prove the second statement for $j > n_1$.

Let $j \le n_1 + n_2$. The statement of Theorem follows from the identity

$$y_j^k \frac{\partial^k}{\partial y_j^k} = \frac{\partial^k}{\partial (\log y_j)^k}. \tag{2.9}$$

Let $n_1 + n_2 < j$. We prove by induction on k that in this case

$$\partial^k y_j = y_j P_k(\partial \log y_j, \partial^2 \log y_j, \dots, \partial^k \log y_j), \tag{2.10}$$

where $P_k(\xi_1, \dots, \xi_k)$ is a polynomial with constant coefficients. We put $v = \log y_j$. Then $y_j = \exp v$ and $\partial y_j = y_j \partial v$. Let Formula (2.10) be true for some k. We prove it for $k + 1$. Indeed, from Equality (2.10) follows the equality

$$\partial^{k+1} y_j = y_j (\partial \log y_i) P_k + y_j \sum_{i=1}^{k} \frac{\partial P_k}{\partial \xi_i} \partial^{i+1} \log y_j,$$

which has the form (2.10) with

$$P_{k+1}(\xi_1, \dots, \xi_{k+1}) = \xi_1 P_k(\xi_1, \dots, \xi_k) + \sum_{i=1}^{k} \frac{\partial P_k}{\partial \xi_i} \xi_{i+1}.$$

According to (2.10), we have

$$y_j^{-1}\partial^k y_j = P_k(\partial \log y_j, \ldots, \partial^k \log y_j),$$

whence it follows the second statement of Theorem for $n_1 + n_2 < j$. □

We note that the multiplication of the polynomial $g(Y)$ by the factor Y^T means the parallel translation of the set $\mathbf{S}(g)$ by the vector T: $\mathbf{S}(gY^T) = \mathbf{S}(g) + T$. In this way, if $\dim \mathbf{\Gamma}(f) < n$, then by the power transformation (2.1) and by the multiplication by Y^T, one can place the support of the polynomial f in the d-dimensional coordinate subspace. That allows to lower the dimension of the truncated problem. The original equation is the regular perturbation of its truncation in the corresponding domain of the X-space, where the truncation is the first approximation. It remains valid after the application of the power transformation as well. But then the domain may be made close to the coordinate subspace in the Y-space. From Section 11 of Chapter 1, it follows

Theorem 2.3 *Let $\hat{f}_k^{(d)}(X)$ be a truncation of the differential polynomial $f(X)$. There exist such a matrix W and a vector T that after the transformation* (2.1) *there are $n - d$ coordinates y_j, which either absent in the polynomial $Y^T \hat{g}_k^{(d)}(Y)$ or present only in the form $\partial \log y_j$, where $\hat{g}_k^{(d)}(Y) = \hat{f}_k^{(d)}(X)$. Besides $\mathbf{S}(Y^T g) \subset \mathbb{R}_+^n$, where $g(Y) = f(X)$ and $\mathbb{R}_+^n = \{S \geq 0\}$.*

Example 2.1 (the continuation of Example 1.2). Under the power transformation $z = x^{-\alpha} y$, we have

$$y' = z'x^\alpha + (\alpha - 1)zx^{\alpha-1}, \quad y'' = z''x^\alpha + 2\alpha z'x^{\alpha-1} + \alpha(\alpha-1)zx^{\alpha-2}, \tag{2.11}$$

where $' \stackrel{\text{def}}{=} d/dx$. According to the definition, $\alpha = (\sigma + 2)/(1 - \mu)$, and we have $\sigma + \mu\alpha = \alpha - 2$. Hence Equation (1.15′) takes the form

$$x^{\alpha-2}[x^2 z'' + 2\alpha x z' + \beta z - a z^\mu] = 0, \tag{2.12}$$

where $\beta = \alpha(\alpha - 1)$. Cancelling $x^{\alpha-2}$, we obtain the equation

$$x^2 z'' + 2\alpha x z' + \beta z - a z^\mu = 0. \tag{2.13}$$

Its support consists of two points: $(0, 1)$ and $(0, \mu)$ (see Fig. 4.3). If we put $t = \log x$, then

$$z' = x^{-1}\dot{z}, \quad z'' = x^{-2}(\ddot{z} - \dot{z}),$$

where $\dot{} \stackrel{\text{def}}{=} d/dt$, and Equation (2.12) takes the form

$$\ddot{z} + (2\alpha - 1)\dot{z} + \beta z - a z^\mu = 0. \square \tag{2.14}$$

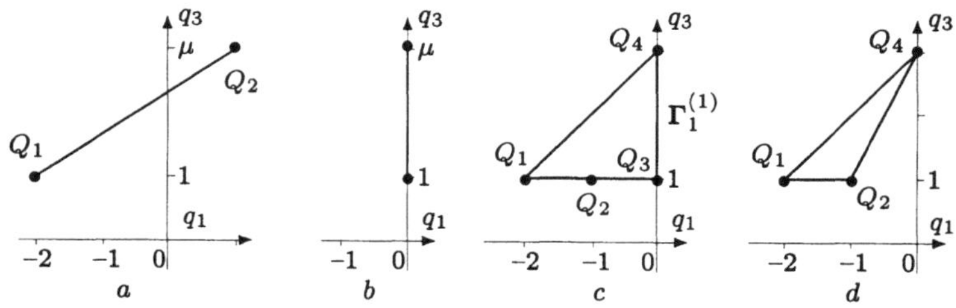

Figure 6.3: Supports and polygons for Equations (1.15′) (a), (2.13) (b), (2.14) with $\beta \neq 0$ (c), and (2.14) with $\beta = 0$ (d)

The co-dimension of the intersection (1.5) is called the *dimension of the truncation* (1.6) *of the aggregate* (1.4). So Theorem 11.1 of Chapter 1 gives

Theorem 2.4 *Let the aggregate of differential polynomials* (1.4) *have the truncation* (1.6) *of the dimension* d. *There exist such a matrix* W *and such vectors* $T_1, \dots, T_m$ *that after the transformation* (2.1) *there are* $n - d$ *coordinates* y_j, *which either absent in polynomials* $Y^{T_i}\hat{g}_i(Y)$, $i = 1, \dots, m$ *or present only in the form* $\partial \log y_j$, *where* $\hat{g}_i(Y) = \hat{f}^{(d_i)}_{ij_i}(X)$. *Besides* $\mathbf{S}(Y^{T_i} g_i) \subset \mathbb{R}^n_+$, *where* $g_i(Y) = f_i(X)$, $i = 1, \dots, m$.

There are two types of the power transformation (2.1): with an arbitrary real matrix W, and with an unimodular matrix W (i.e. $\det W = \pm 1$) with integer elements. The power transformation of the second type gives the one-to-one correspondence between X and Y outside the coordinate subspaces. In order to find the matrix W, it is necessary to use algorithms described in Section 11 of Chapter 1.

3. The logarithmic transformation

Let a differential polynomial $g(Y)$ be such that for some j the coordinate $s_j = 0$ for all $S \in \mathbf{S}(g)$. If y_j is the algebraic coordinate (i.e. a parameter, $j \leq n_1$), then g does not depend on y_j. If y_j is the differential coordinate (i.e. variable, $j > n_1$), then g depends only on $\log y_j$. Let J be the set of such indices $j > n_1$ that $s_j = 0$ for every $S \in \mathbf{S}(g)$. We introduce new variables by means of the *logarithmic transformation*

$$v_j = \log y_j \quad \text{for} \quad j \in J,$$
$$v_k = y_k \quad \text{for} \quad k \notin J.$$

We denote $h(V) = g(Y)$. If g really depends on y_j, $j \in J$, then the support $\mathbf{S}(h)$ has points Q with $q_j \neq 0$. Hence we can find the first approximations for $h(V)$ using its polyhedron $\mathbf{\Gamma}$. Solutions to the corresponding truncated equations may give logarithmic asymptotics of solutions to the original problem. We note that $v_j \to \infty$ when $y_j \to 0$, or $y_j \to \infty$ and $j \in J$.

Example 3.1 (the continuation of Example 2.1). After the transformation $t = \log x$, Equation (2.12) takes the form (2.14). If $\beta \stackrel{\text{def}}{=} \alpha(\alpha - 1) \neq 0$, then its support consists of 4 points: $Q_1 = (-2, 1)$, $Q_2 = (-1, 1)$, $Q_3 = (0, 1)$ and $Q_4 = (0, \mu)$, where $q_1 = \text{ord } t$, $q_2 = \text{ord } z$. Since $t \to \infty$, then for the studied solutions the vector $P = (p_1, p_2)$ has $p_1 \geq 0$. The polyhedron $\mathbf{\Gamma}$ for Equation (2.5) has only one edge $\mathbf{\Gamma}_1^{(1)}$, the normal vector P of which has $p_1 > 0$ (see Fig. 6.3, *c*). The corresponding boundary subset $\mathbf{S}_1^{(1)}$ consists of two points: Q_3, Q_4, and the truncated equation is $\beta z - az^\mu = 0$. Its solutions are the values b cited in Example 1.2.

If $\beta = 0$, i.e. $\alpha = 0, 1$, then the support of Equation (2.14) consists of three points: Q_1, Q_2, Q_4. Again the polyhedron $\mathbf{\Gamma}$ has only one edge, whose normal vector P has the component $p_1 > 0$ (see Fig. 6.3, *d*). But now the corresponding boundary subset consists of points Q_2, Q_4, and the truncated equation is $\gamma \dot{z} - az^\mu = 0$, where $\gamma = 2\alpha - 1$. Its solutions are

$$z = [a(1 - \mu)\gamma^{-1} (\log x + c)]^{1/(1-\mu)}, \quad c = \text{const.}$$

Consequently, for $\alpha = 0, 1$, Equation (1.15′) has solutions with the asymptotics

$$y \sim x^{-\alpha}[a(1 - \mu)\gamma^{-1} \log x]^{1/(1-\mu)}.$$

That coincides with Formulae (6.33), (6.34) in Section 6 of Chapter 3. We note that the found in Example 6.4 of Chapter 3 logarithmic asymptotics (6.32) cannot be the asymptotics of a solution to Equation (1.15′). It relates only to solutions to the system of equations (2.12) of Chapter 3. But not every solution to that system is a solution to Equation (1.15′). □

Example 3.2 (the continuation of Example 1.2). We consider Equation (1.15′) when $\mu = 1$ and $\sigma \neq -2$. Now the points Q_1 and Q_2 have the same second coordinates $q_2 = 1$. Hence we have to cancel y and make logarithmic transformation $v = \log y$ in Equation (1.15′), i.e. $y = \exp v$. Since $y' = yv'$, $y'' = y(v')^2 + yv''$, where $' \stackrel{\text{def}}{=} d/dx$, then Equation (1.15′) with $\mu = 1$ takes the form

$$(v')^2 + v'' - ax^\sigma = 0. \tag{3.1}$$

Its support consists of three points, and the convex hull is the triangle. Only one edge of the triangle has the normal $P = (p_1, p_2)$ with the component $p_2 > 0$. To this edge there corresponds the truncation $(v')^2 = ax^\sigma$ of Equation (3.1). Its solutions are

$$v_{1,2} = c \pm 2\sqrt{a}\, x^{(\sigma+2)/2}/(\sigma + 2),$$

where $c =$ const. To these solutions there correspond asymptotics of solutions to Equation (1.15′)

$$y_{1,2} = c \exp[\pm 2(2+\sigma)^{-1}\sqrt{a}\, x^{(\sigma+2)/2}]. \square$$

Example 3.3. We consider the equation

$$\frac{\partial^2 \psi}{\partial x^2} + \frac{\partial^2 \psi}{\partial y^2} = f(x, y),$$

where f is a homogeneous polynomial in x, y of the degree $m-2$. Here $n_1 = 0$, $n_2 = 2$, $n_3 = 1$, $n = 3$, $X_2 = (x, y)$, $X_3 = \psi$. The support of the equation consists of points $(-2, 0, 1)$, $(0, -2, 1)$, $(k, m-2-k, 0)$, where $0 \le k \le m-2$. All the points lie in the plane $q_1 + q_2 + mq_3 = m - 2$, where $q_1 = \text{ord } x$, $q_2 = \text{ord } y$, $q_3 = \text{ord } \psi$. The power transformation

$$u = x, \quad z = x^{-1}y, \quad \varphi = x^{-m}\psi$$

and the cancellation of u^{m-2} reduce the equation to the form

$$m(m-1)\varphi + 2mu\frac{\partial \varphi}{\partial u} + u^2\frac{\partial^2 \varphi}{\partial u^2} - 2uz\frac{\partial^2 \varphi}{\partial u \partial z} + z^2\frac{\partial^2 \varphi}{\partial z^2} + 2z\frac{\partial \varphi}{\partial z} + \frac{\partial^2 \varphi}{\partial z^2} = f(1, z). \tag{3.2}$$

Its support lies in the plane $s_1 = 0$, where $s_1 = \text{ord } u$, $s_2 = \text{ord } z$, $s_3 = \text{ord } \varphi$. Hence we assume $t = \log u$. Then the equation takes the form

$$m(m-1)\varphi + z^2\frac{\partial^2 \varphi}{\partial z^2} + 2z\frac{\partial \varphi}{\partial z} + \frac{\partial^2 \varphi}{\partial z^2} + (2m-1)\frac{\partial \varphi}{\partial t} - 2z\frac{\partial^2 \varphi}{\partial t \partial z} + \frac{\partial^2 \varphi}{\partial t^2} = f(1, z). \tag{3.3}$$

Its support consists of points $(0, 0, 1)$, $(0, -2, 1)$, $(-1, 0, 1)$, $(-2, 0, 1)$, $(0, k, 0)$, and it has three-dimensional convex hull $\mathbf{\Gamma}$. Now $q_1 = \text{ord } t$, $q_2 = \text{ord } z$, $q_3 = \text{ord } \varphi$. Fig. 6.4 shows the support and its polyhedron $\mathbf{\Gamma}$ for $m = 4$. Since $t \to \infty$, then here we need only faces with the normal $P = (p_1, p_2, p_2)$, which has $p_1 > 0$. Such a face here is unique. To it there corresponds the truncation of Equation (3.3), where the derivatives with respect to t are absent, i.e. for φ we obtain the ordinary differential equation. The same equation is obtained from (3.2) if φ does not depend on u. The corresponding solution ψ to the initial equation is the so called "self-similar solution" (see Chapter 7). $\square$

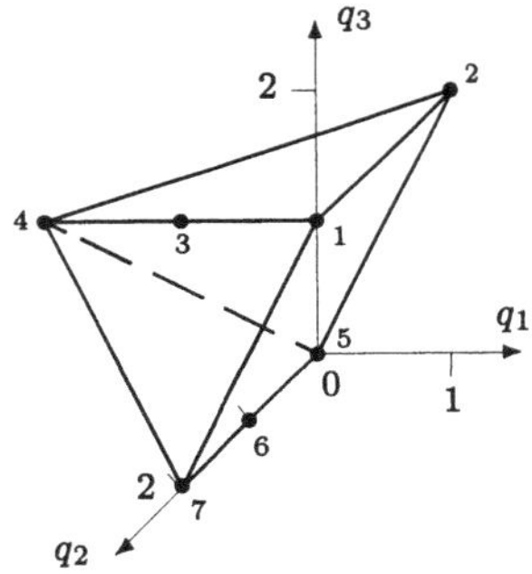

Figure 6.4: The support and the polyhedron for Equation (3.3) of Example 3.3.

4. A big example

Let us find asymptotics of solutions $y(x)$ to the equation

$$y''' - x^{\sigma}y^{\mu} = 0, \quad \sigma, \mu \in \mathbb{R}, \tag{4.1}$$

when $x \to \infty$. The equation is always real when $x, y > 0$. If σ or μ are irrational with an odd denominator, then the equation is also real when $x < 0$ or $y < 0$ respectively. In notation of Sections 1 and 2, in Equation (4.1) $n_1 = 0$, $n_2 = n_3 = 1$, $n = 2$, $X_2 = x$, $X_3 = y$. The left hand side of Equation (4.1) consists of two monomials with the vector powers: $Q = (-3, 1)$, $Q_2 = (\sigma, \mu)$.

Firstly, we consider the case

$$\mu \neq 1. \tag{4.2}$$

Then $Q_1 \neq Q_2$, and the polyhedron $\mathbf{\Gamma}$ is the oblique segment $[Q_1, Q_2]$ (see Fig. 6.5, *a*). The vector $R \overset{\text{def}}{=} Q_2 - Q_1 = (\sigma + 3, \mu - 1)$ is directed along the segment. We put

$$\alpha = -(\sigma + 3)/(\mu - 1). \tag{4.3}$$

Then the normal to the segment $\mathbf{\Gamma}$ is $N = (1, \alpha)$. The segment $\mathbf{\Gamma}$ has three faces: $\mathbf{\Gamma}_1^{(1)} = \mathbf{\Gamma}$, $\mathbf{\Gamma}_1^{(0)} = Q_1$, $\mathbf{\Gamma}_2^{(0)} = Q_2$. We consider the corresponding truncations one by one.

To the edge $\mathbf{\Gamma}_1^{(1)}$ there corresponds truncation, which coincides with the original equation (4.1). We denote $\varkappa \overset{\text{def}}{=} \operatorname{sgn} x = \pm 1$. We are going to find the solution to Equation (4.1) corresponding to the normal vector N, i.e.

$$y = b|x|^{\alpha}, \quad b = \text{const} \tag{4.4}$$

Substituting this expression in (4.1), we obtain

$$\beta \varkappa^{-3} b|x|^{\alpha-3} = \varkappa^{\sigma} b^{\mu} |x|^{\sigma+\alpha\mu},$$

where $\beta = \alpha(\alpha-1)(\alpha-2)$. By virtue of (4.3), we have $\alpha - 3 = \sigma + \alpha\mu$. Reducing the obtained equation by $b|x|^{\alpha-3}\varkappa^{\sigma}$, we obtain the equation

$$\beta^{\mu-1} = \beta\varkappa^{-(\sigma+3)}. \tag{4.5}$$

If

$$\beta \stackrel{\text{def}}{=} \alpha(\alpha-1)(\alpha-2) \neq 0, \tag{4.6}$$

then for each value $\varkappa = \pm 1$, Equation (4.5) may have no more than two real solutions b. Thus, when $\beta > 0$ there are always the real solution b when $\varkappa = \pm 1$. So, when conditions (4.2), (4.6) are satisfied, Equation (4.1) has solutions (4.4), (4.5), which coincide with solutions (8.9), (8.14) found in Section 8 of Chapter 3 under the condition (8.15).

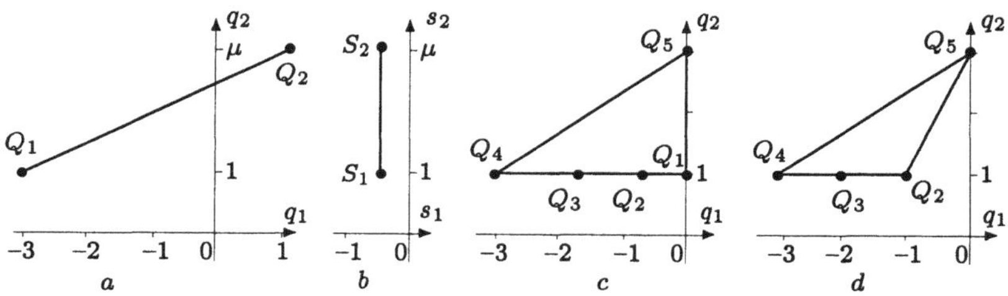

Figure 6.5: Supports and polygons for Equations (4.1), (4.2) (*a*), (4.9) (*b*), (4.11) with $\beta \neq 0$ (*c*), (4.11) with $\beta = 0$ (*d*).

To the vertex $\mathbf{\Gamma}_1^{(0)} \stackrel{\text{def}}{=} Q_1$ there corresponds the truncated equation $y''' = 0$. Its solutions have the form

$$y = cx^{\nu}, \quad c \stackrel{\text{def}}{=} \text{const} \neq 0, \quad \nu = 0, 1, 2. \tag{4.7}$$

The vector order P of such a solution is $P = (1, \nu)$. It lies in the normal cone $\mathbf{U}_1^{(0)}$ of the vertex Q_1 if $\langle P, R \rangle < 0$, i.e.

$$\beta + 3 + \nu(\mu - 1) < 0. \tag{4.8}$$

For $\nu = 0, 1, 2$ the asymptotics (4.7) under the condition (4.8) coincides with asymptotics (8.13), (8.12), (8.11) respectively, which were found in Section 8 of Chapter 3. In fact, asymptotics (8.13) and (8.12) from Section 8 of Chapter 3 include also the second terms of the expansion $y = y_1 + y_2 + \ldots$ over decreasing powers of x. The second terms y_2 of the expansion are obtained by integration from the equation $y_2''' = x^\sigma y_1^\mu$, where y_1 is the corresponding first approximation (4.7).

To the vertex $\mathbf{\Gamma}_2^{(0)} \stackrel{\text{def}}{=} Q_2$ there corresponds the truncated equation $-x^\sigma y^\mu = 0$, which has no non-trivial solutions.

When $\mu \neq 1$, Equation (4.1) has dimension $d = 1 < n = 2$. According to Theorem 2.2, we simplify it by means of the power transformation $y = z|x|^\alpha$. According to the rule of differentiation of a product

$$y''' = z'''|x|^\alpha + 3\alpha\varkappa z''|x|^{\alpha-1} + 3\alpha(\alpha-1)z'|x|^{\alpha-2} + \beta\varkappa z|x|^{\alpha-3}.$$

Substituting this expression in Equation (4.1), we obtain the equation

$$z'''|x|^\alpha + 3\alpha\varkappa z''|x|^{\alpha-1} + 3\alpha(\alpha-1)z'|x|^{\alpha-2} + \\ \beta\varkappa z|x|^{\alpha-3} = \varkappa^\sigma|x|^{\sigma+\mu\alpha}z^\mu, \tag{4.9}$$

the support of which is shown in Fig. 6.5, *b*. Cancelling $|x|^{\alpha-3}\varkappa$ in it and taking into account the equality (4.3), we obtain the equation

$$x^3 z''' + 3\alpha x^2 z'' + 3\alpha(\alpha-1)xz' + \beta z = \varkappa^{\sigma-1}z^\mu. \tag{4.10}$$

According to Section 3, we make now the logarithmic transformation $t = \log x$, $x = e^t$. Denoting the differentiation with respect to t by dot, we obtain

$$z' = x^{-1}\dot{z}, \quad z'' = x^{-2}(\ddot{z} - \dot{z}), \quad z''' = x^{-3}(\dddot{z} - 3\ddot{z} + 2\dot{z}).$$

Substituting these expressions in (4.10) and collecting similar terms, we obtain the equation

$$\dddot{z} + 3(\alpha-1)\ddot{z} + \gamma\dot{z} + \beta z = \varkappa^{\sigma-1}z^\mu, \tag{4.11}$$

where $\gamma \stackrel{\text{def}}{=} 3\alpha^2 - 6\alpha + 1$. For $\beta \neq 0$ the support of this equation consists of five points, and it is shown in Fig. 6.5, *c*. Since $t \to \infty$, we are interested only in truncations, the normal cones of which contain the vector $P = (p_1, p_2)$ with $p_1 > 0$. It is clear from Fig. 6.5, *c*, that there is only one edge (vertical) with that property. To it there corresponds the truncated equation

$$b = \varkappa^{\sigma-1}z^\mu.$$

Its solutions $(z_0, \varkappa_0)$ are the values $(b_0, \varkappa_0)$ satisfying Equation (4.5). To the vertices of the vertical edge there correspond truncated equations, which have only trivial solutions. Consequently, when $\beta \neq 0$ we obtain only asymptotics (4.4), (4.5) found earlier. Let $(b_0, \varkappa_0)$ be a solution to Equation (4.5). When $\varkappa = \varkappa_0$, then $z = b_0$ is the stationary solution to Equation (4.11). In order to find the equation in variations for that solution, we put $z = b_0 + w$ and isolate the linear part in w. We obtain the equation

$$\dddot{w} + 3(\alpha-1)\ddot{w} + \gamma\dot{w} + \beta w = \varkappa_0^{\sigma-1}\mu b_0^\mu w.$$

According to (4.5), $\varkappa_0^{\sigma+1} b_0^{\mu-1} = \beta$, hence that equation is

$$\dddot{w} + 3(\alpha - 1)\ddot{w} + \gamma\dot{w} + \beta(1 - \mu)w = 0.$$

Its eigenvalues λ_i are roots of the equation

$$\lambda^3 + 3(\alpha - 1)\lambda^2 + \gamma\lambda + \beta(1 - \mu) = 0.$$

The discriminant of the equation is

$$D \stackrel{\text{def}}{=} -27\beta^2(1 - \mu)^2 + 54(\alpha - 1)\gamma\beta(1 - \mu) + \\ + 9(\alpha - 1)^2\gamma^2 - 4\gamma^3 - 108(\alpha - 1)^3\beta(1 - \mu).$$

If $D > 0$, then all eigenvalues λ_i are real, and near the solution $z = b_0$ there are no oscillations. If $D < 0$, then there are two complex eigenvalues, i.e. near the solution $z = b_0$ there are oscillatory solutions.

When $\beta = 0$, the support of Equation (4.11) consists of four points, their convex hull is the triangle (Fig. 6.5, *d*). Again only one its edge (the right one) has the normal vector $P = (p_1, p_2)$ with $p_1 > 0$. To the edge there corresponds the truncated equation $\gamma\dot{z} = \varkappa_0^{\sigma-1} z^\mu$. Its solution $t = \gamma\varkappa_0^{\sigma-1}(1-\mu)^{-1}z^{1-\mu}$ or

$$z = [(1 - \mu)\gamma^{-1}\varkappa_0^{-\sigma+1}t]^{1/(1-\mu)} \tag{4.12}$$

gives the logarithmic asymptotics

$$y = |x|^\alpha z \tag{4.13}$$

of solutions to Equation (4.1). Since $\beta = 0$ only when $\alpha = 0, 1, 2$, and for these values α we have $\gamma = 1, -2, 1$ respectively, then the asymptotics (4.12), (4.13) coincides with asymptotics (8.22), (8.21), (8.20) of Chapter 3 respectively. We note that to the ends of the right edge of the triangle in Fig. 6.5, *d* there correspond truncated equations

$$\gamma\dot{z} = 0, \quad \varkappa_0^{\sigma+1} z^\mu = 0. \tag{4.14}$$

The first has solutions $z =$ const. Their vector order $P = (1, 0)$ does not lie in the cone of truncation. The second equation (4.14) has no non-zero solutions.

We consider now the case $\mu = 1$. Here Equation (4.1) is linear

$$y''' = x^\sigma y. \tag{4.15}$$

If $\sigma + 3 \neq 0$, then $Q_1 \neq Q_2$. In that case the segment $\mathbf{\Gamma}$ is placed horizontally (see Fig. 6.6, *a*). According to Section 3, we make the logarithmic transformation $u = \log y$, i.e. $y = e^u$. We compute derivatives

$$y' = u'y, \quad y'' = u''y + (u')^2 y, \quad y''' = u'''y + 3u''u'y + (u')^3 y.$$

Substituting the last expression in Equation (4.15) and cancelling y, we obtain the equation

$$u''' + 3u''u' + (u')^3 = x^\sigma. \tag{4.16}$$

Its support consists of four points, their convex hull is the triangle $\boldsymbol{\Delta}$ (see Fig. 6.6, b). Only at the upper edge of the triangle $\boldsymbol{\Delta}$ the normal $P = (p_1, p_2)$ has $p_2 > 0$. To the edge there corresponds the truncation

$$(u')^3 = x^\sigma \tag{4.17}$$

of Equation (4.16). Equation (4.17) falls apart in three differential equations:

$$u'_k = \varepsilon_k x^{\sigma/3}, \quad k = 0, 1, 2, \tag{4.18}$$

where $\varepsilon_k \stackrel{\text{def}}{=} \exp(2\pi i k/3)$ are roots of the third power from unit: $\varepsilon_k^3 = 1$. Solutions to Equations (4.18) are

$$u_k = \varepsilon_k (1 + \sigma/3)^{-1} x^{1+\sigma/3}, \quad k = 0, 1, 2. \tag{4.19}$$

Consequently, solutions to the linear equation (4.15) have asymptotics of the form

$$y = c_0 \exp u_0 + c_1 \exp u_1 + c_2 \exp u_2$$

composed of solutions (4.19). Since $\operatorname{Im} \varepsilon_1 = -\operatorname{Im} \varepsilon_2 = \sqrt{3}/2 \neq 0$, then among these asymptotics there are oscillatory ones.

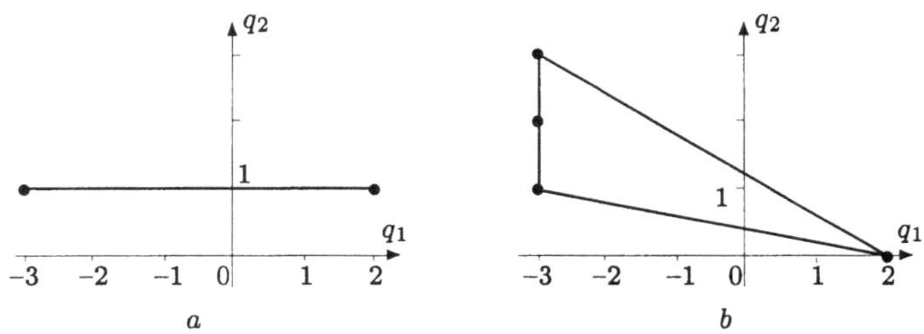

Figure 6.6: Supports and polygons for Equations (4.15) (a) and (4.16) (b).

If $\mu = 1$ and $\sigma + 3 = 0$, then Equation (4.1) is $y''' = x^{-3}y$. Its support consists of one point $Q_1 = Q_2 = (-3, 1)$. After the multiplication by x^3, we obtain the Euler equation $x^3 y''' = y$. Its solutions are

$$y = c_1 x^{\alpha_1} + c_2 x^{\alpha_2} + c_3 x^{\alpha_3},$$

where $c_k = \text{const}$, and α_k are roots of the equation $\alpha(\alpha - 1)(\alpha - 2) = 1$.

Remark 4.1. Using the procedure expounded in this Section for the equation

$$y^{(m)} = x^\sigma y^\mu \tag{4.20}$$

with arbitrary integer $m > 0$, one can find asymptotics of solutions when $x \to \infty$. Here $\alpha = (\sigma + m)/(1-\mu)$, $\beta = \alpha(\alpha-1)\dots(\alpha - m + 1)$. If $\beta = 0$, then $\alpha = k < m$ and $\gamma = (-1)^{m-k-1}k!\,(m-k-1)! \neq 0$. Many formulae of this Section either preserved or changed insignificantly. Since the system of equations $\beta = 0 = \gamma$ has no solutions, solutions to Equation (4.20) have no asymptotics with multiple logarithms.

5. One partial differential equation

The general concepts of Section 1 are described for a partial differential equation with two independent variables x, y and one unknown function ψ in the following way. We define the differential monomial $a(x, y, \psi)$ as the product of powers of coordinates x, y, ψ and derivatives $\partial^{k+l}\psi/\partial x^k \partial y^l$. To the monomial $a(x, y, \psi)$ there corresponds the point (its *vector power*) $Q = Q(a) \in \mathbb{R}^3$: the vector $Q = (q_1, q_2, q_3)$ corresponds to the product $\mathrm{const}\, x^{q_1} y^{q_2} \psi^{q_3}$, the vector $Q = (-k, -l, 1)$ corresponds to the derivative $\partial^{k+l}\psi/\partial x^k \partial y^l$, the sum of vectors Q corresponds to the product of their monomials. A sum of differential monomials is called the *differential polynomial* $f(x, y, \psi)$; to a polynomial f there corresponds the set $\mathbf{S} = \mathbf{S}(f)$ of powers Q of its monomials in $\mathbb{R}^3$. The set $\mathbf{S}$ is called the *support of the polynomial* f. Using the set $\mathbf{S}$, one can build in $\mathbb{R}^3$: the polyhedron $\mathbf{\Gamma}(f)$ as the convex hull of the set $\mathbf{S}(f)$, faces $\mathbf{\Gamma}_j^{(d)}$ of its boundary $\partial\mathbf{\Gamma}(f)$ (here d is the dimension of the face, and j is its number), and the boundary subsets $\mathbf{S}_j^{(d)} = \mathbf{S} \cap \mathbf{\Gamma}_j^{(d)}$. To each face $\mathbf{\Gamma}_j^{(d)}$ in the dual space $\mathbb{R}_*^3$ there corresponds its normal cone $\mathbf{U}_j^{(d)}$. It consists of such $P \in \mathbb{R}_*^3$, for which $\mathbf{H}_P \cap \mathbf{\Gamma} = \mathbf{\Gamma}_j^{(d)}$, where $\mathbf{H}_P$ is the plane, which is supporting to the polyhedron $\mathbf{\Gamma}$ and orthogonal to the vector P.

The truncation $\hat{f}_j^{(d)}$ of the differential polynomial $f(x, y, \psi)$ is the sum of all such monomials $a(x, y, z)$, for which $Q(a) \in \mathbf{S}_j^{(d)}$. The truncation $\hat{f}_j^{(d)}$ is the first approximation to the polynomial f in curves of the form

$$x = b_1\tau^{p_1}(1 + o(1)), \quad y = b_2\tau^{p_2}(1 + o(1)), \quad \psi = b_3\tau^{p_3}(1 + o(1)) \qquad (5.1)$$

with $b_i \neq 0$ and $\tau \to \infty$ if $P = (p_1, p_2, p_3) \in \mathbf{U}_j^{(d)}$.

Let a function $h(\xi)$ be defined when $0 < \xi < \infty$. We put

$$\underline{s} = \underline{\lim} \frac{\log|h(\xi)|}{\log \xi} \quad \text{for} \quad \xi \to 0; \qquad \overline{s} = \overline{\lim} \frac{\log|h(\xi)|}{\log \xi} \quad \text{for} \quad \xi \to \infty.$$

The *asymptotic support* of the functions h is denoted as $\operatorname{supp} h$, and it is defined as the following set in $\mathbb{R}$:

$$\operatorname{supp} h = [\underline{s}, \overline{s}], \quad \text{if} \quad \underline{s} \leq \overline{s}; \qquad \operatorname{supp} h = \mathbb{R} \setminus (\underline{s}, \overline{s}), \quad \text{if} \quad \underline{s} > \overline{s}.$$

We define the *asymptotic support* of the function $x^\rho h(y/x^\sigma)(\log x)^k$ as such a part of the straight line $\mathbf{L}' = \{Q' = (q_1, q_2)\colon q_1 = \rho - c\,\sigma,\ q_2 = c,\ c \in \mathbb{R}\}$ in the plane (q_1, q_2), for which $q_2 \in \operatorname{supp} h$. The straight line $\mathbf{L}'$ is given by the equation $q_1 + \sigma\, q_2 = \rho$, and its normal vector is $P' = (1, \sigma)$. Let $h_k(\xi)$, $k = 0, 1, \ldots, m$ be functions of one variable. The function

$$g_1 = \sum_{k=0}^{m} x^\rho h_k(y/x^\sigma)(\log x)^k$$

is called *pseudo-homogeneous.* The asymptotic support of this function represents the union of supports of all addends, and it lies in the straight line $\mathbf{L}'$. The pseudo-homogeneous function $g_1 = x^\rho h(y/x^\sigma)$ is called the *self-similar* one. If $g(x, y) = g_1 + O(x^{\rho-\varepsilon})$ when $x \to \infty$, $y/x^\sigma = \text{const}$ and $\varepsilon > 0$, then the pseudo-homogeneous function g_1 is the first approximation to the function g with respect to the vector P', i.e. in curves (5.1) with $P' = (p_1, p_2) = (1, \sigma)$. To the expression $\psi = g(x, y)$ there correspond the support of the function g in $\mathbb{R}^2$ and the point $E_3 = (0, 0, 1)$ in $\mathbb{R}^3$. Let $\mathbf{L}_{P'}$ denote the straight line in the plane $\mathbb{R}^2 = \{q_3 = 0\}$ with the normal vector $P' = (p_1, p_2) \in \mathbb{R}^2_* = \{p_3 = 0\}$. Let $\mathbf{M}_P$ be the plane in $\mathbb{R}^3$ passing through the point E_3 and orthogonal to the vector $P = (p_1, p_2, p_3)$. If the plane $\mathbf{M}_P$ passes through the straight line $\mathbf{L}_{P'}$, then

$$P = (P', p_3), \quad p_3 = \langle P', Q' \rangle \quad \text{when} \quad Q' \in \mathbf{L}_{P'}. \tag{5.2}$$

Theorem 5.1 *Let $\psi = g(x, y)$ be a solution to the equation $f(x, y, \psi) = 0$. Let the pseudo-homogeneous function $\hat{g}_{P'}$ be the first approximation to the function g with respect to the vector P', and the support of the function $\hat{g}_{P'}$ be placed in the straight line $\mathbf{L}_{P'}$, and let the vector P defined in* (5.2) *belong to the normal cone $\mathbf{U}_j^{(d)}$. Then $\psi = \hat{g}_{P'}$ is the solution to the truncated equation $\hat{f}_j^{(d)}(x, y, \psi) = 0$.*

Further we assume that the solution ψ to the equation $f(x, y, \psi) = 0$ expands in the asymptotic series in pseudo-homogeneous functions with respect to each $P' \in \mathbb{R}^2_* \setminus \{0\}$.

Let for the equation

$$f(x, y, \psi) = 0 \tag{5.3}$$

we need to find a solution $\psi = \psi(x, y)$ satisfying the boundary conditions of the form

$$\psi = g_i(x, y) \ \text{ when } \ (x, y) \in \mathcal{M}_i, \quad i = 1, \ldots, m, \tag{5.4}$$

where $\mathcal{M}_i$ is the set of points in infinity or in coordinate axes, and g_i are polynomials. In general, we consider in $\mathbb{R}^3$ the support of Equation (5.3) and the asymptotic support of the solution $\psi = g(x, y)$ as well. From conditions

(5.4) one can find its separate parts. We are going to consider Equation (5.3) and boundary conditions (5.4) as a mixed system of differential and algebraic equations. Then to each vector $P = (p_1, p_2, p_3) \in \mathbb{R}^3_* \setminus \{0\}$ there correspond: (a) the truncated equation

$$\hat{f}_P(x, y, \psi) = 0, \tag{5.5}$$

(b) the subsets $\hat{\mathcal{M}}_i$ of sets $\mathcal{M}_i$, which may be reached in curves (5.1) with the vector P (some of subsets $\hat{\mathcal{M}}_i$ may be empty), and (c) truncations $\hat{g}_{iP'}$ of functions g_i with respect to the vector $P' = (p_1, p_2)$. By Theorem 1.1, the first approximation $\psi = \hat{h}_{P'}(x, y)$ to the solution $\psi = h(x, y)$ must satisfy the truncated equation (5.5) and truncated boundary conditions

$$\psi = \hat{g}_i \text{ when } (x, y) \in \hat{\mathcal{M}}_i, \quad i = 1, \ldots, m. \tag{5.6}$$

Here, if the support of the function $\hat{g}_i$ is placed in the straight line $\mathbf{L}_{P'}$ in $\mathbb{R}^2$, then vectors P' and $P = (P', p_3)$ are related by the relation (5.2). In order to consider algorithmically the system of equations (5.3), (5.4), we need for each of the equation to build its own polyhedron, to find its faces and their normal cones. To each non-empty intersection of these cones there corresponds its own *truncated problem*, which consists of the truncated equation and truncated boundary conditions. The solutions to these truncated problem may serve as asymptotics to the solution to the original problem. In the considered three-dimensional case, in order to find the intersection of the normal cones of Equations (5.3), (5.4), one may use the graphical method of Section 3 in Chapter 1, i.e. to consider the intersection of the cones with the planes $p_3 = 1$ and $p_3 = -1$.

Not every boundary condition has the form (5.4), but the general principle remains valid: one needs to seek such solutions to the truncated equation, which satisfy the corresponding "truncated" boundary conditions. This approach is realized in the following section.

6. The viscous fluid flow around a plate

1. The first approximations to the Helmholtz equation. We consider the system of the Navier–Stokes equations describing the stationary flow of the viscous incompressible fluid [Navier 1827, Stokes 1849]:

$$u\frac{\partial u}{\partial x} + v\frac{\partial u}{\partial y} + \frac{1}{\rho}\frac{\partial p}{\partial x} = \nu\left(\frac{\partial^2 u}{\partial x^2} + \frac{\partial^2 u}{\partial y^2}\right),$$

$$u\frac{\partial v}{\partial x} + v\frac{\partial v}{\partial y} + \frac{1}{\rho}\frac{\partial p}{\partial y} = \nu\left(\frac{\partial^2 v}{\partial x^2} + \frac{\partial^2 v}{\partial y^2}\right), \tag{6.1}$$

$$\frac{\partial u}{\partial x} + \frac{\partial u}{\partial y} = 0.$$

Here x, y are rectangular coordinates, u and v are components of the vector of the velocity of the flow with respect to axes x and y, p is the pressure, ρ is the density, ν is the kinematic coefficient of the viscosity.

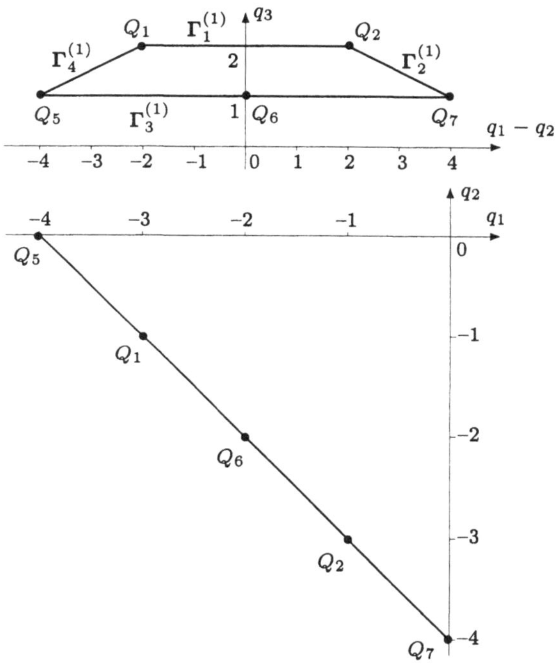

Figure 6.7: The support and the trapezoid $\mathbf{\Gamma}$ of Equation (6.3) in coordinates $q_1 - q_2$, q_3 (above), and its projection on the plane q_1, q_2 (below).

If we exclude the pressure p from the first two equations, and using the third equation we introduce the stream function ψ by formulae

$$u = \frac{\partial \psi}{\partial y}, \quad v = -\frac{\partial \psi}{\partial x}, \tag{6.2}$$

then for the function ψ we obtain the Helmholtz [1858] equation:

$$\frac{\partial \psi}{\partial y}\frac{\partial^3 \psi}{\partial x^3} + \frac{\partial \psi}{\partial y}\frac{\partial^3 \psi}{\partial x \partial y^2} - \frac{\partial \psi}{\partial x}\frac{\partial^3 \psi}{\partial x^2 \partial y} - \frac{\partial \psi}{\partial x}\frac{\partial^3 \psi}{\partial y^3} =$$

$$= \nu\left(\frac{\partial^4 \psi}{\partial x^4} + 2\frac{\partial^4 \psi}{\partial x^2 \partial y^2} + \frac{\partial^4 \psi}{\partial y^4}\right). \tag{6.3}$$

Let us count the powers of all monomials as they appear in the equation:

$$Q_1 = (-3,-1,2), \quad Q_2 = (-1,-3,2), \quad Q_3 = Q_1, \quad Q_4 = Q_2,$$

$$Q_5 = (-4,0,1), \quad Q_6 = (-2,-2,1), \quad Q_7 = (0,-4,1).$$

The support $\mathbf{S}$ consists of five points: Q_1, Q_2, Q_5, Q_6, Q_7. They are all placed in the vertical plane $q_1 + q_2 = -4$. Their convex hull $\mathbf{\Gamma}$ is the trapezoid placed in the same plane. In the upper Fig. 6.7, it is shown in coordinates $q_1 - q_2$, q_3. Its boundary $\partial\mathbf{\Gamma}$ consists of four edges $\mathbf{\Gamma}_1^{(1)}$– $\mathbf{\Gamma}_4^{(1)}$. The lower Fig. 6.7 shows its projection on the plane q_1, q_2.

For each edge $\mathbf{\Gamma}_j^{(1)}$ we isolate its boundary subset $\mathbf{S}_j^{(1)}$, write down the corresponding truncated equation, and compute its normal cone $\mathbf{U}_j^{(1)}$ by formulae of Chapter 1.

1) The edge $\mathbf{\Gamma}_1^{(1)}$. The boundary subset $\mathbf{S}_1^{(1)} = \{Q_1, Q_2, Q_3, Q_4\}$. The truncated equation

$$\frac{\partial\psi}{\partial y}\left(\frac{\partial^3\psi}{\partial x^3} + \frac{\partial^3\psi}{\partial x\partial y^2}\right) - \frac{\partial\psi}{\partial x}\left(\frac{\partial^3\psi}{\partial x^2\partial y} + \frac{\partial^3\psi}{\partial y^3}\right) = 0. \tag{6.4}$$

The normal cone

$$\mathbf{U}_1^{(1)} = \{P : \langle P, Q_1\rangle = \langle P, Q_2\rangle > \langle P, Q_6\rangle\} = \{P : p_1 = p_2,\ p_3 > 0\}.$$

2) The edge $\mathbf{\Gamma}_2^{(1)}$. The boundary subset $\mathbf{S}_2^{(1)} = \{Q_2, Q_4, Q_7\}$. The truncated equation

$$\frac{\partial\psi}{\partial y}\frac{\partial^3\psi}{\partial x\partial y^2} - \frac{\partial\psi}{\partial x}\frac{\partial^3\psi}{\partial y^3} = \nu\frac{\partial^4\psi}{\partial y^4}. \tag{6.5}$$

The normal cone

$$\mathbf{U}_2^{(1)} = \{P : \langle P, Q_2\rangle = \langle P, Q_7\rangle > \langle P, Q_5\rangle\} = \{P :\ p_1 = p_2 + p_3,\ p_1 > p_2\}.$$

3) The edge $\mathbf{\Gamma}_3^{(1)}$. Here $\mathbf{S}_3^{(1)} = \{Q_5, Q_6, Q_7\}$. The truncated equation

$$0 = \nu\left(\frac{\partial^4\psi}{\partial x^4} + 2\frac{\partial^4\psi}{\partial x^2\partial y^2} + \frac{\partial^4\psi}{\partial y^4}\right). \tag{6.6}$$

The normal cone

$$\mathbf{U}_3^{(1)} = \{P : \langle P, Q_5\rangle = \langle P, Q_7\rangle > \langle P, Q_1\rangle\} = \{P : p_1 = p_2,\ p_3 < 0\}.$$

4) The edge $\mathbf{\Gamma}_4^{(1)}$. Here $\mathbf{S}_4^{(1)} = \{Q_1, Q_3, Q_5\}$. The truncated equation

$$\frac{\partial\psi}{\partial y}\frac{\partial^3\psi}{\partial x^3} - \frac{\partial\psi}{\partial x}\frac{\partial^3\psi}{\partial x^2\partial y} = \nu\frac{\partial^4\psi}{\partial x^4}. \tag{6.7}$$

If x and y change places, and the sign of ψ is changed, then Equation (6.7) is transformed in Equation (6.5), and all sets related to it are transformed into the corresponding sets for $\mathbf{\Gamma}_2^{(1)}$.

The points Q_1, Q_2, Q_5 and Q_7 are vertices of the trapezoid $\mathbf{\Gamma}$. In order to simplify notation, we are going to denote their normal cones as $\mathbf{U}_1^{(0)}$, $\mathbf{U}_2^{(0)}$, $\mathbf{U}_5^{(0)}$ and $\mathbf{U}_7^{(0)}$ respectively. Fig. 6.8 shows intersections of the normal cones $\mathbf{U}_j^{(d)}$ with planes $p_3 = 1$ and $p_3 = -1$.

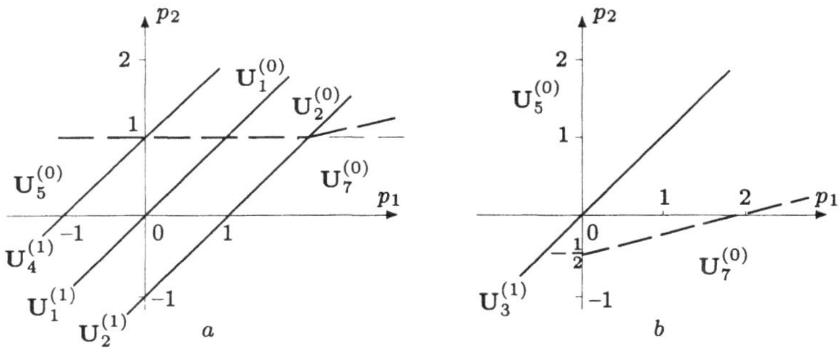

Figure 6.8: Intersections of the normal cones with planes $p_3 = 1$ (a) and $p_3 = -1$ (b) for the Helmholtz equation.

We note that Equation (6.4) is obtained by the substitution (6.2) from the Euler equations for the non-viscous flow [Euler 1755]:

$$u\frac{\partial u}{\partial x} + v\frac{\partial u}{\partial y} + \frac{1}{\rho}\frac{\partial p}{\partial x} = 0, \quad u\frac{\partial v}{\partial x} + v\frac{\partial v}{\partial y} + \frac{1}{\rho}\frac{\partial p}{\partial y} = 0, \quad \frac{\partial u}{\partial x} + \frac{\partial v}{\partial y} = 0.$$

Equation (6.5) is obtained by the substitution (6.2) from the Prandtl equations for boundary layer [Prandtl 1904]:

$$u\frac{\partial u}{\partial x} + v\frac{\partial u}{\partial y} + \frac{1}{\rho}\frac{\partial p}{\partial x} = \nu\frac{\partial^2 u}{\partial y^2}, \quad \frac{\partial u}{\partial x} + \frac{\partial v}{\partial y} = 0, \quad \frac{\partial p}{\partial y} = 0.$$

Equation (6.7) is also obtained from such equations by substitution of (y, x, v, u) instead of (x, y, u, v). Equation (6.6) is obtained by substitution (6.2) from the Stokes equations for the creeping flow [Stokes 1851; Schlichting 1965, Chapter 6]:

$$\frac{1}{\rho}\frac{\partial p}{\partial x} = \nu\left(\frac{\partial^2 u}{\partial x^2} + \frac{\partial u}{\partial y^2}\right), \quad \frac{1}{\rho}\frac{\partial p}{\partial y} = \nu\left(\frac{\partial^2 v}{\partial x^2} + \frac{\partial v}{\partial y^2}\right), \quad \frac{\partial u}{\partial x} + \frac{\partial v}{\partial y} = 0.$$

Therefore for the sake of brevity, we are going to call Equations (6.4), (6.5) and (6.6) as *Euler, Prandtl* and *Stokes* equations respectively.

2. The first approximations of a solution in infinity. Let the plane semi-infinite plate be placed in the half-line $\{x, y: x \geq 0,\ y = 0\}$. We consider the stationary flow of the viscous incompressible fluid around the plate in the positive direction of the x-axis. Such a flow is described by the Navier–Stokes system of equations (6.1) with the boundary conditions

$$u = u_\infty, \quad v = 0 \text{ when } x \to -\infty;$$

$$u = 0, \quad v = 0 \text{ when } y = 0, \quad x \geq 0.$$

That flow is also described by the Helmholtz equation (6.3) with two boundary conditions

$$\partial\psi/\partial x = 0, \quad \partial\psi/\partial y = u_\infty \text{ when } x \to -\infty; \tag{6.8}$$

$$\partial\psi/\partial x = 0, \quad \partial\psi/\partial y = 0 \text{ when } x \geq 0, \quad y = 0. \tag{6.9}$$

Let us make the substitution

$$\psi = \nu\tilde{\psi}, \quad x = \nu\tilde{x}/u_\infty, \quad y = \nu\tilde{y}/u_\infty.$$

Omitting tildes at new variables, we obtain Equation (6.3) with $\nu = 1$, and the boundary conditions (6.8), (6.9) with $u_\infty = 1$.

The boundary condition (6.8) takes the form

$$\psi = y \text{ when } (x, y) = (-\infty, y). \tag{6.10}$$

We consider Equation (6.3) and the condition (6.10) as the system of two equations: the differential and the incomplete algebraic ones. We are going to study this system by methods of Sections 3 and 9 of Chapter 1 (see the previous Section). To the equation $\psi = y$ there corresponds the support $\mathbf{S}_1$ consisting of two points: $E_3 = (0,0,1)$ and $E_2 = (0,1,0)$. Their convex hull is the segment $\mathbf{\Gamma}_2 = [E_3, E_2]$. Its normal plane is $\mathbf{U}_{21}^{(1)} \stackrel{\text{def}}{=} \{P: p_2 = p_3\} \subset \mathbb{R}_*^3$. It intersect the plane $p_3 = 1$ by the straight line $p_2 = 1$ (see Fig. 6.8, *a*).

We consider at first the problem for $x < 0$, where the condition (6.9) is absent. In Fig. 6.8, *b* to the boundary condition (6.8) there corresponds the point at infinity $(p_1, p_2) = (+\infty, 1)$ lying in the intersection $\mathbf{U}_7^{(0)} \cap \mathbf{U}_{21}^{(1)}$. To the vertex Q_7 there corresponds the truncation $0 = \partial^4\psi/\partial y^4$ of Equation (6.3). The function $\psi = y$ satisfies the equation. Consequently, the function is suitable as the first approximation in the whole intersection $\mathbf{U}_7^{(0)} \cap \mathbf{U}_{21}^{(1)}$, i.e. for the whole half-line $p_1 > 3/2$, $p_2 = 1$ of the plane $p_3 = 1$ in Fig. 6.8. The point $(p_1, p_2) = (3/2, 1)$ lies in the normal cone $\mathbf{U}_2^{(1)}$. To the edge $\mathbf{\Gamma}_2^{(1)}$ there corresponds the truncated equation (6.5), which is satisfied by the function $\psi = y$. The interval $p_1 \in (1, 3/2)$, $p_2 = 1$ lies in the normal cone $\mathbf{U}_2^{(0)}$. To the vertex $Q_2 = Q_4$ there corresponds the truncated equation (6.5) with zero instead of the right hand side; it is satisfied by the function $\psi = y$. The point $(p_1, p_2, p_3) = (1, 1, 1) \in \mathbf{U}_1^{(1)}$. To the edge $\mathbf{\Gamma}_1^{(1)}$ there corresponds the truncated equation (6.4), which is satisfied by the function $\psi = y$. Moving further left along the straight line $p_2 = 1$, we pass the points belonging to normal cones of the vertex $Q_1 = Q_3$, of the edge $\mathbf{\Gamma}_4^{(1)}$, and of the vertex Q_5. The function $\psi = y$ satisfies the corresponding truncated equations. To the movement along the straight line $p_2 = 1$ from $p_1 = +\infty$ to $p_1 = -\infty$ there corresponds the shift along $|x| + |y| = \infty$ from $x = -\infty$ to $x = 0$ in the plane (x, y). Our analysis shows that here the function $\psi = y$ is the first approximation to the solution $\psi(x, y)$ everywhere.

Now we consider the problem when $x \geq 0$ moving from $x = 0$ to $x = +\infty$. In the plane $p_3 = 1$ to this movement there corresponds the return along the straight line $p_2 = 1$ towards the increasing p_1. When $x > 0$ there is the boundary condition (6.9). For each $P' = (p_1, 1)$ we will seek the solution to the corresponding truncated problem

$$\hat{f}_P = 0, \quad \psi = y \text{ when } |x| + |y| = \infty, \tag{6.11}$$
$$\partial\psi/\partial x = \partial\psi/\partial y = 0 \text{ when } x \geq 0 = y.$$

For the pseudo-homogeneous solution $\psi = \hat{g}_{P'}$ of the problem (6.11), the support of the function $\hat{g}_{P'}(x, y)$ lies in the straight line passing through the point $(q_1, q_2) = (0, 1)$ and orthogonal to the vector $P' = (p_1, 1)$, i.e. in the straight line $p_1 q_1 + q_2 = 1$. It intersects the axis q_1 at the point $q_1 = 1/p_1 \stackrel{\text{def}}{=} \alpha$. If the function $\hat{g}_{P'}(x, y)$ is a self-similar one, then it has the form

$$\psi = x^\alpha h(\xi), \quad \xi \stackrel{\text{def}}{=} y/x^\alpha, \quad \alpha \stackrel{\text{def}}{=} 1/p_1. \tag{6.12}$$

The boundary conditions in (6.11) for it are

$$\begin{aligned} h &= \xi + o(1) \text{ when } \xi \to \infty, \\ h &= dh/d\xi = 0 \text{ when } \xi = 0. \end{aligned} \tag{6.13}$$

Lemma 6.1 *For $p_1 < 2$, the problem* (6.11) *for the corresponding truncated equations has no self-similar solution* (6.12), (6.13).

Proof. Increasing p_1 from $-\infty$ to 2, we consider the corresponding truncated problems (6.11). When $p_1 < 0$, the point $(p_1, 1, 1) \in \mathbf{U}_5^{(0)}$. The corresponding truncation of Equation (6.3) is $\partial^4\psi/\partial x^4 = 0$. All its solutions are polynomials of the third power in x. If they are self-similar, i.e. have the form (6.12), then

$$\psi = b_0 y + b_1 x y^{1-p_1} + b_2 x^2 y^{1-2p_1} + b_3 x^3 y^{1-3p_1}.$$

Since $-p_1 > 0$, from the first condition (6.13) it follows that $b_1 = b_2 = b_3 = 0$, and from the second it follows that $b_0 = 0$. Consequently, for $p_1 < 0$ the problem (6.11)–(6.13) has no solution.

When $p_1 = 0$, the point $(0, 1, 1) \in \mathbf{U}_4^{(1)}$. The corresponding truncated equation is (6.7), and the solution (6.12) has the form $\psi = yh(x)$. If this function satisfies the first boundary condition in (6.11), then $h(x) \equiv 0$. Again there is no solution.

For the further reasoning we compute partial derivatives of the function (6.12) with respect to x and y through the derivatives of the function h with respect to ξ, which we denote by prime. We obtain

$$\begin{aligned}
&\partial\psi/\partial y = h', \quad \partial^2\psi/\partial y^2 = x^{-\alpha}h'', \quad \partial^3\psi/\partial y^3 = x^{-2\alpha}h''', \\
&\partial^4\psi/\partial y^4 = x^{-3\alpha}h^{(4)}, \quad \partial\psi/\partial x = \alpha x^{\alpha-1}(h - \xi h'), \\
&\partial^2\psi/\partial x^2 = \alpha x^{\alpha-2}[(\alpha-1)(h - \xi h') + \alpha\xi^2 h''], \\
&\partial^3\psi/\partial x^3 = \alpha x^{\alpha-3}[(\alpha-1)(\alpha-2)(h-\xi h') - 3\alpha\xi^2 h'' - \alpha^2\xi^2 h'''], \\
&\partial^3\psi/(\partial x\partial y^2) = -\alpha x^{-\alpha-1}[h'' + \xi h'''], \\
&\partial^3\psi/(\partial x^2\partial y) = x^{-2}[(\alpha^2+\alpha)\xi h''' + \alpha^2\xi^2 h'''].
\end{aligned} \tag{6.14}$$

When $p_1 \in (0, 1)$ the point $(p_1, 1, 1) \in \mathbf{U}_1^{(0)}$. The corresponding truncated equation is (6.7) with zero instead of the right hand side. After the substitution there expressions (6.14), cancelling $\alpha x^{\alpha-3}$ and collecting similar terms, we obtain the equation

$$(\alpha-1)(\alpha-2)hh' - (\alpha-1)(\alpha-2)\xi h'^2 - (\alpha^2+\alpha)\xi hh'' + \\ + (\alpha^2 - 2\alpha)\xi^2 h'h'' - \alpha^2\xi^2 hh''' = 0. \tag{6.15}$$

Let

$$h = b\xi^k + \dots \quad \text{when } \xi \to 0. \tag{6.16}$$

Substituting this expression in Equation (6.14), isolating terms of the smaller power $2k - 1$ in ξ, cancelling ξ^{2k-1}, and collecting similar terms, we obtain the equation

$$-2b^2k(k-1)(k\alpha - \alpha + 1) = 0.$$

It has a solution $b \neq 0$ only when $k = 0$, $k = 1$, and $k = (\alpha - 1)/\alpha = 1 - p_1$. Since $p_1 \in (0, 1)$, for all these solutions $k \leq 1$; and by the second boundary condition (6.13), it must be $k > 1$. Hence Equation (6.15) has no such solution of the form (6.16) with a finite $k > 1$, which satisfies conditions (6.13). Let us check as yet whether Equation (6.15) has a solution (6.16) with $k = \infty$. For that purpose we make the logarithmic transformation $g = \log h$. Then

$$h' = hg', \quad h'' = h(g'^2 + g''), \quad h''' = h(g'^3 + 3g'g'' + g''').$$

Substituting this expression in Equation (6.15) and cancelling h^2, we obtain the equation

$$(\alpha-1)(\alpha-2)g' - (\alpha-1)(\alpha-2)\xi g'^2 - (\alpha^2+\alpha)\xi(g'^2+g'') +$$

$$+ (\alpha^2-2\alpha)\xi^2 g'(g'^2+g'') - \alpha^2\xi^2(g'^3+g'g''+g''') = 0.$$

In the plane $(\text{ord}\,\xi, \text{ord}\,g)$, its support consists of three points: $(-1,1)$, $(-1,2)$, $(-1,3)$. When $\xi\to 0$, we are interested in solutions $g = b\xi^{-k}+\dots$ to the equation, i.e. $p_1\le 0$, $p_2\ge 0$. For $p_1\le 0<p_2$, the boundary subset of the support consists of one point $(-1,3)$. To it there corresponds the truncated equation $(\alpha^2-2\alpha)\xi^2 g'^3 - \alpha^2\xi^2 g'^3 = 0$, i.e. $-2\alpha\xi^2 g'^3 = 0$. It has no non-trivial solutions.

When $p_1 = 1$ the point $(1,1,1)\in \mathbf{U}_1^{(1)}$. The corresponding truncated equation is (6.4). Substituting in it expressions (6.14) with $\alpha = 1$, multiplying by x^2 and collecting similar terms, we obtain the equation

$$\xi^2 h'h'' + h'h'' + 2\xi hh'' + \xi^2 hh''' + hh''' = 0.$$

It has the first integral

$$(\xi^2+1)hh'' = c \stackrel{\text{def}}{=} \text{const.}$$

Since when $\xi\to 0$, its left hand side tends to zero by virtue of the boundary condition (6.13), then $c=0$, i.e. $hh''=0$. Since $h\ne 0$, we obtain the equation $h''=0$. Its solution $h\ne 0$ is a linear function of ξ, and it does not satisfy the second condition (6.13).

When $p_1\in(1,2)$ the point $(p_1,1,1)\in\mathbf{U}_2^{(0)}$. The corresponding truncation of Equation (6.3) is Equation (6.5) with zero instead of the right hand side. Substituting in it expressions (6.14), cancelling α and collecting similar terms, we obtain the equation $hh''' + h'h'' = 0$. It has the first integral $hh'' = c \stackrel{\text{def}}{=} \text{const.}$ In the same way as in the case $p_1 = 1$, it follows from here the absence of the solution (6.12) with the property (6.13). □

Remark 6.1. In the situation of Lemma 6.1, one can prove not only the absence of self-similar solutions but the absence of the pseudo-homogeneous as well. But such a proof is significantly more cumbersome.

Lemma 6.2 *When $p_1 = 2$, the problem* (6.11) *for the corresponding truncated equation has the self-similar solution* (6.12), (6.13).

Proof. When $p_1 = 2$ the point $(2,1,1)\in\mathbf{U}_2^{(1)}$. The corresponding truncation of Equation (6.3) is Equation (6.5). Substituting in it expressions (6.14) with $\alpha = 1/2$, we obtain the equation $hh''' + h'h'' + 2h^{(4)} = 0$. It has the first integral $hh'' + 2h''' = c \stackrel{\text{def}}{=} \text{const}$. According to (6.13), when $\xi\to\infty$ its left hand side tends to zero. Consequently, $c = 0$, and the equation takes the form $hh'' + 2h''' = 0$. It is the Blasius [1908] equation (see also Examples 3.2 and 7.2 of Chapter 3, and Example 7.1 of Chapter 7). The boundary conditions (6.13) determine its unique solution $h = b(\xi)$, which is called the Blasius solution. It was studied analytically and numerically [Blasius 1908; Kochin et al. 1948; Schlichting 1965]. See also Theorem 7.2 of Chapter 7. In particular, it turned out that

$$b''(0) \stackrel{\text{def}}{=} \tilde{\alpha} \text{ and } b(\xi) = \xi + \beta + o(\xi) \text{ when } \xi \to \infty, \tag{6.17}$$

where $\tilde{\alpha} \approx 0.33206$, $\beta \approx -1.72077$. In the plane $(q_1, q_2) = (\text{ord } x, \text{ord } y)$, the support of the function $\sqrt{x}b(y\sqrt{x})$ is placed in the straight line $2q_1 + q_2 = 1$ denoted as $\mathbf{L}_1$ in Fig. 6.9. The Blasius solution describes the asymptotics of the stream function in the *boundary layer*

$$\{x, y\colon x \to +\infty,\ |y\sqrt{x}| < \infty\}.\square \tag{6.18}$$

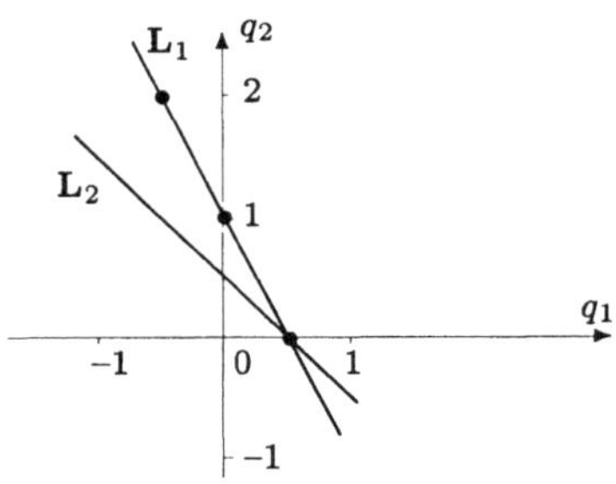

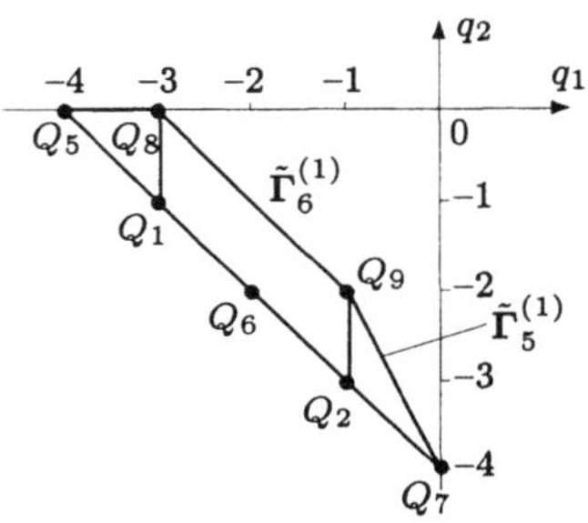

Figure 6.9 (left): Straight lines $\mathbf{L}_1$ and $\mathbf{L}_2$ in the plane (q_1, q_2).

Figure 6.10 (right): The projection of the pentahedron $\tilde{\mathbf{\Gamma}}$ in the plane (q_1, q_2).

From Lemmas 6.1 and 6.2, it follows

Theorem 6.1 *If* $x^2 + y^2 \to \infty$, *then the first approximations to the solution* ψ *to Equation* (6.3) *with boundary conditions* (6.8) *are* $\hat{\psi}_1 = \sqrt{x}\, b\,(y/\sqrt{x})$ *inside the boundary layer* (6.18) *and* $\hat{\psi}_2 = y$ *outside it.*

3. The second approximation to the solution in infinity. To study the second approximation to the stream function ψ outside the boundary layer, we put $\psi = y + \varphi$ in (6.3), and we obtain the equation for φ

$$\left(1 + \frac{\partial \varphi}{\partial y}\right) \frac{\partial^3 \varphi}{\partial x^3} + \left(1 + \frac{\partial \varphi}{\partial y}\right) \frac{\partial^3 \varphi}{\partial x \partial y^2} - \frac{\partial \varphi}{\partial x} \frac{\partial^3 \varphi}{\partial x \partial y^2} - \frac{\partial \varphi}{\partial x} \frac{\partial^3 \varphi}{\partial y^3} = \Delta^2 \varphi, \tag{6.19}$$

where Δ is the Laplace operator, and boundary conditions

$$\partial\varphi/\partial x = \partial\varphi/\partial y = 0 \text{ when } x \to -\infty \quad \text{or} \quad |y| \to \infty. \tag{6.20}$$

The support of Equation (6.19) is $\tilde{\mathbf{S}} = \mathbf{S} \cup \{Q_8, Q_9\}$, where $Q_8 = (-3, 0, 1)$ and $Q_9 = (-1, -2, 1)$ (Fig. 6.10). The convex hull $\tilde{\mathbf{\Gamma}}$ of the support $\tilde{\mathbf{S}}$ is the pentahedron; it has the edge $\tilde{\mathbf{\Gamma}}_6^{(1)} \supset \{Q_8, Q_9\}$ with $\tilde{\mathbf{U}}_6^{(1)} = \{P: 0 < p_1 = p_2 > p_3\}$. According to equalities (6.17), to the Blasius asymptotics $\psi = \sqrt{x} b(y/\sqrt{x})$ when $y/\sqrt{x} \to \infty$ there corresponds $\varphi \approx \beta\sqrt{x}$, i.e. the point $Q_3' = (1/2, 0)$ in the plane $q_3 = 0$. There exists the unique plane $\mathbf{M}_P$ passing through points E_3 and Q_3' with $P \in \tilde{U}_6^{(1)}$. For it $P = (1, 1, 1/2)$ and $\mathbf{L}_P \stackrel{\text{def}}{=} \{Q: q_1 + q_2 = 1/2\} \stackrel{\text{def}}{=} \mathbf{L}_2$ (Fig. 6.10). The plane $\mathbf{H}_P$, which is the supporting one to $\tilde{\mathbf{\Gamma}}$, intersect $\tilde{\mathbf{\Gamma}}$ along the edge $\tilde{\mathbf{\Gamma}}_6^{(1)}$. To the edge there corresponds the truncated equation

$$\partial^3\varphi/\partial x^3 + \partial^3\varphi/\partial x \partial y^2 = 0.$$

We seek its self-similar solution as a function $\varphi = \hat{\varphi} \stackrel{\text{def}}{=} \sqrt{|x|}\, h(y/x)$ with the support in the straight line $\mathbf{L}_2$. The boundary conditions $h(0) = \beta$ and (6.20) are satisfied only by the odd solution

$$\hat{\varphi} = c_1 \operatorname{sgn} y \sqrt{\sqrt{x^2 + y^2} + x} \quad \text{with} \quad c_1 = \beta/\sqrt{2}.$$

Thus, without any physical assumption we have proved

Theorem 6.2 *If $x^2 + y^2 \to \infty$, then the solution ψ to Equation* (6.3) *with boundary conditions* (6.8) *has the asymptotics $\hat{\psi}_1 = \sqrt{x} b(y/\sqrt{x})$ when $x \to +\infty$, $|y/\sqrt{x}| < \infty$ and $\hat{\psi}_2 = y + \beta \operatorname{sign} y \sqrt{(\sqrt{x^2 + y^2} + x)/2}$ in other cases.*

We note that to the lower face (with $q_3 = 1$) of the pentagon $\tilde{\mathbf{\Gamma}}$ there corresponds the Oseen linear approximation [Oseen 1910], frequently used in the flow problems.

In [Imai 1957] there were obtained the second approximation to the solution in the boundary layer and the third approximation outside it. They contain terms with $\log x$. In the book [Goldstein 1960, Chapter 8], the theory of further approximations was constructed, which are pseudo-homogeneous ones, and which include terms with powers of $\log x$. All this is in agreement with the theory of Sections 1–3, 5 of this Chapter.

4. The first approximation to the solution in the origin. We consider the problem of the flow around the plate $\{x \geq 0,\ y = 0\}$ (i.e. Problem (6.3) with $\nu = 1$, (6.9), (6.10)) near the leading edge of the plate, i.e. in the neighborhood of the point $x = y = 0$. The boundary conditions (6.9), (6.10) here take the form

$$\psi(x, y) = -\psi(x, -y), \tag{6.21}$$

$$\partial\psi/\partial y = 0 \text{ when } x \geq 0,\ y = 0, \tag{6.22}$$

$$\partial\psi/\partial y \neq 0 \text{ when } x < 0,\ y = 0. \tag{6.23}$$

In particular, from (6.21) it follows that $\psi(x, 0) = 0$. Since here ψ is small, then when x, y are small the corresponding first approximation to the Helmholtz equation (6.3) is the Stokes equation (6.6), which is a biharmonic one. According to Theorem 5.1, we seek the first approximation to the stream function ψ as a self-similar solution $\psi = x^k h(y/x)$, $k \geq 1$ to the biharmonic equation (6.6). In the polar coordinates r, θ: $x = r\cos\theta$, $y = r\sin\theta$ such a solution is written in the form

$$\psi = r^k g(\theta), \quad k \geq 1. \tag{6.24}$$

Accordingly, the boundary conditions (6.21), (6.22) take the form

$$g(\pi - \theta) = -g(\pi + \theta), \quad 0 \leq \theta \leq \pi, \tag{6.25}$$

$$g'(0) = g'(2\pi) = 0, \tag{6.26}$$

where the prime means the derivative with respect to θ. In general, the boundary condition (6.23) is not necessarily satisfied for the first approximation. But if it is satisfied for the function (6.24), then it has the form

$$g'(\pi) \neq 0. \tag{6.27}$$

Theorem 6.3 *Equation* (6.6) *has Solution* (6.24) *satisfying the boundary conditions* (6.25), (6.26) *only with integer* $k \geq 3$ *and half-integer* $k > 1$. *This solution has*

$$g(\theta) = b[k\sin(k-2)\theta - (k-2)\sin k\theta], \quad k \in \mathbb{Z}, \quad k \geq 3, \tag{6.28}$$

$$g(\theta) = a[\cos(k-2)\theta - \cos k\theta], \quad k = l + 1/2, \quad l \in \mathbb{Z}, \quad l \geq 1, \tag{6.29}$$

where $a, b \stackrel{\text{def}}{=} \text{const} \neq 0$.

Proof. In the polar coordinates the biharmonic equation (6.6) has the form

$$\frac{\partial^4\psi}{\partial r^4} + \frac{2}{r^2}\frac{\partial^4\psi}{\partial r^2\partial\theta^2} + \frac{1}{r^4}\frac{\partial^4\psi}{\partial\theta^4} + \frac{2}{r}\frac{\partial^3\psi}{\partial r^3} - \frac{2}{r^3}\frac{\partial^3\psi}{\partial r\partial\theta^2} +$$

$$+ \frac{4}{r^4}\frac{\partial^2\psi}{\partial\theta^2} - \frac{1}{r^2}\frac{\partial^2\psi}{\partial r^2} + \frac{1}{r^3}\frac{\partial\psi}{\partial r} = 0.$$

Substituting in it the expression (6.24), cancelling r^{k-4} and collecting similar terms, we obtain the ordinary differential equation for $g(\theta)$

$$g^{(4)} + 2(k^2 - 2k + 2)g'' + k^2(k-2)^2 g = 0. \tag{6.30}$$

If we seek a solution in the form $g = \exp(\lambda\theta)$, then for λ we obtain the characteristic equation

$$\lambda^4 + 2(k^2 - 2k + 2)\lambda^2 + k^2(k-2)^2 = 0.$$

Its roots are

$$\lambda_{1,2} = \pm i(k-2), \quad \lambda_{3,4} = \pm ik. \tag{6.31}$$

For $k \geq 1$, among these roots there are multiple ones only when $k = 1$ and $k = 2$. We consider these cases.

$k = 1$. In this case the general solution to Equation (6.30) is

$$g = a_1 \cos\theta + b_1 \sin\theta + a_2\theta \cos\theta + b_2\theta \sin\theta, \tag{6.32}$$

where a_i, $b_i = \text{const}$. From the condition (6.25) it follows that

$$g(0) = g(\pi) = g(2\pi) = 0. \tag{6.33}$$

For the function (6.32) that means that

$$a_1 = 0, \quad a_1 + a_2\pi = 0, \quad a_1 + a_2 2\pi = 0,$$

i.e. $a_1 = a_2 = 0$. The condition (6.26) means that

$$b_1 = 0, \quad b_1 + b_2 2\pi = 0,$$

i.e. $b_1 = b_2 = 0$. Consequently, in this case there is no non-zero solution to Equation (6.30) satisfying conditions (6.25), (6.26).

$k = 1$. In this case the general solution to Equation (6.30) is

$$g = a_1 \cos 2\theta + b_1 \sin 2\theta + a_2 + b_2\theta. \tag{6.34}$$

Conditions (6.33) mean that

$$a_1 + a_2 = 0, \quad a_1 + a_2 + b_2\pi = 0, \quad a_1 + a_2 + b_2 2\pi = 0,$$

i.e. $a_1 + a_2 = b_2 = 0$. Conditions (6.26) mean that $2b_1 + b_2 = = 0$. Consequently, $a_1 + a_2 = b_1 = b_2 = 0$, and the function (6.34) has the form

$$a_1(\cos 2\theta - 1) = -2a_1 \sin^2\theta.$$

This function does not satisfy the condition (6.25). Thus, in this case there is also no required solution.

If $k > 1$ and $k \neq 2$, then all eigenvalues (6.31) are different. Hence the general solution to Equation (6.30) is

$$g = a_1 \cos(k-2)\theta + b_1 \sin(k-2)\theta + a_2 \cos k\theta + b_2 \sin k\theta. \tag{6.35}$$

Conditions (6.33) mean that

$$a_1 + a_2 = 0,$$
$$(a_1 + a_2)\cos k\pi + (b_1 + b_2)\sin k\pi = 0,$$
$$(a_1 + a_2)\cos 2k\pi + (b_1 + b_2)\sin 2k\pi = 0,$$

i.e.

$$a_1 + a_2 = 0, \quad (b_1 + b_2)\sin k\pi = 0. \tag{6.36}$$

Conditions (6.26) mean that

$$b_1(k-2) + b_2 k = 0, \tag{6.37}$$

and

$$-[a_1(k-2) + a_2 k]\sin 2k\pi + [b_1(k-2) + b_2 k]\cos 2k\pi = 0.$$

Taking into account (6.36) and (6.37), from the last equality we obtain

$$2a_1 \sin 2k\pi = 0. \tag{6.38}$$

If $a_1 = 0$, then $a_2 = 0$ by virtue of (6.36), besides $b_1 + b_2 \neq 0$ by virtue of (6.37), i.e. $\sin k\pi = 0$ by virtue of (6.36). Consequently, k is integer and (6.37) is satisfied. Such solution (6.35) is (6.28).

If $a_1 \neq 0$, then $a_2 = -a_1$ by virtue of (6.36), besides the number $2k$ is integer according to (6.38). If k is not integer, then $\sin k\pi \neq 0$ and $b_1 + b_2 = 0$ according to (6.36). Together with (6.37) that means that $b_1 = b_2 = 0$. Such solution (6.35) is (6.29).

If $a_1 \neq 0$ and k is integer, then Solution (6.35) with properties (6.36)–(6.38) is

$$g = a[\cos(k-2)\theta - \cos k\theta] + b[k\sin(k-2)\theta - (k-2)\sin k\theta].$$

It satisfies the condition (6.25) only when $a = 0$, i.e. it is again Solution (6.28). □

Lemma 6.3 *If the solution ψ to Equation* (6.3) *has the first approximation* (6.24), *then it expands in the asymptotic series*

$$\psi = \sum_{m=1}^{\infty} g_m(r,\theta) r^{km}, \tag{6.39}$$

where functions $g_m(r,\theta)$ have the zero order in r and $g_1 \stackrel{\text{def}}{=} g(\theta)$.

Proof. The support of the series (6.39) in r consists of points km with $m \geq 1$. If we substitute the series in Equation (6.3), then the left hand side of the equation will contain terms of powers $km_1 - 1 + km_2 - 3 = k(m_1 + m_2) - 4$, where m_1 and m_2 are natural numbers, i.e. the number $m_1 + m_2 \geq 2$ and integer. The right hand side of the equation (6.3) will contain terms of powers $km-4$, where m is natural. In this way, the supports of both parts of Equation (6.3) lie in the set of numbers $km - 4$, where m is natural. Equating terms with the same power r, we obtain the infinite system of equations for functions g_m, which is solved one-by-one over m. Here the powers of r do not leave the set. □

The more concrete definition of the solution in comparison with Theorem 6.3 can be obtained by studying the second and further approximations to Solution (6.39) satisfying the boundary conditions. Thus, for example, the condition (6.27) is not satisfied for functions (6.28), and it is satisfied for functions (6.29). The function (6.29) in the open interval $\theta \in (0, \pi)$ turns to zero $l - 1$ times; but the stream function must have even number of such zeros. The smallest value k permitted by Theorem 6.3 equals $3/2$. For this case in [Carrier and Lin 1948] the first approximation to Solution (6.29) was suggested, and the second one, which was proved to be wrong. It was corrected in [Van-Dyke 1964] (see there Formula (3.24)). In [Carrier and Lin 1948] it was supposed that the asymptotic expansion near the leading edge (6.39) is directly matched with the asymptotic expansion in the boundary layer. In fact, from the second approximations near the leading edge and in the boundary layer it is clear that these asymptotic expansions near the plate cannot be matched directly. In [Van de Vooren and Dijkstra 1970, McLachlan 1991] the results of computations of the flow around the finite and semi-infinite plate respectively are cited.

The preliminary results of this Chapter were published in [Bruno 1994, 1995a, b, c, 1996b, c, 1997a, b, 1998c, 1999] for Sections 1–4 and in [Bruno and Vasiliev 1995, 1996, 1997, 1998; Bruno 1999] for Sections 5 and 6.

More complicated problems with the boundary layer were considered with methods of Power Geometry by Vasiliev [1998, 1999].

Chapter 7

Self-similar solutions

1. Supports of a function

Let $X = (x_1, \ldots, x_n) \in \mathbb{K}^n$, where $\mathbb{K}$ is $\mathbb{C}$ or $\mathbb{R}$, and $Q = (q_1, \ldots, q_n) \in \mathbb{R}^n$; we are going to denote as X^Q the monomial $x_1^{q_1} \ldots x_n^{q_n}$. To the sum

$$f(X) = \sum f_Q X^Q \text{ over } Q \in \mathbf{S}, \tag{1.1}$$

in which the coefficients $f_Q \in \mathbb{K}$ and all the similar terms are collected, we put in correspondence the set

$$\mathbf{S}(f) = \{Q : f_Q \neq 0\} \tag{1.2}$$

in the space $\mathbb{R}^n$. That set $\mathbf{S}(f)$ is called the *support* of the sum (1.1). Let $\mathbf{SS}(f)$ be the smallest linear manifold in $\mathbb{R}^n$ containing the set $\mathbf{S}(f)$; we call it the *super-support* of the sum (1.1). The number $d(f) \stackrel{\text{def}}{=} \dim \mathbf{SS}(f)$ is called the *dimension* of the sum (1.1).

If the function $f(X)$ is not a sum of the form (1.1), then it may not have the point-wise support (1.2), but it may have the super-support $\mathbf{SS}(f)$. Let $d < n$, vectors $B_1, \ldots, B_d \in \mathbb{R}^n$ be linearly independent, and the vector $R \in \mathbb{R}^n$. Let $g(y_1, \ldots, y_d)$ be an arbitrary smooth function in its arguments, and $\partial g / \partial y_j \not\equiv 0$, $j = 1, \ldots, d$. Then the linear manifold in $\mathbb{R}^n$

$$\{Q : Q = R + \sum_{j=1}^{d} \mu_j B_j, \ \ \mu_j \in \mathbb{R}^n\} \tag{1.3}$$

is called the *super-support of the function*

$$f(X) = X^R g(X^{B_1}, \ldots, X^{B_d}), \tag{1.4}$$

and we denote it as $\mathbf{SS}(f)$. For the sum (1.1), both definitions of the super-support $\mathbf{SS}(f)$ coincide. The number $d(f) \stackrel{\text{def}}{=} \dim \mathbf{SS}(f)$ is called the *dimension of the function* (1.4).

Example 1.1. Let $n = 2$ and $f = x_1 + x_2$. Then the support $\mathbf{S}(f)$ consists of two points $E_1 = (1,0)$ and $E_2 = (0,1)$, and the super-support according to the first definition is the straight line

$$q_1 + q_2 = 1 \tag{1.5}$$

passing through points E_1 and E_2, hence $d(f) = 1$. Its normal vector is $N = (1,1)$. If we write f in the form (1.4) as $f = x_1(1 + x_1^{-1}x_2)$, then $d = 1$, $B_1 = N$, and according to the second definition (1.3), the super-support $\mathbf{SS}(f)$ is the straight line

$$\mathbf{SS}(f) = \{Q : Q = (1,0) + \mu(-1,1), \ \ \mu \in \mathbb{R}\} \tag{1.6}$$

with the directing vector $B_1 = (-1,1)$. When $\mu = 0$ and $\mu = 1$, it passes through the points E_1 and E_2 respectively, i.e. it coincides with the straight line (1.5). Formulae (1.5) and (1.6) are the *dual descriptions* [Chernikov 1968] of the same straight line. □

We consider now the space $\mathbb{R}^n_*$, that is dual to the space $\mathbb{R}^n$. So that for $P = (p_1, \ldots, p_n) \in \mathbb{R}^n_*$ and $Q = (q_1, \ldots, q_n) \in \mathbb{R}^n$ the scalar product $\langle P, Q\rangle = p_1q_1 + \ldots + p_nq_n$ is defined.

Let $\mathbf{K}$ be a linear manifold in $\mathbb{R}^n$, and $\mathbf{N} = \mathbf{N}(\mathbf{K})$ be the maximal linear subspace in $\mathbb{R}^n_*$, which is normal to the manifold $\mathbf{K}$. For a function f of the form (1.4), together with the super-support $\mathbf{SS}(f)$, we will consider its *normal subspace* $\mathbf{N}(f) \stackrel{\text{def}}{=} \mathbf{N}(\mathbf{SS}(f)) \subset \mathbb{R}^n_*$. Let vectors $N_1, \ldots, N_{n-d} \in \mathbb{R}^n_*$ be linearly independent and normal to vectors $B_1, \ldots, B_d$. Then they form the basis of the normal subspace $\mathbf{N}(f)$ of the function (1.4).

2. Supports of a differential polynomial

The prime here marks vectors of the length $n-1$. Thus, $X' = (x_1, \ldots, x_{n-1})$. The *differential monomial* $a(X)$ is the product of powers of coordinates X and the derivatives of the form

$$\partial^{\|L'\|}x_n/\partial x_1^{l_1} \ldots \partial x_{n-1}^{l_{n-1}}, \tag{2.1}$$

where $L' = (l_1, \ldots, l_{n-1}) \in \mathbb{Z}_+^{n-1}$, i.e. all $l_i \geq 0$ and integer, and $\|L'\| = l_1 + \ldots + l_{n-1}$. To each differential monomial $a(X)$ we put in correspondence the point $Q(a) \in \mathbb{R}^n$ (its vector exponent) by the following rule: the point Q corresponds to the monomial $\text{const}X^Q$; the point

$$Q = (-L', 1), \ \text{ i.e. } \ Q' = -L', \ \ q_n = 1, \tag{2.2}$$

corresponds to the derivative (2.1); and the point $Q(ab) = Q(a) + Q(b)$ is put in correspondence to the product of two differential monomials $a(X)$ and $b(X)$. The finite sum of differential monomials

$$f(X) = \sum_{i=1}^{s} a_i(X) \tag{2.3}$$

is called the *differential polynomial.* The set of points $\mathbf{S}(f) \stackrel{\text{def}}{=} \{Q(a_i),\ i = 1, \ldots, s\}$ is called the *support* of the differential polynomial (2.3). The minimal linear manifold $\mathbf{SS}(f)$ containing the support $\mathbf{S}(f)$ is called the *super-support of the polynomial* (2.3). The dimension $d(f) \stackrel{\text{def}}{=} \dim \mathbf{SS}(f)$ and normal space $\mathbf{N}(f) \stackrel{\text{def}}{=} \mathbf{N}(\mathbf{SS}(f)) \subset \mathbb{R}^n_*$ are defined in a similar way.

Example 2.1. We consider the one-dimensional equation describing the combustion process without sources

$$u_t = (u^\sigma u_x)_x, \quad \sigma = \text{const} > 0, \tag{2.4}$$

for the temperature u, the time t, and the unique space coordinate x [Samarskii et al. 1987, Klokov et al. 1997]. Here $n = 3$,

$$x_1 = t, \quad x_2 = x, \quad x_3 = u. \tag{2.5}$$

The support $\mathbf{S}$ consists of two points $Q_1 = (-1, 0, 1)$ and $Q_2 = (0, -2, 1 + \sigma)$. The super-support $\mathbf{SS}$ here is the straight line passing through the point Q_1 with the directing vector $B = Q_2 - Q_1 = (1, -2, \sigma)$, i.e. the straight line $\{Q : Q_1 + \mu B,\ \mu \in \mathbb{R}\}$. Consequently, $d = 1$. The normal subspace $\mathbf{N}$ is two-dimensional with the basis $N_1 = (2, 1, 0)$ and $N_2 = (0, \sigma, 2)$. □

3. The Lie operators

The differential *equation* $f(X) = 0$ corresponding to the polynomial (2.3) *admits the Lie operator*

$$\sum_{i=1}^{n} \lambda_i x_i \frac{\partial}{\partial x_i}, \tag{3.1}$$

if the equation is invariant under the change of coordinates

$$\tilde{x}_i = x_i \mu^{\lambda_i}, \quad \mu \in \mathbb{R}, \quad \mu > 0, \quad i = 1, \ldots, n, \tag{3.2}$$

i.e. it is a quasi-homogeneous equation (see [Ovsiannikov 1978, Ibragimov 1983]).

Theorem 3.1 *The differential equation* $f(X) = 0$ *corresponding to the polynomial* (2.3) *admits the Lie operator* (3.1) *if the vector* $\Lambda = (\lambda_1, \ldots, \lambda_n) \in \mathbf{N}(f)$.

Proof. Under the substitution (3.2), the monomial $X^Q = \mu^{-\langle \Lambda, Q\rangle}\tilde{X}^Q$, the derivative (2.1)

$$\partial^{\|L'\|}x_n/\partial x_1^{l_1}\dots\partial x_{n-1}^{l_{n-1}} = \mu^{-\langle \Lambda, Q\rangle}\partial^{\|L'\|}\tilde{x}_n/\partial\tilde{x}_1^{l_1}\dots\partial\tilde{x}_{n-1}^{l_{n-1}}$$

with the vector Q cited in (2.2); and a product of monomials and derivatives acquire as a factor such a power of μ which is equal to the sum of powers of μ acquired by each of the factors. Hence after the substitution (3.2) the differential monomial $a(X)$ turns into the monomial $a(\tilde{X})\mu^{-\langle \Lambda, Q\rangle}$, where $Q = Q(a)$, and the differential polynomial (2.3) turns into

$$\sum_{i=1}^{s} a_i(\tilde{X})\mu^{-\langle \Lambda, Q_i\rangle}, \tag{3.3}$$

where $Q_i = Q(a_i)$. If the vector $\Lambda \in \mathbf{N}(f)$, then all scalar products $-\langle \Lambda, Q_i\rangle$ in (3.3) are equal to each other for $i = 1, \dots, s$. Hence $f(X) = f(\tilde{X})\mu^{-\langle \Lambda, Q_1\rangle}$, and the equation $f(X) = 0$ turns into itself under the substitution (3.2), i.e. it admits the operator (3.1). □

If $d \stackrel{\text{def}}{=} d(f) < n$, then $\dim \mathbf{N} = n - d > 0$, and the equation $f(X) = 0$ admits $n - d$ linearly independent operators (3.1), which are easily computed from the support $\mathbf{S}(f)$.

The expression $x_n = \varphi(X')$ is called the *solution to the equation* $f(X) = 0$ if $f(X', \varphi(X')) \equiv 0$. The solution is called the *self-similar* one, if it is invariant under the changes of coordinates forming the Lie group [Ovsiannikov 1978, Ibragimov 1983]. In particular, the solution is self-similar, if it admits the operator of the form (3.1), i.e. it is invariant under the changes (3.2). One can seek such self-similar solutions $x_n = \varphi(X')$ to the equation $f(X) = 0$, which admit some of the operators (3.1) admissable by the equation. If the solution $x_n = \varphi(X')$ admits the operator (3.1), then the corresponding super-support $\mathbf{SS}(x_n - \varphi(X'))$ is normal to the vector Λ. For the remainder $x_n - \varphi(X')$, the super-support necessarily passes through the point $Q = E_n = (0, \dots, 0, 1)$ corresponding to the term x_n, and the super-support $\mathbf{SS}(\varphi(X'))$ is placed in the hyperplane $q_n = 0$. Hence the super-support $\mathbf{SS}(x_n - \varphi(X'))$ is placed in the hyperplane

$$\{Q : \langle \Lambda, Q\rangle = \lambda_n\}. \tag{3.4}$$

It intersects the hyperplane $q_n = 0$ by the linear manifold

$$\{Q' : \langle \Lambda', Q'\rangle = \lambda_n\}, \tag{3.5}$$

in which the super-support $\mathbf{SS}(\varphi(X'))$ is placed. If solution $x_n = \varphi(X')$ admits several operators of the form (3.1) with vectors $\Lambda_i = (\lambda_{1i}, \dots, \lambda_{ni})$, $i = 1, \dots, m$, then the super-support $\mathbf{SS}(\varphi(X'))$ lies in the linear manifold

$$\{Q' : \langle \Lambda'_i, Q'\rangle = \lambda_{ni}, \;\; i = 1, \dots, m\}. \tag{3.6}$$

Example 3.1 (the continuation of Example 2.1). Equation (2.4) admits operators (3.1) with $\Lambda = N_1$ and $\Lambda = N_2$. □

4. Self-similar solutions

For the sum (2.3), the sum of some number of its addends $a_i(X)$, where $0 < s_1 \le s$, is called its *proper subsum* $\check{f}$.

Theorem 4.1 *Let $x_n = \varphi(X')$ be such a solution to the equation $f(X) = 0$ corresponding to the polynomial* (2.3) *that no proper subsum $\check{f}$ of the sum* (2.3) *with* $\dim(\check{f}) < \dim(f)$ *is annulled identically on the solution. Then*

$$\mathbf{N}(x_n - \varphi(X')) \subset \mathbf{N}(f). \tag{4.1}$$

Proof. Let the vector $\Lambda \in \mathbf{N}(x_n - \varphi(X'))$, then according to (3.5) $\lambda_n = \langle \Lambda', Q' \rangle$, where Q' runs the super-support of the function $\varphi(X')$. Under the substitution (3.2) the differential polynomial (2.3) assumes the form (3.3). Since $x_n = \varphi(X')$ is the solution to the equation $f(X) = 0$, that is invariant under the substitution (3.2), i.e. $\tilde{x}_n = \varphi(\tilde{X}')$ is the same solution, then in it

$$\sum_{i=1}^{s} a_i(\tilde{X}) \mu^{-\langle \Lambda, Q_i \rangle} = 0. \tag{4.2}$$

Let the different values of scalar products $\langle \Lambda, Q_i \rangle$ be $c_1 < c_2 < \ldots < c_t$. We denote as I_r the collection of all such indices i that $\langle \Lambda, Q_i \rangle = c_r$, $r = 1, \ldots, t$. We define the differential polynomials

$$f_r(X) \stackrel{\text{def}}{=} \sum a_i(X) \text{ over } i \in I_r, \quad r = 1, \ldots, t.$$

Then the equality (4.2) takes the form

$$\sum_{r=1}^{t} \mu^{-c_r} f_r(\tilde{X}) = 0.$$

Since the equality is valid for all $\mu > 0$, then on the solution the equalities

$$f_r(\tilde{X}) = 0, \quad r = 1, \ldots, t$$

are satisfied, and it is the solution to that system of equations. According to the construction of polynomials $f_r(X)$, the vector $\Lambda \in \mathbf{N}(f_r)$ for all $r = 1, \ldots, t$. Consequently,

$$\Lambda \in \mathbf{N}(f_1) \cap \ldots \cap \mathbf{N}(f_t). \tag{4.3}$$

If $d(f_r) = d(f)$ for some r, then $\mathbf{N}(f_r) = \mathbf{N}(f)$, and the property (4.1) follows from (4.3). By the condition of Theorem, $d(f_r) = d(f)$ for all $r = 1, \dots, t$. Hence the property (4.1) is satisfied. But if $d(f_r) < d(f)$ for all r, and the intersection (4.3) does not lie in $\mathbf{N}(f)$, then the inclusion (4.1) may be violated (see below Example 4.3). □

Example 4.1 (the continuation of Examples 2.1 and 3.1). We consider Equation (2.4) in coordinates (2.5). If we draw through the point $E_3 = (0,0,1)$ the straight line parallel to $\mathbf{SS}(f)$, i.e. with the directing vector $B = (1,-2,\sigma)$, then that straight line will cross the plane $q_3 = 0$ in the point $Q_4 = (-1/\sigma, 2/\sigma, 0)$. The property (4.1) means that for the solution $x_3 = \varphi(x_1, x_2)$ to Equation (2.4), the super-support of the function φ in the plane $q_3 = 0$ either consists of the point Q_4, or it is the straight line passing through the point, or it coincides with the whole plane. Therefore solutions related to the first two cases may be found in the form

$$x_3 = c w_1^{-1/\sigma} w_2^{2/\sigma}, \tag{4.4}$$

where $w_i = c_i + \delta_i x_i$, $c_i = \text{const}$, $\delta_i = \pm 1$, $i = 1, 2$. In the first case $c = \text{const}$, and for c we obtain an "algebraic" equation $-\delta_1 = 2(2+\sigma)c^\sigma/\sigma$. In the second case c is a function of an arbitrary monomial $w_1^b w_2$. In the plane $q_3 = 0$, the straight line passing through the point Q_4 with the directing vector $(b,1,0)$ has the form $(q_1, q_2) = (-1/\sigma, 2/\sigma) + \mu(b,1)$, $\mu \in \mathbb{R}$. It intersects the axis q_1 when $\mu = -2/\sigma$ in the point with $q_1 = -(1+2b)/\sigma$. Hence that straight line can be also written in the form

$$(q_1, q_2) = (-(1+2b)/\sigma, 0) + \mu(b,1),$$

and the self-similar solution (4.4), which has the super-support of the right hand side placed in that straight line, can be written in the form

$$x_3 = w_1^{-(1+2b)/\sigma} v(\xi), \quad \xi = w_1^b w_2.$$

Substituting that expression in Equation (2.4) and cancelling $w_1^{-1-(1+2b)/\sigma}$, for the function $v(\xi)$ we obtain the ordinary differential equation

$$\delta_1[bv'\xi - (1+2b)v/\sigma] = v^\sigma v'' + \sigma v^{\sigma-1} v'^2,$$

where prime means the derivative with respect to ξ. In both cases, Solution (4.4) is self-similar one; but it has different dimension $d(x_3 - \varphi)$: the unit in the first case and two in the second one. When $\delta_1 = -1$ and $c_1 > 0$, Solution (4.4) in the first case is the *blow-up solution*, i.e. it goes into infinity in a finite time [Samarskii et al. 1987]. These solutions were studied in [Samarskii et al. 1987, Chapter III] in § 1 and § 2 for the first and the second cases respectively. □

Example 4.2. We consider the equation of the combustion process with a source [Samarskii et al. 1987, Klokov et al. 1997]

$$u_t = (u^\sigma u_x)_x + u^\beta, \quad \sigma = \text{const} > 0, \ \beta = \text{const} > 0. \tag{4.5}$$

Here $n = 3$. In the notation (2.5), the support $\mathbf{S}$ for equation (4.5) consists of three points $Q_1 = (-1, 0, 1)$, $Q_2 = (0, -2, \sigma + 1)$, $Q_3 = (0, 0, \beta)$. We compose remainders

$$Q_2 - Q_1 = (1, -2, \sigma) \stackrel{\text{def}}{=} B_1,$$

$$Q_3 - Q_1 = (1, 0, \beta - 1) \stackrel{\text{def}}{=} B_2$$

and compute their vector product $N \stackrel{\text{def}}{=} [B_1, B_2] = (2 - 2\beta, 1 + \sigma - \beta, 2)$. Since $N \neq 0$ with any values of exponents σ and β, then vectors B_1 and B_2 are linearly independent, and the dimension of the equation $d = 2$. Its super-support in $\mathbb{R}^3$ is the plane with the normal vector N. The property (4.1) means that the super-support of the solution $x_3 = \varphi(w_1, w_2)$ either coincides with $\mathbb{R}^3$ or it is the plane with the normal vector N. Since the plane $\mathbf{SS}(x_3 - \varphi)$ passes through the point $E_3 = (0, 0, 1)$, then its Equation (3.4) is $2(1-\beta)q_1 + (1+\sigma-\beta)q_2 + 2q_3 = 2$, and it intersects the plane $q_3 = 0$ by the straight line (3.5), i.e.

$$2(1-\beta)q_1 + (1+\sigma-\beta)q_2 = 2. \tag{4.6}$$

We consider at first the case

$$\beta \neq 1. \tag{4.7}$$

In this case the straight line (4.6) intersects the axis q_1 at the point $q_1 = 1/(1-\beta)$. Consequently, Equation (4.5) has a self-similar solution of the form

$$u = w_1^{1/(1-\beta)} \psi(w_1^{\varkappa} w_2), \quad \varkappa = -\frac{1+\sigma-\beta}{2(1-\beta)}, \tag{4.8}$$

where the vector $(\varkappa, 1, 0)$ is the directing one for the straight line (4.6), $w_1 = c_1 + \delta_1 t$, $w_2 = c_2 + \delta_2 x$, $c_1, c_2 = \text{const}$, $\delta_i = \pm 1$. The function $\psi(\xi)$ satisfies the ordinary differential equation

$$\delta_1(\psi/(1-\beta) + \varkappa\psi'\xi) = \psi''\psi^\sigma + \delta\psi'^2\psi^{\sigma-1} + \psi^\beta, \tag{4.9}$$

where $\xi = w_1^{\varkappa} w_2$ and the prime means the differentiation with respect to ξ. In [Samarskii et al. 1987, Klokov et al. 1997] the asymptotics and local properties of bounded solutions to Equation (4.9) are either studied at length or they are given without explanation for various values of exponents $\sigma > 0$ and $\beta > 1$. But these properties may be simply studied by methods of

Chapter 6 of this book (see the end of this Example). If the function $\psi(\xi)$ does not tend to zero when $w_1 \to 0$, then Solution (4.8) tends to infinity when $\beta > 1$, i.e. when $c_1 > 0$ and $\delta_1 = -1$ it is a blow-up solution [Samarskii et al. 1987, Klokov et al. 1997]. When $\beta < 1$ and the function $\psi(\xi)$ is bounded, Solution (4.8) does not tend to infinity with the finite t.

Now we consider the case

$$\beta = 1. \tag{4.10}$$

In this case the equation of the straight line (4.6) is $q_2 = 2/\sigma$. Hence Equation (4.5) in the case (4.10) has the self-similar solution

$$x_3 = v(t) w_2^{2/\sigma}. \tag{4.11}$$

Equation (4.5) for the function v gives the ordinary differential equation

$$\dot{v} = \omega v^{\sigma+1} + v, \quad \text{with} \quad \omega = 2(2+\sigma)/\sigma^2, \tag{4.12}$$

where the dot means differentiation with respect to t. Its solutions are

$$\log \frac{v}{(\omega v^\sigma + 1)^{1/\sigma}} = t + c,$$

where c is an arbitrary constant. When $v \to \infty$, the left hand side of the last equality has the asymptotics

$$-\frac{1}{\sigma}\left(\log \omega + \frac{1}{\omega v^\sigma}\right).$$

When $v \to \infty$, we have $v^{-\sigma} \approx -\sigma\omega t + c_1$, $c_1 = \text{const}$, i.e. $v \approx (c_1 - \sigma\omega t)^{-1/\sigma}$. That means that in the case (4.10) Solution (4.11) to Equation (4.5) is the blow-up solution.

That result can be obtained by another means: by methods of Chapter 6 of this book. In this case $n = 2$. In the plane q_1, q_2 the support of Equation (4.12) consists of three points $(-1, 1)$, $(0, \sigma+1)$, $(0, 1)$. Their convex hull is the triangle with vertices in the points. The boundary of the triangle consists of three vertices and three edges. The edge of the triangle that connects the first two vertices has the ray as the normal cone, which is spanned on the vector $(-\sigma, 1)$. To the edge there corresponds the truncated equation $\dot{v} = \omega v^{\sigma+1}$. It has the power solution $v = \alpha w_1^{-1/\sigma}$, where $-\delta_1/\sigma = \omega\alpha$, i.e. $\alpha = -\delta_1/(\sigma\omega)$. That solution to the truncated equation is the asymptotics of the solution to the complete equation when $w_1 \to 0$ and $v \to \infty$, since in the basis vector $(-\sigma, 1)$ of the normal cone the first coordinate is negative and the second one is positive.

The third means is to apply the methods of Chapter 3. In this case $n = 1$. We write Equation (4.12) in the form $(\log v)^{\cdot} = \omega v^\sigma + 1$. Then the support of the system consists of two points $q_1 = \sigma$ and $q_1 = 0$. Their convex hull is the segment $[0, \sigma]$, since $\sigma > 0$. To solutions with $v \to \infty$ there corresponds the first vertex of the segment, i.e. the point $q_1 = \sigma$. To it there corresponds the truncated system $(\log v) = \omega v^\sigma$. Further, it is as it was in the second method.

The fourth method to obtain blow-up solutions was suggested by S.A. Posashkov. When $\beta = 1$ the change $\tilde{u} = ue^{-t}$, $\tilde{t} = (e^{\sigma t} - 1)/\sigma$ reduces Equation (4.5) into Equation (2.4) (see [Samarskii et al. 1987, Chapter II, § 7, Subsection 1]). Here to blow-up solutions to Equation (2.4) there correspond blow-up solutions to Equation (4.5). In particular, Solution (4.4) related to the first (algebraic) case may be taken. □

Theorems 3.1 and 4.1 give such an algorithm for finding all self-similar solutions, which is substantially simpler than the traditional method of computation of the Lie operators admitted by the equation [Ovsiannikov 1978, Ibragimov 1983]. The test of the inclusion (4.1) for the results cited in [Ibragimov 1994] showed there a number of misprints and inaccuracies. Some of them are discussed below in Example 4.3. According to [Ibragimov 1994], among all symmetries of differential equations approximately 70% are the symmetries corresponding to operators of the form (3.1), i.e. power and logarithmic ones.

Example 4.3. The test of the inclusion (4.1) for the results cited in [Ibragimov 1994] showed there the following three cases of its failure.

1. Page 120, line 12 (see also [Ovsiannikov 1960]). The equation $u_t = (k(u)u_x)_x$ with $k = u^{-4/3}$, the solution $u = x^{-3}f(t)$. Here $n = 3$, $x_1 = t$, $x_2 = x$, $x_3 = u$. The support of the equation consists of two points $Q_1 = (-1, 0, 1)$ and $Q_2 = (0, -2, -1/3)$, its dimension $d = 1$. The directing vector of the super-support of the equation is $B \stackrel{\text{def}}{=} Q_2 - Q_1 = (1, -2, -4/3)$. If $f(t)$ is an arbitrary function, then the super-support of the solution is parallel to the vector $E_1 = (1, 0, 0)$; besides, it passes through points $E_3 = (0, 0, 1)$ and $Q_3 = (0, -3, 0)$, i.e. it is parallel to the vector $E_3 - Q_3 = (0, 3, 1) \stackrel{\text{def}}{=} B_2$. The vector product $[E_1, B_2] = (0, -1, 3) \stackrel{\text{def}}{=} N$ is the normal vector to the super-support of the solution. The scalar product $\langle N, B \rangle = -2 \neq 0$, i.e. the property (4.1) is not satisfied. Let us find the equation for the function $f(t)$. Substituting the mentioned solution into the original equation, we obtain the equation $f' = 0$ for $f(t)$. Consequently, $f(t) = \text{const}$. In this case the support of the solution consists of two points E_3 and Q_3, and the directing vector of the super-support of the solution is B_2. It is not collinear to the vector B, and the property (4.1) of Theorem 4.1 is not satisfied. In this solution, the property (4.1) of Theorem 4.1 is not satisfied either, since $u_t \equiv 0$ and $(u^{-4/3}u_x)_x \equiv 0$, i.e. both parts of the original equation vanish. Each of them has the zero dimension, and for each of them the property (4.1) is satisfied. That shows that in Theorem 4.1 the condition on the solution is essential.

2. Page 196, § 11.10. For the equation

$$u_t + uu_x + \beta u_{ttt} = 0 \tag{4.13}$$

the self-similar solution

$$u = t^2 U(t^{-3}x) \tag{4.14}$$

is cited. Here $n = 3$, $x_1 = t$, $x_2 = x$, $x_3 = u$. The support of Equation (4.13) consists of three points $Q_1 = (-1,0,1)$, $Q_2 = (0,-1,2)$, $Q_3 = (-3,0,1)$. Their remainders are

$$B_1 \stackrel{\text{def}}{=} Q_2 - Q_1 = (1,-1,1),$$

$$B_2 \stackrel{\text{def}}{=} Q_3 - Q_1 = (-2,0,0).$$

The vector product $[B_1, B_2] = -(0,2,2)$. Consequently, the vector $N_1 = (0,1,1)$ is the normal one to the super-support of Equation (4.13), and not the vector $\Lambda = (1,3,2)$ cited in [Ibragimov 1994] for the operator X_3, and which is normal to the super-support of Solution (4.14). Let us move coordinates of vectors Q_i in such a way that we have obtained the vector product equal to $-\Lambda$. Indeed, if instead of Q_3 we take $Q_4 = (-3,0,2)$, then

$$B_3 \stackrel{\text{def}}{=} Q_4 - Q_3 = (-2,0,1),$$

and the vector product $[B_1, B_3] = -(1,3,2)$. Consequently, there is a misprint in Equation (4.13): the factor u in the third addend is missing. The correct equation with the self-similar solution (4.14) is

$$u_t + uu_x + \beta uu_{ttt} = 0.$$

Indeed, in the original paper [Kostin 1969] that equation was considered with the number (0.4), and Equation (4.13) is a result of the misprint in [Ibragimov 1994].

3. Pages 299–300, § 13.4.1 (see also [Vinogradov and Vorob′ev 1976]. For the equation $u_{tx} - (uu_x)_x - u_{yy} = 0$ the solution $u^2 = (4x/t - y^2/t^2)/8$ (the second one in the page 300) is cited. Here $n = 4$, $X = (t, x, y, u)$; the support of the equation consists of three points $Q_1 = (-1,-1,0,1)$, $Q_2 = (0,-2,0,2)$, $Q_3 = (0,0,-2,1)$. Their remainders

$$B_1 \stackrel{\text{def}}{=} Q_2 - Q_1 = (1,-1,0,1),$$

$$B_2 \stackrel{\text{def}}{=} Q_3 - Q_1 = (1,1,-2,0)$$

are linearly independent. Consequently, the dimension of the equation $d = 2$. The support of the solution consists of three points $Q_4 = (0,0,0,2)$, $Q_5 = (-1,1,0,0)$, $Q_6 = (-2,0,2,0)$. Their remainders

$$B_3 \stackrel{\text{def}}{=} Q_5 - Q_4 = (-1,1,0,-2),$$

$$B_4 \stackrel{\text{def}}{=} Q_6 - Q_4 = (-2, 0, 2, -2)$$

are linearly independent. Consequently, the dimension of the solution equals two. Here $B_2 = B_3 - B_4$, but the vector B_1 is not expressed linearly through B_3 and B_4. Consequently, the property (4.1) of Theorem 4.1 is not satisfied. But on the solution $(uu_x)_x \equiv 0$, i.e. the condition of Theorem 4.1 is not satisfied. That solution also satisfies the equation $u_{tx} - u_{yy} = 0$. The directing vector of the super-support of the equation B_2 is linearly expressed through vectors B_3 and B_4, i.e. the property (4.1) of Theorem 4.1 is satisfied for each of the two equations of dimensions 0 and 1 respectively. □

5. The power transformation

Let us form the vector $\log X = (\log x_1, \dots, \log x_n)^*$, where the asterisk means the transposition. Let $A = (a_{ij})$ be a square non-singular n-matrix. The change of coordinates

$$\log Y = A \log X \tag{5.1}$$

is called the *power transformation* [Bruno 1962, 1979a]. The inverse change $\log X = A^{-1} \log Y$ is also the power transformation.

Theorem 5.1 *Under the power transformation* (4.1) :

1) *the sum* (1.1) *turns into the sum* $\tilde{f}(Y) = \sum f_Q Y^{\tilde{Q}}$, *where* $\tilde{Q} = A^{*-1}Q$;

2) *the function* $f(X)$ *of the form* (1.4) *turns into the function* $\tilde{f}(Y) = f(X)$ *of the form* (1.4), *besides* $\mathbf{SS}(\tilde{f}) = A^{*-1}\mathbf{SS}(f)$;

3) *the differential polynomial* (2.3) *turns into the ratio of two differential polynomials* $g(Y)/h(Y) = f(X)$, *besides the support* $\mathbf{S}(h)$ *consists of one point* $Q = 0$, *and the support* $\mathbf{S}(g) = A^{*-1}\mathbf{S}(f)$;

4) *the normals to supports are transformed linearly by the rule* $\tilde{N} = AN$.

For the proof of Theorem 5.1, auxiliary statements are needed. Let $\rho(X)$ be the product of logarithmic derivatives of the form

$$\frac{\partial^{\|L'\|} \log x_n}{(\partial \log x_1)^{l_1} \dots (\partial \log x_{n-1})^{l_{n-1}}} \stackrel{\text{def}}{=} \frac{\partial^{\|L'\|} \log x_n}{(\partial \log X')^{L'}}, \tag{5.2}$$

where $L' = (l_1, \dots, l_{n-1})$, $l_i \geq 0$. Such a derivative may be written in the ordinary form as a differential polynomial. Consequently, the product $\rho(X)$ of such derivatives is also a differential polynomial. It is obvious that $\mathbf{S}(\rho) = 0$. We call $\rho(X)$ the *differential logarithmic monomial*. The sum

$$\sum_{i=1}^{s} X^{Q_i} \rho_i(X), \tag{5.3}$$

where $\rho_i(X)$ are differential logarithmic monomials, we call the *logarithmic form*. It is obvious that every logarithmic form (5.3) may be written in the form of the differential polynomial (2.3). The reverse is also true.

Lemma 5.1 *Every differential polynomial* (2.3) *may be written in the logarithmic form* (5.3).

Proof. It would suffice to make it for one differential monomial $a(X)$. As it was shown in the proof of Theorem 2.2 of Chapter 6, the equality

$$\frac{\partial^l x_n}{\partial x_1{}^{l_1} \ldots \partial x_{n-1}{}^{l_{n-1}}} = X'^{L'} x_n \frac{P_l(\partial \log x_n, \ldots, \partial^l \log x_n)}{(\partial \log x_1)^{l_1} \ldots (\partial \log x_{n-1})^{l_{n-1}}} \tag{5.4}$$

is valid, where $l = l_1 + \ldots + l_{n-1}$ and $P_l(\xi_1, \ldots, \xi_l)$ are some polynomials with constant coefficients containing only the terms of the form

$$\text{const}\, \xi_1^{k_1} \ldots \xi_l^{k_l} \text{ with } k_1 + 2k_2 + \ldots + l\, k_l = l, \;\; k_i \geq 0.$$

Since the differential monomial $a(X)$ is the product of an ordinary monomial and some derivatives of the form (5.4), then, according to (5.4), it is written in the form

$$a(X) = X^Q \sum_{i=1}^{t} \rho_i(X),$$

where $Q = Q(a)$ and ρ_i are differential logarithmic monomials. □

Lemma 5.2 *Let*

$$x_n = \varphi(X') \stackrel{\text{def}}{=} \sum \varphi_{Q'} X'^{Q'} \quad \textit{over } Q' \in \mathbb{Z}_+^{n-1} \tag{5.5}$$

be a local expansion with arbitrary coefficients $\varphi_{Q'}$. Let as a result of the affine transformation

$$X = WY, \;\; \det W \neq 0, \;\; W = (w_{ij}) \tag{5.6}$$

the expansion

$$y_n = \psi(Y') \stackrel{\text{def}}{=} \sum \psi_{R'} Y'^{R'} \quad \textit{over } R' \in \mathbb{Z}_+^{n-1} \tag{5.7}$$

is obtained from the expansion (5.5). *Then its coefficients have the form*

$$\psi_{R'} = \beta_{R'}(\{\varphi_{Q'}\})/a^{2||R'||-1},$$

where $\beta_{R'}(\{\varphi_{Q'}\})$ are polynomials in coefficients $\varphi_{Q'}$ with $||Q'|| \leq ||R'||$, and

$$a = w_{nn} - \sum_{i=1}^{n-1} \varphi_{E_i} w_{in}, \tag{5.8}$$

where E_i denotes the i-th unit vector.

Proof. We write the expansion (5.5) in the form $f(X) \stackrel{\text{def}}{=} x_n - \varphi(X') = 0$ and make in the equation the change (5.6). Then it takes the form

$$f(X) = g(Y) \stackrel{\text{def}}{=} \sum_{i=1}^{n} w_{ni}y_i - \varphi((AY)') = 0. \tag{5.9}$$

We consider it as an equation for the implicit function (5.7). It is clear from (5.9) that $a \stackrel{\text{def}}{=} \partial g/\partial y_n$ has the form (5.8) in the origin. Applying to Equation (5.9) Theorem 1.1 of Chapter 2, we obtain the existence and uniqueness of the expansion (5.7), and the mentioned form of its coefficients. □

Corollary 5.1 *Under the change* (5.6), *partial derivatives* $\partial^{||R'||}y_n/(\partial Y')^{R'}$ *are rational functions of partial derivatives* $\partial^{||Q'||}x_n/(\partial X')^{Q'}$ *with* $||Q'|| \leq ||R'||$, *besides, denominators of these rational functions are powers of the sum*

$$a \stackrel{\text{def}}{=} w_{nn} - \sum_{i=1}^{n-1} w_{ni}\partial x_n/\partial x_i.$$

Indeed, according to the Tailor formula

$$\varphi_{Q'} = \frac{1}{Q'!}\frac{\partial^{||Q'||}x_n}{(\partial X')^{Q'}}, \quad \psi_{R'} = \frac{1}{R'!}\frac{\partial^{||R'||}y_n}{(\partial Y')^{R'}},$$

where $Q'! = q_1!\dots q_{n-1}!$, i.e. the derivatives differ from the coefficients of expansions (5.5) and (5.7) by constant factors.

Corollary 5.2 *Let* $A = W^{-1} = (a_{ij})$. *Then in the situation of Lemma 5.2, partial derivatives* $\partial^{||Q'||}x_n/(\partial X')^{Q'}$ *are rational functions of derivatives* $\partial^{||R'||}y_n/(\partial Y')^{R'}$ *with* $||R'|| \leq ||Q'||$, *besides, their denominators are powers of the sum*

$$a_{nn} - \sum_{i=1}^{n-1} a_{in}\partial y_n/\partial y_i. \tag{5.10}$$

Remark 5.1. If the matrix W has the block-triangular form cited in Section 2 of Chapter 6, i.e. $w_{1n} = \dots = w_{n-1\,n} = 0$, then according to (5.8) $a = w_{nn} = \text{const}$, and derivatives $\partial^{||R'||}y_n/(\partial Y')^{R'}$ are polynomials in derivatives $\partial^{||Q'||}x_n/(\partial X')^{Q'}$. The reverse is also true, i.e. the second derivatives are polynomials in the first ones, since in that case $a_{1n} = \dots = a_{n-1\,n} = 0$, and the sum (5.10) is equal to the constant a_{nn}.

Proof of Theorem 5.1. Statement 1 is Property 1 of the power transformation from Section 3 of Chapter 2.

Statement 2 of Theorem 5.1 follows from the equality $X^{B_i} = Y^{\tilde{B}_i}$, where $\tilde{B}_i = A^{*-1}B_i$.

Let us prove Statement 3 of Theorem 5.1. Firstly, according to Lemma 5.1, we write the differential polynomial in the logarithmic form (5.3), where $\rho_i(X)$ are products of derivatives of the form (5.2). After the transformation (5.1), $X^{Q_i} = Y^{\tilde{Q}_i}$, where $\tilde{Q}_i = A^{*-1}Q_i$. We denote

$$\xi_i = \log x_i, \quad \eta_i = \log y_i, \quad i = 1, \dots, n.$$

In that notation the derivative (5.2) is

$$\partial^{||L'||}\xi_n/(\partial\Xi')^{L'}, \tag{5.11}$$

and the transformation (5.1) is an affine transformation $\mathrm{H} = A\Xi$. According to Corollary 5.2 from Lemma 5.2, derivatives (5.11) are rational functions of derivatives $\partial^{||M'||}\eta_n/(\partial\mathrm{H}')^{M'}$ with $||M'|| \le ||L'||$, besides, the denominators there are powers of the sum

$$\tilde{a} = a_{nn} - \sum_{i=1}^{n-1} a_{in}\partial\eta_n/\partial\eta_i.$$

Consequently, differential logarithmic monomial $\rho_i(X)$ is the ratio $\tilde{\rho}_i(Y)/\tilde{a}^m$, where $\tilde{\rho}_i$ is the polynomial in derivatives of the form $\partial^{||M'||}\log y_n/(\partial \log Y')^{M'}$ and $\tilde{a} = a_{nn} - \sum\limits_{i=1}^{n-1} a_{in}\partial \log y_n/\partial \log y_i$; and the whole sum (5.3) is $g(Y)/\tilde{a}^{\tilde{m}}$, where $g(Y)$ is the logarithmic form of some differential polynomial, and $\tilde{m} \ge 0$ is integer. Since the supports of differential polynomials $\tilde{\rho}_i(Y)$ and $\tilde{a}(Y)$ consist of one point (the origin), then the support $\mathbf{S}(\tilde{a}^{\tilde{m}}) = 0$ and the support $\mathbf{S}(g) = A^{*-1}\mathbf{S}(f)$. That completes the proof of the third statement of Theorem 5.1.

The fourth statement follows from the first three, since in order to preserve the scalar product, the transformation (5.1) induces the transformation $\tilde{N} = AN$ in the dual space $\mathbb{R}^n_*$. □

To the multiplication of the differential polynomial $f(X)$ by the monomial X^T there corresponds in $\mathbb{R}^n$ the parallel translation of supports $\mathbf{S}(f)$ and $\mathbf{SS}(f)$ by the vector T. Hence if $d(f) = d < n$, then by a linear transformation of the form $\tilde{Q} = A^{*-1}(Q + T)$ the set $\mathbf{S}(f)$ can be placed in the coordinate subspace

$$\tilde{q}_1 = \dots = \tilde{q}_{n-d} = 0. \tag{5.12}$$

Theorem 5.2 *For the differential polynomial $f(X)$ with $d(f) \stackrel{\text{def}}{=} d < n$ there exist the vector $T \in \mathbb{R}^n$ and the matrix A, such that the power transformation*

(4.1) *reduces the differential equation* $X^T f(X) = 0$ *to the form* $g(Y) = 0$, *where for all points* $\tilde{Q}$ *of the support* $\mathbf{S}(g)$ *the property* (5.12) *holds.*

Proof follows from the statement 3 of Theorem 5.1 and the solution of Problem 3 in Section 11 of Chapter 1. As a vector $-T$, any vector from the super-support $\mathbf{SS}(f)$ can be taken. □

Remark 5.2. If the coordinate x_i is present in the polynomial (2.3) only under the differentiation sign, then the power changes of the form $dx_i = Y^{B_i} dy_i$ are also can be made, where $B_i = (b_{i1}, \ldots b_{in})$ and $b_{ii} = 0$.

Example 5.1 (the continuation of Examples 2.1, 3.1, 4.1). For Equation (2.4), $n = 3$ and $d = 1$. In notation (2.5), we make the power transformation (5.1)

$$y_1 = x_1,$$
$$y_2 = x_2,$$
$$y_3 = X^{B/\sigma} = x_1^{1/\sigma} x_2^{-2/\sigma} x_3$$

with the matrix

$$A = \begin{pmatrix} 1 & 0 & 0 \\ 0 & 1 & 0 \\ 1/\sigma & -2/\sigma & 1 \end{pmatrix}. \tag{5.13}$$

In the original coordinates t, x, u, that transformation is $u = vt^{-1/\sigma}x^{2/\sigma}$. After cancelling $t^{-1-1/\sigma}x^{2/\sigma}$, Equation (2.4) turns into the equation

$$(v_t t) - v/\sigma = (v_{xx}x^2)v^\sigma + \sigma(v_x x)^2 v^{\sigma-1} + 4(\sigma+1)\sigma^{-1}(v_x x)v^\sigma + 2(2+\sigma)\sigma^{-2}v^{\sigma+1}. \tag{5.14}$$

The support of the equation consists of two points

$$(0, 0, 1) \quad \text{and} \quad (0, 0, 1+\sigma). \tag{5.15}$$

Here the vector $T = (1 + 1/\sigma, -2/\sigma, 0)$, and the matrix A is given by Formula (5.13). Instead of t and x, w_1 and w_2 can be taken as x_1, x_2. □

6. The logarithmic transformation

Let all points $\tilde{Q}$ of the support of a differential polynomial $g(Y)$ have the coordinate $\tilde{q}_i = 0$. Then the coordinate y_i belongs to the $g(Y)$ only as powers of the differential $\partial \log y_i$. Hence if we make the *logarithmic transformation* $z_i = \log y_i$, then $g(Y)$ will become the differential polynomial in $y_1, \ldots, y_{i-1}, z_i, y_{i+1}, \ldots, y_n$.

Theorem 6.1 *If for all points $\tilde{Q}$ of the support $\mathbf{S}(g)$ of the differential polynomial $g(Y)$ the property* (5.12) *is satisfied, then after the logarithmic transformation*

$$\begin{aligned} z_i &= \log y_i, \quad i = 1, \dots, n-d, \\ z_j &= y_j, \qquad j = n-d+1, \dots, n \end{aligned} \tag{6.1}$$

the differential polynomial $\tilde{g}(Z) = g(Y)$ is obtained.

Proof. According to Lemma 5.1, we write the differential polynomial $g(Y)$ in the logarithmic form

$$g(Y) = \sum_{k=1}^{t} Y^{\tilde{Q}_k} \tilde{\rho}_k(Y), \tag{6.2}$$

where $\tilde{\rho}_k(Y)$ are differential logarithmic monomials. By the condition of the Theorem, for all vector exponents $\tilde{Q}_i$ the property (5.12) is satisfied, i.e. $y_1, \dots, y_{n-d}$ are absent in monomials $Y^{\tilde{Q}_k}$, which are present in (6.2). These coordinates are present only in $\tilde{\rho}_k(Y)$ in the form $\log y_i$. Hence after the logarithmic change (6.1), e^{z_i} do not appear in the logarithmic form (6.2), i.e. the form (6.2) remains the differential polynomial. □

Example 6.1 (the continuation of Example 5.1). We put $z_1 = \log t$, $z_2 = \log x$. Since $v_t t = \partial v/\partial z_1$, $v_x x = \partial v/\partial z_2$, $v_{xx} x^2 = \partial^2 v/\partial z_2^2 - \partial v/\partial z_2$, then Equation (5.14) takes the form

$$\frac{\partial v}{\partial z_1} - \frac{v}{\sigma} = \frac{\partial^2 v}{\partial z_2^2} v^\sigma + \sigma \left(\frac{\partial v}{\partial z_2} \right)^2 v^{\sigma-1} + \frac{3\sigma + 4}{\sigma} \frac{\partial v}{\partial z_2} v^\sigma + \frac{2(2+\sigma)}{\sigma^2} v^{\sigma+1}. \tag{6.3}$$

The support of the equation consists of points

$$(-1,0,1), (0,0,1), (0,-2,1+\sigma), (0,-1,1+\sigma), (0,0,1+\sigma). \tag{6.4}$$

Consequently, two points (5.15) of the support of Equation (5.14) are blown up into five points (6.4) of the support of Equation (6.3). □

Example 6.2 (the continuation Example 4.2). We consider Equation (4.5) firstly in the case (4.7). We make the power transformation

$$\begin{aligned} t &= t, \\ y &= t^{\varkappa} x, \\ v &= X^{B_2/(\beta-1)} = t^{1/(\beta-1)} u \end{aligned} \tag{6.5}$$

where the number $\varkappa$ is defined in (4.8). Here vectors $(\varkappa, 1, 0)$ and $(1/(\beta - 1), 0, 1)$ form the basis of the linear subspace in $\mathbb{R}^3$, which is parallel to the super-support of Equation (4.5). The inverse transformation to (6.5) is

$$
\begin{aligned}
t &= t, \\
x &= t^{-\varkappa} y, \\
u &= t^{1/(1-\beta)} v.
\end{aligned} \tag{6.6}
$$

We compute the derivatives using (6.5) and (6.6)

$$u_t = \frac{1}{1-\beta} t^{1/(1-\beta)-1} v + t^{1/(1-\beta)} (v_t + v_y \varkappa t^{-1} y), \quad u_x = t^{1/(1-\beta)} v_y t^{\varkappa},$$

$$u^\sigma u_x = t^{\sigma(1-\beta)+1/(1-\beta)+\varkappa} v^\sigma v_y = t^{(\sigma+1+\beta)/2(1-\beta)} v^\sigma v_y,$$

$$(u^\sigma u_x)_x = t^{(\sigma+1+\beta)/2(1-\beta)+\varkappa} (\sigma v^{\sigma-1} v_y^2 + v^\sigma v_{yy}) = t^{\beta/(1-\beta)} (\sigma v^{\sigma-1} v_y^2 + v^\sigma v_{yy}).$$

Substituting these values in Equation (4.5) and cancelling $t^{\beta/(1-\beta)}$, we obtain the equation

$$\frac{1}{1-\beta} v + t v_t + \varkappa y v_y = \sigma v^{\sigma-1} v_y^2 + v^\sigma v_{yy} + v^\beta. \tag{6.7}$$

Now we make the logarithmic transformation $\tau = \log t$ (i.e. $t = e^\tau$). Equation (6.7) transforms into

$$\frac{1}{1-\beta} v + v_\tau + \varkappa y v_y = \sigma v^{\sigma-1} v_y^2 + v^\sigma v_{yy} + v^\beta. \tag{6.8}$$

If v does not depend on τ, i.e. $v_\tau = 0$, then the equation becomes Equation (4.9). However, now we can seek a solution to the complete Equation (4.5) or (6.8) in the form of a series in negative powers of τ:

$$v = \psi(y) + \tau^{-1} \psi_1(y) + \tau^{-2} \psi_2(y) + \ldots,$$

or in the form of a polynomial in τ:

$$v = \psi(y) \tau^m + \psi_1(y) \tau^{m-1} + \ldots + \psi_m(y),$$

where $\psi(\xi)$ is a solution to Equation (4.9). This leads to such expansions of solutions $u(t,x)$ to Equation (4.5) in powers of $(\log t)^{-1}$, which begin with a self-similar solution.

We consider now Equation (4.5) in the case (4.10). We make the power transformation

$$
\begin{aligned}
t &= t, \\
x &= x, \\
v &= x^{-2/\sigma} u,
\end{aligned} \tag{6.9}
$$

where the vectors $(1, 0, 0)$ and $(0, -2/\sigma, 1)$ form a basis of the two-dimensional linear subspace parallel to the super-support of Equation (4.5). Here $u = x^{2/\sigma}v$. We compute derivatives:

$$u_x = \frac{2}{\sigma}x^{2/\sigma-1}v + x^{2/\sigma}v_x, \quad u_x u^\sigma = \frac{2}{\sigma}x^{2/\sigma+1}v^{\sigma+1} + x^{2/\sigma+2}v^\sigma v_x, \quad (u_x u^\sigma)_x =$$

$$\frac{2(2+\sigma)}{\sigma^2}x^{2/\sigma}v^{\sigma+1} + \frac{4(\sigma+1)}{\sigma}x^{2/\sigma+1}v^\sigma v_x + x^{2/\sigma+2}v^{\sigma-1}v_x^2 + x^{2/\sigma+2}v^\sigma v_{xx}.$$

Substituting this expression in Equation (4.5) and cancelling $x^{2/\sigma}$, we obtain the equation

$$v_t = \frac{2(2+\sigma)}{\sigma^2}v^{\sigma+1} + \frac{4(\sigma+1)}{\sigma}xv^\sigma v_x + x^2v^{\sigma-1}v_x^2 + x^2v^\sigma v_{xx}.$$

After the logarithmic transformation $\log x = \eta$ (i.e. $x = e^\eta$) the equation takes the form

$$v_t = \frac{2(2+\sigma)}{\sigma^2}v^{\sigma+1} + \frac{3\sigma+4}{\sigma}v^\sigma v_\eta + v^{\sigma-1}v_\eta^2 + v^\sigma v_{\eta\eta} + v, \tag{6.10}$$

since $v_x = v_\eta/x$, $v_{xx} = (v_{\eta\eta} - v_\eta)/x^2$. If v does not depend on η, then the last equation coincides with the Equation (4.12). We can seek a solution to Equation (6.10) in the form of a polynomial in η or as a series in η^{-1}; here the coefficient at leading term is the solution to Equation (4.12). In the power transformations (6.5), (6.9) and so on, one can use w_1 and w_2 instead of t and x. □

7. The ordinary differential equation

When $n = 2$, the derivative (2.1) is ordinary, and the differential equation $f(X) = 0$ corresponding to the polynomial (2.3) is also ordinary. Theorems 3.1, 4.1, 5.1, 5.2 and 6.1 remain valid for it. Here the inequality $d < n$ means that $d = 0$ or $d = 1$. However for ordinary differential equations one can prove more.

Theorem 7.1 *Let $n = 2$, and an ordinary differential equation $f(X) = 0$ have the order m and the dimension $d(f) < n$. Then the equation may be reduced to the equation of the order $m - (n - d(f))$ by means of power and logarithmic transformations.*

Proof. Let at first $d = 1$. According to Theorem 5.1, we make the power transformation which places the super-support $\mathbf{SS}(f)$ in the horizontal straight line $\tilde{q}_2 = \text{const} = c$. After cancellation of y_2^c, we obtain the equation $g(Y) = 0$, the support of which lies in the axis $\tilde{q}_2 = 0$. Its order equals m.

This equation includes y_2 only in the form $d\log y_2/dy_1$. Taking this derivative as a new dependent variable, we obtain the equation of the order $m-1$.

We consider now the case $d(f)=0$, when the support $\mathbf{S}(f)$ consists of one point Q. Multiplying the equation by X^{-Q}, we obtain the equation with the zero support. Since $q_2=0$, then x_2 belongs to it only in the form $d\log x_2/dx_1$. Taking this derivative as a new dependent variable, we obtain the equation of the order $m-1$. Its support is placed in the vertical axis $q_1=0$. Exchanging the dependent and the independent variables, which corresponds to the power transformation (5.1) with the matrix

$$A=\begin{pmatrix} 0 & 1 \\ 1 & 0 \end{pmatrix},$$

we obtain the equation of the order $m-1$, the support of which is placed in the horizontal axis $\tilde{q}_2=0$. Repeating the means applied above, we lower the order of the equation by one yet again. □

Example 7.1. We consider the Blasius equation (see Examples 3.2 and 7.2 of Chapter 3 and the proof of Lemma 6.2 of Chapter 6)

$$y'''+y''y=0, \tag{7.1}$$

where the prime means the derivation with respect to x. Here $n=2$, $m=3$, $x_1=x$, $x_2=y$. The support of Equation (7.1) consists of two points $Q_1=(-3,1)$ and $Q_2=(-2,2)$ lying in the straight line $q_1-q_2+4=0$, i.e. $d=1$. Consequently, $m-(n-d)=2$. According to Remark 5.2, we make the power transformation $dt=ydx$. Then $dt/dx=y$. We denote by dot the differentiation with respect to t. We have

$$y'=\dot{y}y,\quad y''=\ddot{y}y^2+\dot{y}^2y,\quad y'''=\dddot{y}\,y^3+4\ddot{y}\dot{y}y^2+\dot{y}^3y.$$

Substituting the derivatives in Equation (7.1) and cancelling y^4, we obtain the equation

$$\dddot{y}\,y^{-1}+4\ddot{y}\dot{y}y^{-2}+\dot{y}^3y^{-3}+\dot{y}y^{-1}+\dot{y}^2y^{-2}=0. \tag{7.2}$$

We make the logarithmic transformation $\log y=z$, $y=e^z$ in it. Then

$$\dot{y}=\dot{z}y,\quad \ddot{y}=\ddot{z}y+\dot{z}^2y,\quad \dddot{y}=\dddot{z}\,y+3\ddot{z}\dot{z}y+\dot{z}^3y.$$

Substituting these derivatives in Equation (7.2) and collecting similar terms, we obtain the equation

$$\dddot{z}+7\ddot{z}\dot{z}+6\dot{z}^3+\ddot{z}+2\dot{z}^2=0.$$

Assuming here $\dot{z} = u$, we obtain the equation of the second order

$$\ddot{u} + 7\dot{u}u + 6u^3 + \dot{u} + 2u^2 = 0.$$

The support of this equation consists of five points, and its dimension equals two. □

If the dimension $d(f) < n$, then the equation $f(X) = 0$ and its solution are quasi-homogeneous. Hence the one boundary value problem at two ends of an interval can be reduced to two boundary value problems at the one end of the interval (see [Klamkin 1962, 1970; Na 1979; Ibragimov 1994, Chapter 16]). In the following two theorems we suppose that the solution to any boundary value problem $f(X) = 0$, $x_2^{(k)}(0) = \alpha_k$, $k = 0, \dots, m-1$ exists, and it is unique.

Theorem 7.2 *Let $f(X) = 0$ be an ordinary differential equation of the order m and the dimension 1, and the vector $\Lambda = (\lambda_1, \lambda_2)$ be normal to its super-support. If $l\lambda_1 \neq \lambda_2$, then the solution of the boundary value problem*

$$x_2(0) = x'_2(0) = \dots = x_2^{(m-2)}(0) = 0, \tag{7.3}$$

$$x_2^{(l)}(\infty) = c \neq 0 \tag{7.4}$$

is reduced to the solution of two boundary value problems with conditions only at zero $x_1 = 0$.

Proof. Let $x_2 = \varphi(x_1)$ be a solution to the boundary value problem (7.3), $x_2^{(m-1)}(0) = 1$. Let $\varphi^{(l)}(\infty) \stackrel{\text{def}}{=} \gamma \neq 0, \infty$. Under the substitution (3.2) the equation is transformed into itself, and the solution $x_2 = \varphi(x_1)$ is transformed into the solution

$$x_2 = \mu^{-\lambda_2}\varphi(\mu^{\lambda_1}x_1). \tag{7.5}$$

Its kth derivative is

$$x_2^{(k)} = \mu^{-\lambda_2+k\lambda_1}\varphi^{(k)}(\mu^{\lambda_1}x_1). \tag{7.6}$$

Solution (7.5) obviously satisfies zero boundary conditions (7.3). It satisfies the boundary condition (7.4) if $\mu^{-\lambda_2+l\lambda_1}\gamma = c$, i.e. $\mu = \mu_0 \stackrel{\text{def}}{=} (c/\gamma)^{1/(l\lambda_1-\lambda_2)}$, since according to the condition of Theorem $-\lambda_2 + l\lambda_1 \neq 0$. Here according to (7.6), on Solution (7.5) with $\mu = \mu_0$ we have

$$x_2^{(m-1)}(0) = \mu_0^{-\lambda_2+(m-1)\lambda_1}\varphi^{(m-1)}(0) = (c/\gamma)^{[(m-1)\lambda_1-\lambda_2]/(l\lambda_1-\lambda_2)}. \tag{7.7}$$

Hence, in order to solve the boundary value problem (7.3), (7.4), one must solve firstly the boundary value problem (7.3), $x_2^{(m-1)}(0) = 1$, and find the value $\gamma \stackrel{\text{def}}{=} \varphi^{(l)}(\infty)$; and then to solve the second boundary value problem (7.3), (7.7). □

If the dimension of an equation equals zero, then the equation is invariant under the changes (3.2) with two independent parameters. Hence the mentioned in Theorem 7.2 reduction is possible in the case when there is a unique non-zero boundary condition at $x_1 = 0$. For an equation of the second order this situation is analyzed in [Klamkin 1962, 1970; Ibragimov 1994, Section 16.1].

Theorem 7.3 *Let $f(X) = 0$ be an ordinary differential equation of the order $m \geq 2$ and dimension 0. In generic case, the solution of the boundary value problem*

$$x_2(0) = x'_2(0) = \ldots = x_2^{(m-3)}(0) = 0, \tag{7.8}$$

$$x_2^{(m-2)}(\infty) = a, \quad x_2^{(m-1)}(\infty) = b \tag{7.9}$$

is reduced to the solution of two boundary value problems with conditions only at zero $x_1 = 0$.

Proof. Let $x_2 = \varphi(x_1)$ be a solution to the boundary value problem (7.8),

$$x_2^{(m-2)}(0) = x_2^{(m-1)}(0) = 1, \tag{7.10}$$

and $\varphi^{(m-2)}(\infty) = \alpha \neq 0$, $\varphi^{(m-1)}(\infty) = \beta \neq 0, \infty$. Since the equation $f(X) = 0$ is zero-dimensional, it is invariant under the changes $\tilde{x}_1 = \mu x_1$, $\tilde{x}_2 = x_2$ and $\tilde{x}_1 = x_1$, $\tilde{x}_2 = \nu x_2$, where $\mu, \nu \in \mathbb{R}$, $\mu, \nu > 0$. Therefore with any $\mu, \nu > 0$, the expression $x_2 = \nu\varphi(\mu x_1)$ is also the solution to the equation $f(X) = 0$. It obviously satisfies the boundary conditions (7.8). Since for that solution

$$x_2^{(k)} = \nu\mu^k\varphi^{(k)}(\mu x_1),$$

then it satisfies conditions (7.9) if

$$\nu\mu^{m-2}\alpha = a, \quad \nu\mu^{m-1}\beta = b.$$

From these equations the concrete values of parameters $\mu = \mu_0$, $\nu = \nu_0$ are determined, which satisfy the equations: $\mu_0 = (b/\beta)(a/\alpha)^{-1}$, $\nu_0 = (a/\alpha)\mu_0^{2-m} = (a/\alpha)^{m-1}(b/\beta)^{2-m}$. When $x_1 = 0$, the solution $x_2 = \nu_0\varphi(\mu_0 x_1)$ has

$$x_2^{(m-2)}(0) = \nu_0\mu_0^{m-2} = a/\alpha, \quad x_2^{(m-1)}(0) = \nu_0\mu_0^{m-1} = b/\beta. \tag{7.11}$$

Hence in order to solve the problem (7.8), (7.9), one must at first solve the problem (7.8), (7.10), and find values α, β, and then solve the problem (7.8), (7.11). □

Particular cases of Theorems 7.2 and 7.3 and other types of boundary conditions, and also problems with a finite interval of the independent variable and other examples see in [Klamkin 1970; Na 1979].

8. The system of equations

Let $X' = (x_1, \ldots, x_{n'})$ be independent variables, and $X'' = (x_{n'+1}, \ldots, x_n)$ be dependent variables. In a similar way, we will split each vector Y of the length n into two subvectors Y' and Y'' of the length n' and $n - n'$ respectively. The product of powers of X and derivatives of the form

$$\partial^{\|L'\|} x_j / \partial X'^{L'}, \tag{8.1}$$

where $L' \in \mathbb{Z}_+^{n'}$ and $j > n'$, is called the *differential monomial* $a(X)$. To each differential monomial $a(X)$ we put in correspondence the point $Q(a) \in \mathbb{R}^n$ (its vector power exponent) by the following rule: to the monomial const X^Q there corresponds the point Q; to the derivative (8.1) there corresponds the point

$$Q = (-L', E_j), \text{ i.e. } Q' = -L', \quad Q'' = E_j; \tag{8.2}$$

and to the product of two differential monomials $a(X)$ and $b(X)$ there corresponds the point $Q(ab) = Q(a) + Q(b)$. The finite sum of differential monomials

$$f(X) = \sum_{i=1}^{s} a_i(X) \tag{8.3}$$

is called the *differential polynomial*. The set of points $\mathbf{S}(f) \stackrel{\text{def}}{=} \{Q(a_i), i = 1, \ldots, s\}$ is called the *support* of the differential polynomial (8.3). The affine hull $\mathbf{SS}(f)$ of the support $\mathbf{S}(f)$ is called the *super-support of the polynomial* (8.3). Its *dimension* $d(f) \stackrel{\text{def}}{=} \dim \mathbf{SS}(f)$ and the *normal subspace* $\mathbf{N}(f) \stackrel{\text{def}}{=} \mathbf{N}(\mathbf{SS}(f)) \subset \mathbb{R}_*^n$ are defined similarly.

Now we consider the aggregate of differential polynomials

$$f_1(X), \ldots, f_m(X). \tag{8.4}$$

Each of the polynomials $f_i(X)$ has its own support $\mathbf{S}(f_i)$, the normal subspace $\mathbf{N}(f_i)$, and the dimension $d(f_i)$. However the *normal subspace* $\mathbf{N}$ *of the aggregate* (8.4) is the intersection

$$\mathbf{N} = \mathbf{N}(f_1) \cap \ldots \cap \mathbf{N}(f_m); \tag{8.5}$$

and its codimension d, i.e. $d = n - \dim \mathbf{N}$, is the *dimension of the aggregate* (8.4). If we put in correspondence to the aggregate (8.4) the system of equations

$$f_i(X) = 0, \quad i = 1, \ldots, m, \tag{8.6}$$

then the analogues to Theorems 3.1, 4.1, 5.1, 5.2, 6.1, 7.1–7.3 are valid for the aggregate (8.4) and for System (8.6) as well. We formulate these theorems without proofs, since they repeat almost literally the proofs of the mentioned Theorems.

The *system* of equations (8.6) *admits the Lie operator* (3.1) if each of the equations is invariant under the change (3.2).

Theorem 8.1 *The system of equations* (8.6) *admits the Lie operator* (3.1) *if the vector* $\Lambda = (\lambda_1, \dots, \lambda_n)$ *lies in the intersection* (8.5).

The expression $X'' = \Phi''(X')$ is called the *solution to the system of equations* (8.6) if each equation turns into identity with respect to X' by this substitution. A solution is the *self-similar* one if it admits the operator of the form (3.1), i.e. it is invariant under the changes (3.2). If the solution $X'' = \Phi''(X')$ admits the operator (3.1), then all super-supports $\mathbf{SS}(x_j - \varphi_j(X'))$, $j = n' + 1, \dots, n$ are normal to the vector Λ. Here the super-support $\mathbf{SS}(x_j - \varphi_j(X'))$ passes through the point E_j, and the super-support $\mathbf{SS}(\varphi_j(X'))$ is placed in the coordinate subspace $Q'' = 0$. Hence the super-support $\mathbf{SS}(\varphi_j(X'))$ is placed in the linear manifold

$$\langle \Lambda', Q' \rangle = \lambda_j, \quad Q'' = 0. \tag{8.7}$$

Theorem 8.2 *Let* $X'' = \Phi''(X')$ *be such a solution to System* (8.6) *which does not satisfy any system* $\check{f}_i = 0$, $i = 1, \dots, m$ *formed by proper subsums of the sum* f_i, *and having the dimension smaller than the dimension of System* (8.6). *Then the normal subspace of the solution lies in the normal subspace of the system.*

Example 8.1. We consider System (1.6) from [Klokov et al. 1997, page 102]

$$\begin{aligned} u_t &= (u^{\sigma_1} u_x)_x + u^{\beta_1} v^{\gamma_2}, \\ v_t &= (u^{\sigma_2} v_x)_x + v^{\beta_2} u^{\gamma_1}, \quad \beta_i, \gamma_i \geq 0, \quad \sigma_i > 0. \end{aligned} \tag{8.8}$$

The functions u, v may be the temperatures or the concentrations of two substances in a self-catalytic reaction. Here $X = (t, x, u, v)$, $n = 4$, $n' = 2$. The support of the first equation consists of three points $Q_1 = (-1, 0, 1, 0)$, $Q_2 = (0, -2, \sigma_1 + 1, 0)$, $Q_3 = (0, 0, \beta_1, \gamma_2)$. The support of the second equation consists of three points $Q_4 = (-1, 0, 0, 1)$, $Q_5 = (0, -2, 0, \sigma_2 + 1)$, $Q_6 = (0, 0, \gamma_1, \beta_2)$. We compose the remainders

$$\begin{aligned} B_1 &\stackrel{\text{def}}{=} Q_2 - Q_1 = (1, -2, \sigma_1, 0), \\ B_2 &\stackrel{\text{def}}{=} Q_3 - Q_1 = (1, 0, \beta_1 - 1, \gamma_2), \\ B_3 &\stackrel{\text{def}}{=} Q_5 - Q_4 = (1, -2, 0, \sigma_2), \\ B_4 &\stackrel{\text{def}}{=} Q_6 - Q_4 = (1, 0, \gamma_1, \beta_2 - 1). \end{aligned} \tag{8.9}$$

The normal subspace $\mathbf{N}$ of System (8.8) is normal to four vectors B_1, B_2, B_3, B_4. Its dimension is greater than zero only if $\Delta \stackrel{\text{def}}{=} \det (B_1 B_2 B_3 B_4)^* = 0$. The computations give

$$\Delta = -2[\sigma_1(\beta_2 - 1 - \gamma_2) + \sigma_2(\gamma_1 + 1 - \beta_1)].$$

In order to simplify the computations of the determinant by (8.9), it is convenient to subtract the first line from the third one, and from the fourth line the second one. Consequently, $\mathbf{N} \ni \Lambda \neq 0$ only if $\Delta = 0$, i.e.

$$\frac{\sigma_1}{\sigma_2} = \frac{\gamma_1 + 1 - \beta_1}{\gamma_2 + 1 - \beta_2}.$$

Computing the cofactors of the dimension 3×3 in the matrix $(B_1 B_2 B_3)^*$, we obtain the normal vector

$$\Lambda = (\lambda_1, \lambda_2, \lambda_3, \lambda_4) = (-2(\sigma_2(\beta_1 - 1) + \sigma_1\gamma_2), \sigma_1\sigma_2 - \sigma_1\gamma_2 - \sigma_2(\beta_1 - 1), 2\sigma_2, 2\sigma_1)$$

According to (8.7), for the self-similar solution $u = \varphi(t, x)$, $v = \psi(t, x)$, the super-support of the function φ is placed in the straight line $\lambda_1 q_1 + \lambda_2 q_2 = \lambda_3$, $q_3 = q_4 = 0$, and the super-support of the function ψ is placed in the straight line $\lambda_1 q_1 + \lambda_2 q_2 = \lambda_4$, $q_3 = q_4 = 0$. Consequently,

$$u = t^{m_1} g_1(\xi), \quad v = t^{m_2} g_2(\xi), \tag{8.10}$$

where $m_1 = \lambda_3/\lambda_1$, $m_2 = \lambda_4/\lambda_1$, $\xi = t^{\varkappa} x$, $\varkappa = -\lambda_2/\lambda_1$. Here the functions g_1 and g_2 satisfy the system of ordinary differential equations

$$m_1 g_1 + \varkappa\xi g'_1 = (g_1^{\sigma_1} g'_1)' + g_1^{\beta_1} g_2^{\gamma_2},$$

$$m_2 g_2 + \varkappa\xi g'_2 = (g_2^{\sigma_2} g'_2)' + g_2^{\beta_2} g_1^{\gamma_1},$$

where the prime means the differentiation with respect to ξ. In Formulae (8.10) and further, instead of t and x one may use $w_1 = c_1 + \delta_1 t$ and $w_2 = c_2 + \delta_2 x$. □

Theorem 8.3 *Under the power transformation* (5.1) *the differential polynomial* (8.3) *is transformed into the ratio of two differential polynomials* $g(Y)/h(Y) = f(X)$; *here the support* $\mathbf{S}(h)$ *consists of one point* $Q = 0$, *and the support* $\mathbf{S}(g) = A^{*-1}\mathbf{S}(f)$. *The normals to supports are transformed by the rule* $\tilde{N} = AN$.

Theorem 8.4 *For the aggregate of differential polynomials* (8.4) *with the dimension* $d < n$ *there exist vectors* $T_1, \ldots, T_m \in \mathbb{R}^n$ *and the matrix* A *such that the power transformation* (5.1) *reduces the system of equations*

$$X^{T_i} f_i(X) = 0, \quad i = 1, \ldots, m$$

into the form

$$g_i(Y) = 0, \quad i = 1, \ldots, m,$$

where for all points $\tilde{Q}$ *of supports* $\mathbf{S}(g_i)$ *the property* (5.12) *holds.*

Theorem 8.5 *If for all points* $\tilde{Q}$ *of supports* $\mathbf{S}(g_i(X))$, $i = 1, \ldots, m$ *of differential polynomials* $g_1(Y), \ldots, g_m(Y)$ *the property* (5.12) *holds, then after the logarithmic transformation* (6.1), *the differential polynomials* $\tilde{g}_i(Z) = g_i(Y)$, $i = 1, \ldots, m$ *are obtained.*

When $n' = 1$, the derivative (8.1) is an ordinary one, and the differential equations of System (8.6) are also ordinary. Let for the dependent variable x_j, $j > 1$ the maximal order of the differentiation in System (8.6) be m_j. We call it the *order of System* (8.6) *with respect to the coordinate* x_j. The quantity $m'' = m_2 + \ldots + m_n$ is called the *order of System* (8.6).

Theorem 8.6 *Let* $n' = 1$, *the ODE system* (8.6) *have the order* m'' *and the dimension* $d < n$. *Then by means of the power and logarithmic transformations the system can be reduced to a system of the order* $m'' - (n - d)$.

Theorem 8.7 *Let* $n' = 1$, *the ODE system* (8.6) *have the order* m_j *with respect to the coordinate* x_j, $j = 2, \ldots, n$, *the order of the system be* $m'' = m_2 + \ldots + m_n$ *and the dimension* $d < n$. *Let* $N_1, \ldots, N_{n-d}$ *be a basis of the normal subspace of System* (8.6). *Let a solution to any boundary value problem* (8.6), $x_j^{(k)}(0) = \alpha_{jk}$, $k = 0, \ldots, m_j - 1$, $j = 2, \ldots, n$ *exist and be unique. Let the boundary conditions*

$$x_j^{(k)}(0) = a_{jk} \neq \infty; \quad x_j^{(l)}(\infty) = b_{jl} \neq 0, \infty$$

be given, where for every j *there are altogether* m_j *such conditions and* $0 \leq k, l < m_j$; *besides among* m'' *constants* a_{jk} *and* b_{jl} *no less than* $n - d$ *be different from zero including all* b_{jl}. *We put in correspondence to each constant* $a_{jk} \neq 0$ *the vector* $R(j,k) = -kE_1 + E_j$ *and to the constant* b_{jl} *the vector* $R(j,l) = -lE_1 + E_j$. *Let them be the vectors* $R_1, \ldots, R_s$. *We form the matrix*

$$\mathcal{R} = \begin{pmatrix} \langle N_1, R_1 \rangle & \ldots & \langle N_{n-d}, R_1 \rangle \\ \ldots & \ldots & \ldots \\ \langle N_1, R_s \rangle & \ldots & \langle N_{n-d}, R_s \rangle \end{pmatrix}.$$

If $\operatorname{rank} \mathcal{R} = n - d$, *then the solution of the mentioned boundary value problem with conditions at the zero* $x_1 = 0$ *and at the infinity* $x_1 = \infty$ *is reduced to the solution of two boundary value problems with conditions only at zero.*

Example 8.2. We demonstrate that in the situations of Theorems 7.2 and 7.3, the conditions of Theorem 8.7 are satisfied. In Theorem 7.2 we have $n - d = 1$, and the unique non-zero boundary condition is (7.4). To it there corresponds the vector $R = (-l, 1)$. The scalar product $\langle \Lambda, R \rangle = -l\lambda_1 + \lambda_2$. By the condition of Theorem 7.2, it is different from zero.

In Theorem 7.3 we have $n - d = 2$, the normal subspace is the whole plane $\mathbb{R}^2_*$, the basis there is formed by vectors $N_1 = E_1$ and $N_2 = E_2$. The two non-zero boundary conditions are (7.9). To them there correspond vectors $R_1 = (2 - m, 1)$, $R_2 = (1 - m, 1)$. The matrix $\mathcal{R}$ is

$$\begin{pmatrix} 2 - m & 1 \\ 1 - m & 1 \end{pmatrix}.$$

Its determinant equals one. Consequently, its rank equals two, i.e. $n - d$. □

Chapter 8

On complexity of problems of Power Geometry

1. The levels of complexity

When problems are solved with the help of the Power Geometry, one has to use various objects in the space of power exponents $\mathbb{R}^n$ and in the dual space $\mathbb{R}^n_*$. Here we give a classification of complexity of the problems based upon the classification of complexity of these geometrical objects. Below we introduce that classification in the order of increasing complexity.

Level I. Linear subspaces and manifolds, i.e. linear equalities for the vector powers.

Level II. Linear changes of coordinates.

Level III. Convex polyhedra, their faces and normal cones, i.e. local properties of boundaries of polyhedra described by the linear inequalities.

Level IV. Global properties of convex polyhedra (the volume, the mixed volume etc.).

The problems of the first three levels of complexity are discussed below in Sections 2, 3, 4 for system of equations of the following types. The fourth level is discussed in Remark 4.2 at the end of Section 4.

Here $X = (x_1, \ldots, x_n)$, $Q = (q_1, \ldots, q_n)$, $X^Q = x_1^{q_1} \ldots x_n^{q_n}$.

1. The system of algebraic equations

$$0 = f_i(X) \stackrel{\text{def}}{=} \sum f_{iQ} X^Q \text{ over } Q \in \mathbf{S}_i, \; i = 1, \ldots, m. \tag{1.1}$$

Here each equation has its own support $\mathbf{S}_i = \mathbf{S}(f_i) \subset \mathbb{R}^n$.

2. The system of ordinary differential equations

$$0 = f_i(X) \stackrel{\text{def}}{=} \sum a_{iQ}(X) \text{ over } Q \in \mathbf{S}_i, \; i = 1, \ldots, m. \tag{1.2}$$

Here $x_1, \ldots, x_{n_1}$ are parameters, x_l is the independent variable ($l = n_1 + 1$), $x_{l+1}, \ldots, x_n$ are dependent variables, $n - l \leq m$, and $f_i(X)$ are sums of differential monomials $a_{iQ}(X)$, each of which is the product of the ordinary monomial X^T and the derivatives of the form $d^k x_j / dx_l^k$ with $j > l$, and it has its own power exponent $Q = Q(a)$ (see Section 1 of Chapter 6). Here each equation has its own support $\mathbf{S}_i = \mathbf{S}(f_i) \subset \mathbb{R}^n$.

3. The system of partial differential equations of the form (1.2), where $x_1, \ldots, x_{n_1}$ are parameters, $x_{n_1+1}, \ldots, x_{n_1+n_2}$ are independent variables, $x_{n_1+n_2+1}, \ldots, x_n$ are dependent variables and $n_3 \stackrel{\text{def}}{=} n - n_1 - n_2 \leq m$. Here $f_i(X)$ are sums of differential monomials $a(X)$, each of which is the product of an ordinary monomial X^T and the derivatives of the form $\partial^{\|K\|} x_j / \partial X_2^K$, where $K = (k_{n_1+1}, \ldots k_{n_1+n_2}) \in \mathbb{Z}_+^{n_2}$, $X_2 = (x_{n_1+1}, \ldots, x_{n_1+n_2})$, $j > n_1 + n_2$; and it has its own power exponent $Q(a)$ (see [Bruno 1994] and Section 1 of Chapter 6). Here each equation has its own support $\mathbf{S}_i = \mathbf{S}(f_i)$.

In the problems of Type 2 (systems of ODE), we isolate two subtypes, which are of particular interest for the author.

2*. The autonomous system of the ordinary differential equations resolved with respect to derivatives

$$dX/dt \stackrel{\text{def}}{=} \dot{X} = \Phi(X). \tag{1.3}$$

This system is written in the form

$$(\dot{\log} X) = F(X) \stackrel{\text{def}}{=} \sum F_Q X^Q \quad \text{over } Q \in \mathbf{S} \tag{1.4}$$

(see [Bruno 1962] and Section 2 of Chapter 3). Here the whole system has one support $\mathbf{S} = \mathbf{S}(F)$. There may be parameters among the coordinates x_j, since the parameter x_j satisfies the equation $\dot{x}_j = 0$.

2. The Hamiltonian system** with l degrees of freedom

$$\dot{x}_i = \partial h / \partial x_{i+l}, \quad \dot{x}_{i+l} = -\partial h / \partial x_i, \quad i = 1, \ldots, l, \tag{1.5}$$

where $2l = n$, and the Hamiltonian function

$$h = \sum h_Q X^Q \quad \text{over } Q \in \mathbf{S}(h) \subset \mathbb{R}^n. \tag{1.6}$$

There are two supports corresponding to this system: the support $\mathbf{S}(F)$ of System (1.3) of Type 2* and the support $\mathbf{S}(h)$ of the Hamiltonian function $h(X)$.

In the problems of the type 3 (systems of partial differential equations), we isolate a subtype, for the study of which the methods of Power Geometry were used.

3*. The linear partial differential equation

$$g(D)u = g_0(x_1, \ldots, x_m), \tag{1.7}$$

where $D = (\partial/\partial x_1, \ldots, \partial/\partial x_m)$ is a vector differential operator, $g(Y)$ is the symbolic polynomial

$$g(Y) = \sum a_K Y^K \;\text{ over }\; K \in \mathbf{S} \subset \mathbb{Z}^m \tag{1.8}$$

and g_0 is a sufficiently smooth function. Here $n_1 = 0$, $n_2 = m$, $n_3 = 1$, $X = (x_1, \ldots, x_m, u)$. The support $\mathbf{S}_1$ of the left hand side of Equation (1.7) consists of the points

$$\{Q = (-K, 1) \;\text{ for }\; a_K \neq 0\}.$$

Since all points Q of the support $\mathbf{S}_1$ have the same last coordinate $q_{m+1} = 1$, we can omit it; and to the left hand side of Equation (1.7) we can put in correspondence the support $-\mathbf{S}$, i.e. the support of the symbolic polynomial (1.8) taken with the negative sign.

Further we assume that $X \in \mathbb{R}^n$ or $\mathbb{C}^n$ if all power exponents are integer, and $X \in \mathbb{R}^n_+$ if they are only real.

2. The linear equalities

Let in $\mathbb{R}^n$ the set $\mathbf{S}$ be given, and its affine hull AFF $\mathbf{S}$ (see Section 1 of Chapter 1) have the dimension $d = \dim \text{AFF}\,\mathbf{S} \stackrel{\text{def}}{=} d(\mathbf{S})$. Then in the dual space $\mathbb{R}^n_*$ there is the subspace $\mathbf{N}(\mathbf{S})$ normal to the linear manifold AFF $\mathbf{S}$, and $\dim \mathbf{N}(\mathbf{S}) = n - d$. Let vectors

$$\Lambda_i = (\lambda_{i1}, \ldots, \lambda_{in}), \;\; i = 1, \ldots, n - d \tag{2.1}$$

form a basis of the normal subspace $\mathbf{N}(\mathbf{S})$. Then the manifold AFF $\mathbf{S}$ is determined by the system of equations

$$\{Q : \langle \Lambda_i, Q \rangle = c_i, \;\; i = 1, \ldots, n - d\}.$$

If in $\mathbb{R}^n$ the collection of sets

$$\mathbf{S}_1, \ldots \mathbf{S}_m, \tag{2.2}$$

be given, then each of them has its own normal subspace $\mathbf{N}(\mathbf{S}_i)$. The intersection

$$\mathbf{N} \stackrel{\text{def}}{=} \mathbf{N}(\mathbf{S}_1) \bigcap \ldots \bigcap \mathbf{N}(\mathbf{S}_m) \tag{2.3}$$

is the normal subspace of the collection of sets (2.2). Let $e = \dim \mathbf{N}$, then its co-dimension $d = n - e$ is the dimension of the collection of sets (2.2). Let vectors (2.1) form a basis of the normal subspace (2.3). We suppose that the left minor of the order e of the matrix $(\Lambda_1 \dots \Lambda_e)^*$ is different from zero, i.e.

$$\det \begin{pmatrix} \lambda_{11} & \dots & \lambda_{1e} \\ \dots & \dots & \dots \\ \lambda_{e1} & \dots & \lambda_{ee} \end{pmatrix} \neq 0. \tag{2.4}$$

Now, through the unit points E_{e+j}, $j = 1, \dots, d$, we draw linear manifolds $\mathbf{M}_j$ normal to $\mathbf{N}$. Then each manifold $\mathbf{M}_j$ is determined by the system of equations

$$\{Q : \langle \Lambda_i, Q \rangle = \lambda_{e+j}, \ \ i = 1, \dots, e\}.$$

It crosses the coordinate subspace $q_{e+1} = \dots = q_n = 0$ in the point

$$R_j \stackrel{\text{def}}{=} (r_{j1}, \dots, r_{je}, 0, \dots, 0), \tag{2.5}$$

which is the unique solution to the system of equations

$$\lambda_{i1} r_{j1} + \dots + \lambda_{ie} r_{je} = \lambda_{ie+j}, \ \ i = 1, \dots, e.$$

We split each n-vector X into two subvectors $X' = (x_1, \dots, x_e)$ and $X'' = (x_{e+1}, \dots, x_n)$. Then $R_j = (R_j', 0)$.

1. The system of algebraic equations. In System (1.1) each equation has its own support $\mathbf{S}_i$; and to the whole System (1.1) there corresponds the collection of supports (2.2) and the normal subspace (2.3). If its dimension $e > 0$, then System (1.1) is quasi-homogeneous. Let vectors (2.1) form a basis in $\mathbf{N}$, and the inequality (2.4) is satisfied. Then the system of equations (1.1) has quasi-homogeneous solutions of the form

$$x_{e+j} = b_j X'^{R_j'}, \ \ j = 1, \dots, d, \tag{2.6}$$

where vectors R_j' are the non-trivial parts of vectors (2.5), and b_j are non-zero constants satisfying the system of equations

$$f_i(B', B'') = 0, \ \ i = 1, \dots, m,$$

where $B' = (1, \dots, 1)$, $B'' = (b_{e+1}, \dots, b_n)$. Moreover, for the algebraic system (1.1), quasi-homogeneous solutions (2.6) exhaust all its solutions lying outside the coordinate hyperplanes $x_i = 0$.

Special cases of this situation were known long ago especially for the system of homogeneous equations [Shafarevich 1972].

2. The system of ODE. We denote $n_2 \stackrel{\text{def}}{=} 1$, $n_3 \stackrel{\text{def}}{=} n - n_1 - n_2$, and we suppose that in the system of ODE (1.2) $n_3 \le d < n$, i.e. the dimension of the collection of supports (2.2) is no less than the number of dependent variables. Let vectors (2.1) form a basis of the normal subspace (2.3), and the property (2.4) is satisfied.

If $n_3 = d$, then System (1.2) has quasi-homogeneous solutions of the form (2.6), where b_j are constants satisfying the system of algebraic equations. Examples of such solutions see in [Bruno 1965] and in the end of Example 4.2 of Chapter 7, and in Section 3 of Chapter 6.

If $n_3 < d$, then $e \leq n_1$, and vector X' consists only of parameters. In this case System (1.2) has quasi-homogeneous solutions of the form (2.6) with $j = l+1, \ldots, n$, where b_j are functions of the new independent variable

$$\xi_l = x_l / X'^{R'_{l-e}}$$

satisfying the system of ODE with $n_1 - e$ new parameters

$$\xi_k = x_k / X'^{R'_{k-e}}, \quad k = e+1, \ldots, n_1. \tag{2.7}$$

These functions $b_{l+1}(\xi_l), \ldots, b_n(\xi_l)$ satisfy the system of ODE with a smaller number of parameters $n_1 - e$.

Besides, for the system of ODE (1.2) with $d < n$, the single boundary value problem with conditions at two ends of the interval of the independent variable can be replaced by two boundary value problems at one end of the interval (see Theorems 7.2, 7.3 and 8.7 of Chapter 7).

3. The system of partial differential equations. Let in the system of partial differential equations (1.2) $d \geq n_3$, i.e. $e \leq n_1 + n_2$. Let vectors (2.1) form a basis of the subspace (2.3) of System (1.2), and the inequality (2.4) be satisfied.

If $d = n_3$, i.e. $e = n_1 + n_2$, then System (1.2) has self-similar solutions of the form (2.6), where b_j are constants satisfying a system of algebraic equations.

If $n_1 \leq e < n_1 + n_2$, then System (1.2) has self-similar solutions of the form (2.6) with $j = n_1 + n_2 + 1, \ldots, n$, where b_j are functions of $n_1 + n_2 - e$ new independent variables

$$\xi_k = x_k / X'^{R'_{k-e}}, \quad k = e+1, \ldots, n_1 + n_2. \tag{2.8}$$

These n_3 functions b_j satisfy the system of partial differential equations with $n_1 + n_2 - e$ independent variables and without parameters.

If $e < n_1$, then the vector X' consists of parameters, and System (1.2) has self-similar solutions of the form (2.6) with $j = n_1 + n_2 + 1, \ldots, n$, where b_j are functions of n_2 new independent variables of the form (2.8) with $k = n_1 + 1, \ldots, n_1 + n_2$. These n_3 new functions satisfy the system of partial differential equations with $n_1 - e$ new parameters of the form (2.7) with $k = e+1, \ldots, n_1$. It follows from Theorems 4.1 and 8.2 of Chapter 7.

Examples of such self-similar solutions see in Examples 4.1, 4.2, 8.1 of Chapter 7. Here we cite the smallest possible number of independent variables and parameters of self-similar solutions, but one can seek self-similar solutions with greater number of them. (see Example 4.1 of Chapter 7).

2^*. **The autonomous system of ODE.** If System (1.3) has dimension $d < n$, the vectors (2.1) form a basis of its normal subspace, and the property (2.4) is satisfied, then System (1.3) has particular integrals (i.e. invariant manifolds) of the form (2.6), where b_j are constants satisfying a system of algebraic equations.

To systems of the form (1.3) with $d < n$, there belongs the resonant normal form of a system of the same form in the neighborhood of an elementary stationary point [Bruno 1964, 1971, 1979a]. All vector exponents Q of the normal form (1.3) in notation (1.4) satisfy the equation

$$\langle \Lambda, Q \rangle = 0, \tag{2.9}$$

where vector $\Lambda = (\lambda_1, \dots, \lambda_n) = F_0$ is formed by eigenvalues of the matrix of the linear part of the original system. For the normal form (1.3), its every formal (or analytic at the origin) first integral

$$h = \sum h_Q X^Q \quad \text{over} \quad Q \in \mathbb{Z}_+^n$$

includes only resonant monomials $h_Q X^Q$ satisfying Equation (2.9). In particular, if Equation (2.9) has no solutions $Q \in \mathbb{Z}_+^n$, $Q \neq 0$, then the original system of ODE, for which System (1.3) is its normal form, has no formal (and analytic at the stationary point) first integral. If Equation (2.9) has d linearly independent solutions $Q \in \mathbb{Z}_+^n$, then the original system has no more than d independent first integrals analytic at the stationary point. Hence it follows the analytic non-integrability of systems of the form (1.3) established for various cases.

The other class of Systems (1.3) with $d < n$ constitutes systems invariant with respect to a continuous group of linear transformations. For example, if System (1.3) is invariant with respect to all rotations in the plane x_1, x_2, then in notation (1.4) the relation $q_1 = q_2$ is satisfied for all vectors Q, i.e. the support of System (1.4) satisfies the Equation (2.9) with $\Lambda = (1, -1, 0, \dots, 0)$. Such systems occur in hydrodynamics after the reduction of the evolutionary equation onto the central manifold [Afendikov and Babenko 1988, Afendikov and Mielke 1999a, b].

2^{**}. **The Hamiltonian system.** If the Hamiltonian system (1.5) is the normal form, then its vector Λ has the form $\Lambda = (\Lambda', -\Lambda')$, where $\Lambda' = (\lambda_1, \dots, \lambda_l)$, and the expansion of the Hamiltonian function (1.6) includes only resonant terms satisfying Equation (2.9), i.e.

$$\langle \Lambda', Q' - Q'' \rangle = 0, \tag{2.10}$$

where $Q = (Q', Q'')$. If the equation $\langle \Lambda', Q' \rangle = 0$ has exactly k linearly independent solutions $Q_j^{'} \in \mathbb{Z}^l$, $j = 1, \dots, k$, then the Hamiltonian system (1.5) has $l - k$ independent first integrals of the form

$$\langle K', \rho \rangle, \tag{2.11}$$

where $\langle K', Q_j^{'} \rangle = 0$, $j = 1, \dots, k$, $\varrho_i = x_i x_{i+l}$, $\rho = (\varrho_1, \dots, \varrho_l)$ [Bruno 1971, 1990a].

It turned out that Equality (2.10) is satisfied for the classical Hamiltonian functions (1.6) corresponding to the quantum Hamiltonian functions of some problems of the quantum optics [Vedenyapin and Orlov 1996; 1999, § 2; Vedenyapin and Mingalev 1997]. Here to the quadratic integrals (2.11) there correspond the conservational laws of the combinative dispersion. The integrals (2.11) are quadratic with respect to X but linear with respect to ρ. Vedenyapin and Orlov [1999, § 5] cited the similar to (2.11) linear integrals of discrete models of the Boltzmann equation.

3. The linear transformations

As a result of the power transformation

$$\log Y = A \log X, \quad \det A \neq 0 \tag{3.1}$$

a system of equations in X of some type is transformed into the system of equations in Y of the same type (see Chapter 2, Sections 3, 7; Chapter 3, Section 2; Chapter 4, Section 2; and Chapter 7, Sections 5 and 8). Here the supports of equations undergo the affine transformation $\tilde{Q} = A^{*-1}Q$, and the objects in the dual space $\mathbb{R}_*^n$ undergo the transformation $\tilde{P} = AP$. Hence by means of the power transformation and reductions a system of the dimension d may be reduced to the system the support of which is placed in the d-dimensional coordinate subspace, i.e. all vector power exponents $\tilde{Q} = (\tilde{q}_1, \dots, \tilde{q}_n) \in \tilde{\mathbf{S}}_i$ have $n - d$ coordinates $\tilde{q}_j$ identically equal to zero.

1. The system of algebraic equations. Multiplying each equation of System (1.1) by X^{T_i} and making the required power transformation (3.1), we reduce System (1.1) to the form

$$g_i(y_1, \dots, y_d) = 0, \quad i = 1, \dots, m,$$

where g_i are polynomials (see Chapter 2, Sections 3, 7).

Special cases of power transformations of algebraic equations have been used for a long time. Thus, to resolve singularities the sigma process was used [Walker 1950, Shafarevich 1972, Bruno 1979a]. It is a finite sequence of power transformations (3.1) with matrices of the form $A = E \pm A_{ij}$, where E is the unit matrix, and the matrix A_{ij} has the element in the intersection of the

ith line and jth column equal to one, and other elements equal to zero. The product of such matrices is a unimodular matrix.

2. The system of ODE. For the ODE System (1.2) power transformations (3.1) are convenient to use only with the block-triangular matrix

$$A = \begin{pmatrix} A_{11} & 0 \\ A_{12} & A_{22} \end{pmatrix}, \tag{3.2}$$

where the square matrix A_{11} has the order n_1 corresponding to the number of parameters, in order that new parameters would depend on the old ones and would not depend on the old variables. In this connection, one has to strive firstly, that possibly the maximal number of new parameters belonged to the transformed system in zero powers, i.e. were absent there, and secondly, that possibly the maximal number of new dependent variables y_j with $j > l$ had zero exponents $\tilde{q}_j$ in supports of all equations. Then these coordinates y_j may be replaced by $d \log y_j / dy_l$, and the order of the system with respect to the coordinate may be lowered (see Chapter 7, Theorems 7.1 and 8.6, Example 7.1).

It is worthy to note that, generally speaking, under the power transformation (3.1) the differential polynomial $f_i(X)$ is transformed into the ratio of two differential polynomials $f_i(X) = g_i(Y)/h(Y)$, where the support $\mathbf{S}(h)$ consists of one point, that is zero. Therefore the system of equations (1.2) is transformed into the system $g_i(Y) = 0$, $i = 1, \dots, m$. If the upper line in the matrix A_{22} in (3.2) has the form $(a, 0, \dots, 0)$, i.e. the new independent variable does not depend on the old dependent variables, then $h(Y) \equiv 1$ (see Chapter 6, Section 2; and Chapter 7, Section 5, Remark 5.1).

3. The system of partial differential equations. For it, the power transformation (3.1) must be taken with the block-triangular matrix (3.2). Besides, with the power transformation, one has to strive that supports of the transformed equations had as many zero coordinates corresponding to parameters and independent variables as possible. That reduces the number of parameters and essential independent variables in the transformed system. Since for an independent variable y_j with the zero power exponent in supports, one can make the logarithmic transformation $z_j = \log y_j$, that would not violate the character of equations as differential polynomials (see Chapter 6, Section 3; and Chapter 7, Section 6). The solutions which do not depend on these independent variables z_j are the self-similar ones.

2*. The autonomous system of ODE. With the help of the power transformation (3.1) and the change of time of the form $d\tilde{t} = X^T dt$, System (1.4) may be reduced to the form

$$(\log y_i) = g_i(y_1, \dots, y_d), \quad i = 1, \dots, n. \tag{3.3}$$

From this system, the subsystem for $y_1, \dots, y_d$ splits off. The remaining y_j are obtained from them with the help of quadratures (see Chapter 3, Section 2; [Bruno 1962, 1971, 1979a]).

In particular, for the normal form (1.4), (2.9), the dimension d does not exceed the multiplicity of the resonance. Hence, with the help of the power transformation, its integration is reduced to the solution of a system of the order d (see [Bruno 1964, 1971, 1979a]).

2**. **The Hamiltonian system.** Generally speaking, the power transformation (3.1) is not a canonical one, and it destroys the Hamiltonian structure of System (1.5). In [Bruno 1990a, Chapter I, § 3], the canonical transformation is cited, which reduces the number of degrees of freedom of the Hamiltonian system (1.5) if $\dim \mathrm{AFF}\,\mathbf{S}(h) < n$. Namely, if the equation $\langle \Lambda', Q' \rangle = 0$ has k linearly independent solutions $Q' \in \mathbb{Z}^l$, then the Hamiltonian system (1.5) is reduced to the Hamiltonian system with k degrees of freedom and $l-k$ parameters.

Remark 3.1. The algorithms of computation of the matrix A for the power transformation (3.1), which reduces the system to the mentioned simplifications, are described in Chapter 1, Section 11 of this book.

4. Linear inequalities

Now we consider algebraic and differential equations, the supports of which are not necessarily lie in a linear manifold of a smaller dimension. But by means of the support $\mathbf{S}$ of the equation $0 = f(X) = \sum a_Q(X) = 0$, its convex hull $\mathbf{\Gamma}$ is constructed. The surface $\partial\mathbf{\Gamma}$ of the polyhedron $\mathbf{\Gamma}$ consists of faces $\mathbf{\Gamma}_j^{(d)}$ of various dimensions d. To each face $\mathbf{\Gamma}_j^{(d)}$ there corresponds the boundary subset $\mathbf{S}_j^{(d)} = \mathbf{\Gamma}_j^{(d)} \bigcap \mathbf{S} \subset \mathbb{R}^n$ and the normal cone $\mathbf{U}_j^{(d)} \subset \mathbb{R}_*^n$. Besides, to each face $\mathbf{\Gamma}_j^{(d)}$ there corresponds the truncated equation

$$0 = \hat{f}_j^{(d)}(X) \stackrel{\text{def}}{=} \sum a_Q(X) \quad \text{over } Q \in \mathbf{S}_j^{(d)} \tag{4.1}$$

and the set $\mathcal{U}_j^{(d)}(\varepsilon)$ in the X-space. Here the truncated equation (4.1) is the first approximation to the equation $f(X) = 0$ in the set $\mathcal{U}_j^{(d)}(\varepsilon)$ (see [Bruno 1979a]).

For the complete system of equations, in this way, we extract the truncated system, which has a dimension $d < n$ and its own normal cone, and its own set of the type $\mathcal{U}(\varepsilon)$, where the truncated system is the first approximation to the complete system. In this connection, if the complete system has a solution lying in the set $\mathcal{U}(\varepsilon)$, then its first approximation is a solution to the corresponding truncated system. It allows to find consecutively the asymptotics of solutions and asymptotic expansions of solutions. To find solutions to the truncated systems, one can use the power and logarithmic transformations, as it is described above. Chapters 2–6 of this book were devoted to these questions for various types of equations. Namely, systems of algebraic equations (1.1) were considered in Chapter 2. Systems of ordinary differential equations were considered in Sections 1–4 of Chapter 6. General systems of partial differential equations were considered in Sections 1, 5, 6 of Chapter 6. Systems of equations of the type (1.3) were considered in Chapters 3 and 5, and Hamiltonian systems (1.5) were considered in Chapter 4.

In [Aranson 1996] and in Chapter 1 of this book, the algorithms for computation of the boundary subsets (i.e. truncated systems) and their normal cones are described.

Now we cite the applications of Power Geometry, which were not included in Chapters 2–6 of the present book.

2*. The autonomous system of ODE. For ODE Systems (1.3) of the type 2*, this technique allows to compute asymptotic expansions not only for asymptotic solutions but for neutral (periodic, quasi-periodic etc.) as well [Bruno 1979a]. For this purpose, we find such a power transformation, which reduces the truncated system to the form (3.3). That power transformation is done in the complete system. Then, in the neighborhood of each elementary singular point with $y_{d+1} = \ldots = y_n = 0$, the complete system is reduced to the normal form. If necessary, another power transformation is done in the normal form. In [Sadov 1994, Bruno 1988], this approach was employed for the first time to the analysis of solutions to concrete systems. In [Bruno and Soleev 1998] and in Chapter 5 of this book, a concrete system of the form (1.3) was studied by this method. In [Afendikov and Mielke 1999b], the same approach was used for another system (1.3), which appeared in a hydrodynamic problem.

Besides, in [Bruno 1971], in the proof of divergence of the normalizing transformation for systems of the type (1.3), the concepts of Power Geometry were used for the extraction of the first approximations of these transformations and for explicit estimates of the Tailor coefficients of the extracted first approximations. In [Bruno 1990c], the attempt was made to explain popularly these geometric constructions, but they still remain not understood till now. For two-dimensional real systems (1.3), in [Bruno 1979b, 1982] the proofs of divergence of the normalizing transformation were given, which also used the Power Geometry.

For the linear non-autonomous ODE system of the form $\dot{X} = \Phi(t)X$, the Power Geometry also allows to obtain new results and to interpret the old ones (see [Bruno 1971, Introduction; 1979b, 1982, 2000]).

Remark 4.1. In Chapter 3 of this book, the procedure of computation of asymptotics of solutions to System (1.3) is based upon the consideration in

$\mathbb{R}^n$ of the unique support of the whole system. It is possible that the general approach would be more universal, according to which to each equation of System (1.3) its own support in $\mathbb{R}^{n+1}$ is put in correspondence, and the truncated systems are found by these supports according to general rules (see Example 1.5 of Chapter 6).

3*. The linear partial differential equation (1.7). In the terms of the support of the symbolic polynomial (1.8), the conditions were formulated, which guarantee the local solvability of some problems for Equation (1.7). Since to local properties (in a neighborhood of the point X_2^0) there corresponds the cone of the problem $\{P_2 < 0\}$ for the support $\mathbf{S}_1$ of the differential polynomial $g(D)u$, and the equality $\mathbf{S}(g) = -\mathbf{S}_1$ is valid, then for the support $\mathbf{S}(g)$, to these local properties there corresponds the cone of the problem $\{P_2 > 0\}$. In studies [Mikhailov 1963, 1965, 1967a, b; Friberg 1967; Volevich and Gindikin 1968, 1985; Gindikin 1973, 1974; Volevich 1974; Gindikin and Volevich 1992; Boggiato et al. 1996] the symbolic polynomial is written in the form

$$g(iZ) = a(Z) + ib(Z) = \sum a_K Z^K + \sum b_L Z^L.$$

Obviously, $\mathbf{S}(g) = \mathbf{S}(a) \cup \mathbf{S}(b)$. There are considered the polyhedrons $\mathbf{\Gamma}$, which is the convex hull of the support $\mathbf{S}(a)$, and $\mathbf{\Gamma}_1$, which is the convex hull of the set $\mathbf{S}(a) \cup \{0\}$. In the polyhedron $\mathbf{\Gamma}$, its leading part with respect to the negative orthant $\mathbb{R}_-^m = \{K \le 0\}$ is isolated: $\tilde{\mathbf{\Gamma}} = \mathbf{\Gamma}/\mathbb{R}_-^m$ (see Chapter 1, Section 8); and in the polynomial $a(Z)$ its main part

$$\tilde{a}(Z) = \sum a_k Z^k \quad \text{over } K \in \tilde{\mathbf{\Gamma}}$$

is isolated. We assume that
(a) the set $\tilde{\mathbf{\Gamma}}$ has points in each coordinate axis;
(b) each truncated polynomial $\hat{a}_j^{(m-1)}(Z)$ corresponding to the hyperface $\mathbf{\Gamma}_j^{(m-1)} \subset \tilde{\mathbf{\Gamma}}$ (i.e. with the normal vector $P > 0$) has no real roots outside the coordinate subspaces.

Then for real $Z \to \infty$, the following estimate is fulfilled

$$|a(Z)| \ge c \sum Z^k \quad \text{over } K \in \mathbf{V},$$

where the constant $c > 0$, and $\mathbf{V}$ is the set of vertices of the hypersurface $\tilde{\mathbf{\Gamma}}$. Such an estimate allows to prove the hypoellipticity of Equation (1.7) and the local solvability of some problems for the differential equation (1.7) (see [Mikhailov 1963, 1965, 1967a, b; Friberg 1967; Volevich and Gindikin 1968, 1985; Gindikin 1973, 1974; Volevich 1974; Gindikin and Volevich 1992; Boggiato et al. 1996]). In these studies, also such an equation (1.7) was considered, in which the operator $g(D)$ dependent on coordinates $x_1, \ldots, x_m$. But this dependence was supposed to be weak and not influencing the structure of the polyhedron $\tilde{\Gamma}$. Besides, in [Volevich and Gindikin 1968] a system of linear equations, and in [Denk et al. 1998] problems with a parameter were considered.

Remark 4.2. Concerning the global properties of polyhedra, there are only the results for systems of algebraic equations. For example, in a generic case the number of solutions to the system of n algebraic equations in n variables is equal to the mixed Minkowski volume of the Newton polyhedra of these equations [Bernshtein 1975]. In the studies by Khovanskii and his colleagues, the interesting links were established between the topological structure of algebraic varieties and Newton polyhedra of the equations determining these varieties [Khovanskii 1978a, b, 1983, 1988, 1992, 1997a, b, 1999, 2000; Kazarnovskii 1984, 1997; Danilov and Khovanskii 1986; Pukhlikov and Khovanskii 1992; Gelfond and Khovanskii 1996]. The principal achievement here is the discovery of the relation between the Newton polyhedra and the toric manifolds, i.e. the construction of the toric compactification and the resolution of singularities for sufficiently wide class of algebraic systems of equations with fixed Newton polyhedra. This construction has allowed to apply the theory of toric manifolds to the polyhedra and has lead to the decisive progress in this field. See the review of the studies in Khovanskii [2000].

5. On applications of Power Geometry

Let us enumerate some of the problems of various sciences, in which new results were obtained with the help of Power Geometry.

Mechanics (Robotics). The singular positions of mechanisms were studied in Section 9 of Chapter 2. They are the singularities of a system of algebraic equations.

Celestial Mechanics. Families of periodic solutions to the equation of the plain oscillations of a satellite on the elliptic orbit (the Beletskii [1959] equation) and their singularities were studied [Sadov 1995, 1996a, b, 1997a, b, 1998, 1999; Bruno and Varin 1995, 1996, 1997a, b, 1999; Varin 1996, 1997, 1999a, b]. The problem is described by the ordinary differential equation.

The singularities of the space oscillations of a satellite on a circular orbit were studied in Section 3 of Chapter 4. The problem is described by a Hamiltonian system.

The singularities of families of periodic solutions to the restricted three-body problem were studied in Section 4 of Chapter 4 and in [Bruno 1978, 1999]. In particular, in the problem of Sun-Jupiter-Spaceprobe the family of stable periodic orbits with close approach to Jupiter and Earth was found. The problem is described by a Hamiltonian system.

Theoretical Physics. The singularities of solutions to a finite-dimensional approximation to the Schrödinger equation were studied in [Sadov 1994a, b; Bruno and Sadov 1995]. The problem is described by an autonomous system of ordinary differential equations.

Conservation laws for discrete models of the Bolzmann equation were studied in [Vedenyapin and Orlov 1996, 1999; Vedenyapin and Mingalev 1997].

Hydrodynamics. New families of periodic waves and new families of conditionally periodic waves on the water surface were found in Chapter 5. The problem is described by the autonomous ODE system. Similar problems were studied in [Afendikov and Mielke 1999a, b].

A rigorous mathematical substantiation to the theory of the boundary layer in a simplest flow problem was given in Section 6 of Chapter 6. The initial problem is described by the system of partial differential equations. The reduced problem is described by one ordinary differential equation (see Example 7.1 of Chapter 7). More complicated problems with a boundary layer were considered by Vasiliev [1998, 1999].

Thermodynamics. A simple way to analyze equations of the combustion process was given in Examples 2.1, 4.1, 4.2, 5.1, 6.1, 6.2, 8.1 of Chapter 7. The initial problem is described by a partial differential equation (or by a system of such equations). The reduced problem is described by an ordinary differential equation (or by a system of such equations) and can be easily studied by methods of Chapter 6.

6. Historical remarks

The studies which use methods of Power Geometry are subdivided into three jets, each one with its own origin:

1. The studies by the author and his followers beginning with [Bruno 1962];
2. The studies by V.P. Mikhailov and his followers beginning with [Mikhailov 1963];
3. The studies by V.I. Arnol'd and his followers beginning with [Arnol'd 1974];

Below, from the author's viewpoint, the history of these jets and their mutual influence is expounded.

1. In [Bruno 1962], for System (1.3), there were introduced the polyhedron of the system, its faces, truncated systems, normal cones, power transformations; and they were applied to the computation of asymptotics of the asymptotical solutions (see also [Bruno 1965, 1973a, b, 1979a] and Chapter 3 of this book). In [Bruno 1973a] this approach was extended on the

computation of branches of an algebraic curve near a degenerate critical point (see also [Bruno 1976, 1979a; Soleev 1983; Bruno and Soleev 1990, 1991a, b, 1992] and Chapter 2 of this book). In [Bruno 1973a], to each normal cone in $\mathbb{R}_*^n$ there was put in correspondence its own set $\mathcal{U}(\varepsilon)$ in the X-space, and power series were introduced, the supports of which lie in the given cone (see also [Bruno 1979a]). In [Bruno 1978], with the help of the Newton polyhedron of a Hamiltonian function, the first approximations for a Hamiltonian system near a singularity were isolated (see also [Bruno 1990b, 1999; Bruno and Soleev 1995e, 1996] and Chapter 4 of this book). In [Soleev 1982], the development of an algorithm for the computation of a polyhedron, its faces and normal cones using the given support had begun (see also [Soleev and Aranson 1994, Aranson 1996] and Chapter 1 of this book). In [Bruno and Soleev 1994b] the approach of Power Geometry was extended on the resolving of singularities of arbitrary algebraic varieties (see also Chapter 2 of this book). In [Bruno 1994], it was shown how to isolate first approximations to arbitrary systems of differential equations and to find asymptotics of their solutions with the help of Power Geometry. In particular, it was shown that the methods of papers [Mikhailov 1963, 1965, 1967a, b; Volevich and Gindikin 1968, 1985; Gindikin 1973, 1974; Volevich 1974; Gindikin and Volevich 1992] for Equation (1.7) and the results of the book [Vasil′eva and Butusov 1990] concerning the singular perturbations fall into the general scheme (see also [Bruno 1997a, b, 1998c, 1999] and Section 1 of Chapter 6 of this book). In [Bruno 1996b], it was noted that under the power transformation, the support of a differential polynomial undergoes the linear transformation (see also [Bruno 1997a, b, 1998c, 1999] and the preset book). In [Bruno 1997a], the logarithmic transformation was introduced.

2. For the differential equation (1.7), Mikhailov [1963] proposed to consider the support of its characteristic polynomial and the polyhedron, which is the convex hull of the support and the origin (see the end of the previous Section and also [Mikhailov 1965, 1967a, b; Friberg 1967; Volevich and Gindikin 1968, 1985; Gindikin 1973, 1974; Volevich 1974; Gindikin and Volevich 1992; Boggiatto et al. 1996]). In this connection, in the early studies [Mikhailov 1963, 1965] the polyhedron had no name; in [Mikhailov 1967a, b] and [Volevich and Gindikin 1968] it was called the "characteristic" one; and beginning with [Gindikin 1973] and further, it was called the "Newton polyhedron", although usually the Newton's open polygon was applied in a neighborhood of the origin, and its convex part corresponding to the principal part of a polynomial was directed to the origin. But in problems with the hypoelliptic characteristics polynomials, the convex part of the Newton polyhedron corresponding to the principal part of the polynomial is directed to the infinity.

3. Arnol'd [1974, § 9.9] had noted that in a generic case, some local properties of a function of two variables in a neighborhood of a degenerate point depend only on the size of the domain limited by axis and by the convex part of the Newton's open polygon of the function. That served as an impetus to the writing of papers [Kushnirenko 1975a, b], where the Newton polyhedra of one or several series were already considered. It is interesting to note that in these studies the Newton polyhedron is not simply the convex hull of the support, but it is the convex hull of the support added by the non-negative orthant $\{Q \geq 0\}$, similar to that as in [Gindikin 1973, 1974] the Newton polyhedron is the convex hull of the support added by the negative orthant $\{Q \leq 0\}$. However, similar definitions of the Newton polyhedron were also given in some subsequent studies (see for example [Arnol'd et al. 1984, § 6.2]). The explicit definition of the Newton polyhedron as the convex hull of the support was given in [Bernshtein et al. 1976], and implicit one is in [Bernshtein 1975]. Subsequently, this approach was used mainly for the computation of asymptotics of integrals [Varchenko 1976; Vasil'ev 1977; Arnol'd et al. 1984; Varchenko and Khovanskii 1985] and for the study of the properties of solutions to a system of algebraic equations [Kouchnerenko 1976, Kushnerenko 1976; Khovanskii 1978a, b, 1983, 1992, 1997a, b, 1999, 2000; Danilov and Khovanskii 1986; Pukhlikov and Khovanskii 1992; Gelfond and Khovanskii 1996].

Now we will dwell upon the mutual influence of the mentioned three jets. As far as the author knows, each of the jets 2 and 3 had no noticeable ideological influence on the other jets except the terminological one. Thus, the term *Newton polyhedron*, which appeared in the paper [Gindikin 1973] of the second jet, became generally accepted. The same is true for the term *support* from the third jet. Therefore we will consider the influence of the first jet on the second and the third.

The influence of the first jet on the second one had begun apparently in 70th through L.R. Volevich, who was working then in the Keldysh Institute of Applied Mathematics (KIAM) in the same department with the author. That is especially noticeable in the paper [Volevich and Gindikin 1985] and in the book [Gindikin and Volevich 1992], where there were used (without precise references) the normal cones and the sets $\mathcal{U}(\varepsilon)$ introduced by the author in [Bruno 1962, 1973a, 1979a]. Apparently, for the first time, the both jets have merged in [Bruno 1979a, Subsection 2.4 of Chapter IV]. In [Bruno 1994] and in this book (Chapter 6, Example 1.6 and Subsection 3* of Section 4 of Chapter 8), it is shown how Power Geometry of the second jet fits in the general scheme.

The influence of the first jet on the third one took place in the following way. In 1960 and 1961, the author submitted his studies, which would become the basis for publications [Bruno 1962, 1965, 1973b], for the student's research competition in the Mathematical faculty of Moscow State University. V.I. Arnol'd, who was then a postgraduate student, was a referee on these studies. In 1968, he was an official referee for the doctoral (professorial) thesis by the author, which was subsequently published in [Bruno 1971]. The author keeps the review by V.I. Arnol'd on the manuscripts of these studies, that was written in December of 1969. In the review, V.I. Arnol'd criticizes "the geometry of power exponents" in the three following remarks.

9. On page 14, the author writes about "new concepts of the geometry of power exponents". On page 18, it becomes clear that these concepts are "cones, polyhedra and other geometrical objects, which have no analogues in Lie algebras".

   I must confess that I do not see new concepts in the "the geometry of exponents". I think, it was better to write "new concepts in author's opinion".

11. On page 18, the author writes: "We think that the concepts of the geometry of power exponents are more fruitful" (than the point of view of the Lie algebras). The referee thinks that the opinion expressed here by the author is explained simply by his ill acquaintance with the Lie algebras and with those simplifications which the general group viewpoint could have brought in his expounding of material. It is better to omit the cited criticism of the Lie algebras.

16. On pages 195–198 the author speaks of his studies in the following terms:

   "The proofs ... are based exclusively on methods developed by the author. These methods are essentially based on the geometry of power exponents..." "The power transformations introduced by the author in 1962 make the geometry of power exponents more profound".

   In referee's opinion:

   A. "The geometry of exponents" consists in the fact that the power exponents in the multiple Fourier (Laurent, Taylor) series are depicted by points of the integer grid. That geometry becomes "more profound" when the theorem of reduction of an integer matrix to the standard form with the help of elementary transformations is used. Both notions are well known and usually used without any pomposity.

   B. "Power transformations introduced by the author" are widely known by the name of "multiple sigma- and anti sigma-processes" or "blowing up of singularities" in Algebraic Geometry. They are also used for a long time in differential equations (see for example the memoir by Bendixson, 1901).

> It seems to the referee that the self-appraisal of the author in pages 195–198 should be omitted: the readers (and the author) would only benefit from that.

In the paper [Bruno 1971], the criticized issues remained in the following form. To the remark 9 there relates the paragraph before the formulation of Condition ω in Subsection II of Introduction (page 140). The text cited in the remark 11 was omitted, since at the time, V.I. Arnol'd was the Vice-president of the Moscow Mathematical Society (MMS), and the editorial staff of the Transactions of MMS could not permit to ignore such a referee. The remark 16 relates to the first paragraph of Subsection III, § 6, Chapter III (page 230) and to the paragraph before the formulation of Theorem 10 in the same Subsection (page 231). So, V.I. Arnol'd required 14 years (from 1960 to 1974) in order to realize the usefulness of Power Geometry. Further, the influence of the first jet on the third was exerted through A.G. Khovanskii, who was working in KIAM in the same room with the author in 1973–1976. Thus, Bernshtein [1975] used the power transformation and the related to it linear transformation in the space of power exponents (of cause without references). The power transformations were used also in [Arnol'd et al. 1984, § 8, Subsection 8.1. B] under the name of *monomial transformations* and also without references. In the same book, the analogues of normal cones and other instruments of Power Geometry were used.

Bibliography

Here the publications in Russian are marked as such. The language of other publications coincides with that of the citation. The publications of the same study in different languages are separated by the equal sign "=".

Abyhankar, S.S. 1982: Weighted expansions for canonical desingularization. Lecture Notes in Math. N^o910. Springer, Berlin–Heidelberg–New York.

Afendikov, A.L., Babenko, K.I. 1988: Bifurcation of cycle generation in some problems with symmetry. Doklady Akad. Nauk SSSR **300**, N^o1, 14–18. *(Russian)* = Soviet Math. Doklady **37**, N^o3 (1988) 600–605.

Afendikov, A., Mielke, A. 1999a: Bifurcation of homoclinic orbits to a saddle-focus in reversible systems with SO(2)-symmetry. J. Differential Equations **159**, N^o2, 370–402.

Afendikov, A., Mielke, A. 1999b: Unfolding family of reversible vector fields with SO(2) symmetry and nonsemisimple eigenvalue zero of multiplicity four. Doklady Akad. Nauk **369**, N^o2, 154–157. *(Russian)* = Russian Acad. Sci. Doklady. Mathem. **60**, N^o3 (1999) 332–336.

Aizengendler, P.G. 1966: Some questions of branching solutions of nonlinear equations. Uspekhi Mat. Nauk **21**, N^o1, 182–183. *(Russian)*

Amick, C., McLeod, J. 1991: A singular perturbation problem in water waves. Stab. and Appl. Anal. of Cont. Media. **1**, 127–148.

Aranson, A.B. 1996: Computation of the Newton polyhedron. In: Proceedings of the International Conference and Chebyshev Readings, Celebrated the 175 Anniversary of the P.L. Chebyshev Birthday. MGU, Moscow, v.1, 32–34. *(Russian)*

Arnol'd, V.I. 1971: On matrices depending on parameters. Uspekhi Mat. Nauk **26**, N^o2, 101–114. *(Russian)* = Russian Math. Surveys **26** (1971).

Arnol'd, V.I. 1974: Normal forms of functions in a neighborhood of degenerate critical points. Uspekhi Mat. Nauk **29**, N^o2, 11–49. *(Russian)* = Russian Math. Surveys **29**, N^o2 (1974).

Arnol'd, V. I. 1984: Reversible systems. In: Nonlinear and Turbulent Processes in Physics. Harwood Ac. Publ., New York, v. 3, 1161–1174.

Arnol'd, V. I., Sevryuk, M. B. 1986: Oscillations and bifurcations in reversible systems. In: Nonlinear Phenomena in Plasma Physics and Hydrodynamics. Mir, Moscow, 31–64.

Arnol'd, V.I., Varchenko, A.N., Gusein-Zade, S.M. 1984: Singularities of Differentiable Maps. II. Monodromy and Asymptotics of Integrals. Nauka, Moscow. *(Russian)* = Mono. in Math. N^o83. Birkhäser, Boston–Basel–Berlin, 1988.

Bautin, N.N. 1954: On periodic solutions of a system of differential equations. Prikl. Mat. Mekh. **18**, 128–134. *(Russian)*

Becker, T., Weispfenning, V. 1993: Gröbner Bases. Springer, New York–Berlin etc.

Beletskii, V.V. 1959: On the librations of a satellite. In: Artificial Satellites of the Earth. Izdat. Akad. Nauk SSSR, Moscow, Part 3, 13–31. *(Russian)* = NASA TTF–429(1966) 257–261.

Belitskii, G.R. 1975: The normal forms of local mappings. Uspekhi Mat. Nauk **30**, N^o1, 223. *(Russian)* = Russian Math. Surveys **30**, N^o1 (1975) 223–224.

Belitskii, G.R. 1979a: Normal forms with respect to the filtering action of a group. Trudy Mosk. Mat. Obshch. **40**, 3–46. *(Russian)* = Trans. Moscow Math. Soc. **40** (1979) 3–46.

Belitskii, G.R. 1979b: Normal Forms, Invariants and Local Mappings. Naukova Dumka, Kiev. *(Russian)*

Benest, D. 1976: Libration effects for retrograde satellites in the restricted three-body problem. Celestial Mechanics **13**, N^o2, 203–215.

Berkovich, L.M. 1998: Factorization and transformations of nonlinear differential equations: methods and algorithms. In: Intern. Congress on Nonlinear Analysis and its Applications, Moscow, Sept. 1–5, 1998. Electronic Proceedings, 595–625. See: `http://www.acad-ns. newmail.ru/nlaia98`. *(Russian)*

Bernshtein, D.N. 1975: The number of roots of a system of equations. Funkts. Anal. Prilozh. **9**, *N°*3, 1–4. *(Russian)* = Funct. Anal. Appl. **9** (1975) 183–185.

Bernshtein, D.N., Kushnirenko, A.G., Khovanskii, A.G. 1976: Newton polyhedra. Uspekhi Mat. Nauk **31**, *N°*3, 201–202. *(Russian)* = Russian Math. Surveys **31**, *N°*3 (1976) 201–202.

Bierstone, E., Milman, P.D. 1989: Uniformization of analytic spaces. Journal AMS **2**, *N°*4, 801–836.

Blasius, H. 1908: Grenzschichten in Flüssigkeiten mit kleiner Reibung. Zeit. für Math. und Phys. **56**, 1–37.

Boggiatto, P., Buzano, E., Rodino, L. 1996: Global Hypoellipticity and Spectral Theory. Akademie Verlag, Berlin.

Botashev, A.I. 1976: Finite Methods in the Theory of the Multidimensional Branching. Ilim, Frunze. *(Russian)*

Botashev, A.I. 1985: Analytical methods in the theory of branching. Uspekhi Mat. Nauk **40**, *N°*4, 147–148. *(Russian)* = Russian Math. Surveys **40**, *N°*4 (1985).

Briot, C., Bouquet, T. 1856: Recherches sur les proprietes des equations differentielles. J. l'Ecole Polytechn. **21**, *N°*36, 133–199.

Broer, H. W., Huitema, G. B., Sevryuk, M. B. 1996: Quasi–Periodic Motions in Families of Dynamical Systems. LNM 1645. Springer, Berlin.

Brøndsted, A. 1983: An Introduction to Convex Polytopes. Springer, N.Y.–Heidelberg–Berlin. = Mir, Moscow, 1988. *(Russian)*

Bruno, A.D. 1962: The asymptotic behavior of solutions of nonlinear systems of differential equations. Doklady Akad. Nauk SSSR **143**, *N°*4, 763–766. *(Russian)* = Soviet Math. Doklady **3** (1962) 464–467.

Bruno, A.D. 1964: The normal form of differential equations. Doklady Akad. Nauk SSSR **157**, *N°*6, 1276–1279. *(Russian)* = Soviet Math. Doklady **5** (1964) 1105–1108.

Bruno, A.D. 1965: Power asymptotics of solutions of nonlinear systems. Izvestiya Akad. Nauk SSSR, Ser. Mat. **29**, *N°*2, 329–364. *(Russian)*

Bruno, A.D. 1971: Analytical form of differential equations. Trudy Mosk. Mat. Obshch. **25** (1971) 119–262; **26** (1972) 199–239. *(Russian)* = Trans. Moscow Math. Soc. **25** (1971) 131–288; **26** (1972) 199–239.

Bruno, A.D. 1973a: Elements of Nonlinear Analysis. Course of Lectures. Samarkand. *(Russian)*

Bruno, A.D. 1973b: On the power asymptotics of solutions of nonlinear systems. Preprint N^o54, Inst. Appl. Math., Moscow. *(Russian)*

Bruno, A.D. 1973c: Local invariants of differential equations. Mat. Zametki **14**, N^o4, 499–507. *(Russian)* = Math. Notes **14**, N^o4 (1973) 844–848.

Bruno, A.D. 1974: The sets of analyticity of a normalizing transformation. Preprints N^o97 and 98, Inst. Appl. Math., Moscow. *(Russian)* = [Bruno 1989a, Part II]

Bruno, A.D. 1976: Local methods in the nonlinear analysis. In: Functional Analysis and Some Questions of QTDE (E.V. Voskresenskii, ed.). Mordovian University, Saransk, 77-80. *(Russian)*

Bruno, A.D. 1978: On periodic flights round moon. Preprint N^o91, Inst. Appl. Math., Moscow. *(Russian)* = On periodic flybys of the moon. Celestial Mechanics **24**, N^o3 (1981) 255–268.

Bruno, A.D. 1979a: Local Method of Nonlinear Analysis of Differential Equations. Nauka, Moscow. *(Russian)* = [Bruno 1989a, Part I]

Bruno, A.D. 1979b: Divergence of the real normalizing transformation. Preprint N^o62, Inst. Appl. Math., Moscow. *(Russian)* = Selecta Math. formerly Sovietica **12**, N^o1 (1993) 13–23.

Bruno, A.D. 1982: Divergence of a real normalizing transformation. Mat. Zametki **31**, N^o3, 403-410. *(Russian)* = Math. Notes **31**, N^o3 (1982) 207-211.

Bruno, A.D. 1988: Bifurcation of the periodic solutions in the symmetric case of a multiple pair of imaginary eigenvalues. In: Numerical Solution of Ordinary Differential Equations (S.S. Filippov, ed.). Inst. Appl. Math., Moscow, 161–176. *(Russian)* = Selecta Math. formerly Sovietica **12**, N^o1 (1993) 1–12.

Bruno, A.D. 1989a: Local Methods in Nonlinear Differential Equations. Springer, Berlin etc. = [Bruno 1974, 1979a] *(Russian)*

Bruno, A.D. 1989b: Analysis of a system in the center manifold. In: VII. Allunion Conference "Qualitative Theory of Diff. Eq.". Abstracts. Riga, 43. *(Russian)* = The analysis of bifurcations on a center manifold. In: Intern. Congress of Mathem., Short Comm. Abstracts. Kyoto, 1990, 193.

Bruno, A.D. 1990a: The Restricted 3-Body Problem: Plane Periodic Orbits. Nauka, Moscow. *(Russian)* = Walter de Gruyter, Berlin, 1994.

Bruno, A.D. 1990b: A local analysis of Hamiltonian systems. Preprint M33, I.H.E.S., Paris.

Bruno, A.D. 1990c: On divergence of normalizing transformation. Preprint M72, I.H.E.S., Paris.

Bruno, A.D. 1992: A general approach to the study of complex bifurcations. Prikladnaja Mekhanika **28**, N^o12, 83–86 *(Russian)* = International Applied Mechanics **28**, N^o12 (1993) 849–853.

Bruno, A.D. 1994: First approximations of differential equations. Doklady Akad. Nauk **335**, N^o4, 413-416. *(Russian)* = Russian Acad. Sci. Doklady. Mathem. **49**, N^o2 (1994) 334–339.

Bruno, A.D. 1995a: General approach to the asymptotic analysis of singular perturbations. In: Dynamical Systems and Chaos (N. Aoki, K. Shiraiwa, Y. Takahashi, eds.). World Scientific, Singapore, v. 1, 11–17.

Bruno, A.D. 1995b: The Newton polyhedron in nonlinear Analysis. Preprint N^o48, Inst. Appl. Math., Moscow. *(Russian)*

Bruno, A.D. 1995c: Newton polyhedron in Nonlinear Analysis. Vestnik Mosk. Univ., Ser. 1 Mat. Mekh. N^o6, 45–51. *(Russian)* = Moscow Univ. Mathem. Bulletin **50**, N^o6 (1995) 4–9.

Bruno, A.D. 1995d: Ways of computing of a normal form. Doklady Akad. Nauk **344**, N^o3, 298–300. *(Russian)* = Russian Acad. Sci. Doklady. Mathem. **52**, N^o2 (1995) 200-202.

Bruno, A.D. 1996a: General approach to the asymptotic nonlinear analysis. Vestnik Mosk. Univ., Ser. 1 Mat. Mekh. N^o6, 24–27. *(Russian)* = Moscow Univ. Mathem. Bulletin **51**, N^o6 (1996) 13–15.

Bruno, A.D. 1996b: Algorithms of the nonlinear analysis. Uspekhi Mat. Nauk **51**, N^o5, 186. *(Russian)* = Russian Math. Surveys **51**, N^o5 (1996) 956.

Bruno, A.D. 1996c: New applications of the Newton polyhedron. Problems of Nonlinear Analysis in Engineering Systems, Kazan′, N^o2, 37–45.

Bruno, A.D. 1997a: Power geometry. J. Dynamical and Control Systems **3**, N^o4, 471–492.

Bruno, A.D. 1997b: Algorithms of the local nonlinear analysis. Nonlinear Analysis **30**, N^o7, 4683–4694.

Bruno, A.D. 1998a: Simplification of systems of algebraic and differential equations. Foreword. Mathematics and Computers in Simulation **45**, N° 5–6, 409-411.

Bruno, A.D. 1998b: Normal forms. Ibid. 413-427.

Bruno, A.D. 1998c: Newton polyhedra and power transformations. Ibid. 429-443.

Bruno, A.D. 1999: Power Geometry and four applications. J. Mathem. Sciences **95**, N^o5, 2483–2512.

Bruno, A.D. 2000: Meromorphic reducibility of a linear triangle ODE system. Doklady Akad. Nauk. **371**, N^o5. *(Russian)* = Russian Acad. Sci. Doklady. Mathem. **61**, N^o2 (2000).

Bruno, A.D., Parusnikov, V.I. 1997: Comparison of various generalizations of continued fractions. Matem. Zametki **61**, N^o3, 339–348. *(Russian)* = Mathem. Notes **61**, N^o3 (1997) 278–286.

Bruno, A.D., Petrovich, V.Yu. 1994a: Regularization of satellite's oscillations on a very stretched orbit. Preprint N^o4, Inst. Appl. Math., Moscow. *(Russian)*

Bruno, A.D., Petrovich, V.Yu. 1994b: Computation of periodic oscillations of a satellite. The singular case. Ibid. N^o44. *(Russian)*

Bruno, A.D., Petrovich, V.Yu. 1997: Computation of periodic oscillations of a satellite. Matematicheskoe Modelirovanie (Mathematical Modelling) **9**, N^o6, 82–94. *(Russian)*

Bruno, A.D., Sadov, S.Yu. 1995: Formal integral of a divergence-free system. Matem. Zametki **57**, N^o6, 803–813. *(Russian)* = Math. Notes **57**, N^o6 (1995) 565–572.

Bruno, A.D., Soleev, A. 1990: The local uniformization of branches of an algebraic curve. Preprint M34, I.H.E.S., Paris.

Bruno, A.D., Soleev, A. 1991a: Local uniformization of branches of a space curve, and Newton polyhedra. Algebra i Analis **3**, N^o1, 67–101. *(Russian)* = St. Petersburg Math. J. **3**, N^o1 (1992) 53–82.

Bruno, A.D., Soleev A. 1991b: The local uniformization of branches of a space curve. In: Proc. of the 22 Annual Iranian Math. Conference (M.R.R. Moghaddam, M.A. Pourabdollah, eds.). Ferdowsi University, Mashhad, 72–85.

Bruno, A.D., Soleev, A. 1992: The local uniformization of branches of an algebraic curve. Contemporary Mathematics **131**, N^o3, 361–378.

Bruno, A.D., Soleev, A. 1994a: Classification of singularities of the position's function of mechanisms. Problemy Mashinovedenija i Nadezhnosti Mashin N^o1, 102–108. *(Russian)* = J. Machinery Manufacture and Reliability N^o1 (1994).

Bruno, A.D., Soleev, A. 1994b: First approximations of algebraic equations. Doklady Akad. Nauk **335**, N^o3, 277–278. *(Russian)* = Russian Acad. Sci. Doklady. Mathem. **49**, N^o2 (1994) 291–293.

Bruno, A.D., Soleev, A. 1995a: On singular positions of some mechanisms. Uzbek. Matem. Zhurnal N^o3, 31–39. *(Russian)*

Bruno, A.D., Soleev, A. 1995b: Local analysis of a singularity of a reversible ODE system. Simple cases. Preprint N^o40, Inst. Appl. Math., Moscow. *(Russian)*

Bruno, A.D., Soleev, A. 1995c: Local analysis of a singularity of a reversible ODE system. Complicated cases. Ibid. N^o47. *(Russian)*

Bruno, A.D., Soleev, A. 1995d: Homoclinic solutions of a reversible ODE system. Ibid. N^o54. *(Russian)*

Bruno, A.D., Soleev, A. 1995e: The Hamiltonian truncations of a Hamiltonian system. Ibid. N^o55. *(Russian)*

Bruno, A.D., Soleev, A. 1995f: Local analysis of singularities of a reversible ODE system. Uspekhi Mat. Nauk **50**, N^o6, 169–170. *(Russian)* = Russian Math. Surveys **50**, N^o6 (1995) 1258–1259.

Bruno, A.D., Soleev, A. 1995g: Bifurcations of solutions in a reversible ODE system. Doklady Akad. Nauk **345**, N^o5, 590–592 *(Russian)* = Russian Acad. Sci. Doklady. Mathem. **52**, N^o3 (1995) 419–421.

Bruno, A.D., Soleev, A. 1996: Hamiltonian truncated systems of a Hamiltonian system. Ibid. **349**, N^o2, 153–155. *(Russian)* = Ibid. **54**, N^o1 (1996) 512–514.

Bruno, A.D., Soleev, A. 1997: Local analysis of a reversible ODE system and the Newton polyhedron. Nonlinear Analysis **30**, N^o8, 4833–4838.

Bruno, A.D., Soleev, A. 1998: The local analysis of singularities of a reversible ODE system. Trudy Mosk. Mat. Obshch. **59**, 3–72. *(Russian)* = Trans. Moscow Math. Soc. **59** (1998) 1–64.

Bruno, A.D., Varin, V.P. 1995: The first (second) limit problem for the equation of oscillations of a satellite. Preprints N^o124 and 128, Inst. Appl. Math., Moscow. *(Russian)*

Bruno, A.D., Varin, V.P. 1996: The fractal structure of periodic oscillations of a satellite. In: Proc. Intern. Conference and Chebyshev Readings, Celebrated the 175 Anniversary of the P.L. Chebyshev Birthday, MGU, Moscow, v.1, 75–77. *(Russian)*

Bruno A.D., Varin V.P. 1997a: The limit problems for the equation of oscillations of a satellite. Celestial Mechanics and Dynamical Astronomy **67**, N^o1, 1–40.

Bruno A.D., Varin V.P. 1997b: Singularities of oscillations of a satellite on highly eccentric elliptic orbits. Nonlinear Analysis **30**, N^o4, 2541–2546.

Bruno A.D., Varin V.P. 1999: Generalized periodic solutions to the equation of oscillation of a satellite. ZAMM **79**, Supplement 2, S283–S284.

Bruno, A.D., Vasiliev, M.M. 1995: Newton polyhedra and the asymptotic analysis of the viscous fluid flow around a flat plate. Preprint N^o44, Inst. Appl. Math., Moscow. *(Russian)*

Bruno, A.D., Vasiliev, M.M. 1996: The asymptotic analysis of a flow around a semiinfinite plate by the method of the Newton polyhedron. In: Proc. Intern. Conference and Chebyshev Readings, Celebrated the 175 Anniversary of the P.L. Chebyshev Birthday. MGU, Moscow, v.1, 78–80. *(Russian)*

Bruno, A.D., Vasiliev, M.M. 1997: Asymptotic analysis of the viscous fluid flow around a flat plate by the Newton polyhedra. Nonlinear Analysis **30**, N^o8, 4765–4770.

Bruno, A.D., Vasiliev, M.M. 1998: Newton polyhedron and Prandtl equations for a boundary layer. ZAMM **78**, Supplement 1, 309–310.

Buchberger, B. 1970: Ein algorithmisches Kriterium fur die Losbarkeit eines algebraischen Gleichungssystems. Aequationes Math. **4**, N^o3, 374–383.

Bugaev, N.V. 1888: Applications of the principle of greatest and least power exponents in the theory of algebraic functions. Mat. Sbornik **14**, 553–590. *(Russian)*

Bugaev, N.V. 1891: The principle of greatest and least power exponents in the theory of differential equations. Ibid. **16**, 39–80. *(Russian)*

Burger, E. 1956: Über homogene lineare Ungleichungssysteme. Z. angew. Math. und Mech. **36**, $N^o3/4$, 135–139.

Carrier, G.F., Lin, C.C. 1948: On the nature of the boundary layer near the leading edge of a flat plate. Quart. Appl. Math. **6**, 63–68.

Chang, Y.F., Tabor, M., Weiss, J. 1982: Analytic structure of the Henon-Heiles Hamiltonian in integrable and nonintegrable regimes. J. Math. Phys. **23**, N^o4, 531–538.

Chebotarev, N.G. 1943: The "Newton polygon" and its role in current developments in mathematics. In: Isaac Newton. Izd. Akad. Nauk SSSR, Moscow - Leningrad, 99–126. See also: Collected Works. Izd. Akad. Nauk SSSR, Moscow - Leningrad, 1950, v. 3, 47–80. *(Russian)*

Chernikov, S.N. 1968: Linear Inequalities. Nauka, Moscow. *(Russian)*

Cicogna, G., Gaeta, G. 1999: Symmetry and Perturbation Theory in Nonlinear Dynamics. LNP, M57. Springer, Berlin etc.

Coleman, C.S. 1966: Brjuno, A.D. Power asymptotics of solutions of non-linear systems. Math. Reviews **31**, N^o1, 81, # 441.

Coppel, W.A. 1966: A survey of quadratic systems. J. Differential Equations **2**, 293–304.

Danilov, V.I., Khovanskii, A.G. 1986: Newton polyhedra and an algorithm for computing Hodge-Deligne numbers. Izvestiya Akad. Nauk SSSR, Ser. Mat. **50**, N^o5, 925-945. *(Russian)* = Math. USSR Izvestiya **19**, N^o2 (1987) 279–298.

Denk, R., Mennicken, R., Volevich, L. 1998: The Newton polygon and elliptic problems with parameter. Math. Nachr. **192**, 125–157.

Dimentberg, F.M. 1950: Determination of Position of Space Mechanisms. Akad. Nauk SSSR, Moscow. *(Russian)*

Edneral, V.F. 1998: A symbolic approximation of periodic solutions of the Henon–Heiles system by the normal form method. Mathematics and Computers in Simulation **45**, N^o5–6, 445–463.

Écalle, J. 1993: Six lectures on transseries, analyzable functions and the constructive proof of Dulac's conjecture. In: Bifurcations and Periodic Orbits of Vector Fields (D. Schlomiuk, ed.). Kluwer Academic Publishers, 75–184.

Elphick, C., Tirapegui, E., Brachet, M., Coullet, P., Iooss, G. 1987: A simple global characterization for normal forms of singular vector fields. Physica **29**D, 95–127.

Euler, L. 1755: Principes généraux du mouvement des fluides. Hist. de l'Acad. de Berlin. = De principiis motus fluidorum. Novi. Comm. Acad. Petrop. **14**, N^o1 (1759).

Friberg, J. 1967: Multiquasielliptic polynomials. Ann. Sckola Norm. Super. Pisa **21**, N^o2, 233–260.

Frommer, M. 1928: Die Integralkurven einer gewöhnlichen Differentialgleichung erster Ordnung in der Umgebung rationaler Unbestimmtheitsstellen. Math. Ann. **99**, 222–272 = Uspekhi Mat. Nauk. N^o9 (1941) 212–253. *(Russian)*

Fuks, B.A. 1962: Introduction to the Theory of Analytic Functions of Several Complex Variables. Fizmatgiz, Moscow. *(Russian)* = Transl. of Math. Mono., v. 8, Am. Math. Soc., Providence, R.I., 1963.

Fukuda, T., Aoki, K., Sun, W. Z. 1986: On the number of branches of a plane curve germ. Kodai Math. J. **9**, 178–187.

Gaeta, G. 1994: Normal forms of reversible dynamical systems. Intern. J. Theor. Phys. **33**, N^o9, 1917–1928.

Gaeta, G. 1997: Reduction of Poincare normal forms. Lett. Math. Phys. **42**, N^o2, 103–114.

Gaeta, G. 1999: Poincare renormalized forms. Ann. Inst. Henri Poincare **70**, N^o6, 461–514.

Gantmacher, F.R. 1967: Theory of Matrices. 3rd. ed. Nauka, Moscow. *(Russian)* = Chelsea Publ. Co., New York, 1959.

Gelfond, O.A., Khovanskii, A.G. 1996: Newton polyhedra and Grothendieck residues. Doklady Acad. Nauk **350**, N^o3, 298–300. *(Russian)* = Russian Acad. Sci. Doklady. Mathem. **54**, N^o2 (1996) 700–702.

Gindikin, S.G. 1973: On a class of differential operators admitting the two-sided energy estimates. Uspekhi Mat. Nauk **28**, N^o6, 199–200. *(Russian)* = Russian Math. Surveys **28**, N^o6 (1973) 199–200.

Gindikin, S.G. 1974: Energy estimates and Newton polyhedra. Trudy Mosk. Mat. Obshch. **31**, 189–236. *(Russian)* = Trans. Moscow Math. Soc. **31** (1974) 193–246.

Gindikin, S.G., Volevich, L.R. 1992: The Method of Newton's Polyhedron in the Theory of Partial Differential Equations. Kluver Acad. Publ., Dordrecht.

Goldman, A., Tucker, A. 1956: Polyhedral convex cones. In: Linear Inequalities and Related Systems. Ann. Math. Stud. **38**, 19–40. = IL, Moscow, 1959, 142–161. *(Russian)*

Goldstein, S. 1960: Lectures on Fluid Mechanics. Wiley, New York.

Goursat, E. 1927: Cours d'Analyse Mathématique. 5em ed., Gauthier–Villars, Paris, t. 1. = GTTI, Moscow, Leningrad, 1933, v. 1, pt. 2. *(Russian)*

Grawes, L. M. 1955: Remarks on singular points of functional equations. Trans. Amer. Math. Soc. **79**, 150–157.

Grin', A.N. 1971: To the branching theory of a nonlinear equation in the multidimensional case of branching. Doklady Akad. Nauk SSSR **201**, N^o4, 22–23. *(Russian)* = Soviet Math. Doklady **12**, N^o6 (1971).

Grobman, D.M. 1962: The topological classification of a vicinity of a singular point in the n-dimensional space. Mat. Sbornik **56**, N^o1, 77–94. *(Russian)* = Math. USSR Sbornik **4** (1962).

Hammersley, J.M., Mazarino, G. 1989: Computational aspect of some autonomous differential equations. Proc. Roy. Soc. London. Ser. A. **424**, 19–37.

Hartshorne, R. 1977: Algebraic Geometry. Springer, New York–Heidelberg–Berlin. = Mir, Moscow, 1981. *(Russian)*

Helmholtz, H.L.F. 1858: Über Integrale der hydrodynamischen Gleichungen, welche den Wirbelbewegungen entsprechen. J. Reine und Angewandte Mathematik **55**, 25–55.

Henon, M. 1969: Numerical exploration of the restricted problem. V. Hill's case: periodic orbits and their stability. Astronomy and Astrophysics **1**, 223–238.

Henon, M., Heiles, C. 1964: The applicability of the third integral of motion: some numerical experiments. Astron. J. **69**, N^o1, 73–79.

Hill, G.W. 1878: Researches in the lunar theory. Amer. J. Math. **1**, 5–26, 129–147, 245–260. Also: Collected Mathematical Works. Carnegie Inst., Wash., D.C., 1905, v. 1, 284–335.

Hironaka, H. 1964: Resolution of singularities of an algebraic variety over a field of characteristic zero. Ann. of Math. (2), **79**, 109–326.

Ibragimov, N.H. 1983: Transformation Groups Applied to Mathematical Physics. Nauka, Moscow. *(Russian)* = D. Reidel: Dordrecht, 1985.

Ibragimov, N.H. (ed.) 1994: Lie Group Analysis of Differential Equations. CRC Press, Boca Ration, v. 1.

Imai, I. 1957: Second approximation to the laminar boundary–layer flow over a flat plate. J. of Aeronautical Sci. **24**, 155–156.

Iooss, G. 1995a: A codimension 2 bifurcation for reversible vector fields. Fields Institute Communications **4**, 201–217.

Iooss, G. 1995b: Capillary-gravity water-waves problem as a dynamical system. In: Structure and Dynamics of Nonlinear Waves in Fluids (A. Mielke, K. Kirchgässner, eds.). World Scientific, Singapore, 42–57.

Iooss, G., Adelmeyer, M. 1992: Topics in Bifurcation Theory and Applications. World Scientific., Singapore.

Iooss, G., Kirshgässner, K. 1990: Bifurcation d'ondes solitaires en presence d'une faible tension superficielle. C.R. Acad. Sci. Paris **311**, N^o1, 265–268.

Iooss, G., Kirshgässner, K. 1992: Water waves for small surface tension: an approach via normal form. Proceedings of the Royal Society of Edinburg **122A**, 267–299.

Iooss, G., Pérouème, M.-C. 1993: Perturbed homoclinic solutions in reversible 1:1 resonance vector fields. J. Differential Equations **102**, 62–88.

Kazarnovskii, B. Ya. 1984: Newton polyhedra and roots of systems of exponential sums. Funkts. Anal. Prilozh. **18**, N^o4, 40–49. *(Russian)* = Funct. Anal. Appl. **18** (1984).

Kazarnovskii, B. Ya. 1997: Exponential analytical sets. Funkts. Analiz Prilozh. **31**, N^o2, 15–26. *(Russian)* = Funct. Analys Appl. **31**, N^o2 (1977).

Khinchin, A.Ya. 1961: Continued Fractions. Fizmatgiz, Moscow. *(Russian)* = University of Chicago Press, Chicago–London, 1964.

Khovanskii, A.G. 1978a: Newton polyhedra and genus of complete intersections. Funkts. Anal. Prilozh. **12**, N^o1, 51–61. *(Russian)* = Funct. Anal. Appl. **12**, N^o1 (1978) 38–46.

Khovanskii, A.G. 1978b: Newton polyhedra and the Euler–Jacobi formula. Uspekhi Mat. Nauk **33**, N^o6, 237–238. *(Russian)* = Russian Math. Surveys **33**, N^o6 (1978) 237–238.

Khovanskii, A.G. 1981: Newton polyhedra and the index of a vector field. Uspekhi Mat. Nauk **36**, N^o4, 234. *(Russian)*

Khovanskii, A.G. 1983: Newton polyhedra (resolution of singularities). In: Itogi Nauki i Tekhniki. Sovr. Probl. Mat., v. 22, VINITI, Moscow, 207–239. *(Russian)* = J. Soviet Math. **27** (1984) 2811–2830.

Khovanskii, A.G. 1988: Algebra and mixed volumes. In: Burago, Yu.D., Zalgaller, V.A., Geometric Inequalities. Springer, Berlin, 182–207.

Khovanskii, A.G. 1992: Newton polyhedron, Hilbert polynomial and the number of points in sums of finite sets. Funkts. Anal. Prilozh. **26**, N^o4, 57–63. *(Russian)* = Funct. Anal. Appl. **26**, N^o4 (1992) 276–281.

Khovanskii, A.G. 1997a: Newton polytopes, curves on toric surfaces and inversion of Weil's theorem. Uspekhi Mat. Nauk **52**, N^o6, 113–142. *(Russian)* = Russian Math. Surveys **52**, N^o6 (1997) 1251–1279.

Khovanskii, A.G. 1997b: Consistent partitions of polytopes and polynomial measures. In: Topics in Singularity Theory: V.I. Arnold's 60th Anniversary Collection. Amer. Math. Soc. Transl., Ser. 2, v. 180, Amer. Math. Soc., Providence, RI, 157-166.

Khovanskii, A.G. 1999: Newton polyhedrons, a new formula for mixed volume, product of roots of a system of equations. In: The Arnoldfest, Proceedings of a Conference in Honour of V.I. Arnold for his Sixtieth Birthday. Fields Institute Communications **24**, 325–364.

Khovanskii, A.G. 2000: Geometry of polyhedra and algebraic geometry. Electronic version in: `http://www.math.toronto.edu/askold/askoldpap.html`

Kieffer, J., Litvin, F.L. 1991: Local parametric representation of displacement functions for linkades and manipulators. Mech. Mach. Theory. **26**, N^o1, 41–53.

Kirchgässner, K. 1988: Nonlinearly resonant surface waves and homoclinic bifurcation. Adv. Appl. Mech. **26**, 135–181.

Klamkin, M.S. 1962: On the transformation of a class of boundary value problems into initial value problems for ordinary differential equations. SIAM Rev. **4**, 43–57.

Klamkin, M.S. 1970: Transformation of boundary value problems into initial value problems. J. Math. Anal. Appl. **32**, N^o2, 308–330.

Klein, F. 1872: Vergleihende Betrachtungen über neuere geometrische Forschungen. A. Deichert, Erlangen.

Klokov, Yu.A., Mikhailov, A.P., Ad'yutov, M.M. 1997: Nonlinear mathematical models and nonclassical boundary-value problems for ordinary differential equations. In: Fundamental Basis of the Mathematical Modelling. Nauka, Moscow, 98–197. *(Russian)*

Kochin, N.E., Kibel', I.A., Rose, N.V. 1948: Theoretical Hydrodynamics. Pt. II. Gostekhizdat, Moscow. *(Russian)*

Kogan, A.Yu. 1988: Remote satellite's orbits in the circular restricted three-body problem. Kosmicheskie Issledovaniya **26**, N^o6, 813–818. *(Russian)* = Cosmic Research **26**, N^o6 (1988).

Kostin, V.M. 1969: Some invariant solutions of Koteweg-deVries equations. Zh. Prikl. Mekh. Tekhn. Fiz. N^o4, 69–73. *(Russian)* = J. Appl. Mech. Techn. Phys. **10**, N^o4 (1969).

Kouchnerenko, A.G. 1976: Polyhedra de Newton et nombers de Milnor. Invent. Math. **32**, N^o1, 1–32.

Kozlov, V.V., Furta, S.D. 1996: Asymptotics of Solutions to the Strongly Nonlinear Systems of Differential Equations. MGU, Moscow. *(Russian)*

Kukles, I.S., Gruz, D.M. 1958: On the number of operations connected with the application of the Frommer's method. Izvestiya Akad. Nauk Uzbek. SSR, Ser. Fiz.-Mat. Nauk N^o1, 29-45. *(Russian)*

Kushnirenko, A.G. 1975a: The Newton polyhedron and the number of solutions of a system of k equations in k unknowns. Uspekhi Mat. Nauk **30**, N^o2, 266–267. *(Russian)*

Kushnirenko, A.G. 1975b: The Newton polygon and Milnor numbers. Funkts. Anal. Prilozh. **9**, N^o1, 74–75. *(Russian)* = Funct. Anal. Appl. **9**, N^o1 (1975) 71–72.

Kushnirenko, A.G. 1976: Newton polyhedra and Bezout's theorem. Funkts. Anal. Prilozh. **10**, N^o3, 82–83. *(Russian)* = Funct. Anal. Appl. **10**, N^o3 (1976) 233–234.

Lefshets, S. 1954: Complete families of periodic solutions of differential equations. Comm. Helv. **28**, 341–345.

Lefshets, S. 1957: Differential Equations: Geometrical Theory. Intersience Publishers, N.Y.–London. = IL, Moscow, 1961. *(Russian)*

Lekkerkerker, C.G. 1969: Geometry of Numbers. North-Holland Publ. Comp., Amsterdam.

Lidov, M.L., Vashkov'yak, M.A. 1990: Quasisatellite periodic orbits. In: Analytical Celestial Mechanics (K. V. Kholshevnikov, ed.). Kasan' University, Kasan', 53–57. *(Russian)*

Lidov, M.L., Vashkov'yak, M.A. 1993: Perturbation theory and analysis of evolution of quasisatellite orbits in the restricted three-body problem. Kosmicheskie Issledovaniya **31**, N^o2, 75–99. *(Russian)* = Cosmic Research **31**, N^o2 (1993) 187–207.

Lidov, M.L., Vashkov'yak, M.A. 1994: On quasisatellite orbits for the experiment for a correction of the gravitational constant. Pis'ma v Astron. Zh. **20**, N^o3, 239–240. *(Russian)* = Astronomy Letters **20**, N^o2 (1994) 188–198.

Liouville, J. 1833: Sur la determination des integrales dont la valeur est algebrique. J. l'Ecole Polytechnique N^o22, 124–193.

Loginov, B.V. 1985: Branching Theory for Solutions to Nonlinear Differential Equations with the Group Invariance. FAN, Tashkent. *(Russian)*

Loginov, B.V., Trenogin, V.A. 1975: About utilization of the group invariance in the branching theory. Differents. Uravn. **11**, N^o8, 1518–1521. *(Russian)* = Differential Equations **11** (1975).

Lunev, V.V. 1994a: Meromorphic solutions to the equations of motion of a solid body with a fixed point. Prikl. Mat. Mekh. **58**, N^o1, 30–40. *(Russian)* = J. Appl. Math. Mech. **58**, N^o1 (1994) 31–41.

Lunev, V.V. 1994b: The Integrable Dynamical Systems and the Method of Kovalevskaya–Lyapunov. Institute of Mechanical Engineering Research, Moscow. *(Russian)*

Lunev, V.V., Misyurin, S.Yu. 1993: Singular manifolds of planar and space mechanisms with nonused degrees of freedom. Problemy Mashinovedenya i Nadezhnosti Mashin N^o1, 102–109. *(Russian)* = J. Machinery Manufacture and Reliability N^o1 (1993).

Lunev, V.V., Misyurin, S.Yu. 1994: Application of the Newton polyhedra method to the problem of position of mechanisms. Ibid. N^o1, 109–118. *(Russian)* = Ibid. N^o1 (1994).

Lyapunov, A.M. 1892: The General Problem of the Stability of Motion. Kharkov. Reprints: ONTI, Moscow–Leningrad, 1935. Also: GIT, Moscow–Leningrad, 1950. *(Russian)* = Problème Général de la Stabilité du Mouvement. Ann. Fac. Sci. Toulouse Math. (2), **9** (1907) 204–474. Also: Ann. of Math. Studies **17** (1947). = Stability of Motion. Academic Press, New York–London, 1966.

MacMillan, W. 1912a: A reduction of a system of power series to an equivalent system of polynomials. Math. Ann. **72**, 157–179.

MacMillan, W. 1912b: A method for determining the solutions of a system of analytic functions in the neighborhood of a branch point. Ibid., 180–202.

Markeev, A.P., Sokol'skii, A.G., Chekhovskaya, T.N. 1977: On stability of the conic precession of the dynamically symmetrical solid body. Pis'ma v Astron. Zh. **3**, N^o7, 333–336. *(Russian)* = Astronomy Letters **3** (1977).

McLachlan, R.I. 1991: The boundary layer on a finite flat plate. Phys. Fluids. A, **3**, N^o2, 341–348.

Mel'nikov, V.K. 1963: On the stability of the centre for time-periodic perturbations. Trudy Mosk. Mat. Obshch. **12**, 3–52. *(Russian)* = Trans. Moscow Math. Soc. **12** (1963) 1–56.

Mikhailov, V.P. 1963: On the first boundary value problem for some semi-bounded operators. Doklady Akad. Nauk SSSR **151**, 282–285. *(Russian)* = Soviet Mathem. Doklady **4** (1963) 997–1001.

Mikhailov, V.P. 1965: On the behavior at infinity of some classes of polynomials. Doklady Akad. Nauk SSSR **164**, N^o3, 499–502. *(Russian)* = Soviet Mathem. Doklady **6**, N^o4 (1965).

Mikhailov, V.P. 1967a: Behavior at infinity of a certain class of polynomials. Trudy Mat. Inst. Steklova **91**, 59–80. *(Russian)* = Proc. Steklov Inst. Math. **91** (1967) 61–82.

Mikhailov, V.P. 1967b: The first boundary-value problems for quasi-elliptic and quasi-parabolic equations. Ibid. 81–99. *(Russian)* = Ibid. 83–101.

Milnor, J. 1968: Singular Points of Complex Hypersurfaces. Princeton Univ. Press: Princeton, N.J., 1968. = Mir, Moscow 1971. *(Russian)*

Mishchenko, E.F., Rozov, N.Kh. 1975: Differential Equations with a Small Parameter and Relaxation Oscillations. Nauka, Moscow. *(Russian)* = Plenum Press, N.Y.–London, 1980.

Mishchenko, E.F., Kolesov, Yu.S., Kolesov, A.Yu., Rozov, N.Kh. 1994: Asymptotic Methods in Singularly Perturbed Systems. Consultants Bureau, Plenum Publ. Corp., N.Y.–London. = Periodic Solutions and Bifurcation Processes in Singularly Perturbed Systems. Fizmatlit, Moscow, 1995. *(Russian)*

Moiseev, N.N. 1983: Epilogue. In: Iooss, G., The Elementary Theory of Stability and Bifurcations: Russian translation from English. Mir, Moscow, 291–296. *(Russian)*

Moser, J. 1958: On the generalization of a theorem of A. Liapounoff. Comm. Pure and Appl. Math. **11**, N^o2, 257–271.

Motzkin, T.S., Schoenberg, G. J. 1954: The relaxation method for linear inequalities. Canad. J. Math. **6**, 393–404.

Na, T.Y. 1979: Computational Methods in Engeneering Boundary Value Problems. Academic Press, N.Y.

Navier, C.L.M.H. 1827: Memoir sur les lois du mouvement des fluids. Mem. de L'Ac. Royale de Sc. de L'Institut de France **7**, 375–394.

Nemytskii, V.V., Stepanov, V.V. 1947: Qualitative Theory of Differential Equations. 1st ed. OGIZ, Gostekhizdat, Moscow; 2nd ed. OGIZ, Gostekhizdat, Moscow, 1949. *(Russian)* = Princeton Univ. Press, Princeton, 1960.

Newton, I. 1711: A treatise of the method of fluxions and infinite series, with its application to the geometry of curve lines. In: The Mathematical Works of Isaac Newton (H. Woolf, ed.). Johnson Reprint Corp., New York–London, 1964, v. 1, 27–137. = ONTI, Moscow–Leningrad, 1937, 25–166. *(Russian)*

Oseen, C.W. 1910: Über die Stokessche Formel und über eine verwandte Aufgabe in der Hydrodynamik. Arkiv für Math. Astr. Fys. **6**, N^o29; **7**, N^o1 (1911).

Ovsiannikov, L.V. 1960: Group properties of Chaplygin equation. Zh. Prikl. Mekh. Tekhn. Fiz. N^o3, 120–131. *(Russian)* = J. Appl. Mech. Techn. Phys. **1**, N^o3 (1960).

Ovsiannikov, L.V. 1978: Group Analysis of Differential Equations. Nauka, Moscow. *(Russian)* = Academic Press, N.Y., 1982.

Petrovskii, I.G. 1964: Lectures on the Theory of Ordinary Differential Equations. 5th ed., Nauka, Moscow. *(Russian)* = Ordinary Diff. Equations. Dover Publ., New York, 1973.

Prandtl, L. 1904: Über Flüssigkeitsbewegung bei sehr kleiner Reibung. In: Verhandl. III Kongr., Heidelberg.

Pshenichnyi, V.I. 1980: Convex Analysis and Extremal Problems. Nauka, Moscow. *(Russian)*

Pukhlikov, A.V., Khovanskii, A.G. 1992: A Riemann-Roch theorem for integrals and sums of quasipolynomials over virtual polytopes. Algebra i Analiz **4**, N^o4, 188–216. *(Russian)* = St. Petersburg Math. J. **4**, N^o4 (1993) 789–812.

Puiseux, V. 1850: Recherches sur les fonctions algebriques. J. Math. Pures et Appl. **15**, 365–480.

Reyn, J.W. 1987: Phase portraits of a quadratic systems of differential equations occurring frequently in applications. Nieuw Archief voor Wiskunde (4), **5**, N^o2, 107–155.

Roberts, J. A. G., Quispel, G. R. W. 1992: Chaos and time-reversal symmetry. Order and chaos in reversible dynamical systems. Phys. Rep. **216**, N^o2–3, 63–177.

Rozov, N.Kh. 1985: Emden equation. In: Mathem. Entsiklop. **5**, 997–998. Sov. Entsikl., Moscow. *(Russian)* = Encyclopaedia of Mathematics. Kluver Acad. Publ., Dordrecht, 1989, v. 3, 380.

Sadov, S.Yu. 1994a: A study of a finite-dimensional dynamical system approximating the evolution of quantum averages. In: Hamiltonian Mechanics (J. Seimenis, ed.). Plenum Press, New york, 413–420.

Sadov, S.Yu. 1994b: On a dynamic system arising from a finite-dimensional approximation of the Schrödinger equation. Matem. Zametki **56**, N^o3, 118–133. *(Russian)* = Math. Notes **56**, N^o3 (1994) 960–971.

Sadov, S.Yu. 1995: Normal form of the equation of oscillations of a satellite in a singular case. Matem. Zametki **58**, N^o5, 785–789. *(Russian)* = Math. Notes **58**, N^o5 (1995).

Sadov, S.Yu. 1996a: Higher approximations of the averaging method for the equation of planar oscillations of a satellite. Preprint N^o48, Inst. Appl. Math., Moscow. *(Russian)*

Sadov, S.Yu. 1996b: The averaging method for the equation of oscillations of a satellite for the eccentricity near to one. In: Proc. of Intern. Conference and Chebyshev Readings, Celebrated the 175 Anniversary of the P.L. Chebyshev Birthday. MGU, Moscow, v. 2, 305–307. *(Russian)*

Sadov, S.Yu. 1997a: Planar motion of a near symmetric satellite about its mass center with rational rotation numbers. Preprint N^o31, Inst. Appl. Math., Moscow. *(Russian)*

Sadov, S.Yu. 1997b: Singular normal form for a quasilinear ordinary differential equation. Nonlinear Analysis **30**, N^o8, 4973–4978.

Sadov, S.Yu. 1998: Functions that determine stability of rational rotations of a near symmetric satellite. Mathem. and Computers in Simulation **45**, N^o5–6, 465–484.

Sadov, S.Yu. 1999: Lissajous solutions of the satellite oscillation equation. Nonlinear Phenomena in Complex Systems. Minsk. **2**, N^o2, 96–101.

Saito, M. 1983: Exponents and Newton polyhedra of isolated hypersurfaces singularities. Ann. Inst. Fourier, Grenoble **33**, 1–10.

Samarskii, A.A., Galaktionov, V.A., Kurdyumov, S.P., Mikhailov, A.P. 1987: Blow-up in Quasilinear Parabolic Equations. Nauka, Moscow. *(Russian)* = Walter de Gruyter, Berlin, 1995.

Samovol, V.S. 1972: Linearization of a system of differential equations in the neighborhood of a singular point. Doklady Akad. Nauk SSSR **206**, 545–548. *(Russian)* = Soviet Math. Doklady **13** (1972) 1255–1259.

Samovol, V.S. 1979: Linearization of systems of differential equations in a neighborhood of invariant toroidal manifolds. Trudy Mosk. Mat. Obshch. **38**, 187–219. *(Russian)* = Trans. Moscow Math. Soc. **38**, N^o2 (1979) 190–225.

Samovol, V.S. 1982: Equivalence of systems of differential equations in a neighborhood of a singular point. Trudy Mosk. Mat. Obshch. **44**, 213–234. *(Russian)* = Trans. Moscow Math. Soc. **44**, N^o2 (1982) 217–237.

Sather, D. 1973: Branching of solutions of nonlinear equations. Rocky Mountain J. Math. **3**, N^o2, 203–250.

Schlichting, H. 1965: Grenzschicht-Theorie. 5te Auflage. Verlag G. Braun, Karlsruhe = Nauka, Moscow, 1969. *(Russian)*

Sevryuk, M. B. 1986: Reversible Systems. LNM 1211. Springer, Berlin.

Shafarevich, I.R., 1972: Basic Algebraic Geometry. Nauka, Moscow. *(Russian)* = Springer, Berlin etc., 1977.

Shestakov, A.A. 1960: Asymptotic behavior of solutions of multidimensional systems of ordinary differential equations having a singular point of higher order. Doklady Akad. Nauk SSSR **131**, 1038–1041. *(Russian)* = Soviet Math. Doklady **1** (1960) 394–397.

Shestakov, A.A. 1961: Asymptotic behavior of solutions of higher-dimensional system of differential equations having a singularity of higher order. Sib. Mat. Zh. **2**, N^o5, 767–788. *(Russian)*

Shih, Ch.-W. 1993: Normal forms and versal deformations of linear involutive dynamical systems. Chinese J. Math. **21**, N^o4, 333–347.

Sintsov, D.M. 1898: Rational Integrals of Linear Equations. Uch. Zap. Kazan. Univ. *(Russian)*

Sokol'skii, A.G. 1980: To the problem of stability of the regular precessions of the symmetric satellite. Kosmicheskie Issledovaniya **18**, N^o5, 698–706. *(Russian)* = Cosmic Research **18**, N^o5 (1980).

Sokol'skii, A.G. 1981: On stability of an autonomous Hamiltonian system with two degrees of freedom in case of zero frequencies. Prikl. Mat. Mekh. **45**, N^o3, 441–449. *(Russian)* = J. Appl. Math. Mech. **45**, N^o3 (1981) 321–329.

Soleev, A. 1982: Algorithm of computation of the Newton polyhedron. Doklady Akad. Nauk Uzbek. SSR N^o5, 14–16. *(Russian)*

Soleev, A. 1983: Determining the branches of an analytic curve, and the Newton polyhedra. Doklady Akad. Nauk SSSR **268**, N^o6, 1305–1307. *(Russian)* = Soviet Math. Doklady **27**, N^o1 (1983) 252–254.

Soleev, A., Aranson, A.B. 1994: Computation of a polyhedron and of normal cones of its faces. Preprint N^o36, Inst. Appl. Math., Moscow. *(Russian)*

Soleev, A., Aranson, A.B. 1995: The first approximations of a reversible ODE system. Preprint N^o28, Inst. Appl. Math., Moscow. *(Russian)*

Soleev, A., Bruno, A.D. 1995: Newton polyhedra and Hamiltonian systems. Vestnik Mosk. Univ., Ser. 1 Mat. Mekh. N^o6, 84–86. *(Russian)* = Moscow Univ. Mathem. Bulletin **50**, N^o6 (1995) 36–38.

Stokes, G.G. 1849: On the theories of the internal friction of fluids in motion and of the equilibrium and motion of elastic solids. Trans. Cambr. Philos. Society **8**, 287–319.

Stokes, G.G. 1851: On the effect of internal friction of fluids on the motion of pendulums. Trans. Camb. Philos. Society **9**, part II, 8–106. Also: Math. and Phys. Papers, Cambridge Univ. Press, London–New York, 1901, v. 3, 1–141.

Szafraniec, A. 1992: A formula for the number of branches for one-dimensional semianalitic sets. Math. Proc. Camb. Phil. Soc. **112**, 527–534.

Trenogin, V.A. 1960: The equation of branching and the Newton diagram. Doklady Akad. Nauk SSSR **131**, N^o5, 519–522. *(Russian)* = Soviet Math. Doklady **1**, N^o2 (1960).

Tucker, A. 1956: Dual systems of homogeneous linear relations. In: Linear Inequalities and Related Systems. Ann. Math. Stud. **38**, Princeton Univ. Press. = IL, Moscow, 1959, 127–141. *(Russian)*

Vainberg, M.M., Aizengendler, P.G. 1966: Research methods in the theory of solution's branching. In: Itogi Nauki. Mat. Analiz. VINITI, Moscow, 7–69. *(Russian)*

Vainberg, M.M., Trenogin, V.A. 1969: The Theory of Branching of Solutions of Nonlinear Equations. Nauka, Moscow. *(Russian)* = Wolters-Noordhoff Intern. Publ., Leiden, 1974.

Van de Vooren, A.I., Dijkstra, D. 1970: The Navier–Stokes solution for laminar flow past a semi-infinite flat plate. J. Engineering Math. **4**, N^o1, 9–27.

Van Dyke, M. 1964: Perturbation Methods in Fluid Mechanics. Academic Press, New York–London. = Mir, Moscow, 1967. *(Russian)*

Varchenko, A.N. 1976: Newton polyhedra and the estimation of oscillating integrals. Funkts. Anal. Prilozh. **10**, N^o3, 13–38. *(Russian)* = Funct. Anal. Appl. **10** (1976) 175–196.

Varchenko, A.N., Khovanskii, A.G. 1985: On asymptotic behavior of integrals on vanishing cycles and Newton polyhedra. Doklady Akad. Nauk SSSR **283**, N^o3, 521–525. *(Russian)* = Soviet Math. Doklady **32**, N^o1 (1985) 122–127.

Varin, V.P. 1996: The critical families of periodic solutions to the equation of oscillations of a satellite. Preprint N^o101, Inst. Appl. Math., Moscow. *(Russian)*

Varin, V.P. 1997: The critical subfamilies of the family K_0 of periodic solutions to the equation of oscillations of a satellite. Ibid. N^o20. *(Russian)*

Varin, V.P. 1999a: Localization of degeneracies in families of periodic solutions to an ODE and their regularization. Ibid. N^o22. *(Russian)*

Varin, V.P. 1999b: Periodic solutions to the Beletskii equation and their degeneracies. Ibid. N^o23. *(Russian)*

Vasil'ev, V.A. 1977: Asymptotics of exponential integrals, the Newton diagram and a classification of the minimum points. Funkts. Anal. Prilozh. **11**, N^o3, 1–11. *(Russian)* = Funct. Anal. Appl. **11** (1977).

Vasil'eva, A.B., Butusov, V.F. 1990: Asymptotical Methods in the Theory of Singular Perturbations. Vysshaya Shkola, Moscow. *(Russian)* = Vasil'eva, A.B., Butusov, V.F., Kalachev, L.V. The Boundary Function Method for Singular Perturbation Problems. SIAM, Philadelphia, 1995.

Vasiliev, M.M. 1998: About the asymptotic analysis of the viscous heat conductive gas flow equations. Preprint N^o65, Inst. Appl. Math., Moscow.

Vasiliev, M.M. 1999: About the computation of the boundary layer on the semi-infinite flat plate in the viscous heat conductive gas flow. Preprint N^o43, Inst. Appl. Math., Moscow. *(Russian)*

Vedenyapin, V.V., Mingalev, O.V. 1997: Representations of general commutational relations. Asymptotics of spectrum for three quantum Hamiltonians. Doklady Akad. Nauk **352**, N^o2, 155–158. *(Russian)* = Russian Acad. Sci. Doklady. Mathem. **55**, N^o1 (1997).

Vedenyapin, V.V., Orlov, Yu.N. 1996: Asymptotics of the quantum Hamiltonian spectrum. Doklady Akad. Nauk **351**, N^o4, 444–447. *(Russian)* = Russian Acad. Sci. Doklady. Mathem. **54**, N^o3 (1996).

Vedenyapin, V.V., Orlov, Yu.N. 1999: On conservation laws for polynomial Hamiltonians and for discrete models of the Boltzmann equation. Teoretich. Matematich. Fizika **121**, N^o2, 307–315. *(Russian)* = Theoretical and Mathem. Physics **121** (1999).

Vinogradov, A.M., Vorob'ev, E.M. 1976: Applications of symmetries for finding of exact solutions of Zabolotskaya-Khokhlov equation. Akust. Zh. **22**, N^o1, 23–29. *(Russian)* = Soviet Physics. Acoustics **22**, N^o1 (1976).

Volevich, L.R. 1974: Energy method in Cauchy's problem for differential operators correct in Petrovskii's sense. Trudy Mosk. Mat. Obshch. **31**, 147–187. *(Russian)* = Trans. Moscow Math. Soc. **31** (1974) 151–192.

Volevich, L.R., Gindikin, S.G. 1968: On a class of hypoelliptic polynomials. Mat. Sbornik **75(117)**, N^o3, 400–416. *(Russian)* = Math. USSR Sbornik **75** (1968) 369–383.

Volevich, L.R., Gindikin, S.G. 1985: The Newton polyhedron and the local solvability of partial differential equations. Trudy Mosk. Mat. Obshch. **48**, 211–262. *(Russian)* = Trans. Moscow Math. Soc. **48** (1985).

Walker, R.J. 1950: Algebraic Curves. Princeton, N.J. = IL, Moscow, 1952. *(Russian)*

Wang, S.L., Waeldron, K.J. 1987: A study of the singular configurations of serial manipulators. J. Mechanisms Transmissions and Automatition in Design **109**, N^o1, 14–16.

Weierstrass, K. 1902: Theorie der Abelshen Transcendenten, I. Absch., I. Kop. In: Math. Werke. Mayer und Muller, Berlin, v. 4, 11–45.

Worz-Busekros, A. 1993: Complete classification of all two-dimensional Lotka–Volterra systems. Diff. Equations and Dynamical Systems **1**, N^o2, 101–119.

Yoshida, H. 1983: Necessary condition for the existence of algebraic first integrals. Celestial Mechanics **31**, N^o4, Pt. 1, 363–379; Pt. 2, 281–299.

Yossin, B. 1991: Newton polyhedra without coordinates. Mem. Amer. Math. Soc. **19**, 116–124.

Zabreiko, P.P., Kats, B.P. 1971: On the Nekrasov–Nasarov method of solution of the nonlinear operator equations. Sib. Mat. Zh. **12**, N^o5, 1026–1040. *(Russian)* = Sib. Math. J. **12**, N^o5 (1971).

Zariski, O. 1940: Local uniformization on algebraic varieties. Ann. of Math. **41**, N^o2, 852–896.

Subject index

www.ingramcontent.com/pod-product-compliance
Ingram Content Group UK Ltd.
Pitfield, Milton Keynes, MK11 3LW, UK
UKHW020232290926
15470UKWH00002B/491

9 780444 502971